Introduction to
GENERAL TOPOLOGY

Introduction to
GENERAL TOPOLOGY

K. D. Joshi
Department of Mathematics
Indian Institute of Technology
Bombay

A HALSTED PRESS BOOK

JOHN WILEY & SONS
NEW YORK CHICHESTER BRISBANE ONTARIO SINGAPORE

Published in the Western Hemisphere
by Halsted Press, a Division of
John Wiley & Sons, Inc., New York

Library of Congress Cataloging in Publication Data

Joshi, K. D.

Introduction to general topology.

Includes bibliographical references.

1. Topology. I. Title.

QA611.J67 1982 514′.322 82–15520

ISBN 0–470–27556–1

Printed in India at Urvashi Press, Meerut.

To

My Mother

Preface

The present book, as its title indicates, is meant as an introduction to general topology. It is especially written for the average M.A./M.Sc. or the advanced undergraduate student in educational institutions in India. It can also be used by persons from other areas of mathematics or from computer science, physics etc., who want to get a working knowledge of topology. However, it is not meant as a reference book on the subject.

It was relatively recently that general topology was introduced in the M.A./ M.Sc. syllabi of most of our educational institutions offering postgraduate degrees in mathematics. There is a growing and as yet unfulfilled demand for a good textbook on this subject catering especially to the needs of the students in this country, despite the fact that there is a large number of books written on this subject.

In the present book an attempt is made to present the basic material in general topology as simply as possible, with a special emphasis on motivating the definitions and on discussing the significance of the concepts defined and the theorems proved. The topics covered are all standard, it is the presentation which, I believe, is new. For example, instead of starting with a ready-made definition of a topological space, one whole chapter is devoted to developing a motivation for the definition, that is, to convincing the reader why the definition is natural and important. Similarly, while defining compactness, it is painstakingly pointed out how it is the next best thing to finiteness. The same applies to other important concepts in the book. General topology is usually taught to second year students of M.A./M.Sc. and as such it is assumed that the students are familiar with elementary properties of metric spaces through their first year courses in analysis. However, this is not a stringent prerequisite for reading the book as most of it is quite independent of metric spaces and their important properties are proved here anyway. Prerequisites from set theory and analysis are reviewed in Chapter 2, while Chapter 1 gives a warm-up in logic so that the student can do justice to the deductive nature of the subject.

Many books on topology assume a certain degree of maturity on the part of the reader. My experience shows that for the average M.A./M.Sc. students, a presumption of mathematical maturity is grossly incorrect and leads to disastrous consequences. In this book, therefore, I have made an attempt to develop, along with the material, some of the qualities that constitute maturity. This is done both through the exposition of the material and through the exercises at the end of each section. While some exercises are routine applications of the theorems in the text, or as the reader to fill the gaps in the proofs of some theorems, the others are intended to provoke some thoughts. A few exercises have no clear cut answers as they ask the reader to discuss or comment upon something. Their intentional vagueness should not, however, be taken to imply that they are of peripheral interest.

A word or two would probably be in order regarding the language used and the style of presentation. I have tried to avoid stodgy formalism, without being sloppy or too colloquial. Where the subject matter so demands, I have freely indulged into long discussions stressing some points even at the risk of sounding platitudinous to a reader who is already mature enough. Keeping in mind the type of students for whom this book is meant, I would prefer to be guilty of inclusion of such discussions rather than of their omission.

As the contents of the book reveal, it covers most of the material which is standard in a one-year course in general topology. However, occasionally I have exercised my personal choice. An entire chapter is devoted to category theory in order to acquaint the student with an important language in which current mathematics is expressed. On the other hand, I have omitted such topics as paracompactness of metric spaces and the general metrisation theorem. Although their importance can hardly be denied, these topics turn out to be rather too involved for the average student and are often skipped even when they form a part of the syllabus. It may be argued and with good sense that they deserve a better place in a book on general topology than, say, category theory does. The only defence I can put up against such a charge is that the present book is written more for the sake of the student than for the sake of the subject.

The book contains fifteen chapters, more or less of equal length (except Chapter 9). Each chapter is divided into four (sometimes three) sections. Each section generally concentrates itself on one or two major ideas and theorems. It is expected that in an average class, the gist of the material in a section can be covered in one lecture of sixty minutes, allowing one more lecture to discuss some of the exercises. With this pace, there should be little difficulty in completing the book in one academic year in a class meeting three to four times a week.

The first two chapters are strictly preparatory and may be assigned as a self-study. Chapter 3 provides a motivation for topology through geometry. The core of topology is contained in Chapters 4 to 11. The remaining four chapters are relatively independent of each other; they can be read in any order without a significant loss of continuity. Also any of them may be omitted in the interest of time in a short or a rapid course.

All the definitions, propositions, lemmas and theorems occurring in a section are numbered together in the order in which they appear, the section number always coming first, thus: (2.1) Definition, (2.2) Proposition, etc. These numbers are also used in referring to them in the same chapter. However, if they are referred to in another chapter, they will be numbered as (a.b.c) where a is the number of the chapter in which they occur. Thus Theorem (1.4) of Section 1, Chapter 6 will be designated by Theorem (1.4) in Chapter 6 and by Theorem (6.1.4) in all other chapters. The symbol ▮ is used to mark the end of a proof. Where a proof is either omitted or comes prior to the statement, the symbol ▮ is put immediately after the statement. The abbreviations 'iff' for 'if and only if' and 'w.r.t.' for 'with respect to' are also commonly used. In definitions, the concept to be defined is put in bold face.

Exercises form an integral part of the book. Each section has ten to twelve exercises on the average, ranging from the very routine to the challenging. Many of them are referred to in the body of the text. Quite often the student is asked to complete a proof or to supply the details of an argument by way of an exercise. Sometimes the queries such as 'why?' or 'how?' are made in the text where we want to alert the reader that some justification is needed, even though it may be a routine one. Generous hints are provided and the reader is well advised to try at least the less demanding exercises. Following the usual practice, an asterisk (*) is used to mark a challenging problem, realising of course, that there will be a room for disagreement as to whether a particular problem is worth so marking or not. Occasionally, two stars are used to mark a problem. This is usually the case where the problem is unusually demanding or requires extraneous techniques for its solution. A few exercises are designed to acquaint the reader with certain important concepts which could not be introduced in the body of the text for considerations of space. Miscellaneous exercises are provided at the end.

At the end of nearly every section the reader will find notes and a guide to the literature relevant to the material in that section. They are generally intended to direct the interested reader to appropriate references, for further information on certain concepts which are only briefly touched at or sometimes to acknowledge the credit to the original sources I have borrowed from. No attempt is made to trace the origin of each and every result and counter-example. Occasionally a historical remark is also made. However, as the present book is not meant as a reference book on topology, it is not very fruitful—nor is it really practicable—to give an exhaustive bibliography.

A word is in order about the sequence in which certain topics appear especially where it differs from the conventional one. It is customary in books on topology to put off things like convergence of nets, products and compactness until relatively late and then to give concentrated doses of them. We have deviated from this practice. The average reader of the book is sure to be familiar with convergence of sequences at least for real numbers. We have counted on this familiarity and defined convergence of sequences in topological spaces as early as in Chapter 4. Many results about sequences are proved and all this is used as a motivation for the theory of nets in Chapter 10. Similarly, instead of packing all theorems about compactness in a single chapter, we have defined compactness very early in the book and have spread the results about it over several chapters. We hope this will enable the student to absorb this concept without haste. As for products, although there is no conceptual difference between finite products and arbitrary ones, experience shows that while the student can generally handle finite products, often even the definition of an arbitrary product is not clear to him. In the usual treatment, even before the student has sufficient time to develop a feeling for set-theoretic products, the concept of the weak topology generated by a family of functions is thrust upon him and theorems about product spaces are proved at an alarming rate. As a result, the average student often has a very hazy understanding of the product topology. To remedy this, we have defined products of finitely many spaces very early in the book and have proved a number of theorems about them. It is hoped that this approach will allow the student ample time to grasp the essence of the product topology. When he sees arbitrary products in Chapter 8, he would be in a position to accept them as a natural extension of what he already knows.

This procedure has undoubtedly resulted in some repetitions which could have been avoided otherwise. But we feel the clarity gained more than compensates for it.

As for diagrams, I have tried to strike a balance between the two extremes typified on one side by Kelley (without a single figure) and on the other by Munkres (with almost every proof illustrated with a diagram). It is true that most of the arguments in topology have a euclidean or at least a metrical intuition. As a result the importance of drawing appropriate diagrams as an aid to understanding the proofs in topology is indisputable. However, I feel this is a habit which a reader should develop on his own. If this book is used as a text, the instructor, instead of duplicating the proofs, may illustrate them with diagrams (and a lot of hand-waving). As a result I have confined the diagrams only to drawing some important spaces. They are numbered consecutively in each chapter.

It was Professor M. N. Vartak who suggested that I write a book on Topology. Financial assistance to prepare the manuscript was given by the Curriculum Development Centre of the Quality Improvement Programme at I.I.T. Bombay. My sincere thanks to both.

In a book of this type any claim to originality of the results per se must necessarily fail and save for a few exercises and counter-examples, almost all results are borrowed from various sources. The most prominent among these is the book 'General Topology' by J. L. Kelley. I had my first course in topology through this book and its influence on the present work is inevitable. Other books I have occasionally borrowed from are by Dugundji, Hocking and Young, Willard and Munkres. The case of Munkres' book especially deserves a comment. It appeared after the first version of this book was prepared. The similarity between the two was noticeable although perhaps not very surprising since probably the same convictions have guided Munkres as me. In preparing the second version, I have tried to minimize the overlap with Munkres. However a few of his exercises were too tempting not to borrow. As regards the style, I am influenced by I. N. Herstein. All this is of course not to suggest that any of the authors mentioned here is responsible for deficiencies on my part.

Several colleagues and the referee made important suggestions to improve the book. The typing was done mostly by M. Parameswaran.

K. D. Joshi

Contents

Chapter One

Logical Warm-up

If mathematics is regarded as a language, then logic is its grammar. In other words, logical precision has the same importance in mathematics as grammatical correctness in a language. Both can be acquired either through a formal study or through good practice. In this chapter we do not proceed to study formal logic. Rather, we emphasise those aspects of it which are vital to the study of mathematics. Indeed, the essence of what we are going to say is the same as what any layman would tend to think as commonsense, provided he is in the habit of carefully weighing his statements. Some people are naturally good at these sorts of things just as some privileged persons have a musical ear. The moment they come across a flaw in some reasoning, some 'inner sense' within them keeps them on the alert. Unlike musical expertise, however, anyone can acquire this skill through careful practice.

1. Statements and Their Truth Values

It is by no means an easy task to give a complete and rigorous definition of a statement. Indeed, such a definition would cut deep into formal logic, linguistics and philosophy. For our purpose, a statement is a declarative sentence, conveying a definite meaning which may be either true or false but not both simultaneously. Incomplete sentences, questions and exclamations are not statements.

Some examples of statements are:

(1) John is intelligent.
(2) If there is life on Mars then the postman delivers a letter.
(3) Either grandmother chews gum or missiles are not costly.
(4) Every man is mortal.
(5) All men are mortal.
(6) There is a man who is eight feet tall.
(7) Every even integer greater than 2 can be expressed as a sum of two prime numbers.
(8) Every man with six legs is intelligent.

Notice that (2) and (3) are perfectly valid statements although they sound ridiculous, because there is no rational correlation between their components. Statements (4) and (5) can be logically regarded as same or equivalent statements. Note that they are statements about members of a class (in this case the class of all men) as opposed to the first statement which is about a particular individual called John. Statement (6) is an example of what is known as an **existence statement.** It asserts the existence of something but does not do anything more. For example in the present case, statement (6) does not say anything about the name or the whereabouts of the eight-footer nor does it say that he is the only such man. There may or may not be other men who are eight feet tall. The statement is silent on this point.

Statements must be either 'true' or 'false' but not both. The truth value of a statement is said to be 'True' or T or 1 if it is true and 'False' or F or 0 if the statement is false. The beginner is warned not to be misled by the word 'value' here which has the connotation of numerical value. In particular, 1 and 0 are merely symbols here and not the integers which they usually represent. There is indeed some good reason why they are used here despite the likely confusion, but it is beyond our scope to discuss it here. For our purpose, giving the truth value of a statement is merely a fancy way of saying whether it is true or false. What is important is that there is no in-between stage. There is no such thing as saying that a statement is 'somewhat true' or 'almost true'. True means absolutely and completely true, without qualifications. This is especially important to understand in the case of statements such as (4) and (5) which deal with members of a class. Such statements are to be regarded as false even when there is just one instance in which they fail. Indeed this is where a logician or a mathematician differs from a layman for whom the exception proves the rule. In mathematics, even a single exception (a **counter-example** as it is called) renders false a statement about members of a class.

Of course, the truth value of a statement may depend upon the way it is interpreted. For example in statement (1) above, the truth value depends upon what one means by 'intelligent'. But once intelligence is clearly defined then the statement must be declared as either completely true or false. Actually unless such a precise definition is implied and understood, (1) is not a valid statement. Truth values of statements, such as (2) or (3) will be discussed later.

Statement (7) is the famous conjecture of Goldbach. **Conjecture** means a statement whose truth value is not known at present. So far nobody has proved the statement nor has anyone disproved it. (Note that since the statement is about a class of integers, in order to disprove it, that is, to prove that it is false, all one has to do is to exhibit just one even integer greater than 2 which cannot be expressed as a sum of two primes. But no one has found such an integer so far.) Thus, at present no one knows whether (7) is true or false. However, this does not prevent it from being a statement. For it does have a definite truth value, the only trouble is that we do not quite know what it is. This is indeed a fine point. As was remarked earlier, the existence of something is to be distinguished from other knowledge about the thing. This observation applies to the truth value of a statement.

Another interesting point is presented by statement (8), which sounds as a statement from mythology or from fairy tales. We know no man has six legs. And, therefore, the statement (8) (and indeed any statement about a six-legged man) is logically true. Such a statement is said to be **vacuously true.** This certainly appears very puzzling to a layman. A moment's reflexion will, however, clear the mystery. For, if statement (8) is not true, it must be false, as there is no other possibility. But the only way statement (8) can be shown to be false is by finding (that is, proving the existence of) a six-legged man who is not intelligent. Since no man has six legs, this is impossible, the question of intelligence does not arise at all. So the statement (8) cannot be proved to be false. Therefore, we are forced to conclude that it is true. (This argument is not convincing to all persons. Such persons have founded their own school of mathematics called the **constructivist** school. The main point where they differ from the classical mathematical logic is that they do not accept proofs by contradiction, or the so-called *reductio ad absurdum* arguments of which the present one is an example. We shall, however, agree to bend our logical conscience if necessary in order to accept that (8) is true.)

As a practical application of vacuous truth, a student who has not appeared for any subject in an examination can boast that he secured a first class in every subject that he appeared for. Logically, his claim is absolutely correct. It is easy to set him right, however, if one merely notes that in such a case it is equally true to say that he has failed miserably in every subject he appeared for. Both statements are equally true and there is no logical contradiction, because both statements deal with something non-existent. We shall return to this point later when we shall discuss logical implications.

To conclude this section we give some phrases or sentences which are not statements.

(1) Will it rain?
(2) O! Those heavy rains.
(3) I am telling you a lie.

The first two are not statements because they are not declarative sentences. The third one is not a statement because it cannot be assigned any truth value. If it is true then according to its contents it is false. On the other hand, if it is false then according to its contents it is true. The trouble with this sentence is that it refers to itself. Many paradoxical situations arise because of self-referencing and the reader must have come across some popular puzzles which are tacitly based upon self-reference or some variations thereof.

Exercises

1.1 Which of the following expressions are statements? Why?

(1) If it rains the streets get wet.
(2) If it rains the streets remain dry.
(3) It rains.
(4) If it rains.

(5) John is intelligent and John is not intelligent.
(6) If John is intelligent then John is intelligent.
(7) If John is intelligent then John is not intelligent.
(8) For every man there is a woman who loves him.
(9) There exists a woman for whom there exists no man who loves her.
(10) There exists a woman such that no man loves her.
(11) This sentence is false.
(12) There is life outside our solar system.

1.2 On both the sides of a piece of paper it is written, 'The sentence on the other side is false.' Are the two sentences so written statements? Why? What if on one side 'The sentence on the other side is true' is written and on the other side 'The sentence on the other side is false'?

1.3 A barber in a village makes an announcement, 'I shave those (and only those) persons in this village who do not shave themselves.' Is this announcement a statement? Why?

2. Negation, Conjunction, Disjunction and Truth Tables

There are three ways of manufacturing new statements from given ones. Any complicated statement can be shown to be obtained from some very elementary or simple statements (such as 'it rains' or 'John is intelligent'). Let us study one by one the three ways of generating new statements.

(i) Negation: To negate a statement is to make another statement which will be opposite to the original one in terms of truth value. That is, we want that the negation be true precisely when the original statement is false, and vice versa. The simplest way to achieve this would be to merely prefix the phrase 'it is not the case that' before the original statement. For example the negation of 'John is intelligent' would be 'It is not the case that John is intelligent'. However, this is a very mechanical way and sentences formed like this tend to be linguistically clumsy. It is much better to say that 'John is not intelligent' which conveys exactly the same meaning. We can even go further and abbreviate 'not intelligent' to 'unintelligent', or to, say, 'dumb', provided we agree that 'intelligent' and 'dumb' are antonyms, that is, words whose meanings are opposite to each other. Similarly, 'X is rich' may be negated as 'X is poor' and so on.

In mathematical statements symbols are often used for brevity. The negation of such statements is expressed by putting a slash (/) over that symbol which incorporates the principal verb in the statement. Thus '$x = y$' (read 'x is equal to y') is negated as '$x \neq y$' (read 'x is not equal to y'). Similarly, '$x \notin A$' is the negation of '$x \in A$', (read 'x belongs to A').

Things are not so simple when we come to more complicated sentences. For example suppose the original statement is 'John is very intelligent'. Of course one way to negate it is simply to say 'It is not the case that John is very intelligent'. But this is hardly satisfying. A much more reasonable answer is 'John is not very intelligent'. In practice we often regard such a statement as a polite way of saying that John is dumb (in fact very dumb). In mathematics this is not so. The original statement is about the degree of John's intelligence. There are many degrees of intelligence ranging from very high to very low. Just because John lacks a very high degree does not automatically mean that the level of his intelligence is very low. It could be that he is just average. In this case, therefore, the correct logical negation is merely 'John is not very intelligent' and not 'John is very dumb'. A similar remark applies in all cases where there are more than two possibilities. For example to say that a book is not red does not necessarily mean that it must be of some other specific colour, say, green (although they are complementary colours). Hence, the logical negation of 'The book is red' is simply 'The book is not red' and not 'The book is green' or 'The book is blue'.

Even greater care is necessary when we come to statements such as 'Every man is mortal'. A layman is most apt to negate this statement either as 'Every man is immortal' or 'No man is mortal' or perhaps even as 'Every woman is mortal'. A logician will however negate it as 'There exists a man who is not mortal' or as 'There is an immortal man'. Recall that the statement in question is about a whole class, namely, the class of all men. To say it is false simply means that it is false in at least one instance. This easily leads to the correct negation. Note that the negation of a statement beginning with 'every' is an existence statement. We can also state the negation as 'Not every man is mortal'. But it is better to avoid saying 'Every man is not mortal' as it is likely to be confused with 'Every man is immortal'.

When we come to the negation of existence statements, the tables are turned around. As expected, the negation will be a statement about a class, asserting that every member of that class fails to be something whose existence is asserted by the original statement. For example the logical negation of 'There exists a rich man' would be that 'Every man is poor' or 'No man is rich', and not 'There exists a poor man'. The negation of 'There exists a woman such that no man loves her' is 'For every woman there exists a man who loves her'. The reader is strongly urged to master these types of sentences thoroughly, that is, to interpret them precisely as well as to negate them correctly as mathematics is full of such statements.

Sometimes statements are represented by symbols p, q, r, etc. just as in algebra one uses the symbol $x, y, z, \ldots$, etc. to denote some variable quantities. With this notation there is a special symbol, $\neg$ (read as 'not'), for negation. For example if p stands for the statement 'John is strong' then $\neg p$ is read as 'not p' and denotes the statement 'John is not strong'. Note that, by the very definition, the truth values of p and $\neg p$ are always opposite of each other, no matter what p stands for. Although $\neg p$ is standard, the notations $\sim p$ or p' are also commonly used for the negation of p.

(ii) Conjunction: When two statements are joined together by the word 'and', the resulting statement is called their conjunction. For example the conjunction of 'John is rich' and 'Bob is weak' is 'John is rich and Bob is weak'. Where possible, we can paraphrase the conjunction so that it looks better English. For example the conjunction of 'John is intelligent' and 'John is rich' can be stated less mechanically as 'John is both intelligent and rich'. Also, instead of 'and' we can use 'but' when we want to stress an implied contrast between the two statements. For example we may say 'Bombay is big but Delhi is beautiful'. Mathematically, it means the same as 'Bombay is big and Delhi is beautiful'.

The standard notation for conjunction is $\wedge$, read as 'and'. Symbolically, if p and q are statements then their conjunction is denoted by $p \wedge q$ and is read as 'p and q'. The reason for introducing the special symbol $\wedge$ instead of simply writing the word 'and' is that we use the word 'and' in ordinary language so often that its use as a symbol for conjunction is sometimes likely to cause confusion. Note that we can form the conjunction of any two statements whatsoever. There need not be any semantic connection between the two, although in practice such a connection is usually understood. That is why a statement such as 'Missiles are costly and grandma chews gum' sounds ridiculous, although mathematically it is a perfectly valid statement obtained by the process of conjunction.

When we come to the truth value of a conjunction of two statements, it should be obvious to anyone that it is true when both the statements are true. If either one of them or both of them are false, then the conjunction would be false. As there is no such thing as 'half true' in mathematics, we must accept the conjunction as false even when one of its constituents is true and the other false.

(iii) Disjunction: The third way of forming new statements out of old ones is by taking the disjunction of two statements. For this, merely put the word 'or' between the statements. Thus the disjunction of 'John is intelligent' and 'There is life on Mars' is 'John is intelligent or there is life on Mars'. Sometimes we put the word 'either' before the first statement to make the disjunction sound nice, but it is not necessary to do so, so far as a logician is concerned. The symbolic notation for disjunction is $\vee$ (read as 'wedge' or 'or'). If p, q are two statements their disjunction is denoted by $p \vee q$.

A word of warning regarding the use of 'or' is very important. In practice, when one says 'I shall spend my vacation either in Bombay or in Pune' it is generally implied that the person will not spend it both in Bombay and in Pune. Indeed, often the very nature of the two statements is such that they cannot hold simultaneously, for example the statement, 'This book is either red or green'. Occasionally, both the possibilities are also implied together. In such a case the words 'or both' are added to emphasize it as in the statement 'A person with such a handwriting must be either a doctor or a crook or both'. In logic or in mathematics, however, it is not necessary to specify 'or both'. The word 'or' as it is used in mathematics always implies either one or both the two alternatives. This is known as the *inclusive sense* of the word 'or'. When only one of the possibilities is intended and the simultaneous holding of both of them is not automatically precluded by the context, it is explicitly stated by 'but not both'. For example consider the statement, 'For each x, we take either the positive square-root or the negative square-root but not both'. In symbolic terms, $p \vee q$ always means either p or q (or both). The logical expression for 'either p or q but not both' would be $(p \vee q) \wedge (\neg(p \wedge q))$ or equivalently $(p \wedge \neg q) \vee ((\neg p) \wedge q)$.

Another word of caution is necessary when we come to statements about members of a class. Note that the statement 'John is either a man or a woman' can be taken to be the same as the disjunction of the two statements 'John is a man' and 'John is a woman'. However, the statement 'Every human being is either a man or a woman' is definitely not the same thing as saying, 'Either every human being is a man or every human being is a woman'. The essential difference is that unlike the first statement, the second one is about members of a whole class, in this case the class of human beings. Although for each individual member of the class one of the two possibilities holds, it does not follow that the *same* possibility should hold for all members. Quite a few cases of logical fallacies arise because of failure to appreciate this point. A similar observation holds in the case of the conjunction of two existence statements, as would be clear from one of the exercises at the end of this section (Exercise 2.4).

In view of the inclusive sense of the word 'or', it should be obvious that a disjunction $p \vee q$ is true when at least one of the two statements is true and false when both p and q are false.

What is said so far about truth values can be conveniently represented by a device known as *truth tables*. In algebra we consider functions, such as $\sin x$, $x^2 - y^2$, $x \tan (yz)$ etc. of one or several variables. We then prepare tables showing the values of these expressions for various values of the variables x, y, z, etc. Similarly, we can regard statements $\neg p$, $p \wedge q$, $\neg p \vee (q \wedge r)$, etc. as functions (or formulas) of the variables p, q, r, etc. and express their truth values corresponding to various possibilities that may arise by assigning different truth values to p, q, r, etc. Actually, the things are much simpler here because an algebraic variable x, y or z can assume infinitely many values, but each variable p, q, r, etc. assumes only two truth values, 1 or 0. This means that if there are n independent variables in our expression, then the complete truth table for it would need 2^n rows. The following is a truth table for the statement $p \wedge (q \vee \neg r)$ of three variables p, q, r.

The columns for $\neg r$ and $q \vee \neg r$ are auxiliary columns. The reader is urged to check that the table is correct. As an example in the first row p, q, r are all true (i.e. have truth value 1). So $\neg r$ is false. But $q \vee \neg r$ is true and finally $p \wedge (q \vee \neg r)$ is true.

When two statements are obtained as functions of the same variables, they are said to be **logically equivalent** if they have the same truth values for all possible combinations of truth values of these variables (which in terms of truth tables means that the columns under them are identical). As a simple example consider $\neg(\neg p)$ where p is any statement. Logically, this is the

Row	p	q	r	$\neg r$	$q \vee \neg r$	$p \wedge (q \vee \neg r)$
1	1	1	1	0	1	1
2	1	1	0	1	1	1
3	1	0	1	0	0	0
4	1	0	0	1	1	1
5	0	1	1	0	1	0
6	0	1	0	1	1	0
7	0	0	1	0	0	0
8	0	0	0	1	1	0

negation of the negation of p. Note that it is not quite the same as p on the nose although in practice we would like to identify it with p. We can do so because it is easy to verify that $\neg(\neg p)$ and p are logically equivalent. As less trivial examples the reader is urged to verify the equivalence between $p \wedge (q \vee \neg r)$ and $(p \wedge q) \vee (p \wedge \neg r)$ and between $(p \vee q) \wedge (\neg(p \wedge q)$ and $(p \wedge \neg q) \vee ((\neg p) \wedge q)$. Note also that the statement $p \vee \neg p$ is always true and $p \wedge \neg p$ is always false. A statement which is always true is called a **tautology** while a statement which is always false is called a **contradiction.**

An interesting thing happens when we take the negation of a conjunction or a disjunction. For example let p be the statement 'John is rich' and q the statement 'John is intelligent'. Then $p \vee q$ is the statement 'John is either rich or intelligent'. The negation of this, that is to say that it is false, amounts to saying that John is neither rich nor intelligent. But that means that the negations of 'John is rich' and of 'John is intelligent' are both true. Thus, we expect that $\neg(p \vee q)$ should be logically equivalent to $(\neg p) \wedge (\neg q)$. The reader can indeed verify this to be the case by writing the truth tables for both.

The perceptive reader must have been reminded of the well-known De Morgan's laws in set theory at this point. Indeed, most likely he has already noted some similarity between negation and complements, between conjunction and intersection of sets and between disjunction and union of sets. Such a similarity is not coincidental. It is not hard to explain its full significance if one studies what is known as a **truth set of a statement**. This is the bridge needed to go from statements to sets. However, we do not pursue this line further.

Exercises

2.1 Write down the logical negations of all examples of statements given in the last section (except statements involving 'if . . . then . . .').

2.2 Verify that the statements in each of the following pairs are logically equivalent. Also, in each case explain why the equivalence is obvious or reasonable to expect.

(i) $p \vee (q \wedge r)$ and $(p \vee q) \wedge (p \vee r)$

(ii) $(p \vee q) \wedge (\neg(p \wedge q))$ and $(p \wedge \neg q) \vee ((\neg p) \wedge q)$
(iii) $\neg(p \wedge q)$ and $(\neg p) \vee (\neg q)$.

2.3 We can define the conjunction of three statements p, q and r either as $p \wedge (q \wedge r)$ or as $(p \wedge q) \wedge r$. Are these the same? Are they logically equivalent? Justify your answers.

2.4 Suppose X has three friends, A, B and C who are, respectively, good at cricket, music and mountaineering but not at any other fields. What is the fallacy in the following? "The statement 'One of X's friends is a cricketer' is true and so is the statement 'One of X's friends is a musician'. So their conjunction is true, that is, the statement 'One of X's friends is both a cricketer and a musician' is true. But X has no such friend."

3. Logical Implications

As far as mathematics is concerned, the most important aspect of logic is the statements of the form 'If . . . then . . .'. These are called implications. Theorems in mathematics are commonly expressed as statements of this form. For example the theorem that if two triangles are congruent then they are also similar. Some theorems do not appear in this form on the surface but can easily be paraphrased into it. For example the theorem 'The sum of the three angles of any triangle equals 180 degrees' can be put as 'If ABC is a triangle then $\angle A + \angle B + \angle C = 180$ degrees.' The importance of implication statements in mathematics will be discussed in detail a little later. But let us first get down to the precise meaning of such statements. They bear the form 'If p then q' where p and q are some statements. This statement is also read as 'p implies q' and also written as '$p \rightarrow q$' or '$p \Rightarrow q$'. Such a statement is called a **logical implication.** The statements p and q are called, respectively, the **hypothesis** and the **conclusion** of the implication. The meaning of $p \rightarrow q$ is that whenever p is true, q is also true, or in other words that the truth of p ensures (or forces) the truth of q. But it says nothing in the contingency where p is false. In particular, it does not say that if p is false then q is false. This is a very vital point because in practice we tend to attach this extra meaning to implication statements. For example when a person says to his friend 'If Monday is a holiday then I shall attend your wedding', in practice, we think that if Monday is not a holiday then the person will not attend the wedding. To a mathematician, however, the present statement is completely silent as far as the case of Monday not being a holiday is concerned. In such a case whether the person attends the wedding or not, he is not violating his commitment. If the person had something in mind in the event that Monday was not a holiday, he should have made it explicit by another implication statement. For example if he had said, 'I shall attend your wedding if Monday is a holiday, but not otherwise', or 'I shall attend your wedding if, and only if, Monday is a holiday' then this amounts to making two separate statements, 'If Monday is a holiday then I shall attend your wedding' and 'If Monday is not a holiday then I shall not attend your wedding'.

The logic employed here may seem strange in the beginning, but we have already come across something similar to it when we discussed vacuously true statements in Section 1. Recall that the statement there was 'Every man with six legs is intelligent'. This can be given the form of an implication statement by 'If a man has six legs then he is intelligent'. So, if we consider a man who does not have six legs, then whether he is intelligent or not, the statement is true in this case, because it is inapplicable and hence cannot be shown to be false. The only way the implication statement can be false if there is a six-legged man who is not intelligent. Since there is no such man, the statement is true vacuously. In general, an implication $p \Rightarrow q$ is consideried to be true whenever p is false, that is, whenever the hypothesis is not satisfied. When p is true, the truth value of $p \Rightarrow q$ will depend upon whether q is also true or not.

With this rule in mind it is easy to construct the following truth table for $p \Rightarrow q$:

Row	p	q	$p \Rightarrow q$
1	1	1	1
2	1	0	0
3	0	1	1
4	0	0	1

If we construct the truth table for $(\neg p) \vee q$ we see that it is identical to the one above. Therefore, $p \Rightarrow q$ is logically equivalent to $(\neg p) \vee q$. Indeed, this is often taken as the very definition of $p \Rightarrow q$. With this approach, however, it becomes necessary to explain why the statement 'either q holds or else p fails' conforms to our intuitive understanding of logical implication.

Using truth tables, the reader can verify that $p \Rightarrow q$ is logically equivalent to $(\neg q) \rightarrow (\neg p)$. Verbally, this means that 'p implies q' is equivalent to saying 'if q fails then p fails'. This certainly sounds very reasonable. Indeed, this is the very basis of the so-called proof by contradiction or a *reductio ad absurdum* argument. We use this type of argument in every-day practice too. Take for example the statement, 'If it rains the streets get wet'. If the streets are not wet, we conclude on the basis of this statement that it did not rain. The statement $(\neg q) \rightarrow (\neg p)$ is said to be the **contrapositive** of the statement $p \rightarrow q$. Both are logically equivalent to each other.

The contrapositive of an implication statement should not be confused with its converse. The **converse** of $p \rightarrow q$ is defined to be the statement $q \rightarrow p$. In other words, the hypothesis of the original implication statement is the conclusion of the converse and vice versa. For example the converse of the statement 'If two triangles are congruent then they are similar' would be 'If two triangles are similar then they are congruent'. There is no correlation between the truth values of an implication and its converse. Numerous examples can be given where both are true, both are false or where one is true and the other false.

Sometimes, in the statement $p \rightarrow q$ the hypothesis, that is, p, is itself the negation of some statement, as for example, the statement 'If it does not rain the crops will die'. In such a case it is customary to replace the phrase 'if not' by the single word 'unless'. With this change, the present statement would become 'Unless it rains, the crops will die'. We warn once again that this statement says nothing whatsoever about the survival of the crops in the event it does rain. Here again, a logician differs from a layman who would interpret this present statement to mean that if it rains crops will be saved. The safest way to correctly interpret statements involving 'unless' is to substitute for it 'if not'.

In view of the immense importance of implication statements in mathematics, let us consider some other ways of paraphrasing them. Suppose p and q are any statements. Then $p \rightarrow q$ can be read in any of the following ways:

(i) p implies q.
(ii) q follows from p.
(iii) q is a (logical) consequence of p.
(iv) If p is true then q is true.
(v) If q is false then p is false.
(vi) p is false unless q holds.
(vii) p is a sufficient condition for q.
(viii) q is a necessary condition for p.
(ix) p is true only if q is true.

Item (i) is just the definition, while (ii), (iii) and (iv) are its paraphrases. As we have seen before, (v) is the contrapositive of (i) and (vi) a rephrasing of (v). The last three are the only versions which call for a comment. Of these (vii) is fairly straightforward. For example to say 'If it rains the streets get wet' clearly amounts to saying that 'Raining is a sufficient condition for the streets to get wet', or that 'In order that the streets get wet, it suffices if it rains'. Thus, the use of the word 'sufficient' here conforms to its ordinary meaning.

It is a little confusing to use version (vii) in the case of some statements. For example in the example just given, the statement would read 'Wetting of streets is a necessary condition for it to rain'. This sounds absurd. The trouble is with the word 'condition'. In practice, it has the connotation of a prerequisite, that is, something which is to exist prior to the happening of some event. In the present case the question of streets getting wet arises only after the rain and that is why it is hard to swallow that wetting of streets is a necessary condition for it to rain. Perhaps, another example would clarify the situation. Consider the statement, 'If two triangles are congruent then they are similar'. This means that in order that two triangles be congruent they must at least be similar to each other. Congruency can never occur if similarity does not hold. In other words, similarity of the triangles is a necessary condition for them to be congruent. Whether it is sufficient or not is not the concern of the statement, it is the business of the converse statement. Necessity and sufficiency should never be confused with each other. In a sense, they are converse to each other.

About the last version, 'p is true only if q is true' it is once again necessary to distinguish a layman from a logician. When a layman says 'I shall come only if I am free', he generally means that he will come if he is free but not otherwise. A logician, however, makes no such commitment when he makes the same statement. All he is saying is that his being free is a necessary condition for his coming, that is, his coming will be impossible if he is not free. He is saying nothing at all as to what he will do if he is free. Here, too, it is vital to distinguish between 'if' and 'only if'.

There is one exception to the preceding remarks. When something is defined in terms of a condition, it is customary to cite this condition as sufficient, even though it is in fact sufficient as well as necessary. Thus, when we say 'A triangle is called **equilateral** if all its sides are equal', it also means that a triangle all whose sides are not equal will not be called equilateral. In other words, here 'if' means 'if and only if'. This usage is unfortunate but very standard. Fortunately, it appears exclusively in definitions and nowhere else.

In mathematics it often happens that we combine together an implication statement along with its converse. For example take the well-known theorem, 'The sum of opposite angles in a cyclic quadrilateral is 180 degrees and conversely'. If we let p be the statement '$ABCD$ is a cyclic quadrilateral' and q the statement '$\angle A + \angle C = 180$ degrees' then the statement of the theorem is the conjunction of $p \rightarrow q$ and $q \rightarrow p$, that is, the statement $(p \rightarrow q) \wedge (q \rightarrow p)$ or $(p \Rightarrow q) \wedge (q \Leftarrow p)$. These types of statements come up so frequently that it is convenient to have a shorter notation for them. The most natural choice is to use arrows in both directions, that is, to use $p \leftrightarrow q$ or $p \Leftrightarrow q$. Here again it is convenient to list down a number of versions of this statement.

(i) p and q imply each other.
(ii) p and q are equivalent to each other.
(iii) p holds if and only if q holds.
(iv) q is a characterisation of p. (This version is generally used only when p, q express some properties of the same object.)
(v) q holds if p does and conversely.
(vi) q holds if p does, but not otherwise.
(vii) if p is true then q is true and if p is false so is q.
(viii) q is a necessary as well as a sufficient condition for p.

Of course many other formulations are possible in view of the symmetry of p and q. Such statements are called 'if and only if' statements. The expression 'if and only if' appears so often in mathematics that it is customary to abbreviate it to 'iff'. Thus, the geometric theorem quoted above can be stated as 'A quadrilateral is cyclic iff the sum of its opposite angles is 180 degrees'.

A theorem of this sort is really equivalent to two separate theorems which are converses of each other. If we write the statement symbolically as $p \leftrightarrow q$ (or as $p \Leftrightarrow q$) then the implication $p \rightarrow q$ is called the **direct implication** or the **'only if' part** of the theorem while the other way implication, $q \rightarrow p$ is called **converse implication** or the **'if' part** of the theorem. In general, separate proofs are needed for both the parts. Occasionally, it so happens that the steps used in the proof of the direct implication are all reversible. In such a case, the converse is said to follow by reversing the proof of the direct implication. It is by no means the case that both the implications are of the same degree of difficulty. There are many theorems in which one of the implications is simple almost to the point of being trivial, but the other way implication is fairly involved. As an example take the well-known remainder theorem which states, 'Let $f(x)$ be a polynomial in the variable x. Then a real number b is a root of f (i.e. $f(b) = 0$) iff $(x - b)$ is a factor of $f(x)$'. In this case, the 'if' part is trivial, but the 'only if' part is fairly hard to prove.

The concept of implication leads naturally to that of comparison of relative strengths of statements. In practice, we say that a certain statement or piece of information is stronger than another if the knowledge of the former subsumes knowledge of the other. For example we say it is stronger to say that a certain person lives in Kerala than to say that he lives in India. This is so because anyone can infer the latter from the former by sheer commonsense, provided of course, that he knows that Kerala is a part of India.

Mathematically, we say that a statement p is **stronger** than a statement q (or that q is **weaker** than p) if the implication statement $p \rightarrow q$ is true. A few comments are in order. First of all, 'stronger' does not necessarily mean 'strictly stronger'. Note for example that every statement is stronger than itself. The apparent paradox here is purely linguistical. If we want to avoid it we should replace the word 'stronger' by the phrase 'stronger than or possibly as strong as'. However, the use of the word 'stronger' in this context is fairly standard. If $p \rightarrow q$ is true but its converse is false, then we say that p is **strictly stronger** than q (or that q is **strictly weaker** than p). For example it is strictly stronger to say that a given quadrilateral is a rhombus than to say it is a parallelogram. Second, given two statements p and q it may happen that neither is stronger than the other. Indeed, the two statements may not be related at all. In such a case we say that their strengths are not comparable to each other. For example the statement '$ABCD$ is a rectangle' and the statement '$ABCD$ is a rhombus'. The word 'sharper' is used sometimes for 'stronger'. This usage is common when the two statements deal with estimates or approximation of something.

What happens if out of two statements, each is stronger than the other? As we have already noted, in such a case we say that the two statements have the **same (or equal) strength** or that they are (mutually) **equivalent.** For example the statement, '*ABCD* is a cyclic quadrilateral' and the statement, '*ABCD* is a quadrilateral in which $\angle A + \angle C = 180$ degrees' are equivalent to each other. In the last section we defined the logical equivalence of two statements which were constructed from the same statements. It is obvious that in such a case they are of equal strength.

A large part of mathematics is concerned with the determination of relative strengths of statements, that is, with the comparison of strengths of statements. This is a task of varying degree of difficulty. In some cases the comparison of strengths or the equivalence of two statements is a matter of common sense or of using synonymous expressions. For example the statement 'I own this house' is equivalent to the statement 'This house belongs to me'. In some cases, on the other hand, the equivalence may not be obvious and needs to be established by some proof (as in the case of the statements '*ABCD* is a cyclic quadrilateral' and the statement '*ABCD* is a quadrilateral in which $\angle A + \angle C = 180$ degrees').

Exercises

3.1 Using truth tables show that $p \rightarrow q$ and $(\neg q) \rightarrow (\neg p)$ are logically equivalent to each other. Also, verify that $p \leftrightarrow q$ is logically equivalent to $(p \wedge q) \vee [(\neg p) \wedge (\neg q)]$.

3.2 What is the negation of $p \rightarrow q$? Is it the same as the converse of $p \rightarrow q$? Are the two logically equivalent? Justify your answer and illustrate it with examples.

3.3 Take a few theorems you know and cast them in the various forms of an implication statement. Also, state their converses. Give examples of 'iff' theorems in various versions.

3.4 Give an example of an 'iff' theorem where the converse implication is proved merely by reversing the proof of the direct implication.

3.5 Give examples of theorems which are true but whose converses are false.

3.6 Give examples of theorems whose converses are also true but where the two differ considerably in their depth.

3.7 If p, q, r are statements, p is stronger than q and q is stronger than r, show that p is stronger than r. [This is known as the **law of syllogism.** It may sound obvious and indeed trivial. However, a rigorous proof would require that for any three statements p, q, r, the statement $[(p \rightarrow q) \wedge (q \rightarrow r)] \rightarrow (p \rightarrow r)$ is a tautology. Construct a truth table for this purpose.]

3.8 Let p, q, r be statements and suppose q is stronger than r. Show that the implication statement $p \rightarrow q$ is stronger than the implication statement $p \rightarrow r$ (i.e. $(p \rightarrow q) \rightarrow (p \rightarrow r)$ is true given that $q \rightarrow r$ is true). In other words, among two implication statements with the same hypothesis, the one with the stronger conclusion is stronger. What can be said about the relative strengths of two implication statements having the same conclusion but different hypotheses?

3.9 In each of the following pairs of statements determine which of the two statements is stronger. If the two statements are not comparable, or are equivalent, justify why.

(i) Statement p: For every man there exists a woman who loves him.
Statement q: There exists a woman who loves all men.

(ii) p: The diagonals of a parallelogram bisect each other.
q: The diagonals of a rhombus bisect each other.

(iii) p: The diagonals of a rhombus bisect each other.
q: The diagonals of a rhombus bisect each other at right angles.

(iv) p: One of my friends is an actor and one of my friends is a cricketer.
q: One of my friends is an actor and a cricketer.

(v) p: If a man is rich he is also intelligent.
q: If all men are rich then all men are also intelligent.

(vi) p: If Monday is a holiday I shall come.
q: If Monday is not a holiday I shall not come.

(vii) p: π is an irrational number.
q: There exists an irrational number.

(viii) p: This glass is half filled.
q: This glass is half empty.

4. Deductions and Logical Precision

What exactly is the difference between a mathematician, a physicist and a layman? Let us suppose they all start measuring the angles of hundreds of triangles of various shapes, find the sum in each case and keep a record. Suppose the layman finds that with one or two exceptions the sum in each case comes out to be 180 degrees. He will ignore the exceptions and say 'The sum of the three angles in a triangle is 180 degrees.' A physicist will be more cautious in dealing with the exceptional cases. He will examine them more carefully. If he finds that the sum in them is somewhere between 179 degrees to 181 degrees, say, then he will attribute the deviation to experimental errors. He will then state a law, 'The sum of the three angles of any triangle is 180 degrees'. He will then watch happily as the rest of the world puts his law to test and finds that it holds good in thousands of different cases, until somebody comes up with a triangle in which the law fails miserably. The physicist now has to withdraw his law altogether or else to replace it by some other law which holds good in all the cases tried. Even this new law may have to be modified at a later date. And this will continue without end.

A mathematician will be the fussiest of all. If there is even a single exception he will refrain from saying anything. Even when millions of triangles are tried without a single exception, he will not state it as a theorem that the sum of the three angles in *any* triangle is 180 degrees. The reason is that there are infinitely many different types of triangles. To generalise from a million to infinity is as baseless to a mathematician as to generalise from one to a million. He will at the most make a conjecture and say that there is a strong evidence suggesting that the conjecture is true. But that is not the same thing as proving a theorem. The only proof acceptable to a mathematician is the one which follows from earlier theorems by sheer logical implications. For example, such a proof follows easily from the theorem that an external angle of a triangle is the sum of the other two internal angles.

The approach taken by the layman or the physicist is known as the **inductive** approach whereas the mathematician's approach is called the **deductive** approach. In the former, we make a few observations and generalise. In the latter, we deduce from something which is already proven. Of course, a question can be raised as to on what basis this supporting theorem is proved. The answer will be some other theorem. But then the same question can be asked about the other theorem. Eventually, a stage is reached where a certain statement cannot be proved from any other proved statement and must, therefore, be taken for granted to be true. Such a statement is known as an **axiom** or a **postulate**. Each branch of mathematics has its own postulates or axioms. For example one of the axioms of geometry is that through two distinct points there passes exactly one line. The whole beautiful structure of geometry is based on five or six axioms such as this one. Every theorem in geometry can be ultimately deduced from these axioms.

What is even more important is that not only is a mathematician's approach deductive but, in fact, his sole interest is in the process of logical deductions. He cares more about how certain statements imply certain others and not really about the truth or falsehood of the statements *per se*. For example if a mathematician is given two statements, 'All rich men are intelligent' and 'John is a rich man', it would matter little to him whether John is in fact rich or whether there is any rational correlation between richness and intelligence. His sole interest would be to deduce from these two statements (called 'premises') the conclusion 'John is intelligent'. Again, the truth or otherwise of the conclusion makes no difference to the mathematician. He is merely asserting that in case the premises are true, so is the conclusion. He will state his argument as 'Assuming that all rich men are intelligent and assuming that John is a rich man, it follows that John is intelligent'. This hardly sounds very brilliant. But, actually, every mathematical proof consists of a chain of such simple bits of deductive reasoning. Genius is needed not for these bits, but for combining them suitably.

In order that the reader be in a position to do justice to the deductive aspect of the rest of the book, let us examine a little closely the concept of logical validity of an argument. An **argument** is really speaking nothing more than an implication statement. Its hypothesis consists of the conjunction of several statements, called **premises**. In giving an argument, its premises are first listed (the order being immaterial), then a line is drawn and then the conclusion is given. Thus, a typical example of an argument is:

Premises: If it rains, the streets get wet.
If the streets get wet, accidents happen.
Accidents do not happen.

Conclusion: It does not rain.

Symbolically, let us denote the premises of an argument by $p_1, p_2, \ldots, p_n$ and its conclusion by q. Then the argument is the statement $(p_1 \wedge p_2 \wedge \ldots \wedge p_n) \to q$. If this implication statement is true then the argument is said to be **valid** otherwise it is called **invalid**. In other words, an argument is valid if its conclusion is necessarily true under all circumstances in which each of its premises is true and it is invalid if we can find at least one instance in which each of its premises holds but the conclusion fails.

Trivial examples of valid arguments are those in which the conclusion is itself one of the premises or consists of the conjunction of some of the premises. As a truly illustrative example, let us check the validity of the argument given above. Let us denote the three premises by p_1, p_2, p_3 and the conclusion by q. The premises themselves are not simple statements but are formed from other statements. Let us denote by r, s, t, respectively, the statements 'It rains', 'The streets get wet' and 'Accidents happen'. We then have $p_1 = r \to s$, $p_2 = s \to t$ and $p_3 = \neg t$, and $q = \neg r$. We then have to check whether the simultaneous truth of p_1, p_2, p_3 always forces q is to be true. One way of doing this is to write down the truth table of the statement $(p_1 \wedge p_2 \wedge p_3) \to q$ in terms of the truth values of the statements r, s, t. If the column under it consists entirely of 1's, the argument is valid, otherwise it is not. This is always a sure, albeit a little tedious way of testing validity. There are less mechanical ways. For example in the present case we have to deal with only those possibilities in which all the premises are true. Now, the third premise p_3 is true precisely when t is false. So we may confine ourselves only to those cases where t is false. As for the second premise p_2, it is an implication statement, $s \to t$ and so it is true precisely when either s is false or when both s and t are true. Since we are interested only in the case when t is false, we ignore the second possibility. Thus, p_2 and p_3 can hold simultaneously only when both t and s are false. We now bring in p_1, which is also an implication statement, and apply the same reasoning. Then we see that p_1, p_2, p_3 hold simultaneously only when r, s, t are all false. But in this case the conclusion q $(= \neg r)$ is true and thus the argument is valid. As another example, let us consider the argument:

Premises: $p_1 =$ If it rains then missiles are costly.
$p_2 =$ Missiles are costly.

Conclusion: $q =$ It rains.

In the first premise, there is no rational correlation between rain and the cost of the missiles. But this is no reason to declare the argument as invalid because it is none of our business to question the premises. To test the validity of this argument, we let $r \equiv$ 'It rains' and $s \equiv$ 'Missiles are costly'. Then p_2 is the same as s and p_1 is the implication statement $r \rightarrow s$. With the same reasoning as in the last example, we see that both the premises hold simultaneously whenever s is true, regardless of whether r is true or false. In another words, there exists a situation in which both p_1, p_2 hold and the conclusion q (which is the same as r) fails. Hence, the argument is invalid. This does not mean that the conclusion is false. It may very well be true; all we are saying is that its truth cannot be guaranteed from that of the premises. Note, also, that if we modify this argument by retaining its premises but by changing its conclusion to 'It does not rain', the new argument is still invalid. This may sound paradoxical to a beginner who is most apt to ask, 'Surely, either it rains or it does not rain. So at least one of the two arguments must be valid. How can both be invalid?' The answer is that the validity of an argument (with premises $p_1, \ldots, p_n$ and conclusion q, say) does not depend upon the truth of the conclusion q as such, but rather upon the truth of the implication statement $(p_1 \wedge p_2 \wedge \ldots \wedge p_n) \rightarrow q$. The statement $(p_1 \wedge p_2 \wedge \ldots \wedge p_n) \rightarrow \neg q$ is *not* the negation of $(p_1 \wedge p_2 \wedge \ldots \wedge p_n) \rightarrow q$ and so there is nothing wrong if both of them are false simultaneously.

When the number of premises is large, it is hardly practicable to check the validity of an argument either by the truth table or by a reasoning similar to the above. In such cases we resort to what is called the **chain rule** about validity. Simply stated, it says that by chaining together two (or more) valid arguments, we get a valid argument. A precise formulation is as follows. Let A_1 be an argument with premises $p_1, \ldots, p_k$ and conclusion q_1. Let A_2 be an argument with premises $q_1, p_{k+1}, \ldots, p_n$ and conclusion q_2. Let A_3 be the argument with premises $p_1, \ldots, p_k, p_{k+1}, \ldots, p_n$ and conclusion q_2. Then, if A_1 and A_2 are valid, so is A_3. The proof of the chain rule is obvious and left to the reader.

The chain rule is used so frequently in mathematics that an explicit mention is rarely made. Whenever in a proof we cite a theorem previously proved, we are implicitly using the chain rule. By repeated applications of the chain rule a valid argument can be split into a series of some very simple arguments. Among these, probably the most frequently used argument is what is called **modus ponens.** Formally, this is an argument with two premises, one of the form $p \rightarrow q$ and the other p and whose conclusion is q (where p, q are any statements whatever). Verbally, modus ponens is the argument that if an implication statement holds and its hypothesis is true, then so is its conclusion. This is certainly consistent with common sense and it is a triviality to verify the validity of modus ponens. Using modus ponens and the chain rule let us now establish the validity of the argument given on page 17. We follow the notation there and present the reasoning in the form most commonly adopted in mathematical proofs, as a sequence of steps beginning with the premises, ending with the conclusion and defending each step.

(1) $r \rightarrow s$ (given as a premise)

(2) $s \rightarrow t$ (given as a premise)

(3) $\neg t$ (given as a premise)
(4) $\neg t \rightarrow \neg s$ (equivalent to (2))
(5) $\neg s$ ((3), (4) and modus ponens)
(6) $\neg s \neg r$ (equivalent to (1))
(7) $\neg r$ ((5), (6) and modus ponens).

Hence, the argument is valid. Note that the desired conclusion appears at the end. But every step along the way is itself the conclusion of a valid argument. These arguments are often called **minor or subordinate** arguments. Sometimes, the conclusion of a minor argument is itself of some independent interest (such as step (5) above). In such a case it is isolated and called a **lemma** while the whole argument is called a **proposition** or a **theorem** depending upon its degree of depth and utility. Sometimes, it is interesting to go beyond the theorem, either with or without an additional premise. The resulting argument is then called a **corollary** of the theorem. Of course, these are relative terms. Strictly speaking, there is no difference between a lemma, a proposition, a theorem and a corollary from a purely logical viewpoint. The distinction rests on some extrinsic aspects, such as depth, utility and beauty. A lemma is usually useful in a limited context and is too technical to have an aesthetic appeal, a proposition is like a mini-theorem, a true theorem carries with it certain succinctness, while a corollary is like a bonus.

We often come across arguments of the following type:

Premises: Every man is mortal.
John is a man.

Conclusion: John is mortal.

The validity of this argument is obvious. But can it be said to be a case of modus ponens? Strictly speaking, the answer is 'no'. If the first premise were 'If John is a man, then John is mortal' then this argument would be modus ponens. It is true that the first premise, as it stands, has some force of an implication. It in fact asserts that if x is a man then x is mortal where x can be anything at all. Here x is like a variable and if we give it a particular value 'John', then we get the statement 'If John is a man then John is mortal'. Now we can apply modus ponens. Thus, the argument above involves, in addition to modus ponens, an element of substituting a particular value or taking a particular instance. Technically, this is called the **principle of instantiation.** Of course when it is applied, each occurrence of the variable must be replaced by the same value, otherwise disastrous results are obtained.

For our purpose it is not the logical technicalities but the logical precision that really matters. There is simply no substitute for logical rigour. Because of its strict reliance upon deductive logic, mathematics is remarkably free of individual opinions and consequent controversies. There may be disputes *about* mathematics, that is, about its value, etc. but there are no disputes within mathematics. There are some differences on a few fundamental issues.

For example as we mentioned earlier, there are some people who question the validity of a proof by contradiction. Also, sometimes there is a controversy as to whether a particular axiom should be included in a particular branch or not, a classic example of this being the Playfair's axiom about parallel lines. The point, however, is that once a particular discipline of logic (and mostly there is only one such discipline, the one we are following) and a particular set of axioms are agreed upon, there can be no controversy after that. If a proof is logically valid, it has to be accepted regardless of how shocking its conclusion may sound to our intuition; otherwise, it has to be rejected. There is no question of individual taste involved here. Things, such as space-filling curves or bounded plane regions with infinite boundaries, indeed defy our intuition. But a mathematician accepts them coolly, as they are the logical outcome of his axioms.

The importance of avoiding pitfalls of logic cannot be over-emphasized. There are well-known cases where a whole research work, often a published one, had to be withdrawn because a single mistake was detected. A discerning mathematician has an eye for occurrences of logical fallacies, a skill which can also be acquired by practice. What is basically needed is a habit of cautiously weighing the statements one reads, that is, of interpreting them precisely. For example a moment's reflexion shows that 'Not every man is mortal' is not the same thing as saying 'Every man is immortal'. If we compare their relative strengths, the second statement is stronger than the first one (except in one very exceptional event, can you think which it would be?). Similarly, when one expresses oneself it is very important to make sure that the expressed statement is the same as the intended statement, or at least that the two are logically equivalent.

Common instances of logical fallacies arise because of failure to distinguish between statements. Examples of this include incorrectly negating a statement, confusing an implication statement with its converse (or, in other words, confusing necessity with sufficiency), and so on. Often, one statement is stronger than the other. The stronger statement may imply some third statement. In such a case it is a likely mistake to deduce the third statement from the weaker one. This does not mean that the conclusion (that is, the third statement) is false. What is wrong is to say that it follows from the weaker statement.

The reader is especially advised to be cautious about statements involving the phrases 'there exists' (or 'there is') and 'for every' (or 'for all'). These phrases are called **quantifiers**, the former the **existential quantifier** and the latter the **universal quantifier**. The symbols $\exists$ and $\forall$ are sometimes used to abbreviate them. The phrase 'for any' is confusing and should be avoided by a beginner. Whenever used in this book it would always mean 'for every'.

The reason for warning the reader about statements involving quantifiers is multifold. For one thing, such statements will appear very frequently in this book. Second, in mathematics in the interest of unambiguity, they are worded far more clumsily than in practice. And finally, the order of the quantifiers in a statement is often crucial.

We proceed to illustrate these remarks. It will be convenient to use the language of sets here. A reader not familiar with it may pick it up from the first section in the Chapter Two and then return to this material.

We have already seen two statements involving quantifiers. They were 'There is a man who is eight feet tall' and 'All men are mortal'. In mathematics, the statements involving quantifiers are rarely so simple. Moreover, a mathematician is apt to word them more clumsily. Thus, our statements would read 'There exists a man such that (or with the property that) he is eight feet tall' and 'For every man, it is the case that he is mortal'. One can even go further. Let M denote the set (or the class) of all men. Then these statements may be written as '$\exists\ x \in M$ such that x is eight feet tall' and '$\forall\ x \in M$, x is mortal'. Here x is a dummy variable taking values in the set M. We could have as well replaced it by any other symbol not previously used in the particular context (for example we cannot replace x by M). In technical terms, the variable x is said to be **bound** by the quantifiers. A sentence such as '$x \in M$, x is mortal' is meaningless. The expressions '$\exists\ x \in M$' or '$\forall\ x \in M$', so to speak, serve to 'introduce' the variable x. Strictly speaking, such introduction must be made before anything is said about the variable introduced. However, where confusion is not likely, it is customary to defer it a little. Thus, the present statements could be written 'x is eight feet tall for some $x \in M$' and 'x is mortal $\forall\ x \in M$'.

The introduction of a variable bound by a quantifier cannot, however, be postponed at will. A quantifier cannot be shifted beyond another quantifier unless the variables governed by them are completely independent of each other. We illustrate this with examples. Consider the statement 'For every man there is a woman who loves him'. Letting M, W denote, respectively, the sets of all men and women, respectively, we can write this statement as '$\forall\ x \in M$, $\exists\ y \in W$ such that y loves x', or less clumsily as, $\forall\ x \in M$, y loves x for some $y \in W$'. But we cannot write it as '$\exists\ y \in W$ such that $\forall\ x \in M$, y loves x'. This latter statement would mean that there exists a woman who loves each and every man, something not asserted by the original statement. Here the order of quantifiers is crucial since the variable y may depend on a particular value of x.

On the other hand, take the statement 'Every man loves every woman'. We can write it as '$\forall\ x \in M$ and $\forall\ y \in W$, x loves y' or as '$\forall\ x \in M$, x loves y $\forall\ y \in W$' or even as 'x loves y $\forall\ x \in M$ and $\forall\ y \in W$'. Here the variables x, y are independent of each other and so no harm arises by interchanging the order of the quantifiers. Since both the quantifiers are of the same type, it is customary to write '$\forall\ x \in M, y \in W$' instead of '$\forall\ x \in M$ and $\forall\ y \in W$'. Equivalently, one could consider the cartesian product $M \times W$ and write '$\forall\ (x, y) \in M \times W$'.

As a last example take the statement 'A woman who loves a brave man does not love any other man'. Here we let B be the set of all brave men and for each $x \in M$ we let W_x be the set of all women who love x. Then the statement is '$\forall\ x \in B$, $\forall\ y \in W_x$ and $\forall\ z \in M - \{x\}$, y does not love z'. Here the variations of y and z depend on x and so the order of quantifiers cannot be changed.

The need for logical precisions applies not only to proofs but also to the various definitions we come across in mathematics. In giving mathematical definitions, one must not use such vague terms as 'usually', 'reasonable', 'generally speaking' or 'ordinarily', although such terms are commonly used in the definitions one comes across in, say, psychology or law. Another restriction which is followed most scrupulously in mathematics is to avoid what is known as a vicious circle. An example of a **vicious circle** is to define 'long' as 'not short' and 'short' as 'not long' or to define a fool as an idiot, an idiot as a lunatic and a lunatic as a fool! A mathematical definition must never be in terms of something whose definition relies ultimately on the thing to be defined in the first place. All the terms in the definition must have already been defined. For example if we define a triangle as a figure bounded by three line segments, this definition is meaningless unless the terms 'bounded', 'figure' and 'line segments' are already defined without using the word 'triangle'. Of course this process cannot be carried out indefinitely. Eventually, one reaches a stage where certain terms cannot be further defined. Such terms are called **primitive terms.** For example in geometry, the primitive terms are 'point', 'line' and 'incident'. Nothing is said by way of their definitions except whatever can be inferred from the axioms. These terms are not to be defined but rather to be understood privately by the individual. Obviously, their interpretation may change from person to person. For example, a point may mean a small dot and a line a thin straight segment to someone. But one is equally entitled to think of a point as a lock, a line as a key and the relation 'a point is incident upon a line' as the relation 'a lock can be opened by a key'. If a collection of locks and keys satisfies the axioms of geometry with this interpretation, then any theorem of geometry will be applicable to such a system. Indeed, this is one of the many ways mathematics can be put to 'practical' use.

It is well to end the chapter with a discussion of a possible criticism that we are perhaps overstressing the role of deductive logic in mathematics. There is indeed a view that mathematics is nothing but an extension of logic, or that it is the purest form of applied logic. The main demerit of this view is that it deals with mathematics in its finished state, but ignores it in its formative state. It is indeed true that when a mathematician expresses his results (with proofs) he has to follow strictly the discipline of deductive logic. But how does a mathematician obtain his results in the first place? To this there is no single answer. Often, he has to resort to inductive logic, to trial and error and to intelligent guesswork. Sometimes, it is a stroke of genius or of sheer luck that enables one to arrive at the right answer. These things are not revealed by the deductive proof of the results obtained. For example it may take considerable ingenuity to solve a certain differential equation. But to prove that the asserted solution, in fact, satisfies the differential equation is a matter of routine verification. From a deductive point of view, there is no difference between a 'verification' and a 'proof'. Similarly, a strict deductivist will not be impressed by the hard work of a mathematician who has factorised a huge positive integer. The mathematician will have to formulate his result as 'The number . . . has the following factors . . . ', a deductive 'proof' of which consists merely in multiplying the factors together, hardly a matter of a few minutes.

Examples such as these show that it is an over-simplification and indeed a mockery to say that deductive logic is the soul of mathematics. If anything, it is the skeleton of mathematics.

At this point we take the reader back to where we began this chapter. There we said that logic is to mathematics what grammar is to a language. When we think, we rarely do so in terms of complete, grammatically correct sentences. It is when we express ourselves or when we try to understand what others say that we resort to the discipline of grammar. This discipline is necessary for a clear and unambigious communication, but it is no substitute for literary qualities. Shakespeare would turn in his grave if the only appreciation we grant him is that he is grammatically correct. A similar remark would apply for the role of logic in mathematics. Logical precision is an indispensible part of a mathematical communication, but it is no substitute for mathematical talent. Nor is a proof valued simply because it is logically flawless. Using the language developed in this chapter we may say that logical precision is a necessary but not a sufficient condition for being a good mathematician.

Carrying the analogy a bit further, one may ask if occasional lapses of logic are permitted in mathematics just as occasional errors of grammar are condoned in a language. The answer is 'yes' as long as it is clear what you mean. The permissible lapses are of two types. The more common are those where a number of obvious justifications is omitted from a proof, much the same way as some words are left understood when we speak. A proof in which every little step is justified down to the level of 'by modus ponens' would be tedious both for the writer and the reader. Recourse is then taken to phrases such as 'clearly', 'it is obvious that', 'consequently', etc. and the reader is called upon to fill the missing links. The other type of deviation from precision occurs when one knowingly accepts an erroneous version in order to avoid clumsiness or to bow to some well-established convention. Thus, we say that 'A group is finite' when what we really mean is that the underlying set of that group is finite. This lapse of precision causes no confusion because everybody understands the correct meaning either from the context or through some convention.

Thus, insistence upon rigour is relaxed only for cosmetic purposes. It cannot be relaxed where such a lapse would thwart an argument. For example in no case can one replace an implication statement by its converse. It is suggested that in the initial stages the reader adhere to rigour even at the risk of a little loss of elegance. As he acquires maturity, he would know how to express himself elegantly without sacrificing rigour.

Exercises

4.1 Write the correct negation of the following statement in a reasonable way (i.e. without using the phrase 'it is not the case' or its equivalent, except, possibly, while negating the very last part). 'For every real number $E > 0$, there exists a real number $\delta > 0$ such that for every x, $|x - x_0| < \delta \rightarrow |y - y_0| < E$'. (Later on we shall see that this is the definition of continuity of a function $f(x)$ at the point x_0. However, for the present exercise it is unnecessary to know the meaning of the terms involved.)

4.2 Decide which of the following arguments are logically valid. In each case you are given some statements as premises and some statements as a conclusion. Your problem is to decide whether the truth of the conclusion necessarily follows from the truth of all the premises simultaneously.

(i) Premises: No man is rich unless he is intelligent.
John is a rich man.
Conclusion: John is intelligent.

(ii) Premise: Every man is mortal.
Conclusion: John is mortal.

(iii) Premises: If it rains the streets get wet.
It does not rain.
Conclusion: The streets do not get wet.

(iv) Premises: If it rains the streets get wet.
If the streets get wet, accidents happen.
It rains.
Conclusion: Accidents happen.

(v) Premises: Same as in (iv), except the third premise.
Conclusion: If it rains, accidents happen.

(vi) Premises: The streets get wet only if it rains.
It rains.
Conclusion: The streets get wet.

(vii) Premises: If it rains the streets get wet.
If it rains, accidents happen.
The streets get wet.
Conclusion: Accidents happen.

4.3 Point out the logical fallacy in the following *proof* which shows that every cyclic quadrilateral is a rectangle.

Proof: Let $ABCD$ be a cyclic quadrilateral. Draw a circle with the diagonal AC as a diameter. Then B, D lie on this circle as $ABCD$ is given to be cyclic. But then, $\angle B$, $\angle D$ are both angles in a semi-circle and so each is a right angle. Similarly, drawing a circle with BD as a diameter we see that $\angle A$ and $\angle C$ are also right angles. So $ABCD$ is a rectangle.

4.4 Take an English-into-English dictionary (any other language will also do). Start with any word and note down any word occurring in its definition as given in the dictionary. Take this new word and note down any word appearing in its definition. Repeat the process with this new word until a vicious circle results. Prove that a vicious circle is unavoidable no matter which word one starts with. (Caution: The vicious circle may not always involve the original word.)

Notes and Guide to Literature

The present chapter gives only the basic rudiments of logic which the reader should know in order to do justice to the rest of the book. The reader will come across virtually no computations, differential equations and so on in this book. Instead, he will find deductive arguments throughout the book, and to appreciate them one must have a feeling for logical thinking.

What we have called as statements are often known as propositions, and what we have said about their negation, conjunction and disjunction comes under what is known as propositional calculus. This is a rather naive branch of logic. The more formal mathematical logic begins with what is known as predicate calculus. A reader interested in mathematical logic may profitably read Mendelson [1].

We mentioned the constructivist approach to logic in Section 1. This is a relatively recent development and is slowly finding followers. An influential book in this respect is that of Bishop [1].

Exercise (1.3) is based on a popular paradox, the so-called Russell paradox. It has philosophical implications and revolutionized the development of formal set theory. We shall not deal with the axiomatic approach to set theory. A readable account may be found in the appendix to Kelley [1].

The lock-and-key interpretation of points and lines is especially important in the case of finite geometries, that is, where the number of points is finite. They are studied in what is known as *combinatorial mathematics*, a very rapidly growing branch of mathematics. A readable reference on it is Liu [1].

Finally, we must mention the book *How to Solve It* by Polya [1], which, although not a manual for problem solving, gives a charming account of the type of thinking a mathematician has to do to come up with an answer.

Chapter Two

Preliminaries

In this chapter we review the preliminaries that will be needed in this book. It should not be supposed, however, that they are strict prerequisites even for beginning to read the rest of the book. Nor is it the case that they will all be needed equally often. A reader who has had one elementary course in real analysis (and we assume that the majority of the readers will be of this type) would be in a position to proceed at least up to Chapter Eight. If and when necessary, he can refer to the relevant preliminaries.

The preliminaries are divided into three sections. We have not been exhaustive in any of them. Thus, for example besides the preliminaries from real analysis listed here, the reader will have to know some basic facts about measure and integration in working out some exercises. However, such instances will be few and no loss of continuity would result if a reader simply skips such exercises.

1. Sets and Functions

Cantor defined a set as a plurality conceived as a unity. In other words, the concept of a set involves mentally putting together a number of things (or 'objects' as they are often called) and assigning to the things so put together an identity as a whole, that is, an identity separate from that of each of these things. The things are called **elements** or **members** of the set obtained by conceiving them together. Thus, a set consists of or is comprised of its members but is itself a different entity from any of its members. For example a flock (i.e., a set) of birds is not equal to any one of the birds, it is not a bird at all. In fact, the flock is a concept which has no material existence. Of course, we may assign to the flock certain attributes in terms of the corresponding attributes of its members, either by making a convention or by common sense implications. For example we may define 'A flock of birds is flying' to mean that each of its members is flying. It should be noted, however, that the flock has certain attributes which cannot be described in terms of any attributes of individual members. For example 'A large flock of birds' is not the same as 'A flock of large birds'.

It is obvious that sets of material objects are as old as human thought. Things begin to get a little subtle when we form sets of abstract objects, such as integers, real numbers, lines, etc. Indeed, there is nothing to prevent us from forming sets whose objects are themselves sets of some other objects.

However, a little care is necessary in handling such sets. For example a set of sets of lions is not a set of lions much the same way as a set of lions is not a lion.

If we can form sets whose elements are themselves sets, a question naturally arises whether we can form the set of all sets, that is, all possible sets at all. Note that this set will be extraordinary in the sense that it will, unlike a set of lions or a set of real numbers, be a member of itself! Let us agree to call a set as ordinary if it is not a member of itself, and extra-ordinary otherwise. Most of the sets that we come across are ordinary, but some (e.g. the set of sets just considered) are extraordinary. Now let S be the set of all ordinary sets. A deceptively innocuous question is whether the set S is ordinary. We are doomed to get a contradiction whether the answer is affirmative or negative. Such a situation is known as a **paradox**. This particular paradox is due to the philosopher-mathematician Russel. It has revolutionized the approach to set theory for it clearly shows that a set cannot admit such a simplistic definition as 'a collection of objects'. Either we must put some restriction on these objects (e.g. by requiring that they be material objects) or else we have to admit that most, but not all collections of objects are sets. The second alternative is the lesser of the evils.

The approach to set theory in which an attempt is made to define a set and postulate a number of axioms about sets so as to avoid paradoxes is known as the **axiomatic set theory**. Although of considerable interest in its own rights, we shall not follow this approach, because our interest in sets is primarily as building blocks from which we can construct everything else. It turns out that most of the concepts in mathematics can be expressed very succinctly and precisely in terms of sets. Even the integers (and hence the real numbers as well as the complex numbers) as well as their addition, multiplication, etc. can be defined in terms of sets. Although we shall not go this far, we shall have ample evidence to show the utility of sets in giving precise expressions to what we can describe only intuitively otherwise. It is no exaggeration to say that sets are the alphabets of modern mathematics.

Because of our interest in sets as a means rather than as an end, we follow the so-called **naive** approach to **set theory**. In this approach, the paradoxes are avoided by confining ourselves only to certain sets. The most common choice is to fix some set, say U (to be called the **universal set** or the **universe**) and to agree that all the sets under consideration would consist only of elements of U. The choice of the universe need not be specified, but of course must be large enough to include all the symbols, expressions and concepts that we deal with. The existence of a universal set with sufficiently nice properties is a matter of axiomatic set theory.

After these general remarks about set theory, let us now list, specifically, the preliminaries we shall need about sets. For our purpose, 'set' will be a primitive term and the terms 'set', 'family', 'aggregate' and 'collection' will be synonymous. The term 'class' will be of a wider import, in that it will be used not only for sets but also when we put together too many things to form a set. For example we shall speak of 'the class of all sets' rather than 'the set of all sets'. A 'member' or an 'element' of a set will be another primitive term, and the relation of being a member will be denoted by $\in$. The expression '$a \in S$' will be read variously as 'a is a member (or an element) of S' or as 'a belongs to S' or as 'a is contained in S' or as 'S contains a'. The last two versions, however, are not particularly recommended to a beginner as they are likely to be confused with set inclusions, which we are going to define. Sets will generally be denoted by capital letters and their elements by small letters. To denote sets whose elements are themselves sets, we shall generally (but not always) use script letters, such as $\mathcal{A}$, $\mathcal{B}$, $\mathcal{E}$, $\mathcal{F}$, $\mathcal{T}$, $\mathcal{S}$, $\mathcal{U}$ etc. and we shall describe such sets as 'families' of or 'collections' of sets rather than as sets of sets. Occasionally, (and in fact frequently in this book) the word 'point' is also used instead of the word 'element' or 'member' and, in the same vein, '$a \in S$' and '$a \notin S$' are sometimes expressed, respectively, by 'a lies or is in (or inside or within) S' and 'a is or lies outside S'.

A set is completely determined by its elements. This means that two sets having exactly the same elements must be identical. There are two ways of specifying a set directly in terms of its elements. The first is to list all the elements together within curly brackets. Thus, {1, 2, dog} is the (unique) set whose elements are 1, 2 and dog. In so doing, neither the order of appearance of the elements nor the repetition of any elements makes any difference. For example the sets {1, 2, dog}, {2, 1, 2, dog}, {dog, 2, 2, 1} are all identical. It is, of course, not necessary to write down all the elements if it is possible (either from the context or because of some convention) to infer all the elements from those that are actually listed. Thus, $\{1, 2, 3, \ldots, 19, 20\}$ is clearly the set of all integers from 1 to 20. Similarly, $\{2, 4, 6, 8, 10, 12, 14, \ldots, 2n, \ldots\}$ is the set of all positive even integers.

Another method of specifying a set by specifying its elements, is by specifying what is known as a characteristic property of the set. A **characteristic property** of a set is a property which is satisfied by each member of that set and by nothing else. The same set may have more than one characteristic property. Each set has a trivial characteristic property, namely, the property of belonging to that set. This property is, of course, useless if we want to describe the set in the first place. If, however, we know some other characteristic property (and usually such a property is present when we conceive a set) of a set, then the set can be described in terms of it. The standard notation for a set so described is $\{x : \quad \}$ or $\{x \mid \quad \}$. Here x is a dummy symbol and could have been replaced by any other dummy symbol. The space between : (or $\mid$) and } is to be filled by a statement to the effect that x has the property in question. For example the set of cows owned by a farmer F can be denoted by $\{x : x$ is a cow and is owned by $F\}$ or $\{x \mid x$ is a cow and is owned by $F\}$. Similarly, $\{y :$ there exists a positive integer x such that $y = x^2\}$ is the set whose elements are 1, 4, 9, 16, 25, . . . , etc. In such cases it is customary to abbreviate the notation as $\{y : y = x^2$ for some positive integer $x\}$. As in the present case, it often happens that all the elements of a set can be expressed in a certain common form in terms of one (or more) variable. In such a case it is customary to denote the set by writing such a form in curly brackets and specifying what values the variable can take. For example the set just described could have been written as $\{x^2 : x \text{ a positive integer}\}$ or as $\{x^2 : x = 1, 2, 3, \ldots\}$. Note that here too x is a dummy symbol.

When a set is described by specifying some characteristic property and the statement expressing this property is the conjunction of several statements, the word 'and' is often dropped and only commas are used to denote the conjunction. For example $\{x : x \text{ real}, x > 0, x^2 < 2\}$ is the set of those real numbers which are greater than 0 and whose squares are less than 2. As in the present case, it often happens that the elements of the set in question are required to come from some other set. In such cases it is customary to write this requirement before the colon (:) rather than after the colon. For example if we denote by $\mathbb{R}$ the set of real numbers and by $\mathbb{R}^+$ the set of positive real numbers, then the present set could have been denoted by $\{x \in \mathbb{R} : x > 0, x^2 < 2\}$ or by $\{x \in \mathbb{R}^+ : x^2 < 2\}$.

In the same vein, when a set is characterised by a number of conditions which are of a similar form and can be written in terms of a dummy variable, it is customary to write only the values assumed by this dummy variable and to omit a specific mention of the requirement that all the conditions obtained by assigning the various values to the dummy variable are to be simultaneously satisfied. For example let S be the set of those integers which are divisible by every integer from 1 to 20. Then S can be written as $S = \{x : x \text{ an integer}, x \text{ is divisible by } i \text{ for all } i = 1, 2, \ldots, 20\}$ or as $S = \{x : x \text{ an integer}, x \text{ is divisible by } i, i = 1, 2, \ldots, 20\}$. Note that in the latter expression the words 'for all' are omitted and are to be understood. A beginner is warned against interpreting such omission as 'for some'. The set $\{x : x \text{ an integer}, x \text{ is divisible by some } i, i = 1, 2, \ldots, 20\}$ will include many elements not present in the set S. The distinction between the quantifiers 'for all' and 'for some' is vital and confusing the two with each other may be disastrous. The situation is admittedly difficult for a beginner since some authors tend to omit both the quantifiers and leave it to the reader to infer from the context which of them is intended. To avoid confusion, in this book the existential quantifier ('for some' or 'for at least one') will always be mentioned specifically.

It is possible to conceive a set with no elements at all. Such a set is variously known as an **empty** set or a **void** or a **vacuous** or a **null** set. Examples of such sets are the set $\{x : x \text{ an integer and } x^2 = 2\}$ or the set of all six-legged men. One can of course give many other examples. Note, however, that all these sets are equal because they consist of identical elements (viz. no elements at all). To say that two empty sets are unequal would mean that at least one of them contains an element which is not present in the other. Since such an element does not exist, we have to agree that they are equal. By the same logic, any statement to the effect that a certain property holds for every member of the empty set is true, albeit vacuously so. The unique empty set is denoted by 0 or by $\emptyset$. These notations are so standard that they are used even when the same symbols may also represent something else. No confusion results because of such a double use.

Given two sets S and T we say that S is a **subset** of T (or T is a **superset** of S, or S is **contained** in T or T **contains** S) if every element of S is also an element of T. When this is the case we write $S \subset T$ or $T \supset S$. If $S \subset T$ but $S \neq T$, then we say that S is a **proper subset** of T and write $S \subsetneq T$. The reader is cautioned that some authors use the notations $\subseteq$ and $\subset$, where we use $\subset$ and $\subsetneq$, respectively. Obviously, two sets are equal if and only if each is a subset of the other. This is indeed the most straightforward way of proving that two sets are equal. The words 'subfamily' and 'subcollection' are synonymous with subset. Words, such as 'subaggregate', 'superfamily' could be defined but are rarely used. The term 'subclass' will not be defined formally, but has the same relationship with a class as a subset has with a set. This relationship is known as **inclusion**. We shall define set inclusion formally when we define a relation.

Given two sets S and T, the **complement** of S in T (or with respect to T) denoted by $T - S$ (or by $T \sim S$ or $T \setminus S$) is the set of such elements which are in T but not in S. Thus $T - S = \{x : x \in T \text{ and } x \notin S\}$ or $T - S = \{x \in T : x \notin S\}$. Note that we are not requiring here that S be a subset of T, although the complement $T - S$ is always a subset of T. When the set with respect to which complements are considered is understood, the complement of S is denoted by S', $\sim S$ or by $c(S)$.

If S is a set then the set of all subsets of S is called the **power set** of S and will be denoted by $P(S)$. Note that the empty set and the set S itself are always members of the power set $P(S)$. In particular, a power set is never empty. It is easy to show that if S has n (distinct) elements then $P(S)$ has 2^n elements and this is the reason for the name 'power set'. Elements of the power set are in general quite different from those of the original set and the two should not be confused with each other. If x is an element of the original set S then $\{x\}$ will be an element of the power set, that is a subset of the original set; whereas x itself may or may not be an element of $P(S)$.

If S and T are two sets then their **union** (sometimes called **join**) is the set $\{x : x \in S \text{ or } x \in T\}$. It is commonly denoted by $S \cup T$. The notation $S + T$ was also popular till recently. One can define more generally the **union of any family of sets**. Let $\mathcal{F}$ be such a family. Then $\cup \mathcal{F}$ is the set $\{x : x \in F \text{ for some } F \in \mathcal{F}\}$. It is also denoted by $\bigcup_{F \in \mathcal{F}} F$. In the former notation the set F could of course change from element to element of $\cup \mathcal{F}$. In the latter notation, F is nothing but a dummy index (similar to the index of summation written in the sigma notation).

Another important concept is that of intersection of sets. If S, T are two sets, their **intersection** (or **meet**) is the set $\{x : x \in S \text{ and } x \in T\}$. It is denoted by $A \cap B$ or by $A \cdot B$ or by AB. More generally, if $\mathcal{F}$ is a family of sets then its **intersection,** denoted by $\cap \mathcal{F}$ or by $\bigcap_{F \in \mathcal{F}} F$ is defined to be the set $\{x : x \in F \text{ for all } F \in \mathcal{F}\}$. It is interesting to note that if the family $\mathcal{F}$

itself is empty then the condition '$x \in F$ for all $F \in \mathcal{F}$' is vacuously satisfied by all x's. In such a case the intersection, $\bigcap_{F \in \mathcal{F}} F$ therefore, is the whole universe. Often we deal with only families which are subsets of the power set of some set X and by convention we require that the intersections of such families also be subsets of X. In such a case the intersection of the empty family of subsets of X is the entire set X.

A family $\mathcal{F}$ of sets is said to be **closed** under unions (intersections) if for all $\mathcal{G} \subset \mathcal{F}$, $\bigcup \mathcal{G} \in \mathcal{F}$ (respectively $\bigcap \mathcal{G} \in \mathcal{F}$).

If A, B are two sets, we say that A is **disjoint** from B (or that A and B are **mutually disjoint**) if $A \cap B$ is the empty set. Otherwise we say that A **intersects** (or **meets**) B or that A and B **intersect** (or **meet**). A family $\mathcal{F}$ of sets is said to be **pairwise disjoint** if every two distinct members of it are mutually disjoint.

We trust that the reader is familiar with the basic properties of complements, unions and intersections. A few of the more important ones will be given as exercises. We now turn to another important concept, that of a function. First we need the notion of an ordered pair. If x and y are some objects, how should we define the ordered pair (x, y)? It is tempting to define it simply as the set $\{x, y\}$. But if we accept this then we would not be able to distinguish (x, y) from (y, x) for the sets $\{x, y\}$ and $\{y, x\}$ are identical. This difficulty would be overcome if, in addition to the set $\{x, y\}$, we also specify which element we want to consider first. Thus, we define the ordered pair (x, y) to be the set $\{\{x, y\}, x\}$. The reader may wonder why such a simple concept as an ordered pair is defined in such a clumsy way. The answer lies in our insistence upon precision and our commitment to define every mathematical concept in terms of sets. It often happens that we want to define something, say A, which we know intuitively but cannot describe rigorously. Yet, we are able to find something else, say B, which can be described precisely and which is so inexorably linked with A that knowing B is as good as knowing A. We then define A as B. For convenience, we refer to this as the 'definition trick'. The definition of an ordered pair just given is an illustration of the definition trick. We shall see many other instances of this trick. In the same vein, one can define an **ordered triple** (x, y, z) either as the set $\{\{x, y, z\}, \{x, y\}, x\}$ or as the ordered pair $((x, y), z)$ and more generally an **ordered n-tuple** $(x_1, \ldots, x_n)$ for any positive integer n.

Now let A, B be any sets. Then their **cartesian product** (or simply **product**) is defined to be the set $\{(x, y) : x \in A, y \in B\}$. It is denoted by $A \times B$. The name as well as the notation is suggestive in that if A has m elements and B has n elements then $A \times B$ indeed has mn elements. Note that $A \times B$ is in general a different set than $B \times A$. One can, similarly, define the cartesian product of three (or more) sets in terms of ordered triples (or ordered n-triples). The sets whose cartesian product is formed are called the **factor sets** or the **factors.** It is clear that the cartesian product of a finite number of sets is empty, if and only if at least one factor is empty. When we form products, it is not necessary that the factors be distinct sets. The product of the same set A taken n times (or in other words, the product of n copies of A) is called the **nth power** of A and is denoted by A^n. Thus A^2 is the set $A \times A$, A^3 is the set $A \times A \times A$, etc.

We are now ready for the definition of a function, a concept of paramount importance in mathematics. Intuitively, if X and Y are sets then a function, say f, from X to Y is a rule of correspondence which assigns to every element of X a unique element of the set Y. If x stands for an arbitrary element of X then the unique element of Y which is assigned to x by f is denoted by $f(x)$. The difficulty in defining a function this way is that the terms 'rule of correspondence' and 'assign' are not defined earlier and so the definition, although intuitively appealing and clear, is not logically precise. To overcome this difficulty we resort to the definition trick mentioned earlier. Suppose a function f from X to Y (in the intuitive sense) is given. We form the cartesian product $X \times Y$ and let G_f be the subset $\{(x, y) \in X \times Y : y = f(x)\}$. This set is called the **graph** of f. The graph G_f is inexorably related to the function f in the sense that either one of them completely determines the other. This suggests that instead of defining a function from X to Y vaguely, as a certain rule of correspondence, we might as well define it as a certain subset of $X \times Y$. Of course not every subset of $X \times Y$ is the graph of some function from X to Y. It is easy to see that a subset S of $X \times Y$ is the graph of some function from X to Y if and only if S has the property that for every $x \in X$, there is a unique $y \in Y$ such that $(x, y) \in S$. This, now, becomes the definition of a function. In other words, we identify a function with its graph. Formally, a function from a set X to a set Y is a subset, say, f of $X \times Y$, with the property that for each $x \in X$, there is a unique $y \in Y$ such that $(x, y) \in f$. The fact that f is a function from X to Y is expressed by writing $f: X \to Y$ or $X \xrightarrow{f} Y$. If $x \in X$ then the unique element, say y, of Y for which $(x, y) \in f$ is called the **value** of f at x and is denoted by $f(x)$. This is also expressed by saying that f **takes** (or **maps**) x to $f(x)$. The set X is called the **domain** and the set Y a **codomain** of f. A variable which ranges over the domain is called an **argument** of the function. If $y = f(x)$ where $x \in X, y \in Y$, we say that x is a **pre-image** of y.

Although the definition given above is standard, it suffers from a technical drawback in that the same function may have more than one codomain. For example let $\mathbb{N}$, $\mathbb{R}$ denote, respectively, the set of natural numbers and real numbers and let $f = \{(n, n^2) : n \in \mathbb{N}\}$. Then f may be thought of either as a function from $\mathbb{N}$ to $\mathbb{N}$ or from $\mathbb{N}$ to $\mathbb{R}$. This is sometimes undesirable. To avoid this, a **function** is now-a-days defined as an ordered triple (X, Y, f), where X, Y are sets and f is a subset of $X \times Y$ with the property that for each $x \in X$ there is a unique $y \in Y$ such that $(x, y) \in f$. The sets X, Y and f are called, respectively, the **domain,** the **codomain** and the **graph** of the function (X, Y, f). The notations $f: X \to Y$ and $X \xrightarrow{f} Y$ are still valid. Note, however, that with this new definition, the codomain of a function is uniquely determined. Two functions with the same domains and the same graphs are different if they do not have the same codomain as well. In the example just given, $(\mathbb{N}, \mathbb{N}, \{(n, n^2) : n \in \mathbb{N}\})$ is not the same function as $(\mathbb{N}, \mathbb{R}, \{(n, n^2) : n \in \mathbb{N}\})$ even though these two functions have the same graph and, therefore, would have been regarded as identical according to the earlier definition in which a function was identified with its graph. Even with the new approach, a function and its graph are often denoted by the same notation. This abuse of language causes no confusion when the codomain is clearly understood.

Of course, when a mathematician conceives or handles a function he rarely does so in terms of an ordered triple or of a subset of a cartesian product. These devices are meant only to give a precise formulation to the intuitively clear notion of a single-valued correspondence from one set to another. As long as one knows how to express oneself precisely if need be, it is permissible, and indeed preferable, not to deviate unnecessarily from intuition. Thus, in practice, a function is specified by giving its value at a typical element of the domain. It is not necessary that this value be expressed by some formula, and even when so expressed the formula should not be confused with the function itself. The specification of a function is not complete until its domain and codomain are clearly specified. In defining a function it is better to write the arrow notation before giving the formula (if any) which describes the function. Thus instead of saying 'consider the function $f(x) = \sin(x^2)$' it is far better to say 'consider the function $f : \mathbb{R} \to \mathbb{R}$ defined by $f(x) = \sin(x^2)$ for $x \in \mathbb{R}$'. This is especially stressed at this stage, because in an elementary course in analysis one generally confines oneself to functions whose domains and codomains are standard (often subsets of the set of real or complex numbers or their powers) and hence one tends to ignore a specific mention of the domain and the codomain. On the other hand, we shall often deal with functions whose domains and codomains are 'abstract' sets and for whom there are no concrete formulae.

The words, **'transformation'**, **'operator'**, **'map'** and **'mapping'** are really synonyms of 'function' although by convention they are used only in some specific contexts. We shall use 'function' as the general term and reserve the terms 'map' and 'mapping' for certain special type of functions.

If $f : X \to Y$, $g : Y \to Z$ are functions, their **composition** or **composite** is denoted $g \circ f$ (or sometimes by gf) and is defined to be the function from X to Z given by $g \circ f(x) = g(f(x))$ for $x \in X$. In order that the composite of a function f followed by g be defined it is necessary that the codomain of f and the domain of g coincide. It is clear that if f, g, h are functions for which $g \circ f$ and $h \circ g$ are both defined then $(h \circ g) \circ f$ and $h \circ (g \circ f)$ are also defined and are equal to each other. We denote this function by $h \circ g \circ f$.

The simplest functions are the so-called **constant** functions. They assume the same value for all values of the argument and are often denoted by this common value. For any set X, the function $1_X : X \to X$ (or $id_X : X \to X$) defined by $1_X(x) = x$ for all $x \in X$ is called the **identity function** on X. More generally, if Y is a superset of X then the function $i : X \to Y$ defined by $i(x) = x$ for $x \in X$ is called the **inclusion function** of X into Y. If $f : X \to Y$

is a function and $A \subset X$ then the **restriction** of f to A, denoted by $f/A : A \to Y$ is the function defined by $f/A(x) = f(x)$ for all $x \in A$. Equivalently, it is the composite with f of the inclusion function of A into X. A function $f: X \to Y$ is said to be **injective** (or **one-to-one**) if for all $x_1, x_2 \in X$, $f(x_1) = f(x_2)$ implies $x_1 = x_2$. In other words, a function is injective iff it takes distinct points of the domain to distinct points of the codomain. A function $f: X \to Y$ is said to be **surjective** (or **onto**) if for each $y \in Y$ there is some $x \in X$ such that $f(x) = y$. A function which is both injective and surjective is called a **bijective** function or a **bijection**. A bijection of a set onto itself is called a **permutation** of that set. It is easy to show that a function $f: X \to Y$ is bijective iff there exists a function $g : Y \to X$ such that $g \circ f = id_X$ and $f \circ g = id_Y$. When such a function exists it is unique and called the **inverse** function of f. It is denoted by f^{-1}. Note that the inverse function is also a bijection. The term 'one-to-one correspondence' is sometimes used for 'bijection' and is thus different from 'one-to-one function'.

Let $f: X \to Y$ be a function and suppose A, B are subsets of X, Y, respectively. Then the **direct image** (or simply, **image**) of A under f, denoted by $f(A)$ is defined as the set $\{f(x) : x \in A\}$. The set A is said to be **taken** (or **mapped**) by (or under) f onto $f(A)$. The **inverse image** of B under f, denoted by $f^{-1}(B)$ is defined as the set $\{x \in X : f(x) \in B\}$. The inverse image is defined even where f is not a bijection. In case f is a bijection, it coincides with the direct image under the inverse function f^{-1} and so the notation causes no ambiguity. The set $f(X)$ is called the **range** of f. It is evident that a function is onto iff its range is the whole codomain.

We leave as exercises, elementary facts regarding direct and inverse images. We proceed here to show how the concept of a function allows us to define certain concepts succinctly. Take the concept of a sequence. Intuitively, this is an infinite succession of terms. To make this precise, given a sequence, say $\{a_n\}$, we form a function f with domain $\mathbb{N}$, the set of positive integers and defined by $f(n) = a_n$ for $n \in \mathbb{N}$. The codomain of this function is to be the set from which the terms of the sequence come. It is clear that the function f and the sequence $\{a_n\}$ completely determine each other. We, therefore, define a **sequence** as a function whose domain is $\mathbb{N}$. Thus, a sequence of real numbers is a function from $\mathbb{N}$ to $\mathbb{R}$ (the set of real numbers); a sequence of monkeys is a function from $\mathbb{N}$ into the set of all monkeys and so on. Despite this formal definition, we continue to denote sequences by the usual symbols $\{a_n\}$ or $\{a_n\}_{n=1}^{\infty}$ or $\{a_n\}_{n \in \mathbb{N}}$. These notations should not be confused with the set $\{a_n : n \in \mathbb{N}\}$ which is in fact the range of the sequence. A sequence $\{a_n\}$ is said to be **eventually constant** if there exists $m \in \mathbb{N}$ such that $a_n = a_m$ for all $n \geqslant m$.

Another application of functions is in defining an indexed family which is a generalization of a sequence. Instead of $\mathbb{N}$, let us fix an arbitrary set I to be called an **index set**. Let X be any set. Then, by a **family** of elements of X **indexed** over I, we simply mean a function $x : I \to X$. For $i \in I$, we denote $x(i)$ by x_i and write the indexed family as $\{x_i\}_{i \in I}$ or as $\{x_i\}$. We often consider indexed families of sets. In such cases it is tacitly assumed that these sets are subsets of some set and the codomain of the indexed family is understood to be the power set of this set. If $\{A_i\}_{i \in I}$ is an indexed family of sets, then its **union** is defined to be the set $\{x : x \in A_i \text{ for some } i \in I\}$, and is denoted by $\bigcup_{i \in I} A_i$ or by $\sum_{i \in I} A_i$. Similarly, its intersection is defined to be the set $\{x : x \in A_i \text{ for all } i \in I\}$ and is denoted by $\bigcap_{i \in I} A_i$. When the index set is $\mathbb{N}$, the notations $\bigcup_{i=1}^{\infty}$ and $\bigcap_{i=1}^{\infty}$ are also used commonly instead of $\bigcup_{i \in \mathbb{N}}$ and $\bigcap_{i \in \mathbb{N}}$ respectively.

Although a family of sets is conceptually different from an indexed family of sets, the distinction is sometimes unnecessarily fussy. If $\mathcal{F}$ is a family of sets we can view it as an indexed family of sets by taking $\mathcal{F}$ itself as an index set and defining A_F to be F for all $F \in \mathcal{F}$. Conversely, when the indexing is not important, an indexed family may be identified with its range and thus be treated as just an ordinary family.

Now let $\{A_i\}_{i \in I}$ be an indexed family of sets such that $A_i \cap A_j = \emptyset$ for all $i \neq j \in I$. An interesting question is whether we can form a set by picking exactly one element from each A_i, that is a set S such that for each $i \in I$, $S \cap A_i$ consists precisely of one element. Obviously, if some A_i is empty this would be impossible. But what if one is given that each A_i is non-empty? In this case the answer would seem to be 'obviously yes'. But this is not so. Of course, if each A_i contained some distinguished element say x_i, then we may simply let $S = \{x_i : i \in I\}$. When such distinguished elements are not given, the problem boils down to making a simultaneous choice of one element each from the A_i's. That such a choice is possible is one of the fundamental axioms of set theory, the so-called **axiom of choice.** Although it looks self-evident, it has many equivalent formulations which are not obvious. A few such formulations will be mentioned later without proof.

The axiom of choice is applicable even when the original family $\{A_i\}_{i \in I}$ is not pairwise disjoint. In such a case we consider $\{B_i\}_{i \in I}$ where for each $i \in I$, $B_i = A_i \times \{i\}$. If A_i is non-empty so is B_i for $i \in I$. Also, for $i \neq j \in I$, $B_i \cap B_j = \emptyset$ whether $A_i \cap A_j = \emptyset$ or not. By the axiom of choice we get a set S such that for each $i \in I$, $S \cap B_i$ has exactly one element. We now let $T = \{x : \text{for some } i \in I, (x, i) \in S\}$. Then T contains at least one element from each A_i. The technique employed here in switching from $\{A_i\}_{i \in I}$ to $\{B_i\}_{i \in I}$ is known as **disjunctification.**

We conclude this section by showing how functions can be used to give a precise meaning to the notion of the size of a set. Clearly, the smallest set is the empty set. Next, there are the **singleton** sets. By definition, these sets have only one element each. It is easy to see that if X is a given singleton set then for any other set Y, there exists a bijection $f : X \to Y$ if and only if the set Y is also singleton. More generally, if X, Y are sets with m, n elements, respectively, where m, n are positive integers, then there exists a bijection $f : X \to Y$ if and only if $m = n$. If $m < n$ then there exists an injective function $f: X \to Y$ but no surjective function from X to Y.

These simple observations lead to the following definitions:

A set X is **finite** if either $X = \emptyset$ or there exists a bijection $f: \{1, 2, \ldots, n\} \to X$ for some positive integer n. Otherwise it is **infinite.** It is **denumerable** (or **enumerable**) if there exists a bijection $f: \mathbb{N} \to X$ where $\mathbb{N}$ is the set of all positive integers (such a bijection is called an **enumeration** of X). A set which is either finite or denumerable is said to be **countable,** otherwise it is **uncountable.** Two sets X and Y are said to be **equipollent** to each other if there exists a bijection $f: X \to Y$ (or equivalently a bijection $g: Y \to X$). It is easy to show that if X, Y are equipollent and Y, Z are equipollent then X, Z are equipollent. If X is a set, then the class of all sets equipollent to X is called the **cardinal number** of X or the **cardinality** of X (or occasionally, the **pollency** of X) and is variously denoted by $n(X)$, $\#(X)$, $|X|$, $\bar{\bar{X}}$ or card (X). It is clear that for any two sets X and Y, X is equipollent to Y iff $|X| = |Y|$. The cardinal number of a finite set is customarily denoted by 0 (in the case of the empty set) or by that unique positive integer n such that the given set is equipollent to the set $\{1, 2, \ldots, n\}$. The cardinal number of $\mathbb{N}$ (and hence of any denumerable set) is denoted by $\aleph_0$ (read 'aleph naught' or 'aleph zero', $\aleph$ being the first letter of the Hebrew alphabet). The cardinal number of the set of real numbers is denoted by c.

A common misuse of the quantitative adjectives 'finite', 'denumerable', 'uncountable' etc. is noteworthy. Suppose $\mathcal{F}$ is a family of sets. By a **finite union** of members of $\mathcal{F}$, what is meant is either the empty set or a set of the form $F_1 \cup F_2 \cup, \ldots, \cup F_n$ for some $n \in \mathbb{N}$ and some members $F_1, \ldots, F_n$ of $\mathcal{F}$. Equivalently, a finite union in $\mathcal{F}$ means a set of the form $\cup \mathcal{G}$ where $\mathcal{G}$ is a finite subfamily of $\mathcal{F}$. Note that here it is the finiteness of the family $\mathcal{G}$ that matters. Whether its union, that is the set $\cup \mathcal{G}$ is finite or not is irrelevant here. In other words, a finite union may very well be an infinite set. And this is where a beginner is likely to get confused. Unfortunately, the usage is too standard to be changed and we trust that with a little care the reader can master it. It may be helpful to recall that in analysis the terms 'infinite sum' or 'infinite product' refer to the fact that the number of summands (or factors) is infinite. The sum or the product itself may very well be finite.

Terms, such as 'finite intersection', 'countable union' etc. are defined analogously. A finite product of sets means the cartesian product of a finite number of sets. (Of course, so far these are the only products we have considered. Infinite products will be defined much later). To say that a family $\mathcal{F}$ of sets is **closed under finite unions** means that for every finite subfamily $\mathcal{G}$ of $\mathcal{F}$, $\cup \mathcal{G} \in \mathcal{F}$. Similarly, one defines closure under finite intersections, countable unions, etc.

A **finite sequence** is defined as a function whose domain is a set of the form $\{1, 2, \ldots, n\}$ (or of the form $\{0, 1, 2, \ldots, n\}$) for some positive integer n. This phrase is not to be interpreted as if 'finite' were an adjective of the noun 'sequence', because a finite sequence is not a sequence at all!

The definition of a cardinal number may seem artificial but a little reflexion will show that it evolves out of an attempt to give a precise expression to the notion of the size of or the number of elements present in a set. Although for finite sets the definition reasonably conforms to our intuition, this is not so with infinite sets. For example let E be the set of all even positive integers. Intuitively, we would like that $|E|$ is half of $|\mathbb{N}|$ or at any rate that the two are not equal since E is a proper subset of $\mathbb{N}$ (and the complement $\mathbb{N} - E$ is itself an infinite set). However, the function $f: \mathbb{N} \to E$ defined by $f(n) = 2n$, $n \in \mathbb{N}$ is clearly a bijection and so $|\mathbb{N}| = |E|$. It may be shocking that the set $\mathbb{N}$ has as many elements as its proper subset E. But we have to swallow this as it is a logical outcome of our definitions. This does not mean that there is no way of somehow giving a mathematical expression to the 'fact' that there are half as many even positive integers as there are positive integers (one such way will be pointed out in the exercises). It only means that we have to be careful in handling infinite cardinals (i.e. cardinal numbers of infinite sets) by adhering more to the definitions than to intuition.

Cardinal arithmetic is that branch of set theory which deals with elementary operations, such as additions and products of cardinal numbers. The definitions are motivated by the finite cardinals. For example let α, β be two cardinals (that is cardinal numbers of some sets). Find two sets X, Y such that $|X| = \alpha$ and $|Y| = \beta$. These sets are said to **represent** the cardinals α, β, respectively. Then the **product** of α and β, dentoted by $\alpha \cdot \beta$ or $\alpha\beta$ is defined to be $|X \times Y|$. An immediate verification which must be made with this sort of a definition is to check that it is well-defined in the following sense. On the face of it, the definition depends on the sets X, Y which we have taken to represent α, β, respectively. Suppose we chose some other sets say X_1 and Y_1 to represent these cardinals. Then our definition of $\alpha\beta$ would have been $|X_1 \times Y_1|$. For an unambiguous definition it is necessary to ensure that $\alpha\beta$ defined this way is the same as $\alpha\beta$ defined earlier. In other words we have to prove that if $|X| = |X_1|$ and $|Y| = |Y_1|$ then $|X \times Y| = |X_1 \times Y_1|$. This verification is easy and left to the reader. Similarly, to define $\alpha + \beta$, the **sum** of two-cardinal numbers α and β, we choose representatives X, Y, respectively. It is tempting to define $\alpha + \beta$ as $|X \cup Y|$ but as one can see from the finite case, $|X \cup Y|$ would fall short of $\alpha + \beta$ unless $X \cap Y = \emptyset$. To ensure this we disjunctify X and Y, that is, we replace them by $X \times \{a\}$ and $Y \times \{b\}$ where a, b are any two distinct symbols. We now define $\alpha + \beta$ as $|(X \times \{a\}) \cup (Y \times \{b\})|$. Again we have to verify that this is independent of X, Y, a and b. One can similarly define the sum of an indexed family of cardinals.

Finally, we compare cardinals. Let α, β be cardinals. We say $\alpha \leqslant \beta$ (read α **is less than or equal to** β) or that $\beta \geqslant \alpha$ (read β is **greater than or equal to** α) if given any representatives X, Y of α, β respectively, there exists an injective function $f: X \to Y$. If $\alpha \leqslant \beta$ and $\alpha \neq \beta$ we write $\alpha < \beta$ and say that α is **strictly smaller** than β (or write $\beta > \alpha$ and say that β is **strictly greater** than α). It turns out to be sufficient that the condition in this definition be satisfied for some representatives of the cardinals. Two points are noteworthy about comparison of cardinals. First, given two cardinals it can never happen that each is less than or equal to the other unless the two are equal. In terms of sets it means that if X, Y are sets such that there exists injective functions $f : X \to Y$ and $g : Y \to X$ then X and Y are equipollent. This theorem is popularly known as the **Schröder-Bernstein theorem,** a proof of which will be excluded. Another point to be noted about comparison of cardinals is that given two cardinals, they are always comparable one way of the other. In terms of sets this means that given two sets X and Y, either there exists an injective function from X to Y or there exists an injective function from Y to X. This seems obvious when at least one of the two sets is finite. When both the sets are infinite, the proof requires the use of the axioms of choice and will be excluded. The two facts just mentioned can be succinctly summarized as: given any two cardinals α, β exactly one of the three possibilities holds, namely, either $\alpha < \beta$ or $\alpha = \beta$ or $\alpha > \beta$.

Exercises

1.1 Let X be a set, A a subset of X, $\{B_i\}_{i \in I}$ an indexed family of subsets of X and σ a permutation of the index set I. Prove the following identities:

(i) $\bigcup\limits_{i \in I} B_i = \bigcup\limits_{i \in I} B_{\sigma(i)}$ and $\bigcap\limits_{i \in I} B_i = \bigcap\limits_{i \in I} B_{\sigma(i)}$

(ii) $A \cap (\bigcap\limits_{i \in i} B_i) = \bigcap\limits_{i \in I} (A \cap B_i)$ and $A \cup (\bigcup\limits_{i \in I} B_i) = \bigcup\limits_{i \in I} (A \cup B_i)$

(iii) $A \cap (\bigcup\limits_{i \in I} B_i) = \bigcup\limits_{i \in I} (A \cap B_i)$ and $A \cup (\bigcap\limits_{i \in I} B_i) = \bigcap\limits_{i \in I} (A \cup B_i)$

(iv) $(A - \bigcup\limits_{i \in I} B_i) = \bigcap\limits_{i \in I} (A - B_i)$ and $(A - \bigcap\limits_{i \in I} B_i) = \bigcup\limits_{i \in I} (A - B_i)$

[The property (iii) is expressed by saying that intersections and unions are distributive over each other. The property (iv) is the **De Morgan's laws** and is expressed by saying that the complement of a union is the intersection of the complements and the complement of an intersection is the union of the complements.]

1.2 Let $f : X \to Y$ and $g : Y \to Z$ be functions. Prove that

(i) for any $A \subset X$, $(g \circ f)(A) = g(f(A))$.

(ii) for any $B \subset Z$, $(g \circ f)^{-1}(B) = f^{-1}(g^{-1}(B))$.

(iii) for any $A \subset X$, $f^{-1}(f(A)) \supset A$ and for any $B \subset Y$, $f(f^{-1}(B)) \subset B$ and $f^{-1}(Y - B) = X - f^{-1}(B)$.

(iv) for any two subsets A_1, A_2 of X and B_1, B_2 of Y,

$f(A_1 \cup A_2) = f(A_1) \cup f(A_2)$ and $f(A_1 \cap A_2) \subset f(A_1) \cap f(A_2)$

$f^{-1}(B_1 \cup B_2) = f^{-1}(B_1) \cup f^{-1}(B_2)$ and $f^{-1}(B_1 \cap B_2) = f^{-1}(B_1) \cap f^{-1}(B_2)$

$A_1 \subset A_2 \Rightarrow f(A_1) \subset f(A_2)$ and $B_1 \subset B_2 \Rightarrow f^{-1}(B_1) \subset f^{-1}(B_2)$.

1.3 (a) Let $f : X \to Y$ be a function. Prove that the following statements are equivalent to each other:

(i) f is injective

(ii) for every $A \subset X$, $f^{-1}(f(A)) = A$

(iii) for every $A_1, A_2 \subset X$, $f(A_1 \cap A_2) = f(A_1) \cap f(A_2)$

(iv) for any set Z and for any two functions $g, h : Z \rightarrow X$, $f \circ g = f \circ h$ implies $g = h$ (in other words, f can be cancelled from the left)

(v) either $X = \emptyset$ or there exists a function $g : Y \rightarrow X$ such that $g \circ f = id_X$. (such a function g is called a **left inverse** of f.)

(b) Obtain similar characterisations of surjective functions.

(c) Find necessary and sufficient conditions for the composite of two functions to be injective, surjective, bijective.

1.4 (a) For any three sets X, Y, Z prove that there exists a bijection from $(X \times Y) \times Z$ to $X \times (Y \times Z)$ and also a bijection from $X \times Y$ to $Y \times X$.

(b) Let $X_1, \ldots, X_n$ be non-empty sets and let X be their cartesian product $X_1 \times X_2 \times, \ldots, \times X_n$. For $i = 1, 2, \ldots, n$ define $\pi_i : X \rightarrow X_i$ by $\pi_i(x_1, \ldots, x_n) \equiv x_i$ for $(x_1, \ldots, x_n) \in X$. Prove that π_i is a surjection. [Strictly speaking, we should write $\pi_i ((x_1, \ldots, x_n))$ instead of $\pi_i (x_1, \ldots, x_n)$. But where functions whose domains are subsets of cartesian products are concerned, double parentheses are generally not written. The function π_i is called the ***i*th projection** or the projection of X onto X_i. If $x \in X$, $\pi_i(x)$ is called the ***i*th coordinate** of x.]

(c) With X as in (b) suppose Y is a set and for each $i = 1, \ldots, n$ $f_i : Y \rightarrow X_i$ is a given function. Prove that there exists a unique function $f: Y \rightarrow X$ such that for each $i = 1, 2, \ldots, n$, $f_i = \pi_i \circ f$.

1.5 X, Y are sets, let Y^X denote the set of all functions from X to Y. In case X, Y are finite, how many elements are there in Y^X?

1.6 (a) Let $\mathbb{Z}_2$ be the set $\{0, 1\}$ and X any set. For $A \subset X$, the function $f_A : X \rightarrow \mathbb{Z}_2$ defined by $f_A(x) = 1$ if $x \in A$ and $f_A(x) = 0$ if $x \notin A$ for $x \in X$ is called the **characteristic function** of A. If $A, B \subset X$, express $f_{A \cap B}$, $f_{A \cup B}$ and f_{X-A} in terms of f_A and f_B. Prove that there is a bijection from the power set $P(X)$ to the set $\mathbb{Z}_2^X$.

(b) Using (a) and the last exercise, prove that if X has n elements then $P(X)$ has 2^n elements.

*1.7 For any set X prove that $|X|$ is strictly smaller than $|P(X)|$. [This is evident when X is finite. When X is infinite the proof is surprisingly short but tricky.]

1.8 Let X, Y be non-empty sets. Prove that $|X| \geqslant |Y|$ if and only if there exists a surjective function $f : X \rightarrow Y$.

1.9 (a) Prove that a subset of a finite set is finite and that a subset of a countable set is countable.

(b) Prove that the product of two and hence any finite number of countable sets is countable. [Hint: Note that the function $f: \mathbb{N} \times \mathbb{N} \rightarrow \mathbb{N}$ defined by $f(x, y) = \frac{1}{2}(x + y - 1)(x + y - 2) + y$ for $x, y \in \mathbb{N}$ is a bijection.]

(c) If $\{A_i\}_{i \in I}$ is an indexed family of countable sets and the index set

I is countable, prove that $\bigcup_{i \in I} A_i$ is countable. [Hint: First assume that A_i's are mutually disjoint, denumerable and apply (b). Then apply (a) and the last exercise. The present result is expressed by saying that a countable union of countable sets is countable.]

(d) Prove that the set of all finite subsets of a countable set is countable.

(e) Prove that every infinite set contains a denumerable subset.

1.10 Let $A \subset \mathbb{N}$. For $n \in \mathbb{N}$, let $\sigma_n(A) = \frac{1}{n}$ card $(A \cap \{1, 2, \ldots, n\})$. If $\lim_{n \to \infty} \sigma_n(A)$ exists, it is called the **density** (or the **limiting density**) of the subset A. We denote it by $\sigma(A)$.

(a) Prove that the density of the set of even positive integers is $\frac{1}{2}$.

(b) Prove or disprove that a subset $A \subset \mathbb{N}$ is finite iff $\sigma(A) = 0$.

1.11 Let $f : X \to X$ be a function. The functions $f \circ f, f \circ f \circ f, \ldots$ are denoted respectively by $f^2, f^3, \ldots$, etc. By convention, f^0 denotes 1_X. A point $x_0 \in X$ is called a **fixed point** of f if $f(x_0) = x_0$. Prove that a fixed point of f is also a fixed point of f^2. Does the converse hold?

Notes and Guide to Literature

By far, a most recommendable book on set theory is that of Halmos [1]. The axiomatic approach to set theory may be found in the Appendix to Kelley [1]. When the reader consults other literature, a word of caution is in order regarding the definitions and notations about functions. Some authors do not require a function to be single valued while some do not require that it be defined at all the points of the domain set. The term 'natural domain' of a function is sometimes used to denote the set of points where some formula (expressing that function) makes sense. For example the natural domain of the function $\frac{1}{\sqrt{4 - x^2}}$ is the set of real numbers between -2 and 2. We do not follow this practice because it is imprecise and tends to confuse a function with its formula. The word 'range' is often used in the place of 'codomain' which is of relatively recent origin.

If $f : X \to Y$ is a function and $x \in X$, the value of f at x is denoted by some authors by xf rather than $f(x)$. With this convention, the composite of functions is denoted the opposite way. One advantage of this notation is that the order in which the functions appear is also the order in which they act. It is probably for this reason that the new notation is increasingly followed. Nevertheless, we stick to the old notation.

2. Sets with Additional Structures

It was mentioned in the last section that sets are the building blocks of mathematics. Starting from sets, mathematics can be built up in two ways. One of them is to take some very specific sets such as $\emptyset$, $\{\emptyset\}$, $\{\emptyset, \{\emptyset\}\}$ and so on. It turns out that natural numbers can be defined and their properties can be proved in terms of these sets. Once natural numbers are defined, one can successively define integers, rational numbers, real numbers, complex numbers and euclidean spaces as certain sets. One then goes on to prove deeper and deeper properties of these sets. Number theory, euclidean geometry, classical analysis are examples of this type of mathematics. Note that in all these branches, we deal with some very concrete sets.

In the other type of mathematics, on the other hand, we start with an 'abstract' set, say, S. In applications or in giving examples, S will be replaced by a concrete set, for example a set of real numbers, a set of functions or a set of elements. We prove theorems about the abstract set S and when we substitute for S a concrete set, we get results about that concrete set. Of course there is very little one can prove with an abstract set alone. In order to prove non-trivial theorems, we must have, in addition to the given set, certain 'additional structure' on it. This additional structure is usually in the form of certain other sets or functions associated with that set. A few properties of this additional structure are assumed as axioms (or equivalently, they are made a part of the definition of that particular additional structure) and the rest are proved as theorems. The richer the structure, the more non-trivial theorems one can prove.

The simplest additional structure that can be put on a set is that of a binary relation. If S is a set, a **binary relation** on S is defined simply as a subset of $S \times S$. We can similarly define a ternary or more generally an n-ary relation (where n is a positive integer) as a subset of the appropriate power of S. But we shall rarely consider n-ary relations when $n \neq 2$. Hence a relation would always mean a binary relation unless indicated otherwise. If R is a relation on a set S then for $a, b \in S$ we write aRb to mean $(a, b) \in R$ and read it as 'a is related to b under R' or 'a is R-related to b' or simply, as 'a is related to b' where R is understood.

A relation R on a set S is said to be **reflexive** if for all $a \in S$, aRa. It is **symmetric** if for all $a, b \in S$, aRb implies bRa. It is **transitive** if for all $a, b, c \in S$, aRb and bRc implies aRc. It is **anti-symmetric** if for all $a, b \in S$, aRb and bRa implies $a = b$. If T is a subset of S, then the **restriction** of R to T or the relation on T **induced** by R is defined to be the set $R \cap (T \times T)$ (which is a subset of $T \times T$). It is denoted by R/T or occasionally by R itself when such a double use is not likely to cause confusion. If R is a relation on S and $A \subset S$, by $R[A]$ we mean $\{y \in S : xRy \text{ for some } x \in A\}$. For $x \in S$, $R[\{x\}]$ will be denoted by $R[x]$.

A relation is said to be an **equivalence relation** if it is reflexive, symmetric and transitive. If R is an equivalence relation on S and $x \in S$ then $R[x]$ is called the **equivalence class** under R or R-equivalence class containing x. It is clear that for any $x, y \in S$ either $R[x] = R[y]$ or $R[x] \cap R[y] = \emptyset$ and $S = \bigcup_{x \in S} R[x]$. Note also that each equivalence class is non-empty. Moreover the equivalence relation R can be recovered from the equivalence classes by observing that for $x, y \in S$, xRy if and only if x and y are in the same equivalence class. This suggests an alternate way of looking at equivalence relations. A **decomposition** $\mathscr{D}$ of a set S is defined as a family of non-empty subsets of S which is pairwise disjoint and whose union is S. There is then a one-to-one correspondence (i.e. a bijection) between the set of all equivalence relations on a set and the set of its decompositions. The decomposition corresponding to an equivalence relation R on a set S is denoted by S/R and is called the **quotient set** of S by R or the set of **quotient classes modulo** R. The function $p : S \to S/R$ defined by $p(x) = R[x]$ for $x \in S$ is called the **projection** or the **quotient function.**

Besides equivalence relations, the other important type of relations is that of order relations which we now define. The motivations for the various definitions we give, obviously come from the usual order relation for real numbers.

A relation which is reflexive, transitive and anti-symmetric is called a **partial order or ordering.** Partial orders are commonly denoted by symbols such as $\leq$, $\underline{\propto}$, $\geq$ or $\underline{\infty}$ and read in the usual manner. Thus, $a \leq b$ is read as 'a is less than or equal to b'. A **strict order** on a set S is a relation R which is transitive and for which $(a, a) \notin R$ for all $a \in S$. Evidently, such a relation is antisymmetric. Let ΔS be the set $\{(a, a) : a \in S\}$. ΔS is called the **diagonal** on S. It is clear that if R is a partial order on S then $R - \Delta S$ is a strict order while if R itself is a strict order then $R \cup \Delta S$ is a partial order. If a partial order is denoted by $\leq$, the corresponding strict order is denoted by $<$. Occasionally, a partial order is itself denoted by a symbol such as $<$, $\subset$ or $\propto$. In such a case the corresponding strict order is denoted by $\underset{\neq}{<}$, $\underset{\neq}{\subset}$ or $\underset{\neq}{\propto}$. We have already come across an instance of this. Let S be the power set of some set X. Then $\subset$ defines a binary relation on S, called **set inclusion** where for $A, B \in S$, $A \subset B$ means A is a subset of B. It is clear that $\subset$ is a partial order on S.

A **partially ordered set** (or a **poset**) is an ordered pair $(S, \leq)$ where S is a set and $\leq$ is a partial order on S. We say that S is partially ordered by $\leq$. The partial order $\leq$ is said to be **total** (or **linear** or **simple**) if for all $a, b \in S$ either $a \leq b$ or $b \leq a$. This property is known as the **law of dichotomy** and is expressed by saying that every two elements of S are comparable (under $\leq$). Translated in terms of the corresponding strict order $<$, it says that for all $a, b \in S$, either $a < b$ or $a = b$ or $b < a$. In this formulation the property is known as the **law of trichotomy.** A partially ordered set $(S, \leq)$ in which $\leq$ is a total order is called a **chain** (or a **nest** or a **tower**). More generally, if $(S, \leq)$ is a partially ordered set, then a subset T of S is called a **chain** in $(S, \leq)$ if the induced relation $\leq/T$ is a linear order.

A partially ordered set is our first formal example of a set with an additional structure. If $(S, \leq)$ is a partially ordered set, then the set S is called its **underlying set**. It may appear a little clumsy to define a poset as an ordered pair. But it is necessary to do so, because when we talk of a partially ordered set, the binary relation is as important as the underlying set. If there are two distinct partial orders say $\leq_1$ and $\leq_2$ on a set S then the partially ordered sets $(S, \leq_1)$ and $(S, \leq_2)$ are distinct from each other even though they have the same underlying set. A minor but grammatically interesting point is that the entire expression 'partially ordered set' stands for a single gadget, namely, a certain ordered pair. The expression is not to be construed as if the phrase 'partially ordered' were an adjective of the noun 'set'. Thus, a partially ordered set is different from a set which is partially ordered, not a confusing distinction if we keep in mind that a mother-in-law is not the same as a mother in law! Perhaps we should have used the hyphenated expression partially-ordered-set but even without it no confusion need arise once the point is clearly understood. Still, by abuse of language we use terms, such as a subset of a poset, when we really mean a subset of the underlying set of that poset.

If $(S, \leq_1)$ and $(T, \leq_2)$ are two partially ordered sets and $f: S \to T$ is a function then f is said to be **order-preserving** (or **monotonic** or **isotone**) if for all $a, b \in S$, $a \leq_1 b$ implies $f(a) \leq_2 f(b)$. f is said to be **strictly order-preserving** if for all $a, b \in S$, $a <_1 b$ implies $f(a) <_2 f(b)$. If S is a chain, such a function is necessarily injective. If f is a bijection and f as well as its inverse f^{-1} are both order preserving then f is called an **order-isomorphism** or **order-equivalence** from $(S, \leq_1)$ to $(T, \leq_2)$ and when such an order-equivalence exists the two posets are said to be **order-equivalent** or of the same order-type. The **order type** of a poset is the class of all posets order-equivalent to it.

Let $(S, \leq)$ be a partially ordered set and suppose $A \subset S$. An element $x \in S$ is said to be an **upper bound** for (or of) A if for all $a \in A, a \leq x$. A set which has at least one upper bound is said to be **bounded above.** The terms **'lower bound'** and **'bounded below'** are defined similarly. A set which is both bounded above and bounded below is said to be **bounded.** All these definitions are of course relative to the given partial order $\leq$. When the entire set S is bounded (above or below) we say that $\leq$ is bounded or that $(S, \leq)$ is **bounded (above** or **below**). If X is any set and $f: X \to S$ is a function then f is said to be **bounded (above** or **below**) if its range $f(X)$ is bounded (above or below).

Let $(S, \leq)$ and A be as before. An element $x \in A$ is called a **maximal** element of A if it is not strictly less than any element of A, that is, if for each $b \in A$, $a < b$ is false. This falsehood may in some cases be because of the non-comparability of a and b under $\leq$. Consequently, it is possible that the same set may have more than one maximal element. The concept of a minimal element of a set is defined similarly. An element x of A is said to be the **largest** (or **greatest** or **maximum**) element of A if for all $a \in A \sim \{x\}$, $a < x$. It is obvious that a set can have at most one largest element and when it exists, it is also the only maximal element of that set. On the contrary, the existence of even a unique maximal element does not imply the existence of a largest element. In the case of chains, the distinction between a maximal and a maximum element disappears. The concepts of a **minimal** and a **least** (or **smallest** or **minimum**) element of a set are defined similarly. Further, if we have a function $f: X \to S$ then we say it is **maximal** at some $x_0 \in X$ if $f(x_0)$ is a maximal element of the range $f(X)$ (w.r.t. $\leq$, of course). Other terms are defined analogously. The maximum and the minimum elements of a set A are denoted (in case they exist) respectively, by max A and min A.

The least element, if any, of the set of all upper bounds of a set A is called the **least upper bound (l.u.b.** or **supremum**) of A and is denoted by sup (A) or by sup A. The concept of the **greatest lower bound (g.l.b.** or the **infimum**) of a set A is defined analogously and is denoted by inf (A) or by inf A. A partial order is said to be **complete** if every non-empty subset which is bounded above has a sumpremum. It is easy to show that this condition is equivalent to the existence of an infimum for every non-empty subset which is bounded below. A partially ordered set $(S, \leq)$ is said to be a **lattice** if every two-element subset of S (i.e. every subset of the form $\{a, b\}$ where a, b are distinct elements of S) has both a supremum and an infimum. These elements are, respectively, called the **join** and the **meet** of the two elements of that subset.

A subset A of a poset $(S, \leq)$ is said to be **well-ordered** if every non-empty subset of A has a least element. For example if $\leq$ is the usual partial order on $\mathbb{R}$, the set of real numbers, then the set of positive integers is well-ordered w.r.t. it, even though the entire set $\mathbb{R}$ is not so. When the set S is well-ordered, $\leq$ is said to be a **well-ordering** of S. Evidently, every well-ordering is linear and complete. The order type of a well-ordered set is called an **ordinal**. There is a good deal of literature about ordinals. They are also important in topology as sources of interesting counter-examples. Nevertheless, we prefer to work without them because they tend to be rather involved for a beginner; and anyway, the counter-examples require only a minimal knowledge about well-ordering. These counter-examples will be given at the end of the book.

We conclude the discussion of partial orders by mentioning three statements each of which is equivalent to the axiom of choice mentioned in the last section. They are :

(1) **Well-Ordering Principle:** Every set can be well-ordered, that is, on every set there exists a well-ordering.

(2) **Hausdorff Maximal Principle:** Every partially ordered set contains a maximal chain. In other words, if $(S, \leq)$ is a poset then there exists a subset A of S such that $\leq/A$ is a linear order and for any $A \subsetneqq B \subset S$, $\leq/B$ is not a linear order.

(3) **Zorn's Lemma:** A partially ordered set in which every chain has an upper bound has at least one maximal element.

The equivalence of (2) and (3) is not hard to establish. The other equivalences are not so easy and we shall take them for granted. In applications, we shall find it convenient to use Zorn's lemma rather than the well-ordering or the maximal chain principle. The peculiarity of these statements (and other versions of the axiom of choice) is that they all assert the existence of something without actually giving any method of finding it. For example although the well-ordering principle asserts the existence of a well-ordering for every set, nobody has so far exhibited a well-ordering for such a familiar set as that of real numbers. For this reason, a proof based on the axiom of choice is called an **existence** proof or a **non-constructive** proof.

Another type of an important additional structure on a set is that of a binary operation. Although we shall need such structures mostly for examples, we list a few definitions. A **binary operation** on (or in) a set S is a function from $S \times S$ to S. Similarly, we define an n-ary operation for any positive integer n. If $\between$ is a binary operation on a set S and $a, b \in S$ then it is customary to denote $\between(a, b)$ by $a \between b$ and, when the operation is denoted by $\bullet$, by ab as well. A binary operation $\between$ on a set S is said to be **commutative** if for all $a, b \in S$, $a \between b = b \between a$ and it is said to be **associative** if for all $a, b, c \in S$ $a \between (b \between c) = (a \between b) \between c$ (this element can then be denoted unambiguously by $a \between b \between c$). An element $e \in S$ is called a **right** (or **left**) **identity** for $\between$ if for each $a \in S$, $a \between e = a$ (respectively, $e \between a = a$). An element which is both a right identity and a left identity for $\between$ is called an **identity** or **neutral element** of $\between$. Note that such an element, if it exists, is unique. If e is the identity of a binary operation $\between$ on a set S, and $a \in S$ then an element b of S is called a **right** (or **left**) inverse of a if $a \between b = e$ (respectively $b \between a = e$). An element which is both a right and a left inverse of a is called an **inverse** of a. It is easy to see that in the case of an associative operation, such an element is unique. An element which has an inverse is said to be **invertible**. An element a of S is said to be **cancellable from left** (or **right**) if for all $x, y \in S$, $a \between x = a \between y$ implies $x = y$ (respectively, $x \between a = y \between a$ implies $x = y$).

Given two binary operations say $\between$ and $\bullet$ on the same set S, $\bullet$ is said to be **left** (or **right**) **distributive** over $\between$ if for all $a, b, c \in S$, $a \bullet (b \between c) = (a \bullet b) \between (a \bullet c)$ (respectively, $(b \between c) \bullet a = (b \bullet a) \between (c \bullet a)$). If $\bullet$ is both left and right distributive over $\between$, it is said to be **distributive** over $\between$.

With these definitions, we are now in a position to define some simple objects. A **monoid** is an ordered pair $(S, \bullet)$ where S is a set and $\bullet$ is an associative binary operation on S which has an identity. This identity is called the **unit element** of the monoid. If further, every element of S is invertible then $(S, \bullet)$ is called a **group**. When $\bullet$ is understood we also say S is a group. If $\bullet$ is commutative, the group $(S, \bullet)$ is called as **abelian group**. For abelian groups, it is customary to denote the group operation by $+$ and to denote the inverse of an element x by $-x$. The identity element is customarily denoted by 0 (read as 'zero') for abelian groups. If $(G, \bullet)$ is a group, then a non-empty subset H of G is said to be a **subgroup** if for all $x, y \in H$, $x \bullet y \in H$ and $x^{-1} \in H$ where x^{-1} is the inverse of x in G. In this case $\bullet$ induces a binary operation on H w.r.t. which H is also a group in its own right. A subgroup H is called **normal** if for all $x \in H$, $g \in G$, $g^{-1}xg \in H$.

If G, H are groups, a function $f: G \to H$ is called a **homomorphism** (of groups) if for all $x, y \in G$, $f(x \bullet y) = f(x) \bullet f(y)$. Note that here on the left hand side $\bullet$ denotes the binary operation on G, while on the right hand side it denotes the binary operation on H. A homomorphism which is also a bijection is called an **isomorphism**. Two groups are said to be **isomorphic** if there exists an isomorphism from one of them to the other. A property of groups is said to be **group theoretic** if it is invariant under isomorphisms in the sense that if a group G has that property then so does every group isomorphic to G. For example the property of being an abelian group is a group theoretic property. But the property of being a subgroup of a given group is not a group theoretic property. Group theory may be defined as the study of group theoretic properties of groups.

Suppose $(R, +)$ is an abelian group and $\cdot$ is another binary operation on R which is associative and distributive over $+$ then the triple $(R, +, \cdot)$ is called a **ring**. If $\cdot$ has an identity, it is denoted by 1 (read as 'one') and is called the **unit element** or the identity of the ring. If $\cdot$ is commutative, the ring is said to be commutative. The operations $+$ and $\cdot$ are commonly known as **addition** and **multiplication**. A commutative ring with identity (different from 0) in which every non-zero element has an inverse is called a **field**. Terms such as homomorphism or isomorphism of rings are defined so as to require preservation of both addition and multiplication. The definitions of a subring and subfield are also left to the reader to formulate, being analogous to that of a subgroup.

The structures of binary operations and of a binary relation can sometimes be combined, as for example in the case of an ordered field. Formally, an **ordered field** is a quadruple $(F, +, \cdot, <)$ where $(F, +, \cdot)$ is a field and $<$ is a strict, total order on F which is compatible with the addition and multiplication in the following sense: for all $a, b, c \in F$, $b < c$ implies $a + b < a + c$ and if $0 < a$ then $b < c$ implies $ab < ac$. An element x of F is called **positive** if $0 < x$ and **negative** if $x < 0$. For $x \in F$, the magnitude or the **absolute value** of x, denoted by $|x|$ is defined as $\max \{x, -x\}$. Note that for all $x \in F$, $0 \leq |x|$ and equality holds iff $x = 0$. Also, for all $x, y \in F$ we have that $|x \cdot y| = |x| \cdot |y|$ and $|x + y| \leq |x| + |y|$ the latter often being called the **triangle inequality**. The proofs of these facts are left to the reader. By an **isomorphism of ordered fields** we mean a bijection which is both a field isomorphism (that is, a ring isomorphism) and an order isomorphism.

In summary, we observe that in general a set with an additional structure is an ordered pair (or a triple, quadruple etc.) the first entry of which is a set and the remaining entries are certain sets or functions associated with that set. An isomorphism of two sets having the same sort of structure is a bijection between those sets which is compatible with (or preserves) the additional structure. We urge the reader to illustrate with examples the various concepts defined in this section. In this book we shall be concerned with ordered pairs of the form $(X, \mathcal{T})$ where X is a set and $\mathcal{T}$ is a family of subsets of X (i.e., $\mathcal{T}$ is a subset of $P(X)$, or equivalently, an element of $P(P(X))$) satisfying certain conditions. The additional structure here is in the form of a family of subsets. When the additional structure consists of one (or several) binary (or n-ary in general) operations, it is called an **algebraic** structure.

Exercises

2.1 Let R be an equivalence relation on a set S and let $p : S \to S/R$ be the quotient function. If X is any set, a function $f : S \to X$ is said to be **invariant under** R (or to **respect** R) if for all $a, b \in S$, aRb implies $f(a) = f(b)$. Prove that f respects R iff there exists a function $g : S/R \to X$ such that $f = g \circ p$ (this is also expressed by saying that f **factors through** p). Prove that there is a one-to-one correspondence between the set of all functions from S/R to X and the set of those functions from S to X which are invariant under R.

2.2 Let $\mathbb{Z}$ be the set of all integers and n a fixed positive integer. For $a, b \in \mathbb{Z}$ let aRb iff n divides $a - b$ (i.e. iff $a - b = nx$ for some $x \in \mathbb{Z}$). aRb is also written as $a \equiv b$ (modulo n) and read as 'a is congruent to b modulo n.'

(a) Prove that R is an equivalence relation on $\mathbb{Z}$.

(b) Prove that there are n distinct equivalence classes under R and that each equivalence class contains precisely one element of the set $\{0, 1, \ldots, n-1\}$. [These equivalence classes are called **congruence classes** modulo n and the set $\mathbb{Z}/R$ is denoted by $\mathbb{Z}_n$].

(c) Prove that $(\mathbb{Z}, +, \cdot)$ where $+$ and $\cdot$ denote, respectively, the usual addition and multiplication of integers is a commutative ring with identity.

(d) For all $a, b, c, d \in \mathbb{Z}$ prove that aRb and cRd implies $(a + c)R(b + d)$ and $acRbd$.

(e) Prove that it is possible to define addition and multiplication of congruence classes modulo n in such a way that $\mathbb{Z}_n$ is a commutative ring with identity and the quotient function $f : \mathbb{Z} \to \mathbb{Z}_n$ is a ring homomorphism. Prove also that the ring $\mathbb{Z}_n$ is a field if and only if n is a prime.

(f) Use (a) and (d) to test whether the number $2 \cdot 3 \cdot 5 \cdot 7 \cdot 11 \cdot 13 \cdot 17 + 1$ is divisible by 19. Check your answer by direct computation also.

2.3 Let R be a reflexive and transitive relation on a set S. For $a, b \in S$ define aTb iff aRb and bRa. Prove that T is an equivalence relation on S. If $a \in S$, let $[a]$ denote $T[a]$, the equivalence class under T containing a. Define a binary relation $\leq$ on S/T by $[a] \leq [b]$ iff aRb for $a, b \in S$. Prove that $\leq$ is a partial order on S/T. Illustrate this in the case when $S = \mathbb{Z}$ and for $a, b \in \mathbb{Z}$, aRb means a divides b. In this case also verify that $(S/T, \leq)$ is a lattice. [Hint: consider the least common multiple and the greatest common divisor.]

2.4 Let R be a binary relation on a set S. Then its **inverse relation** denoted by R^{-1} is defined by $aR^{-1}b$ iff bRa for all $a, b \in S$. Prove that R is symmetric iff $R = R^{-1}$. Prove also that R is a partial order iff R^{-1} is so. [A maximal element w.r.t. R is a minimal element w.r.t. R^{-1} and vice versa. Because of results of this sort, once a statement is proved for maximal elements, the corresponding statement about minimal elements need not be proved again. For example Zorn's lemma could have equally well been stated as every poset in which every chain has a lower bound has a minimal element. If a partial order is denoted by $\leq$, $\underline{\propto}$, etc. its inverse order is denoted by $\geq$, $\underline{\backsim\!\!\!\!\propto}$, etc.]

2.5 Justify, by means of proofs or counter-examples, the statements about the existence and uniqueness of maximal and maximum elements, made immediately after their definitions.

2.6 Prove that a partial order $\leq$ on a set S is complete iff every non-empty subset of S which is bounded below has a greatest lower bound.

2.7 Let $(S_1, \leq_1)$ and $(S_2, \leq_2)$ be two partially ordered sets. Let $S = S_1 \times S_2$. For $(a_1, a_2), (b_1, b_2) \in S$ define $(a_1, a_2) \leq (b_1, b_2)$ iff either $(a_1 < b_1)$ or $(a_1 = b_1$ and $a_2 \leq b_2)$. Prove that $\leq$ is a partial order on S. Examine which properties of partial orders carry over from $\leq_1$ and $\leq_2$ to $\leq$. For example if $\leq_1$ and $\leq_2$ are total orders is $\leq$ necessarily total? [For obvious reasons, the ordering $\leq$ is called the **lexicographic** or the **dictionary ordering** induced by $\leq_1$ and $\leq_2$. It is a good source of counter-examples.]

2.8 A subset A of a poset $(S, \leq)$ is said to be **dense** (or **order-dense**) in S if for all $a, b \in S$, $a < b$ implies that there exists $c \in A$ such that $a < c < b$ (i.e. $a < c$ and $c < b$). Prove that if $(S, \leq)$ is a simple, denumerable poset which is dense in itself and bounded neither above nor below then $(S, \leq)$ has the same order type as the set of rational numbers with the usual ordering. What if $(S, \leq)$ were either bounded above or bounded below?

2.9 Let $S = P(X)$ for some set X. For $A, B \in S$ define $A \triangle B$ to be the set $(A - B) \cup (B - A)$. It is called the **symmetric difference** of A and B. Prove that $(S, \triangle, \cap)$ is a commutative ring with identity. Prove also that the set inclusion $\subset$ is a complete partial order on $P(X)$ and that $(P(X), \subset)$ is a lattice.

2.10 Prove the equivalence of Zorn's lemma and the Hausdorff maximal principle. [Hint: If Zorn's lemma is given, apply it to the family of all chains in a given poset partially ordered by inclusion of chains and get a maximal chain. For the converse, note that an upper bound of a maximal chain is necessarily a maximal element of the poset.]

2.11 Let $(G, \bullet)$ be a group and H a subgroup of G. For $x, y \in G$ define xRy iff $xy^{-1} \in H$.

(a) Prove that R is an equivalence relation on G. [The equivalence classes under this relation are called the **right cosets** of H in G.]

(b) If H is a normal subgroup of G, prove that for all $x, y, z, w \in G$, xRy and zRw implies $xzRyw$.

(c) In (b), prove that the quotient set G/R can be given the structure of a group in such a way that the quotient function $p : G \to G/R$ is a group homomorphism. [In such a case, the group so formed is denoted by G/H and is called the **factor group** or the **quotient group** and p is called the **natural** (or **canonical** or **quotient**) **homomorphism**. What is really natural about it will be discussed in Chapter Thirteen.]

2.12 Prove the statements made in the text regarding the absolute values of elements of an ordered field. Prove also that the square of every non-zero element in an ordered field is positive.

2.13 Let $(G, \bullet)$ be a group, S any set and X the set of all functions from S to G. For $f, g \in X$ define $f \bullet g \in X$ by $f \bullet g(s) = f(s) \bullet g(s)$ for $s \in S$. Prove that $(X, \bullet)$ is a group and that it is abelian when G is abelian. [This result is typical of those which show how a certain structure on a set induces a structure of a similar type on the set of functions from some set into that set. In the present case the binary operation in X is said to be obtained by **point-wise application** of the binary operation in G.]

2.14 Let $(\mathbb{Z}, +)$ be the abelian group of integers with usual addition and S be any set. Let $(X, +)$ be the abelian group of all functions from S to $\mathbb{Z}$, formed in the last exercise. Let $F(S) = \{f \in X :$ there exists a finite subset T of S such that $f(s) = 0$ for all $s \in S - T\}$. In other words, $F(S)$ consists of those functions which vanish outside a finite subset of S. (Caution: This subset may change from function to function.) Prove that $F(S)$ is a subgroup of $(X, +)$. It is called the **free abelian group** on the set S. It is customary to identify an element, say s, of S with the unique function from S to $\mathbb{Z}$ which is 1 at s and 0 elsewhere. With this identification regard S as a subset of $F(S)$. Prove that every element of $F(S)$ can be written uniquely as $\Sigma n_k s_k$ where for each k, $n_k \in \mathbb{Z}$, $s_k \in S$ and all except finitely many n_k's are non-zero. If H is any abelian group, prove that a function $f : S \to H$ uniquely determines a group homomorphism $\theta : F(S) \to H$ such that $\theta/S = f$.

2.15 Let S be the set of all functions from a set X into itself. Then composition of functions gives rise to a binary operation $\bullet$ on S. Prove that $(S, \bullet)$ is a monoid. Prove further that under this operation, the set of all permutations (see page 34) of X is a group. It is called the **permutation group** of X.

Notes and Guide to Literature

For a thorough discussion of the axiom of choice (and a proof of the equivalence of its various versions) see Halmos [1]. A proof of the result in Exercise (2.8) may be found, for example, in Goffmann [1].

Regarding the definition of a partial order, there is considerable variation among authors. Kelley [1], for example, requires only that it be transitive, which of course covers both partial orders and strict orders in our sense. In view of Exercise (2.3), it is immaterial whether anti-symmetry is assumed or not in the definition of a partial order, for one can always pass to the quotient set. Although we shall make only passing references to lattices, they arise naturally in such diverse branches of mathematics, that they are well worth a systematic study. A classic reference in this direction is Birchoff [1]. Incidentally, the word 'lattice' is also used in an entirely different context (to denote certain subsets of the cartesian plane $\mathbb{R} \times \mathbb{R}$ or of $\mathbb{R}^n$) in a branch of mathematics called geometric number theory (See Cassels [1]).

A word about the exercises. The results of many of them will be needed only in exercises or in a few examples. Nevertheless, they are given as a drill in handling abstract concepts. In some exercises we have referred to $\mathbb{Z}$, the set of integers and also to the set of rational numbers. These will be formally defined (in the exercises) in the next section. To work out the exercises given here the reader may assume as known their basic properties.

3. Preliminaries from Analysis

In the last section we broadly categorized mathematics as 'concrete mathematics' in which we deal with some very specific sets (mostly the set of real numbers, its subsets or some other sets derived from it) and 'abstract mathematics' in which we deal with any arbitrary sets and study them with reference to some additional structures put on them. Classical real (or complex) analysis is an example of concrete mathematics while our subject matter, topology, properly belongs to the second type of mathematics. Nevertheless, the real numbers have an important role to play in the study of topology. For, on one hand, concepts and theorems in real analysis have often served as motivations for certain definitions in topology. Topology (or at least that aspect of it which this book covers) may indeed be thought of as an abstraction of real analysis. Second, although in topology we shall deal with 'abstract' spaces, we shall nevertheless be concerned for considerable time with the question of describing these abstract spaces in terms of the concrete space of real numbers. For this reason, it is important to know at least some elements of real analysis. In this section we list them down. Throughout this book we shall use $\mathbb{R}$, $\mathbb{Q}$, $\mathbb{Z}$, $\mathbb{N}$ to denote, respectively, the sets of real numbers, rational numbers, integers and positive integers. We shall also call positive integers as natural numbers. Sometimes 0 is included as a natural number, but we shall not do so.

The construction of real numbers is relatively unimportant for our purpose. Indeed, it is possible to start with a flat assumption that on $\mathbb{R}$ there exist two binary operations $+$ and $\cdot$ and a complete, linear, strict order $<$ such that $(\mathbb{R}, +, \cdot, <)$ is an ordered field. One can then define integers, rational numbers, etc. and use them systematically. Effectively, this amounts to taking a real number as a primitive term and the existence of $+$, $\cdot$ and $<$ (along with the defining properties of an ordered field) as axioms. One is then at liberty to assign, privately, whatever intuitive meaning to real numbers and to $+$, 1, $<$, etc. one chooses (provided that the axioms are obeyed). At the other extreme, we can actually construct real numbers starting from some very specific sets such as $\emptyset$, $\{\emptyset\}$, $\{\emptyset, \{\emptyset\}\}$. This construction is usually carried out in four steps. First we construct natural numbers, then integers, then rational numbers, and finally the real numbers. Although we shall not attempt to define natural numbers in terms of sets, we shall indicate by way of exercises how the remaining three steps, namely the transitions from $\mathbb{N}$ to $\mathbb{Z}$, from $\mathbb{Z}$ to $\mathbb{Q}$ and finally from $\mathbb{Q}$ to $\mathbb{R}$ are carried out. Although only of peripheral interest to us, we shall also indicate how the set of complex numbers (to be denoted by $\mathbb{C}$) can be formed from $\mathbb{R}$.

Before proceeding with real numbers, it is convenient to define a few of the frequently occurring terms. Let $a, b \in \mathbb{R}$. The sets $\{x \in \mathbb{R} : a < x < b\}$, $\{x \in \mathbb{R} : a \leq x \leq b\}$ are denoted, respectively, by (a, b) and $[a, b]$ and are called, respectively, the **open interval** from a to b and the **closed interval** from a to b. It is a little unfortunate that the notation (a, b) also denotes an ordered pair but the usage is now too established to change. Some authors do, however, denote the open interval from a to b by $]a, b[$. We stick to the notation (a, b) and where a confusion is likely to result otherwise, we shall state explicitly whether we mean an open interval or an ordered pair. The sets $\{x \in \mathbb{R}: a \leq x < b\}$ and $\{x \in \mathbb{R}: a < x \leq b\}$ are called **semi-open intervals,** the former is said to be **closed at left** (or at a) and **open at right** (or at b), the latter is described analogously. If $b < a$, all the four sets (a, b), $[a, b]$, $[a, b)$ and $(a, b]$ are empty while if $a = b$, the second one is a singleton and the remaining three are empty. These are called **degenerate** intervals. When we speak of intervals there will be a tacit assumption that they are non-degenerate whenever it will be needed to avoid trivial exceptions. The numbers a, b are called, respectively, the **left** and **right ends** of any of the four intervals considered, the number $\frac{1}{2}(a + b)$ is called their **centre** and the positive number $b - a$ is called their **length.** Note that for all $x \in \mathbb{R}$ and $r > 0$, the set $\{y \in \mathbb{R}: |y - x| < r\}$ coincides with $(x - r, x + r)$ which is an open interval of length $2r$ and centre x. The interval $[0, 1]$ will be called the **unit interval**, denoted by I.

All these intervals are bounded. We now define unbounded intervals. For $a \in \mathbb{R}$ the sets $\{x \in \mathbb{R}: x > a\}$ and $\{x \in \mathbb{R}: x \geq a\}$ are called, respectively, the open interval from a to ∞ and the closed interval from a to ∞ and are denoted, respectively, by (a, ∞) and $[a, \infty)$. Similarly, $(-\infty, a)$ and $(-\infty, a]$ are defined. It should be noted that ∞ (read as 'infinity') and $-\infty$ (read as 'minus infinity') are not real numbers. Actually, they are non-entities, in the sense that they by themselves mean nothing, nor do they represent anything. While dealing with real numbers, ∞ and $-\infty$ will always come only as parts of some notation and the word 'infinity' will always come only as a part of some phrase. These notations and phrases will be defined in their entirety and to understand them it is unnecessary to know what ∞ and infinity mean much the same way as it is unnecessary to know what is 'law' in order to know what is mother-in-law! By convention $(-\infty, \infty)$ is the entire set $\mathbb{R}$. Expressions involving $[-\infty$ or $\infty]$ are meaningless.

By far the most important concept in real analysis is that of convergence. Convergence (of various types) is indeed the heart of calculus. A sequence $\{a_n\}$ of real numbers is said to **converge to** a real number L or to have L as its **limit** if for every $\epsilon > 0$, there exists $m \in \mathbb{N}$ such that for all $n \in \mathbb{N}$, $n \geq m$ implies $|a_n - L| < \epsilon$. When this holds we write $a_n \to L$ as $n \to \infty$ (the arrow read as 'tends to') or $\lim\limits_{n\to\infty} a_n = L$. A sequence is said to be **convergent** if it converges to some real number (which is then easily seen to be unique). We say $a_n \to \infty$ as $n \to \infty$ if for every $N \in \mathbb{R}$, there exists $m \in \mathbb{N}$ such that for all $n \in \mathbb{N}$, $n \geq m$ implies $a_n > N$. When this holds we say that the sequence $\{a_n\}$ **diverges to** ∞. The definition of divergence to $-\infty$ is similar and left to the reader. A sequence which diverges either to ∞ or to $-\infty$ is said to be **divergent.** We trust that the reader is familiar with simple properties of limits (results of the form 'the limit of a sum of sequences equals the sum of their limits'). Such results are left as exercises.

Convergent sequences can be characterised in various ways. We mention two of them. Recall from section 1 that a sequence is a function from $\mathbb{N}$. Now let f be a sequence. Then a **subsequence** of f is defined as a sequence of the form $f \circ g$ where $g : \mathbb{N} \to \mathbb{N}$ is a strictly monotonically increasing function. (Actually it is enough that g is merely monotonically increasing and $g(n) \to \infty$ as $n \to \infty$. Whichever definition is followed, the same results are obtained.) If the original sequence is denoted by $\{a_n\}_{n=1}^{\infty}$ and $g(k) = n_k$ for $k \in \mathbb{N}$, then the subsequence is denoted by $\{a_{n_k}\}_{k=1}^{\infty}$. A convergent subsequence means a subsequence which, when regarded as a sequence by itself is convergent. It is easy to show that every subsequence of a convergent sequence is convergent and has the same limit as the original sequence. A convergent sequence may be characterised as a sequence which is bounded and all of whose convergent subsequences converge to the same limit.

Another important characterisation of convergent sequences is due to Cauchy. A sequence $\{a_n\}$ of real numbers is said to be a **Cauchy sequence** if for every $\epsilon > 0$, there exists $p \in \mathbb{N}$ such that for all $m, n \in \mathbb{N}$, $m \geq p$ and $n \geq p$ implies $|a_m - a_n| < \epsilon$. It is trivial to show that every convergent sequence is a Cauchy sequence. The converse is non-trivial and is equivalent to the order completeness of $\mathbb{R}$.

A **series** of real numbers is a notation of the form $\sum_{n=1}^{\infty} a_n$ (or simply Σa_n) where $\{a_n\}$ is a sequence of real numbers. For each n, let $b_n = a_1 + a_2 + \ldots + a_n$. b_n is called the ***n*th partial sum** of the series Σa_n. All the definitions regarding convergence, limits etc. of a series Σa_n are in terms of the sequence $\{b_n\}$ of partial sums. Occasionally, we consider series of the form $\sum_{n=k}^{\infty} a_n$, where k is an integer other than 1, in which case the definitions are suitably modified. A number of tests for convergence of series are known. We shall need only two of these. First, the geometric series $\sum_{n=1}^{\infty} r^n$ is convergent iff $|r| < 1$ and second the comparison test which says that if Σa_n, Σb_n are two series with $|a_n| \leqslant b_n$ for all $n \in \mathbb{N}$ then the convergence of Σb_n implies that of Σa_n.

We now define convergence in a more general context. Suppose X is a set and f, g are two real-valued functions on X. Let c, L be real numbers. We say $f(x) \to L$ as $g(x) \to c$ if for every $\epsilon > 0$, there exists $\delta > 0$ such that for all $x \in X$, $0 < |g(x) - c| < \delta$ implies $|f(x) - L| < \epsilon$. When this happens we call L a **limit** of $f(x)$ as $g(x)$ tends to c and write $\lim_{g(x) \to c} f(x) = L$. If for every $\epsilon > 0$, there exists $\delta > 0$ such that for all $x \in X$, $c < g(x) < c + \delta$ implies $|f(x) - L| < \epsilon$, then we say $f(x) \to L$ as $g(x) \to c^+$ (read as $g(x)$ tends to c plus, or to c from **above** or from **right**). In this case we call L as a **right** limit or **upper** limit (or limit from right) of $f(x)$ as $g(x)$ tends to c and write $\lim_{g(x)\to c^+} f(x) = L$. We leave it to the reader to define $f(x) \to L$ as $g(x) \to c^-$ as well as to make appropriate modifications in the definition of $f(x) \to L$ as $g(x) \to c$ when either c or L is replaced by ∞ or by $-\infty$.

In most cases of limits, X is a subset of $\mathbb{R}$ and g is simply its inclusion function into $\mathbb{R}$. A point $c \in \mathbb{R}$ is called a **limit point** or a **point of accumulation** of X if for every $\epsilon > 0$, there exists $y \in X$ such that $0 < |y - c| < \epsilon$. c is called a **right** (or **left**) **accumulation** point (or **right** (or **left**) **limit point**) of X if for every $\epsilon > 0$, there exists $y \in X$ such that $c < y < c + \epsilon$ (respectively, $c - \epsilon < y < c$). It is clear that $\lim_{x\to c^+} f(x)$ is unique (provided it exists) if c is a right accumulation point of X and that similar results hold for $\lim_{x\to c} f(x)$ and $\lim_{x\to c^-} f(x)$. The concepts of various limits are of interest only when c is an accumulation point (or a right accumulation point etc.) of X, since otherwise any number L will satisfy the requirement in the definition vacuously. Note, also, that a limit is a local concept in that it depends only on the values of a function near the point at which it is taken. To be precise, suppose $f, g : X \to \mathbb{R}$ are functions, $c \in \mathbb{R}$ and there exists $r > 0$ such that for all $x \in X$, $0 < |x - c| < r$ implies $f(x) = g(x)$. Then $\lim_{x\to c} f(x)$ exists iff $\lim_{x\to c} g(x)$ exists and the two are equal. Note further that the value of f at c is entirely irrelevant as far as limits of f at c are concerned.

Now suppose $X \subset \mathbb{R}$, $f : X \to \mathbb{R}$ is a function and $c \in X$. We say f is **continuous at** c if for every $\epsilon > 0$ there exists $\delta > 0$, such that for all $x \in C$, $|x - c| < \delta$ implies $|f(x) - f(c)| < \epsilon$. f is said to be **continuous** if it is continuous at every point of X. f is said to be **uniformly continuous** if for every $\epsilon > 0$ there exists $\delta > 0$ such that for all $x, y \in X$, $|x - y| < \delta$ implies $|f(x) - f(y)| < \epsilon$. Suppose for some $r > 0$, $[c, c + r) \subset X$. Then if $\lim_{x\to c^+} [(f(x) - f(c))/(x - c)]$ exists, it is called the **right-handed derivative** of f at c and is denoted by $f'_+(c)$ and we say f is **right-differentiable** at c. Left-handed derivatives and left differentiability are defined similarly. If both $f'_+(c)$ and $f'_-(c)$ exist and are equal, their common value is denoted by $f'(c)$ and is called the **derivative** of f at c. When it exists, f is said to be **differentiable** at c. A function is said to be differentiable on an interval if it is differentiable at every point of it other than the end-points and is appropriately left or right differentiable at an end point (in case it belongs to that interval). Differentiability will not be important for our purpose except in Chapter Twelve, where we shall prove the existence of a continuous function on $[0, 1]$ which is not differentiable at any point.

We now turn to the euclidean spaces. Let n be a fixed positive integer and consider $\mathbb{R}^n$, the cartesian product of n copies of $\mathbb{R}$. Suppose $x = (x_1, \ldots, x_n)$ and $y = (y_1, \ldots, y_n)$ are elements of $\mathbb{R}^n$. We define $x + y$ to be $(x_1 + y_1, \ldots, x_n + y_n)$. It is then easy to see that $(\mathbb{R}^n, +)$ is an abelian group. Elements of $\mathbb{R}^n$ are called **vectors** and the binary operation just defined is called **vector addition.** In this context, elements of $\mathbb{R}$ are called **scalars.** If $a \in \mathbb{R}$ and $x \in \mathbb{R}^n$ then $\alpha \bullet x$ or αx denotes the vector $(\alpha x_1, \ldots, \alpha x_n)$. This way we get a function $\bullet : \mathbb{R} \times \mathbb{R}^n \to \mathbb{R}^n$ which is called the **scalar multiplication.** $\bullet$ has the properties that for all $\alpha, \beta \in \mathbb{R}$ and $x, y \in \mathbb{R}^n$, $(\alpha + \beta)\bullet x = \alpha \bullet x + \beta \bullet x$, $\alpha \bullet (\beta x) = (\alpha \bullet \beta)x$, $1 \bullet x = x$ and $\alpha \bullet (x + y) = \alpha \bullet x + \alpha \bullet y$. The triple $(\mathbb{R}^n, +, \bullet)$ is an example of what is known as a vector space over the field $\mathbb{R}$. If $x, y \in \mathbb{R}^n$, their **dot** product (or **inner product** or **scalar product**), denoted variously by $x \bullet y$ or by $\langle x, y\rangle$ is defined as the number $\sum_{i=1}^{n} x_i y_i$. Note that for any $x \in \mathbb{R}^n$, $\langle x, x\rangle \geq 0$ and equality holds iff $x = 0$ where 0 denotes the zero vector $(0, 0, \ldots, 0)$. The non-negative square root of $\langle x, x\rangle$ is denoted by $\|x\|$ and is called the **norm** or the **length** of the vector x. For $x, y \in \mathbb{R}^n$, $\|x - y\|$ is called the **euclidean distance** or simply **distance** between x and y. Simple properties of the inner product and the norm will be given as exercises. The sets $\{x \in \mathbb{R}^n : \|x\| \leqslant 1\}$, $\{x \in \mathbb{R}^n : \|x\| < 1\}$ and $\{x \in \mathbb{R}^n : \|x\| = 1\}$ are called respectively the **closed unit n-ball** (or simply the **unit n-ball**), the **open unit n-ball** and the unit **$(n - 1)$-sphere.** They will be denoted by D^n, B^n and S^{n-1}, respectively. The set $\{x \in S^{n-1} : x_n = 0\}$ is called the **equator** of S^{n-1} and the point $(0, 0, \ldots, 0, 1) \in S^{n-1}$ the **north pole.** A vector of length 1 (equivalently an element of S^{n-1}) is called a **unit vector.** Every non-zero vector in $\mathbb{R}^n$ can be uniquely expressed as a positive scalar multiple of a unit vector, and this unit vector is called its **direction.** If $x \in \mathbb{R}^n$ and u is a unit vector in $\mathbb{R}^n$ then the set $\{x + tu : t \geq 0\}$ is called the **half ray** from x in the direction u. If $x, y \in \mathbb{P}^n$ and $x \neq y$, the set $\{x + t(y - x)\} : t \in \mathbb{R}\}$ is called the **straight line** or simply the **line** through the points x and y while the subset $\{x + t(y - x) : 0 \leq t \leq 1\}$ is called the **line segment** joining x and y. It is denoted by $\overline{xy}$. $\mathbb{R}^n$, with the additional structure of $+$, $\bullet$ and $\langle , \rangle$ is called the **n-dimensional euclidean space.**

Many concepts such as those of convergence, limits, continuity can be extended from real numbers to euclidean spaces, provided we replace the difference of real numbers with the distance between points. As a typical illustration let $X \subset \mathbb{R}^n$, $c \in X$ and $f: X \to \mathbb{R}$ a function. Then f is said to be **continuous** at c if for every $\epsilon > 0$ there exists $\delta > 0$ such that for all $x \in X$, $\|x - c\| < \delta$ implies $|f(x) - f(c)| < \epsilon$. Similarly, one can define an accumulation point of a set. Note, however, that there is no such thing as a left or a right accumulation point of a subset of $\mathbb{R}^n$ if $n > 1$. The appropriate analogue of these concepts would be an accumulation point along a particular direction. Similarly, one can define directional derivatives of real-valued functions. But they will be of little interest to us. Continuity of functions into a euclidean space can be defined in two ways. Suppose $f: X \to \mathbb{R}^m$ is a function where $X \subset \mathbb{R}^n$ for some n and $c \in X$. Then f is said to be **continuous** at c if for every $\epsilon > 0$, there exists $\delta > 0$ such that for all $x \in X$, $\|x - c\| < \delta$ implies $\|f(x) - f(c)\| < \epsilon$. Note that here the same symbol $\| \;\; \|$ is used to denote the norm on $\mathbb{R}^n$ as well as on $\mathbb{R}^m$. Alternatively, let for $i = 1, 2, \ldots, m$, $\pi_i : \mathbb{R}^m \to \mathbb{R}$ be the ith projection function. Then f is **continuous** at c iff for each $i = 1, \ldots, m$, the composite $\pi_i \circ f$ is continuous at c. It is easy to show that the two definitions are equivalent. Similarly, one can define convergence of sequences of vectors or limits of vector-valued functions.

When $n = 2$, the euclidean space $\mathbb{R}^n$ is called the **euclidean plane.** It deserves a special mention because its points can be identified with complex numbers. The sets D^2, B^2 and S^1 are called the **closed unit disc**, the **open unit disc** and the **unit circle**, respectively.

Exercises

3.1 The construction of $\mathbb{Z}$ from $\mathbb{N}$ is motivated from the requirement that every integer can be expressed as a difference of two positive integers. This expression is of course not unique. Define a binary relation R on $\mathbb{N}\times\mathbb{N}$ by $(a, b)R(c, d)$ iff $a + d = b + c$ for $a, b, c, d \in \mathbb{N}$. Prove that R is an equivalence relation. The equivalence classes under R are called **integers.** The quotient set $\mathbb{N}\times\mathbb{N}/R$ is denoted by $\mathbb{Z}$. If $a, b \in \mathbb{N}$ by $[(a, b)]$ we denote the equivalence class (under R) containing (a, b). If $[(a, b)], [(c, d)] \in \mathbb{Z}$ we define $[(a, b)] + [(c, d)]$ to be $[(a + c, b + d)]$ and $[(a, b)]\cdot[(c, d)]$ to be $[(ac + bd, bc + ad)]$. (If this appears clumsy to you, think of $[(a, b)]$ as $a - b$.)

(a) Prove that $+$ and $\cdot$ are well-defined binary operations on $\mathbb{Z}$, that both are commutative and associative and that $\cdot$ is distributive over $+$.

(b) For any $a, b \in \mathbb{N}$, prove that $[(a, a)] = [(b, b)]$ and $[(a + 1, a)] = [(b + 1, b)]$. Prove further that $[(1, 1)]$ is the identity for $+$ and $[(2, 1)]$ is the identity for $\cdot$ on $\mathbb{Z}$.

(c) Prove that $(\mathbb{Z}, +, \cdot)$ is a ring and further that the cancellation law holds for $\cdot$ (that is to say, every non-zero element is cancellable w.r.t. $\cdot$ which amounts to showing that the product of two nonzero integers is non-zero).

(d) For $[(a, b)], [(c, d)] \in \mathbb{Z}$ define $[(a, b)] \leq [(c, d)]$ iff $a+d \leq b+c$ w.r.t. the usual order $\leq$ on $\mathbb{N}$. Prove that $\leq$ is a total order on $\mathbb{Z}$ which is compatible with the binary operations $+$ and $\cdot$.

(e) Define $f: \mathbb{N} \to \mathbb{Z}$ by $f(n) = [(n + 1, 1)]$ for $n \in \mathbb{N}$. Prove that f is injective and preserves $+$, $\cdot$ and $\leq$. [Because of this we may regard $\mathbb{N}$ as a subset of $\mathbb{Z}$ for all practical purposes.]

(f) For every $c \in \mathbb{Z}$ prove that there exists $a, b \in \mathbb{N}$ such that $a - b = c$.

3.2 The construction of $\mathbb{Q}$ from $\mathbb{Z}$ is motivated from the requirement that every rational number be expressed as the ratio of two integers. Again, this expression is not unique. So we consider $\mathbb{Z} \times (\mathbb{Z} - \{0\})$ and define R by $(a, b)R(c, d)$ iff $ad = bc$ for $a, b, c, d \in \mathbb{Z}$, $b, d \neq 0$. Show that R is an equivalence relation. The set $\mathbb{Z} \times (\mathbb{Z} - \{0\})/R$ is denoted by $\mathbb{Q}$ and its elements are called **rational numbers**. Define addition and multiplication of rational numbers by $[(a, b)] + [(c, d)]$ to be $[(ad + bc, bd)]$ and $[(a, b)]\cdot[(c, d)]$ to be $[(ac, bd)]$ where

$a, c \in \mathbb{Z}$ and $b, d \in \mathbb{Z} - \{0\}$. Taking steps similar to those in the last exercise prove that $(\mathbb{Q}, +, \cdot, \leq)$ is an ordered field and that $\mathbb{Z}$ may be regarded as a subset of $\mathbb{Q}$. Prove that the order $\leq$ is total and $\mathbb{Q}$ is dense in itself and bounded neither above nor below. Also, prove that $\mathbb{Z}$, $\mathbb{Q}$ are denumerable sets.

3.3 The construction of $\mathbb{R}$ from $\mathbb{Q}$ is motivated by the requirement that every real number is uniquely determined by the set of rational numbers less than it. Thus, we call a subset L of $\mathbb{Q}$ a **real number** if L is non-empty, bounded above, has no maximum element and has the property that for all $x, y \in \mathbb{Q}$, $x < y$ and $y \in L$ implies $x \in L$. If L_1, L_2 are real numbers, we define $L_1 \leq L_2$ iff $L_1 \subset L_2$. Then $\leq$ is a total order on $\mathbb{R}$. For $q \in \mathbb{Q}$, we let $L_q = \{x \in \mathbb{Q} : x < q\}$. Then L_q is a real number, the function $f : \mathbb{Q} \to \mathbb{R}$ defined by $f(q) = L_q$ is one-one-one and order preserving. So we regard $\mathbb{Q}$ as a subset of $\mathbb{R}$. Note that it is an order-dense subset. Prove that $\leq$ is a complete order on $\mathbb{R}$. [Hint: If $\mathcal{L}$ is a non-empty subset of $\mathbb{R}$ which is bounded above, show that $\bigcup_{L \in \mathcal{L}} L$ is a real number and that it is the supremum of $\mathcal{L}$.] As regards the algebraic structure on $\mathbb{R}$, there is no difficulty in defining $+$. If $L_1, L_2 \in \mathbb{R}$ we let $L_1 + L_2 = \{x_1 + x_2 : x_1 \in L_1, x_2 \in L_2\}$. It is not hard to show that $L_1 + L_2$ is also a real number and that $+$ gives a commutative and associative binary operation on $\mathbb{R}$ whose identity is 0 (that is the set $\{x \in \mathbb{Q} : x < 0\}$). Note that the additive inverse of a real number L is the set L_{-q} if L is of the form L_q for some $q \in \mathbb{Q}$. Otherwise it is the set $\{x \in \mathbb{Q} : -x \notin L\}$. In defining the product $L_1 \cdot L_2$ a little care is necessary depending upon the signs of L_1 and L_2. One then shows that $(\mathbb{R}, +, \cdot, <)$ is a complete, linear ordered field containing $\mathbb{Q}$ as a subfield.

3.4 A **complex number** is defined as an ordered pair of real numbers, that is, an element of $\mathbb{R}^2$. Addition of complex numbers is defined the same way as for vectors. For multiplication, if (a, b), (c, d) are complex numbers, $(a, b) \cdot (c, d)$ is defined as the complex number $(ac - bd, ad + bc)$. The set of complex numbers is denoted by $\mathbb{C}$. [Thus, actually, $\mathbb{C} = \mathbb{R}^2$ but the former notation is preferred when points of $\mathbb{R}^2$ are viewed as complex numbers rather than as vectors.] It is easy to show that $(\mathbb{C}, +, \cdot)$ is a field and the function $f : \mathbb{R} \to \mathbb{C}$ defined by $f(x) = (x, 0)$ is one-to-one and compatible with $+$ and $\cdot$. So we regard $\mathbb{R}$ as a subset of $\mathbb{C}$ for all practical purposes. It is customary to denote the complex number $(0, 1)$ by i. Note that $i^2 = -1$. Every complex number z can be expressed uniquely as $x + iy$ where x, y are real. They are, respectively, called the **real** and **imaginary parts** of z.

3.5 Let $\{a_n\}$ be a sequence in a euclidean space $\mathbb{R}^k$. A point $x \in \mathbb{R}^k$ is called a **limit** point (or a **cluster** point) of $\{a_n\}$ if for every $\epsilon > 0$ and $m \in \mathbb{N}$, there exists $n \in \mathbb{N}$, such that $n \geqslant m$ and $\|a_n - x\| < \epsilon$. Prove that x is a limit point of $\{a_n\}$ iff some subsequence of $\{a_n\}$ converges to x. Define a Cauchy sequence in $\mathbb{R}^k$ and show that a Cauchy sequence in $\mathbb{R}^k$ can have at most one limit point.

3.6 Assume that $(\mathbb{R}, +, \cdot, <)$ is a linearly ordered field. Prove that the following statements are equivalent:

(1) The order $<$ is complete.

(2) Let $I_1, I_2, \ldots, I_n, \ldots$ be a descending sequence of closed and bounded intervals (this means that for $n \in \mathbb{N}$, $I_n \supset I_{n+1}$). Then $\bigcap_{n=1}^{\infty} I_n$ is non-empty.

(3) Every monotonic and bounded sequence in $\mathbb{R}$ is convergent.

(4) Every bounded sequence in $\mathbb{R}$ has at least one limit point.

(5) Every Cauchy sequence in $\mathbb{R}$ is convergent.

[Hint: It is convenient to prove (1) $\Leftrightarrow$ (2) and then prove the equivalence of (2) with each of the remaining statements either directly or in a cyclic manner.]

3.7 Let b be a fixed positive integer. Let S be the set $\{0, \ldots, b-1\}$. Let E be the set of all sequences in S. Prove that for any such sequence $a = \{a_n\}$, the series $\sum_{n=1}^{\infty} \frac{a_n}{b^n}$ converges to a number $f(a) \in [0, 1]$. Prove that the function $f : E \to [0, 1]$ is onto. Prove that a number $x \in (0, 1)$ can never have more than two preimages under f and that it has two preimages iff it can be written in the form p/b^q for some positive integers p, q. [Such numbers are called **b-adic rational** numbers. A preimage of a number x under f is called a **b-adic expansion** of x. For $b = 10, 2$ or 3 the b-adic expansions are also known as **decimal, binary** (or **dyadic**) and **ternary** (or **triadic**) expansions, respectively.]

*3.8 Let α be an infinite cardinal. Prove that

(a) If β is a finite cardinal then $\alpha + \beta = \alpha$. [Hint: Use Exercise (1.9e).]

(b) $\alpha + \alpha = \alpha$. [Hint: It suffices to show that for some set S of cardinality α, there exists a bijection from $S \times \mathbb{Z}_2$ onto S where $\mathbb{Z}_2 = \{0, 1\}$. Let X be a set with $|X| = \alpha$. Let $\mathcal{F}$ be the set of bijections of the form $f : A \times \mathbb{Z}_2 \to A$ where $A \subset X$. If $f, g \in \mathcal{F}$, define $f \leq g$ iff domain $(f) \subset$ domain g and $f(x) = g(x)$ for all $x \in$ domain (f). In view of Exercise (1.9e), $\mathcal{F}$ is non-empty and $\leq$ is seen to be a partial order on $\mathcal{F}$. Apply Zonrn's lemma and get a maximal element say $h : S \times \mathbb{Z}_2 \to S$ of $\mathcal{F}$. Prove that $X - S$ is finite (use maximality of h along with Exercise (1.9e)). It then follows from (a) that $|S| = |X|$.]

(c) If β is a cardinal and $\beta \leq \alpha$ then $\alpha + \beta = \alpha$.

(d) $\alpha \cdot \alpha = \alpha$. [Hint: As in (a), it suffices to find some set S and a bijection from $S \times S \to S$ such that $|S| = \alpha$. Let X be a set of cardinality α. Consider a suitable family of bijections and get a maximal bijection $h : S \times S \to S$ where $S \subset X$. Now, if $|S| < \alpha$, then by (c) above, $|X - S| = \alpha > |S|$. So $X - S$ contains a subset T with $|T| = |S|$. Now the sets $S \times T$, $T \times S$ and $T \times T$ have the same cardinality as T and so does their union by (b). Hence, it is possible to get a bijection $k : (S \cup T) \times (S \cup T) \to S \cup T$ such that $k > h$ contradicting maximality of h.]

(e) If a set S has cardinality α, so does the set of all finite subsets of S.

3.9 (a) Prove that there is a bijection between any open interval and $\mathbb{R}$.

(b) Prove that all non-degenerate intervals have the same cardinal number as $\mathbb{R}$ (this cardinal number is denoted by c.) [Hint: Use the Schröder-Bernstein theorem.]

(c) Prove that $c > \aleph_0$ or in other words, $\mathbb{R}$ is an uncountable set. [It suffices to show that the unit interval I is uncountable. If not, let $f : \mathbb{N} \to I$ be a bijection. Trisect I in equal parts. At least one of the three intervals $[0, \frac{1}{3}]$, $[\frac{1}{3}, \frac{2}{3}]$ and $[\frac{2}{3}, 1]$ fails to contain $f(1)$. Call one such interval I_1. Trisect I_1 again and let I_2 be a sub-interval such that $f(2) \notin I_2$. Continue this process and apply (2) of Exercise (3.6). Then $\bigcap_{n=1}^{\infty} I_n$ contains a point which cannot be in the range of f, a contradiction. A proof using the decimal expansions of numbers in $[0, 1]$ is also possible. Let $x \in I$ be a number which differs from $f(n)$ in the nth place of decimal for all $n \in \mathbb{N}$. Then x is not in the range of f.]

(d) Prove that $2^{\aleph_0} = c$. [Hint: Let E be the set of all functions from $\mathbb{N}$ to $\{0, 1\}$. From Exercise (3.7) show that $|E| \geq c$. Also, the function f there gives a bijection from E minus a countable set onto I minus a countable set.]

(e) Prove that for all $n \in \mathbb{N}$, the sets $\mathbb{R}^n$, D^n, B^n and S^{n-1} (if $n > 1$) all have the same cardinality c. [Hint: Apply the results of the last and present exercise and the Schröder-Bernstein theorem.]

3.10 For any $x, y, z \in \mathbb{R}^n$ and any $\alpha, \beta \in \mathbb{R}$, prove that

(a) $\langle x, y\rangle = \langle y, x\rangle$ (This is known as **symmetry** of the inner product.)

(b) $\langle \alpha x + \beta y, z\rangle = \alpha\langle x, z\rangle + \beta\langle y, z\rangle$ (This is known as the **linearity** of the inner product in the first argument. By symmetry, linearity also holds in the second argument.)

(c) $\|x + y\| \leq \|x\| + \|y\|$ (This is known as the **triangle inequality**. The proof requires the next result.)

(d) $|\langle x, y\rangle| \leq \|x\| \, \|y\|$ with equality holding iff one of x and y is a scalar multiple of the other. [Hint: Consider the function $f : \mathbb{R} \to \mathbb{R}$ defined by $f(t) = \|x + ty\|^2$. Note that $f(x)$ is a quadratic polynominal in t. Since f never assumes negative values, the discriminant of this quadratic must be at most 0. This famous inequality is known as the **Cauchy-Schwarz inequality**. Using it one can define the concept of an angle between non-zero vectors and then those of perpendicularity and parallelism.]

3.11 Prove that the functions $+: \mathbb{R}\times\mathbb{R}\to\mathbb{R}$, $\cdot: \mathbb{R}\times\mathbb{R}\to\mathbb{R}$, $-: \mathbb{R}\to\mathbb{R}$ defined by $-(x) = -x$ for $x \in \mathbb{R}$, $r : \mathbb{R} - \{0\} \to \mathbb{R}$ defined by $r(x) = 1/x$, the identity function and all constant functions are continuous. Deduce that the sums, differences, products and quotients (wherever defined) of two continuous functions are continuous. Identifying

$\mathbb{R}^n \times \mathbb{R}^n$ with $\mathbb{R}^{2n}$ prove that the function $+: \mathbb{R}^n \times \mathbb{R}^n \to \mathbb{R}^n$ is continuous. Prove also that the inner product is continuous as a function from $\mathbb{R}^n \times \mathbb{R}^n$ to $\mathbb{R}$. Deduce that the norm function from $\mathbb{R}^n$ to $\mathbb{R}$ is continuous.

3.12 Let $m, n \in \mathbb{N}$ and suppose $m < n$. Define $f: \mathbb{R}^m \to \mathbb{R}^n$ by $f(x_1, \ldots, x_m) = (x_1, \ldots, x_m, 0, 0, \ldots, 0)$, for $(x_1, \ldots, x_m) \in \mathbb{R}^m$. Prove that f is an injective function which preserves vector addition, scalar multiplication, and the inner product. Also, prove that f is continuous. (Because of this result we regard $\mathbb{R}^m$ as a subset of $\mathbb{R}^n$ for $m < n$. An alternate approach is to let $\mathbb{R}^\infty$ be the set of all sequences of real numbers which vanish after some stage and let $\mathbb{R}^n$ be the subset consisting of those which vanish after the nth stage. With this approach $\mathbb{R}^1 \subset \mathbb{R}^2 \subset \mathbb{R}^3 \subset \ldots \subset \mathbb{R}^n \subset \mathbb{R}^{n+1} \subset \ldots$ and $\mathbb{R}^\infty = \bigcup_{n=1}^{\infty} \mathbb{R}^n$. $\mathbb{R}^\infty$, with the obvious inner product structure on it, is called the **infinite dimensional euclidean space.**

Notes and Guide to Literature

The construction of real numbers given in Exercise (3.3) is due to Dedekind. An alternate construction, due to Cantor, is motivated from the requirement that every real number is the limit of a sequence of rational numbers. In Chapter 12, we shall carry out an analogous construction for the completion of a metric. For details of Cantor's construction (as well as for a proof that the Dedekind's and Cantor's constructions ultimately yield the same results) see, for example, Goffmann [1].

The results of Exercise (3.8) are taken from Halmos [1]. Exercise (3.9) shows that many subsets of $\mathbb{R}$ (and of $\mathbb{R}^n$) have the same cardinality as $\mathbb{R}$. Using the fact that $\mathbb{Q}$ is countable it is also possible to find many other countable subsets of $\mathbb{R}$. An interesting question is whether $\mathbb{R}$ contains subsets of other cardinalities. In other words, if $S \subset \mathbb{R}$, is it necessary that either $|S| \leq \aleph_0$ or $|S| = c$? This question has not been answered so far. An affirmative answer to this question is known as the **continuum hypothesis.** Although we shall have little occasion to deal with it, the continuum hypothesis is of considerable importance in many constructions.

The euclidean spaces, especially those of dimensions upto three have been studied exhaustively in classical mathematics. They have also served as sources of motivation for many lines of development of the so-called modern mathematics. For example, one may ignore the structure induced by the inner product and study the euclidean spaces only with reference to the structures of vector addition and scalar multiplication. One can then replace the field of real numbers by an abstract field and define vector spaces over it. Their study comes under a branch of mathematics called linear algebra. Similarly, one can ignore the inner product (and consequently the concept of length) but retain the concept of a straight line. A surprisingly large amount of classical geometry can still be salvaged and this study is known as projective geometry. Thirdly, one can ignore the algebraic structure and be concerned solely with the distance between points. This leads to the theory of metric spaces, and we shall take it up in the next chapter.

Finally, a word about the development of real numbers. People have been handling them long before the notions of equivalence relation, ordering and field were formulated. The development given here only shows how these concepts can be utilized in giving a precise construction of real numbers, starting from natural numbers which we assume as known. This approach thus conforms to a saying, 'God gave natural numbers and man constructed everything else.' We used natural numbers as early as the definition of a finite set. The proofs of properties of finite sets often require the so-called principle of mathematical induction. We have not stated it formally as we expect the reader to be familiar with proofs based on it. Note, by the way, that although called 'induction', from a logical point of view the principle is really a case of deduction. As already noted, logical induction has little place in mathematical arguments.

With the advent of axiomatic set theory, even natural numbers could be constructed from sets and their properties (which were hitherto taken as axioms, popularly known as the Peano axioms) proved strictly from the axioms of set theory. This construction may be found in Kelley [1].

Chapter Three

Motivation for Topology

1. What is Topology?

The word 'Topology' is derived from two Greek words, *topos* meaning 'surface' and *logos* meaning 'discourse' or 'study'. Topology thus literally means the study of surfaces. This etymology is, however, inadequate for a clear understanding of the spirit of the subject as it stands today. Indeed, to derive such an understanding is by no means possible with a single terse definition. Perhaps the best way to appreciate what topology is about, is by comparing it with geometry, the two being akin to each other in so many respects. Let us, therefore, take a close look at what geometry is, or at least to that aspect of it, namely, euclidean geometry, with which we are all familiar.

Here, too, the etymological meaning ('the measurement of earth') has long been discarded through the course of time and practice. Geometry, as it stands today, can be slickly defined as the *study of geometric properties of objects*. Such a definition of course immediately raises the questions as to what one means by 'objects' and by 'geometrical properties'. Let us proceed to answer these.

Any schoolboy knows that in geometry he deals with such things as lines, planes, circles, cubes, cylinders, various types of curves and surfaces, and so on. These are, therefore, among the 'objects' to be studied in geometry. But can this be taken as a precise definition? No, because the phrase 'and so on' is too vague to be included in a rigorous mathematical definition. Note that all the 'objects' listed above are sets, indeed they are subsets of the euclidean two dimensional plane or the three-dimensional space. Can we then say that the 'objects' of our study are just sets? Well, we can, but in doing so we shall lose the very spirit and charm of geometry. As you may recall from elementary set theory, there is a one-to-one and onto correspondence (that is, a bijection) between any two objects of the list above (cf. Exercise 2.3.9 (e)). This means, for example, that there is no difference between a plane circle and, say, a solid cube, a conclusion which a geometer will find as disastrous!

So, we cannot afford to be so loose as to declare our objects as mere sets. Some additional structure which will be rigid enough to distinguish between, say, a square and a triangle is necessary. This 'additional structure' will have to comprise of certain other attributes (such as 'distance') between various points defined on the underlying set and required to obey certain conditions or 'axioms'. A geometric object can then be defined not as a mere set, but rather as a set with this 'additional structure' (or, if one wants to be fussy, as an ordered pair consisting of a set and of the additional structure). This situation is quite common in mathematics. In group theory, for example, a group is defined as a set along with a binary operation satisfying certain axioms. The binary operation and the axioms it obeys is what constitutes the 'additional structure' in this case.

Unfortunately, it will take us too far afield to define the additional structure required to convert a set into a geometric object. Although such a definition will be interesting in its own right, we abandon it here because our interest is largely confined to euclidean geometry and there too, only as a motivation to an understanding of the spirit of topology. Our geometric objects will therefore be subsets of euclidean spaces and the 'additional structures' on them will be those they derive from the 'additional structure' on the ambient euclidean space (whose ingredients are concepts of distance and collinearity). Indeed for the rest of this chapter, it will suffice if we merely have an intuitive understanding of geometric objects as everyday physical bodies with definite shape and size. So, without further ado, we turn to the more interesting question of what one means by 'geometric properties' of an object.

In geometry we deal with things like the area, volume and the curvature etc. of the object under study. In other words, these are among the 'geometric properties' of that object. By contrast, we never worry about things like the colour, smell, the melting point or the ownership of these objects in geometry (although we may worry about them in other branches of knowledge). These attributes of an object are therefore not geometric in nature. Exactly on what basis, then, does one call certain properties as geometric and others as not? A somewhat loose answer to this question is that those properties which ultimately depend only on the distance between various points of an object are geometric in character. For example, the volume of a rectangular box is the product of the lengths of its sides, these lengths being the distances between various vertices of the box. Hence volume is a geometric attribute of the box. Its colour, on the other hand, has nothing to do with its shape or its dimensions and hence is not a geometric property.

However, can this criterion be made precise? It is helpful here to recall the concept of congruence from high-school geometry. Although it is customary to define it only for triangles, the idea behind it applies equally well to any pair of objects. Thus we say that two objects are congruent to each other if each can be placed over the other in such a way that every point of one of them corresponds to a unique point of the other and vice versa. Using the terminology of functions we can state this more elegantly: two objects A and B are said to be **congruent** to each other if there exists a bijection $f : A \to B$ which preserves all distances in the sense that $d(x, y) = d(f(x), f(y))$ for all pairs $\{x, y\}$ of points in A, d being used to denote the distance between points. Such a bijection, when it exists, is called a

congruence (or an **isometry**).

It is trivial to verify that being congruent to each other is an equivalence relation on the class of all geometric objects and consequently they are grouped together into disjoint equivalence classes, which may be called congruence classes. A property (or an attribute) of an object is said to be **invariant under or preserved by** congruence if whenever an object has that property (or the attribute) so does every object congruent to it. For example, the area of a triangle is such an attribute. Similarly the property of being an isosceles triangle is preserved under congruence. On the other hand the colour of an object or its smell (if any) are not preserved under congruence since two congruent objects need not have the same colour (or smell). Moreover it is clear that since congruence preserves distance, it also preserves 'anything that ultimately depends upon distances'.

It is now a simple matter to answer what one means by 'geometric properties' of an object. The answer is that they are precisely those properties which are invariant under congruence. With this definition we indeed find that volume, area and curvature are geometric attributes while colour, smell and melting point are not. We have thus given a precise meaning to the notion of 'geometric properties' through a suitable equivalence relation, congruence in the present case. In geometry we are interested only in the geometric properties of an object.

This situation is typical in mathematics. One often deals with certain 'objects', whose definition changes from one branch of study to another. These objects are classified into disjoint classes by a suitable equivalence relation (whose definition again depends on the type of study). One then worries about only those properties of the objects which are invariant under this classification. Probably the simplest example of this scheme is set theory. Here the objects are sets, just bare sets with no 'additional structure'. The fundamental classifying relation is that of **equipollency,** two sets being declared **equipollent** if there exists a bijection between them. A property which is invariant under this equivalence is called a **set-theoretic property**. The number of elements in a set (or its cardinality as it is called) is such a property. The actual nature of its elements is not a set theoretic property—since a set of monkeys can very well be equipollent to a set of real numbers!

As another example, with which most of our readers will be familiar, let us take group theory. Here the objects under study are groups, which as remarked earlier are sets with some additional structures. We then classify all groups by means of isomorphisms; two groups are considered equivalent if they are isomorphic to each other. **Group-theoretic properties** are then defined as those invariant under isomorphisms. Group theory is the study of such properties.

The approach of topology fits perfectly into this general scheme. The 'objects' one deals with in topology are called 'topological spaces'. These will formally be defined in the next chapter, where they will turn out to be sets with some 'additional structure'. For the moment, we regard them loosely as everyday physical bodies, as we did in the case of geometric objects. In other words we do not distinguish between geometric objects and topological objects for the time being. However the basis of classification is no longer congruence, but what is known as a homeomorphism. A formal definition of this concept will be given much later. However in the next section we shall give it an intuitive content and thereby bring out the essential similarity and difference between geometry and topology.

Once the concept of a homeomorphism is defined, it is routine to check that being homeomorphic to each other is an equivalence relation on the class of all topological spaces (or 'objects'). We then define a **topological property** as a property which is invariant under homeomorphism. And finally this leads to a one-line definition of **topology** as the study of topological properties.

The perceptive reader must have noticed the structural resemblance between the four branches of mathematics mentioned so far, viz., geometry, set theory, group theory and topology. In each case one deals with certain 'objects', which are then classified according to some rule and studies properties which are invariant under this classification. It is natural to ask whether a unified theory can be built so as to have all these as special cases. The answer is 'yes'. Such a theory is known as *category theory*. Although in this book we shall deal with it later, suffice it to say here that category theory, originally founded by Eilenberg and MacLane around 1950, has done a tremendous job of unification of ideas from a large number of apparently diverging branches of modern mathematics.

Exercises

1.1 Two geometric objects A and B are said to be **similar** to each other if there is a bijection $f: A \to B$ and a positive constant m such that $d(f(x), f(y)) = md(x, y)$ for all x, y in A where d again denotes the usual distance in a euclidean space.

Prove that similarity is an equivalence relation and that it is **weaker** or **coarser** than congruence (this means that the equivalence classes under congruence are contained in the equivalence classes under similarity).

1.2 If 'similarity' replaces 'congruence' in the definition of geometry, which of the following properties or attributes of geometric objects will be geometric in nature?

(a) volume
(b) the property of being an isosceles triangle
(c) curvature
(d) length
(e) shape
(f) the ratio of the sides (of a rectangle)
(g) the ratio of the area to the perimeter of a rectangle
(h) the centroid of a lamina or volume.

1.3 Make similar studies replacing congruence by other equivalence relations on the class of all geometric objects (for example, consider the binary relations of being images of each other under a homothety, a linear transformation, a linear fractional transformation etc.).

1.4 For each of the following branches of mathematics, what are the 'objects' dealt with and how are they classified? Give examples of some properties invariant under the respective classifying relations in each case.
(a) ring theory
(b) the theory of partially ordered sets
(c) graph theory (in case you happen to know it).

Notes and Guide to Literature

The 'topology' to be studied in this book is more specifically known as **general topology** or the **topology of point sets** and forms only a part of the whole range of topology. The fundamental ideas in general topology are those of convergence and continuity. Although both have appeared in implicit forms for a long time, it was only with the advent of analysis that precise definitions were given for these concepts. It turned out that a good deal of analysis rested on these two concepts. This part was then isolated and came to be known as topology.

There are numerous treatises on the subject and we shall list only a few. Although written nearly a quarter of a century ago, Kelley's [1] book still continues to be an authentic one. However, despite its systematic presentation and wealth of examples, the beginner may have a hard time with it if he lacks the maturity or the motivation to appreciate it. For such a reader we recommend Thron [1] which is easier to digest than Kelley. The books of Kuratowski [1] and Bourbaki [1] are classic, the latter as usual being the most encyclopaedic of all. Two other excellent books on topology are Hocking and Young [1] and Dugundji [1]. Simmon's book [1] also contains a very readable account of most of the standard topics to be covered in an introduction to general topology. Two recent additions to the list of recommendable books on topology are Munkres [1], Engelking [1] and Willard [1], the first for diagrams and the other two for recent results and references.

Euclidean geometry is of course one of the most classical branches of human knowledge. For an account of modern geometry the reader may consult Seidenberg [1].

2. Geometry and Topology

In the last section we remarked that topology, like geometry, deals with certain 'objects', classifies them according to some equivalence relation and then studies those properties of the objects which are invariant under this classification. However, the analogy between the two goes much further. Indeed, if this were the only feature common to them, it would hardly be necessary to dwell on geometry in such detail, for an analogy with, say, group theory would have illustrated the point equally well. The main reason to emphasize the role of geometry in understanding the spirit of topology is that we may assume, at least for the time being that they both deal with the same objects, viz., subsets of euclidean spaces. The difference lies in the fundamental equivalence relation by which these objects are classified.

As we say, in geometry objects are classified by means of congruence while in topology they are classified by means of what is known as a homeomorphism. Let us get down to the definition of this fundamental concept. Unfortunately, the definition leans heavily on the notion of continuity, which in turn would force us to define 'topological spaces' (which are the objects of study in topology). This will be done systematically later on. For the moment, however, since our objects are confined to subsets of euclidean spaces, it suffices to repeat the definition from Chapter Two.

(2.1) Definition: Let A, B be subsets of euclidean spaces, $f : A \to B$ a function and $x_0 \in A$. We say f is **continuous at** x_0 if for each $\epsilon > 0$, there exists $\delta > 0$ such that $d(f(x), f(x_c)) < \epsilon$ for all $x \in A$ for which $d(x, x_0) < \delta$; where, as usual, the same symbol d is used to denote the distance between points of A as well as B. Further, we say f is **continuous**, if it is continuous at all points of A.

Although we have stated this definition in full for the sake of completeness, we trust that the reader is well familiar with it, from his background in analysis. Thus we take for granted some elementary properties of continuous functions, such as the continuity of the composite of two continuous functions. In case the reader has not seen them before, they are to be taken as exercises.

Simplest examples of continuous functions are constant functions and the identity functions. If the set B is a subset of the n-dimensional euclidean space R^n then for $x \in A$ we can write $f(x) = (f_1(x), f_2(x), \ldots, f_n(x))$ where $f_1, f_2, \ldots, f_n$ are real valued functions on A. It is customary to say that $f_1, f_2, \ldots, f_n$ are the **coordinate functions** associated with f. They are of course the composites of f with the projections onto the various coordinates. The continuity of f is equivalent to that of each f_i, $1 \leq i \leq n$. This reduces the problem of checking the continuity of a function to the case where the function is real-valued.

A continuous function is often called a **map**.

A continuous function need not be invertible, that is, need not be a bijection. Even when it is so, its inverse need not be continuous. A simple example of this occurs if we let A be the semi-open interval $[0, 1)$ that is the set $\{x \in \mathbb{R} : 0 \leq x < 1\}$, B the unit circle, $\{(x, y) \in \mathbb{R}^2 : x^2 + y^2 = 1\}$ and define $f : A \to B$ by $f(\theta) = (\cos 2\pi\theta, \sin 2\pi\theta)$, $\theta \in A$. We ask the reader to prove rigorously in this case that f is a continuous bijection whose inverse is not continuous.

Having defined continuity we are now in a position to give the much awaited definition of a homeomorphism.

(2.2) Definition: Let A, B be subsets of euclidean spaces. A **homeomorphism** from A to B is a bijection $f : A \to B$ such that both f and its inverse are continuous. When such f exists, A and B are said to be **homeomorphic** to each other.

In a slightly different form this definition states that a function $f : A \to B$ is a homeomorphism if and only if there exists a continuous function $g : B \to A$ such that the composites $g \circ f$ and $f \circ g$ are equal, respectively, to the identity maps on A and B respectively. The equivalence is trivial to establish. Identity map on any object is the simplest example of a homeomorphism. The inverse of a homeomorphism and the composite of two homeomorphisms are easily seen to be homeomorphisms again. Consequently we see that 'being homeomorphic to each other' is an equivalence relation on the class of all objects. This is the fundamental relation according to which objects are classified in topology. A property or an attribute of an object, which is invariant under homeomorphisms, is said to be topological in character. We shall give a few examples of such properties shortly. They are sometimes called **topological invariants.**

It is now clear that the comparison of geometry and topology boils down to the comparison of congruence and homeomorphism. How are the two related? It is immediate from the definitions that a congruence is always a homeomorphism; the condition that all distances be preserved trivially implies continuity. Consequently if two objects are equivalent for a geometer, they are certainly so for a topologist. Putting it in another form, the geometric classification of objects is **finer** or **stronger** than the topological classification. If a property is invariant under homeomorphism, it is automatically so under congruences. Thus all topological properties are also geometric in nature.

What about the converse? That is, is it true that every homeomorphism is a congruence? If the answer were yes then there would probably be no need to distinguish between geometry and topology. Actually the answer is far from yes. As a simple example, let A, B be respectively two straight pieces of wire of lengths 1 and 2 say. Mathematically we may represent them by the closed intervals $[0, 1]$ and $[2, 4]$ (or any other straight line segments of appropriate lengths). Now define $f : A \to B$ by $f(x) = 2 + 2x$. This function f is clearly continuous and it has an inverse $g : B \to A$ defined by $g(y) = \frac{1}{2}(y - 2)$, which is also continuous. Thus f is a homeomorphism and A, B, are homeomorphic to each other. It is easy to see however that f is not a congruence; indeed there cannot exist any congruence between A and B. Thus two objects which look distinct to a geometer may look the same to a topologist. Intuitively, the homeomorphism stretches the first piece of wire to the second, while its inverse does the opposite, namely, shrinks the second piece to the first. Loosely speaking, stretching and shrinking do not affect the topological properties of the wire although they undoubtedly change its geometric properties (such as length). One can actually do more weird things than stretching and shrinking. For example, the function $h(x) = \left(\cos \frac{\pi}{2} x, \sin \frac{\pi}{2} x\right)$ establishes a homeomorphism between the unit interval [0, 1] and the first quadrant arc of the unit circle. By obvious modifications of this example we see that bending a wire and indeed twisting it will not change it topologically, although so many of its geometric properties such as curvature and torsion get affected in this process. Thus length, curvature and torsion are geometric properties but not topological and hence not studied in topology.

Let us move to some two-dimensional objects such as a sheet of rubber in the form of a circular disc. This sheet can be stretched to varying degrees in various directions so that it assumes the form of a square. This suggests that a circular disc is homeomorphic to a square. An explicit homeomorphism is not hard to construct. Choose any interior points, O, O' as shown in Figure 1. Each ray through O will intersect the boundary of the disc in a unique point, such as A. Let the parallel ray through O' intersect the boundary of the square in A'. The segments OA and $O'A'$ need not be of equal length. However there is a homeomorphism from OA to $O'A'$ which carries O to O'. Moreover as A varies, this homeomorphism can be made to vary 'continuously' so as to ensure that the resulting bijection from the disc to the square will be a homeomorphism. A modification of this technique will show that the disc is also homeomorphic to any triangle, quadrangle, ellipse (that is the area enclosed by and including the ellipse) and indeed any lamina such as the last one in the second row of Figure 1. Note that any homeomorphism between any two of these objects will always carry points on the boundary of one to those on the boundary of the other. This fact is fairly obvious intuitively, but a rigorous proof is highly non-trivial.

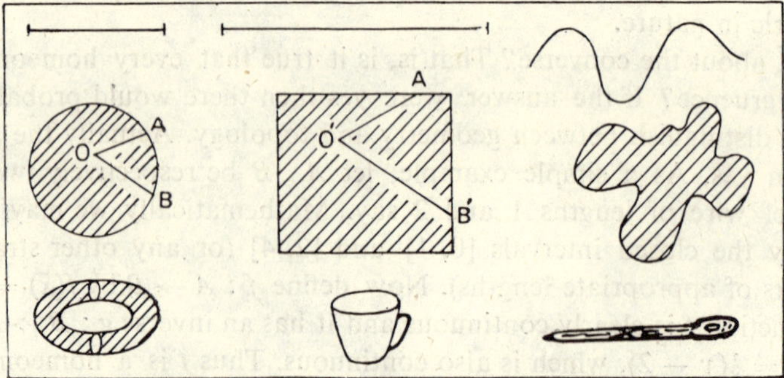

Fig. 1 Homeomorphic objects (objects in each row are mutually homeomorphic).

The surprises from homeomorphisms increase even more as we pass to solid bodies. We invite the reader to convince himself that the three objects in the third row of Figure 1 are mutually homeomorphic. They are respectively a solid torus (the ring with which you play or a doughnut), a tea cup and a needle. Analytical formulation of homeomorphisms will depend upon analytical definitions of the various sets.

The startling revelation from these examples is that many important geometric properties such as length, curvature, area, shape and volume are not topological in character and hence of little interest in topology. It is remarkable that despite this, the dimension of an object remains a topological property. This means, for example, that a solid body can never be homeomorphic to a planar lamina or to a curve. This is yet another fact which looks obvious but which requires a very deep study. Indeed even a rigorous definition of dimension is by no means elementary.

There are many other interesting properties which are topological. We shall describe only a few here. We remark that in the case of some of these, it is by no means easy to establish that they are topological, that is, invariant under homeomorphisms.

Note first of all that inasmuch as every homeomorphism is a bijection to start with, every set-theoretic property is automatically a topological invariant. As a more interesting example, take the property of being a single whole piece, that is not being broken into several parts. The technical name for this property is connectedness and we shall study it formally later on. For the moment we see intuitively that all objects pictured in Figure 1 are connected. Indeed our concept of an object generally requires it to consist of a single piece. However, since we are letting arbitrary subsets of euclidean spaces be our objects, there is no reason why, say, a union of two disjoint line segments cannot be considered as a single object by itself. When we do so, we see that it is not connected and hence cannot be homeomorphic to a connected object; for connectedness is a topological property.

This example shows that although stretching or twisting a piece of wire does not change its topological properties, cutting it at any point definitely does so. The same thing is true of gluing together some portions of the original object. For example, if the two end points of a piece of wire are glued together we get a figure homeomorphic to a ring or a circle. The removal of any point (other than the end points) of the original segment disconnects it, that is to say, the complement of that singleton set is not connected. It is not hard to show that existence of such a point (sometimes called a **cut point**) is a topological property. Clearly the circle possesses no such point. Thus the circle is not homeomorphic to a line segment. When we move to higher dimensions, piercing an object, or cutting a hole into it drastically changes its topological properties. No two objects in Figure 2 are homeomorphic to each other. This may seem obvious but is not easy to establish for all possible pairs. Indeed given two objects it is in general not easy to decide if they are mutually homeomorphic. If they are, the explicit construction of a homeomorphism will settle the issue. But when one wants to prove the non-existence of a homeomorphism between two given objects, there is no canonical method. One has to look for a suitable topological property which is enjoyed by one of the objects but not by the other. Such a property is said to **distinguish** the two objects topologically. In general, if a property distinguishes between one pair of objects, it may not work for some other pair. For example, although the first two objects in Figure 2 can be distinguished by connectedness, the same is not true of, say, the solid sphere and the solid torus both of which are connected.

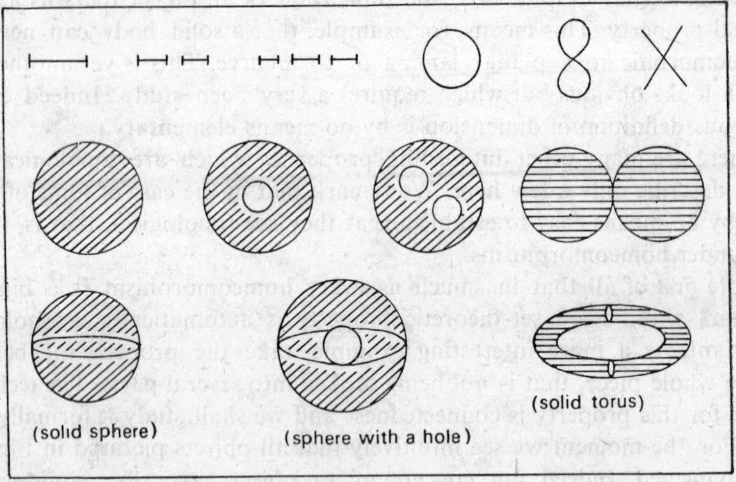

Fig. 2 Non-homeomorphic objects (No two objects shown in this figure are homeomorphic to each other).

The trouble is that the number of topological properties that are needed in order to distinguish between all possible pairs of objects is not finite. In practical terms this means that even if we had an extremely efficient computer which will immediately tell whether a given object has a certain topological property, we cannot write an algorithm or a computer program which will decide whether two given objects are mutually homeomorphic in a finite amount of time. One has to resort to ad hoc methods. Technically this observation is expressed by saying that the problem of classification of objects according to homeomorphism types is **undecidable.** Such questions are treated in mathematical logic and to discuss them further would take us too far afield.

Coming back to the comparison of geometry and topology, we already remarked that the topological classification of objects is much more crude as compared to the geometric classification. A good deal of vital information which a geometer can give about an object, such as its area, volume and curvature, is completely meaningless as far as a topologist is concerned. It is then natural to ask why one should at all study things from a topological viewpoint. There are two good answers to this question besides the intellectual pleasure one derives from studies. One is that although we have restricted our objects to be the same as geometric objects for the time being, the actual class of topological objects is much wider and includes many objects for which the methods of geometry would be inapplicable. Secondly, even when we look at ordinary objects which are covered by euclidean geometry, it is sometimes useful to focus our attention away from geometrical properties as this may help us gain some new insight into the nature of objects. Indeed the idea of a shape of an object comes into prominence only when congruence is replaced by the weaker relation of similarity. Topology goes even further. Given a solid torus, a tea cup and a needle, a geometer will dismiss them as having nothing in common. A topologist will note, however, that they have identical topological properties. He will realise that if we can thread together a bunch of needles, the same can be done with a bunch of tori or of tea cups. While this is also obvious to the layman, a geometer is likely to overlook it if his mind is too preoccupied with areas and volumes!

Exercises

In the following exercises, where 'proofs' are asked it is understood that a completely rigorous argument is not always possible at this stage. The main idea is to develop topological intuition. However, in doing so, try to be as precise as you can.

2.1 Prove that the objects in each row of Figure 1 are mutually homeomorphic.

2.2 Prove that the property of 'being a single piece' (that is, connectedness) is a topological invariant. [Hint: Show that the image of a single piece under a continuous function must also be a single piece].

2.3 Prove that the annulus $\{(x, y) \in \mathbb{R}^2 : 1 \leq x^2 + y^2 \leq 4\}$ is homeomorphic to the curved surface of a right circular cylinder.

2.4 Prove that the curved surface of a right circular cone is homeomorphic to a unit disc.

2.5 Prove that the real line is homeomorphic to any open interval, for example $(-1, 1)$. Generalise to higher dimensional euclidean spaces.

2.6 Can the semi-open interval $[0, 1)$ be homeomorphic to the open interval $(0, 1)$; the closed interval $[0, 1]$; the semi-open interval $(0, 1]$?

2.7 Prove that no two objects in Figure 2 are homeomorphic to each other.

2.8 Consider a knot for example the trefoil knot K in Fig. 3. Prove that K is homeomorphic to the unit circle, $C\{(x, y) \in \mathbb{R}^2 : x^2 + y^2 = 1\}$.

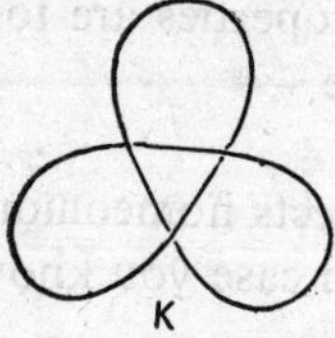

Fig. 3 A trefoil knot

Can we obtain K from C by mere stretching, without cutting and gluing?

2.9 Topology is sometimes defined as the study of those properties of an object which remain unchanged when that object is subjected to stretching, shrinking and twisting without tearing, piercing or gluing. Comment on the definition.

2.10 Prove that the figure eight curve (the fifth object in Figure 2) can be obtained from the unit interval [0, 1] by gluing together certain points in it. Show that it can also be obtained from a pair of disjoint circles or even from one circle. What happens if the points of the form $1/n$, n a positive integer, are all glued together with 0 in [0, 1]?

2.11 As further wonders which gluing can produce, take a square and glue points on one of its sides to the corresponding points on the other in the same sense (that is A and A' will be glued together, B will be glued with B', C with C' etc.) as shown in Figure 4(a). What object do you get?

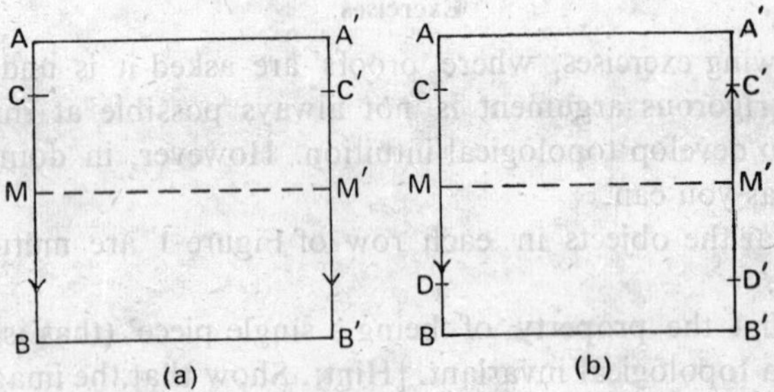

Fig. 4 Objects resulting from gluing points in a square.

If however we glue points on AB with those on $A'B'$ in opposite sense (that is A with B', B with A', C with D', D with C' and so on) the resulting object is popularly known as a Möbius band or strip. It has many weird properties, for example it has only one side. Now glue the side AB with side $A'B'$ as in (a), except this time give a complete twist to the square before matching these two sides. (You will find it is much easier to work with a long rectangular strip than with a square. Topologically this makes no difference. Why?) Now let M, M' be the mid-points of the two sides respectively. The dotted line MM' will turn into a circle in all three cases of gluing. Cut the resulting object along this circle and find what happens in each case.

2.12 Which of the following properties are topological?
(a) having a finite volume
(b) boundedness
(c) being knotted (for objects homeomorphic to a circle)
(d) simple connectivity (in case you know what it means).

Notes and Guide to Literature

The main purpose of this section is to stress the similarities and differences between geometry and topology. The reader is advised to go through it and work out the exercises until he feels reasonably sure of intuitively telling whether two objects are homeomorphic to each other.

Topology from an intuitive and naive point of view has been treated in the chapter on topology in the Time book on mathematics.

A precise topological definition of dimension (as well as a rigorous proof that the euclidean n-dimensional space indeed has dimension n) can be found in the very readable 'Dimension Theory' by Hurewicz and Wallman [1]. A modern treatise on this subject is by Nagata [1].

Knots are not merely some objects of play but form an interesting and fairly well-developed branch of topology, called 'Knot Theory'. The most comprehensive reference is Crowell and Fox [1]. Further information on the concept of decidability may be found in Rogers Jr. [1] or Davis [1].

3. From Geometry to Metric Spaces

Historically topology developed as an outgrowth of analysis rather than of geometry. However, as we saw in the last sections, topology resembles geometry very closely in its spirit and indeed can be considered as an abstract form of geometry. It will therefore be instructive to see how the definition of a topological space (which is the object of study in topology) can be approached in a very natural way starting from geometry. This definition is of fundamental importance and it is vital to create a strong motivation for it lest it sound too arbitrary.

The journey from euclidean geometry to topological spaces can be conveniently broken into two parts. The midway point is metric spaces. We trust that the reader is already familiar with the definition as well as with some of the elementary properties of a metric space. In this section we too shall define metric spaces formally. But the emphasis will not be so much on the definition per se as on discussing why that particular definition is a reasonable one. In doing so some general remarks will be made regarding the isolation and development of one branch of mathematics from another. These remarks will be applicable in the next chapter where we shall see how metric spaces lead naturally to the definition of a topological space.

If we take a close look at euclidean spaces we see that their entire structure is based upon two basic concepts, that of the distance between two points and that of a straight line. (It may be argued that a straight line is not an independent concept since collinearity can be easily characterised in terms of distances; however, in view of its importance we continue to regard it as a basic concept). Every definition in geometry can be made in terms of these concepts. The distances and straight lines are assumed to obey certain postulates or axioms and every theorem in euclidean geometry is proved starting from these axioms by a strictly deductive argument. Of course, not all axioms are needed in every theorem. Indeed, a suprisingly large number of theorems can be proved using only the axioms pertaining to lines and points and having no reference to the distance between points (for example the axiom that any two distinct points lie on a unique line). It is then natural to build a more general theory than classical euclidean geometry by assuming only those of the axioms of the latter which do not involve the distance between points. Such a branch of mathematics is known as projective geomtery.

This is a fairly common phenomenon in mathematics. When a branch of mathematics gets fairly rich in terms of the depth and applicability of its theorems, people begin to think whether some of the basic axioms of that branch can be either dropped totally or at least be replaced by some weaker ones. In doing so, one of course has to pay some price in that the theorems that can be proved with the new axioms will not be as deep and as numerous as those with the original axioms. For example, if we merely retain the axiom that two distinct points lie on a unique line and drop all other axioms from geometry, then there is hardly any theorem worth the name that can be proved solely on the basis of this lone axiom. On the other hand, assuming as few axioms as possible has an advantage in terms of generality. Whenever a certain theory is to be applied somewhere, it must first of all be shown that all its axioms are satisfied in that particular situation. The less the number of axioms and the less restrictive they are, the better are the chances that they are all satisfied. This makes the theory more and more applicable.

Thus, there is an essential contrast between depth and range of applicability. A very deep theory is not likely to be applicable outside a limited compass because its axioms tend to be too restrictive. On the other hand if one strives too much for generality, it can only be done by relaxing or dropping some of the axioms, resulting in a sacrifice of profundity. Depending upon whether one strives for generality or for depth in preference to the other, the particular branch is described as 'soft' or 'hard' mathematics. Number theory and the theory of analytic functions are examples of hard mathematics while category theory and to some extent topology itself can be called soft mathematics. Of course 'hard' and 'soft' are relative terms and it is clear that both the extremes would be equally unappealing. In practice one generally tries to achieve some balance between these two essentially conflicting virtues. That is, one tries to see that the axioms used in the development of a new theory are substantial enough to prove non-trivial theorems but, at the same time, not so restrictive as to disallow interesting applications. Of course, individual tastes are likely to differ as to what is the most optimum balance. Ultimately it is only through the test of time that a theory gets recognition as a worthwhile and separate branch of mathematics.

The theory of metric spaces has well stood the test of time. We shall present the definition of a metric space in such a way as to illustrate the general remarks we made above regarding the formation of a new branch of mathematics from an old one. Let us recall once again that the fundamental concepts of euclidean geometry are that of a line (i.e. a straight line) and that of the distance. Let us ignore the line and concentrate on that aspect of euclidean geometry which can be developed solely with the concept of the distance. It is this aspect which we want to isolate and generalise.

The first step in the generalization is to replace the euclidean space by an abstract set X and the euclidean distance function by an abstract distance function. This will be a real-valued function d on the product $X \times X$; if $x, y \in X$ then $d(x, y)$ would be the (abstract) distance between x and y. With these basic gadgets X and d, let us now ask the key question, "What properties of d should we assume as axioms in order to have a 'reasonable' generalisation of euclidean spaces?" The definition of a metric space evolves naturally as one attempts to answer this key question.

First of all, whatever properties of the abstract distance function d we choose to assume as axioms must all be true in the case of the usual distance function in euclidean spaces. For otherwise, euclidean spaces will not be special cases of metric spaces. So, to begin with, we list down a few properties of the euclidean distance function (which we also denote by d).

(M1) For any points x and y, $d(x, y) \geq 0$ and equality holds iff x and y are identical.

(M2) For any points x and y, $d(x, y) = d(y, x)$.

(M3) For any points x, y and z, $d(x, y) + d(y, z) \geq d(x, z)$.

(M4) For any two points x and y there exists a unique point m such that $d(x, m) = d(y, m) = \frac{1}{2}d(x, y)$.

We could continue this list further but let us pause to consider the properties already listed. For obvious reasons they are respectively called **positivity, symmetry**, the **triangle inequality** and the **mid-point property**. In defining metric spaces we have to assume some of these properties or some others if we like as axioms. Exactly which ones should we pick, keeping in mind that we are trying for a reasonable balance between depth and generality?

Of course a question like this has no unanimous answer. Nevertheless we can learn a little from experience. Let us suppose we choose to make only (M1) and (M2) as the axioms of a metric space. This leaves us ample freedom in defining the distance function, since the restrictions (M1) and (M2) are easy to satisfy. Consequently we can give a large number of examples of metric spaces. For example we could let X be the set of all participants in a league tournament and for two such participants x and y let $d(x, y)$ be the number of matches played between them (assuming that at least one match is played between every pair of distinct participants). The axioms (M1) and (M2) are trivially verified, and thus the system (X, d) will be a perfectly valid example of a metric space. Consequently any theorems in metric spaces will hold for the league tournaments, a remarkable triumph of applications of mathematics!

The only trouble is that there is not much one can prove using exclusively (M1) and (M2), except some utter trivialities. When such 'theorems' are translated in terms of the league tournaments, they would mean nothing more (and perhaps a lot less) than can be said by sheer common sense. In other words, with our definition of a metric space, although the range of applications will be impressive, they will necessarily be very superficial.

In order to achieve some depth, let us add (M3) as another axiom in the definition of a metric space. Note that the set of participants in the league tournaments is not a metric space any more. This means that the theory of metric spaces will no longer be applicable here. There is no help for this. Loss of generality is the price one pays for greater depth. Fortunately, however, it turns out that a good number of fairly non-trivial theorems can be proved using (M1), (M2) and (M3). This, in fact, is the most commonly accepted definition of a metric space today and we shall adopt it. More formally, a **metric space** is a pair (X, d) where X is a set, $d : X \times X \to \mathbb{R}$ is a function and conditions (M1), (M2) and (M3) are satisfied for all points x, y, z in X.

What if we include (M4) in the definition of a metric space? Surely we can prove a few more theorems than we could otherwise. But there are many metric spaces for which (M4) fails. So the theory based upon the axiom system (M1) to (M4) would be inapplicable to them. Experience shows that the resulting gain in depth because of inclusion of (M4) is not worth the loss of generality. So the definition of a metric space today stands where it is. The function d is called a **metric** on the set X. By abuse of language we sometimes say X is a metric space when the metric d is understood.

Even when dealing with two distinct metric spaces at a time it is customary to use the same symbol (generally d) for their distance functions, unless this is likely to cause confusion. A confusion would indeed result if the two metric spaces have the same underlying set.

The two main concepts of topology, namely those of convergence of sequences and continuity can be easily defined for an abstract metric space in exactly the same way as for euclidean spaces with the only change that the abstract distance function replaces the euclidean distances. The definition of continuity of a function was already given in the last section. For the sake of completeness we also include a definition of convergence of a sequence.

(3.1) Definition: A sequence $\{x_n\}$ in a metric space $(X; d)$ is said to **converge** to a point y in X if for every $\epsilon > 0$ there exists a positive integer N such that for all integers $n \geq N$, $d(x_n, y) < \epsilon$.

It is a routine matter to generalise theorems regarding continuity, convergence, limit points etc. from euclidean spaces to abstract metric spaces. We trust that either the reader has already done so or he will work out the exercises at the end of this section. He will also find some examples of metric spaces in the exercises.

Before concluding this section it is perhaps important to note that not all people agree that the definition of a metric space as given here is a 'reasonable' generalization of euclidean spaces. They contend that the condition that the distance between distinct points be strictly positive is rather too restrictive as it does not hold in many 'naturally occurring' situations and thereby prevents them from being metric spaces. There is good merit to this argument because it turns out that most of the nice properties of metric spaces continue to hold if (M1) is replaced by the weaker condition. (M′1) for any points x, y, $d(x, y) \geq 0$ and $d(x, x) = 0$. Thus the price one pays in replacing (M1) by (M′1) is not high and the generality gained is worth-while. However, the term 'metric space' is now too well-established to be used for anything else. So this new object is called a pseudo-metric space. Formally, a **pseudo-metric** space is a pair (X, d) where X is a set, $d : X \times X \to \mathbb{R}$ is a function satisfying (M'1), (M2) and (M3).

As the name indicates, a pseudo-metric space is almost like a metric space. The difference between the two is more technical than conceptual. It is further obviated by the fact that there is a canonical way of obtaining a metric space from a pseudo-metric space as shown in the exercises. In this book we shall deal with metric spaces. It is a general exercise for the reader to check which results continue to hold for pseudo-metric spaces as well.

Exercises

3.1 Prove that in presence of (M2) and (M3), (M1) is equivalent to the weaker condition that $d(x, y) = 0$ iff $x = y$. In other words, in the definition of a metric space it is unnecessary to *assume* that $d(x, y) \geq 0$ for all x, y. (This means there is some redundancy in the definition of a metric space as given. In a logically tight definition such redundancies are avoided. For this, one shows that each axiom is independent of the others by giving an example where all other axioms are satisfied except the one in question. For example, the league tournament example shows that the triangle inequality is independent of positivity and symmetry together).

3.2 Let X be any set and λ a positive real number. Define $d : X \times X \to \mathbb{R}$ by $d(x, y) = 0$ or λ according as $x = y$ or $x \neq y$ for $x, y \in X$. Prove that (X, d) is a metric space. (In such a case d is called a **discrete metric** on X. It is customary to let $\lambda = 1$. In such a case d is called the discrete metric on X.)

3.3 Let S be any set and X be the set of all bounded real-valued functions on S. For $f, g \in X$ define $d(f, g) = \sup. \{|f(s) - g(s)| : s \in S\}$. Prove that d is a metric on X.)

3.4 Given two metric spaces (X_1, d_1) and (X_2, d_2), the cartesian product $X_1 \times X_2$ can be made into a metric space in many ways. We can define $d : (X_1 \times X_2) \times (X_1 \times X_2) \to \mathbb{R}$ by any one of the following formulae:

(i) $d((x_1, x_2), (y_1, y_2)) = \{[d_1(x_1, y_1)]^2 + [d_2(x_2, y_2)]^2\}^{1/2}$

(ii) $d((x_1, x_2), (y_1, y_2)) = d_1(x_1, y_1) + d_2(x_2, y_2)$

(iii) $d((x_1, x_2), (y_1, y_2)) = \max. \{d_1(x_1, y_1), d_2(x_2, y_2)\}$.

Prove that each formula defines a metric on $X_1 \times X_2$. (The first metric is called the **Pythagorean metric**).

**3.5 Consider metric spaces as in (3.4) and let $p \geq 1$ be any real number. Define $d : (X_1 \times X_2) \times (X_1 \times X_2) \to \mathbb{R}$ by $d((x_1, x_2), (y_1, y_2)) = \{[d_1(x_1, y_1)]^p + [d_2(x_2, y_2)]^p\}^{1/p}$. Prove that d is a metric on $X_1 \times X_2$. (Note that (i) and (ii) of the last problem are special cases of this problem.)

3.6 Let X be the set of all real-valued integrable functions on the unit interval $[0, 1]$; i.e. functions $f : [0, 1] \to \mathbb{R}$ which are measurable w.r.t.

the usual Lebesgue measure and for which $\int_0^1 |f(t)|\, dt$ is finite. For f, $g \in X$ define $d(f, g) = \int_0^1 |f(t) - g(t)|\, dt$. Prove that d is a pseudo-metric on X. Why is it not a metric on X?

3.7 Generalise the result of (3.6) for any real number $p \geq 1$, the way (3.5) is a generalisation of (3.4) (ii). (The resulting pseudo-metric spaces are called the **L^p-spaces.**)

3.8 For any set X let $d : X \times X \to \mathbb{R}$ be identically zero. Prove that d is a pseudo-metric on X. (This is clearly the extreme opposite to that of a discrete metric; it is therefore called the **indiscrete pseudo-metric** on X.)

3.9 Let (X, d) be a pseudo-metric space. Define a relation on X by saying $x \sim y$ iff $d(x, y) = 0$ for $x, y \in X$.

(a) Prove that $\sim$ is an equivalence relation.

(b) Let Y be the set of all equivalence classes of X under $\sim$. For A, $B \in Y$, define $e(A, B) = d(x, y)$ where $x \in A$ and $y \in B$. Prove that e is a well-defined metric on Y.

(c) Let $p : X \to Y$ be the natural projection, i.e. for $x \in X$, $p(x)$ is the equivalence class (under $\sim$) containing x. Prove that p is an isometry (i.e. preserves the distances).

3.10 Let (X, d) be a metric space. Then $X \times X$ can be made into a metric space by (3.4). Prove that the function $d : X \times X \to \mathbb{R}$ is continuous w.r.t. the usual metric on $\mathbb{R}$ and any one of the metrics on $X \times X$.

3.11 Let (X, d) be a pseudo-metric space and A a non-empty subset of X. For $x \in X$ define $d(x, A) = \text{inf.}\, \{d(x, y) : y \in A\}$. In other words $d(x, A)$ is the 'shortest distance of x from A'. Prove that $d(x, A)$ is a continuous function from X to $\mathbb{R}$.

3.12 If A, B are nonempty subsets of a metric space X we define $d(A, B) = \text{inf.}\, \{d(x, y) : x \in A, y \in B\}$. Prove that if $A \cap B \neq \phi$ then $d(A, B) = 0$. Prove or disprove the converse.

3.13 Prove that in a metric space no sequence can converge to two distinct points. What if we have a pseudo metric space?

3.14 Let $f : X \to Y$ be a function where X, Y are metric spaces and let $x_0 \in X$. Prove that f is continuous at x_0 iff for every sequence $\{x_n\}$ in X converging to x_0 the sequence $\{f(x_n)\}$ converges to $f(x_0)$ in Y.

Notes and Guide to Literature

The definition and properties of metric spaces are all very standard and can be found in almost any book on modern analysis. However, for a deeper study the reader may consult Sierpinski [1].

Although topological spaces will be treated as a generalisation of metric spaces, certain concepts such as completeness of a metric or uniform continuity admit no generalisation to arbitrary topological spaces. A satisfactory generalisation of metric spaces which retains these features is provided by the so-called uniform spaces. They will be discussed much later.

The idea of a distance breeds the concept of closeness or of proximity.

A systematic theory of proximity spaces (another generalisation of metric spaces) has been developed recently. A standard reference is the book by Naimpally and Warrack [1].

The triangle inequality for the metric in Exercise (3.5) is a special case of the well-known Minkowski inequality. For a proof see for example Royden [1].

Chapter Four

Topological Spaces

In this chapter we give the much-delayed (albeit intentionally) definition of a topological space. We develop it from properties of a metric space. In the second section we give a few examples of topological spaces. In the third and the fourth sections we give two general constructions for topological spaces.

1. Definition of a Topological Space

In the last chapter we began the journey from euclidean spaces to topological spaces and completed it halfway. Let us now do the other half, namely, the transition from metric spaces to topological spaces. This will provide another illustration of the general remarks made in the last chapter regarding the creation of a new branch of mathematics from an old one.

A metric space is a set endowed with an additional structure, namely, the metric or the distance function. It is clear, therefore, that any concept in the theory of metric spaces, unless it is a purely set-theoretic concept, will have to be defined in terms of the metric. For example, the fundamental concepts of continuity and convergence, as defined in the last chapter, rely heavily upon the metric concerned. There is, however, a device by means of which the direct mention of the metric can be avoided frequently. It is known as an 'open' set. Let us define this concept which is of paramount interest. Throughout, $(X; d)$ will be a metric space.

(1.1) Definition: Let $x_0 \in X$ and r be a positive real number. Then the **open ball** with **centre** x_0 and **radius** r is defined to be the set $\{x \in X : d(x, x_0) < r\}$. It is denoted either by $B_r(x_0)$ or by $B(x_0; r)$. It is also called the **open** r**-ball around** x_0. When we want to stress the metric d, we denote it by $B_d(x_0; r)$.

It is obvious that the open ball $B_d(x_0; r)$ depends not only on x_0 and r but on the metric d as well. Let d be the discrete metric on a set X. Then for any $x_0 \in X$, $B(x_0; r)$ consists of X or $\{x_0\}$ depending upon whether $r > 1$ or $r \leq 1$. For the usual metric on the real line, an open r-ball is precisely an open interval of length $2r$ while for the usual euclidean metric on the plane $\mathbb{R}^2$, the open r-balls are open discs of radii r. If, however, we define the metrics d_1 and d_2 on $\mathbb{R}^2$ by

$$d_1((x_1, x_2), (y_1, y_2)) = |x_1 - y_1| + |x_2 - y_2|$$

and $\quad d_2((x_1, x_2), (y_1, y_2)) = \max \{|x_1 - y_1|, |x_2 - y_2|\}$

for $((x_1, x_2), (y_1, y_2)) \in \mathbb{R}^2 \times \mathbb{R}^2$ then the open balls are no longer discs. They are squares with sides parallel to the lines $y = \pm x$ in the case of d_1 and squares with sides parallel to the co-ordinate axes in the case of d_2. The verification is an easy exercise in co-ordinate geometry. Note further that in all these cases, an open r-ball around a point in $\mathbb{R}^2$ is obtained from the open r-ball around the origin by a parallel translation. This is so because the metrics d_1, d_2 (as well as the usual metric on all euclidean spaces) are translation invariant; i.e. for any $x, y, z \in \mathbb{R}^2$ we have $d_1(x+z, y+z) = d_1(x, y)$ and so on. But this need not be the case with all metrics on euclidean spaces. For example, we invite the reader to check that $d_3 : \mathbb{R} \times \mathbb{R} \to \mathbb{R}$ given by $d_3(x, y) = |x - y| + |x^2 - y^2|$ defines a metric on $\mathbb{R}$ and that it is not translation invariant. Note also that the open balls are still open intervals but they are generally not symmetric about the centre; moreover, for a fixed r, their length decreases as we go away from the origin.

A subset A of X is said to be **bounded** if the function $d/A \times A$ is bounded. If A is a non-empty bounded set, its **diameter**, denoted by $\delta(A)$ is the number $\sup \{d(x, y) : x \in A, y \in A\}$. Note that every open ball is bounded but its diameter may be less than twice its radius.

Now we come to the basic concept of an open set in a metric space.

(1.2) Definition: A subset $A \subset X$ is said to be **open** if for every $x_0 \in A$, there exists some open ball around x_0 which is contained in A, that is, there exists $r > 0$ such that $B(x_0; r) \subset A$.

We warn the reader (and we hope a similar warning will not be necessary in future) that the number r in this definition could change as the point x_0 varies over X. The definition does not say that there exists $r > 0$ such that for every $x_0 \in A$ $B(x_0; r) \subset A$. Such a condition would be far stronger.

Before doing anything with open balls and open sets it would be nice to know that open balls are indeed open sets. This follows trivially from the definitions and the triangle inequality.

The fundamental concepts of topology, namely, those of convergence and continuity can be easily stated in terms of open sets as the following two propositions show. The proofs are left as exercises.

(1.3) Proposition: Let $\{x_n\}$ be a sequence in a metric space $(X; d)$. Then $\{x_n\}$ converges to y in X iff for every open set U containing y, there exists a positive integer N such that for every integer $n \geq N$, $x_n \in U$.

(1.4) Proposition: Let $f : X \to Y$ be a function where X, Y are metric spaces and let $x_0 \in X$. Then f is continuous at x_0 iff for every open set V in Y containing $f(x_0)$, there exists an open set U in X containing x_0 such that $f(U) \subset V$.

Of course, there do exist a few concepts which cannot be expressed in terms of open sets, for example, completeness and uniform continuity of a function. But still quite a few concepts can be expressed using open sets only (that is, without direct reference to the metric). What is more, most of the theorems involving such concepts can be proved using a few properties of open sets and without directly involving the metric. We summarise a few such properties in the next theorem. Of course to prove that open sets have these properties does require the use of the metric. But once they are established, they can be used in so many theorems without recourse to the metric directly.

(1.5) Theorem: Let $(X; d)$ be a metric space. Then, (i) the empty set ϕ and the entire set X are open, (ii) the union of any family of open sets is open, (iii) the intersection of any finite number of open sets is open, and (iv) given distinct points $x, y \in X$ there exist open sets U, V such that $x \in U$, $y \in V$ and $U \cap V = \phi$.

Proof: (i) and (ii) are trivial consequences of the definition of open sets. For (iii) first consider the case of the intersection of two open sets say A_1 and A_2. Let $x \in A_1 \cap A_2$. Then $x \in A_1$ and $x \in A_2$. Since A_1 is open, there exists $r_1 > 0$ such that $B(x; r_1) \subset A_1$. Similarly since A_2 is open there exists $r_2 > 0$ such that $B(x; r_2) \subset A_2$. Now let $r = \min\{r_1, r_2\}$. Then clearly $B(x; r) \subset B(x; r_1) \cap B(x; r_2) \subset A_1 \cap A_2$. Thus $A_1 \cap A_2$ is open. One can either generalise this argument or use induction to settle the general case. The exceptional case of the intersection of an empty family of open sets is already covered under (i).

For (iv) let $x, y \in X$ and $x \neq y$. Then $d(x, y) > 0$. Choose r so that $0 < r < \frac{d(x, y)}{2}$ and let $U = B(x, r)$, $V = B(y, r)$. Then clearly U, V are open sets containing x, y respectively. Also they are mutually disjoint, for if $z \in U \cap V$ then $d(x, z) < r$ and $d(z, y) < r$ whence $d(x, y) < 2r$ by triangle inequality, a contradiction. ∎

In view of its importance for our discussion, let us reformulate this theorem slightly.

(1.6) Theorem: Let $\mathcal{T}$ be the collection of all open sets in a metric space X. Then $\mathcal{T}$ has the following properties:

(i) $\phi \in \mathcal{T}$ and $X \in \mathcal{T}$

(ii) $\mathcal{T}$ is closed under arbitrary unions.

(iii) $\mathcal{T}$ is closed under finite intersections.

(iv) Given distinct points x, y in X, there exist $U, V \in \mathcal{T}$ such that $x \in U$, $y \in V$ and $U \cap V = \phi$. ∎

It turns out that many (although not all) of the theorems about metric spaces can be proved using these properties of open sets. As an example, we invite the reader to prove the uniqueness of limits of convergent sequences using Proposition (1.3) above and property (iv) in Theorem (1.5).

Now we come to the key step. Since many interesting concepts in the theory of metric spaces can be defined in terms of open sets (without involving the metric directly) and since many non-trivial theorems can be deduced from the four properties of open sets given in Theorem (1.5), it is natural to look for an abstraction of metric spaces exactly the same way as metric spaces themselves are an abstraction of euclidean spaces. This is precisely the genesis of topological spaces.

So, in order to define a topological space we take a set X, a certain family $\mathcal{T}$ of its subsets and require that $\mathcal{T}$ satisfy some (or all) of the properties listed in Theorem (1.6). Once again we are faced with the problem of achieving an optimum balance between depth and generality, while selecting the properties we want to assume as true. There is little disagreement about the first three properties as they are constantly needed, and consequently any definition of a topological space ought to include them. However, there is some difference of opinion regarding the inclusion of property (iv) in the definition of a topological space. This property is certainly quite important in developing the theory of metric spaces. As a result some people do want to include it in the definition in order that all the nice theorems derived from it carry over to all topological spaces. However, it turns out that (iv) is a bit too restrictive inasmuch as there are many 'naturally occurring' instances wherein (i), (ii) and (iii) hold but (iv) does not. The consequent loss of generality outweighs the gain in terms of depths of theorems that would result by its inclusion. So the official definition of a topological space is as follows.

(1.7) Definition: A **topological space** is a pair $(X; \mathcal{T})$ where X is a set and $\mathcal{T}$ is a family of subsets of X satisfying:

(i) $\phi \in \mathcal{T}$ and $X \in \mathcal{T}$,

(ii) $\mathcal{T}$ is closed under arbitrary unions,

(iii) $\mathcal{T}$ is closed under finite intersections. [Strictly speaking (i) is redundant as it follows by applying (ii) and (iii) to the empty subfamily.]

The family $\mathcal{T}$ is said to be a **topology** on the set X. Members of $\mathcal{T}$ are said to be **open** in X or open subsets of X.

Propositions (1.3) and (1.4) indicate how the concepts of convergence and continuity can be defined for arbitrary topological spaces. We shall not formally state these definitions here.

It is a little unfortunate that the word topology which was used so far to describe a certain branch of mathematics is also used for something else, namely, the family of open sets in a topological space. Still no confusion is likely as the context will make it clear which meaning is intended whenever the word is used.

Clearly, the foremost examples of topological spaces come from metric spaces. Theorem (1.6) shows how each metric space gives rise to a topology on the underlying set. A topological space is said to be **metrisable** if its topology can be obtained from a suitable metric on the underlying set. It may happen that two distinct metrics on a set yield the same topology. A trivial case of this occurs when the two metrics are scalar multiples of each other by a constant factor. Less trivial examples will be found in the exercises. Of course not all topological spaces are metrisable. Indeed as we saw, a topological space in which (iv) fails can never be metrisable. However, (iv) alone is far from a sufficient condition for metrisability. The problem of characterising metrisability of a space $(X, \mathcal{T})$ in terms of intrinsic properties of $\mathcal{T}$ is highly non-trivial and has been solved relatively recently. The full solution is beyond the scope of this book. However, an important sufficient condition for metrisability will be proved in a later chapter.

Before concluding this section we mention three conventions which will be adopted in this book. First, by a standard abuse of language, we shall call the underlying set as a topological space when the topology on it is understood. Secondly, since topological spaces will be our primary objects of study from now on, we shall call them merely as 'spaces'. Finally, although a metric space is not quite the same as the topological space that arises out of it, we shall often ignore the distinction. Thus, while discussing a property of topological spaces, when we say that the property holds in all metric spaces we mean that it holds in the corresponding topological spaces.

A word of explanation is probably in order regarding the term 'open sets'. In a metric space, we defined open sets in terms of the distance function and then verified that their collection indeed forms a topology on the underlying set. In an abstract topological space on the other hand, 'open set' is just another name given to the members of the topology. This double use of the term is sometimes confusing to a beginner. His typical question is 'But, what do you mean by open sets in an abstract topological space?' The answer is that in an abstract space, no more meaning is attached to the term 'open set' than 'a member of the topology'. In any concrete example of a topological space, open sets will have to be specified in terms relevant to that particular example. Perhaps we should not call them as 'open sets' until we verify that their collection is indeed a topology. The confusion should disappear as the reader comes across more and more examples of topological spaces.

Exercises

1.1 Verify the remarks made regarding the nature of open balls in various metric spaces following Definition (1.1).

1.2 Prove that the open balls in a metric space are open sets.

1.3 Prove proposition (1.3).

1.4 Prove proposition (1.4).

1.5 Determine the topology induced by a discrete metric on a set.

*1.6 Prove that if X_1, X_2 are metric spaces then all the three metrics on $X_1 \times X_2$ as defined in Exercise (3.4) of Chapter Three yield the same topology on $X_1 \times X_2$.

1.7 Let X be the set of positive integers. Let d_1 be the usual metric on X (induced by the usual metric on the set of real numbers) and let d_2 be the discrete metric on X. Define $d_3 : X \times X \rightarrow \mathbb{R}$ by $d_3(m, n) = \left|\frac{1}{m} - \frac{1}{n}\right|$ for $m, n \in X$. Prove that d_3 is also a metric on X and that d_1, d_2, d_3 all induce the same topology on X. [Note that d_2 and d_3 are bounded but d_1 is not and also that d_1, d_2 are complete while d_3 is not. (In case you are not familiar with complete metrics, you will find the definition in Chapter Twelve.) This example shows that boundedness or completeness of a metric cannot be characterised in terms of open sets.]

1.8 Prove that a pseudo-metric space also gives rise to a topological space exactly the way a metric space does. Does such a topological space satisfy (iv) in Theorem (1.6)?

1.9 Prove that a subset A of a metric space X is open iff no sequence in $X-A$ converges to a point of A. (Proposition (1.3) shows that convergence can be defined in terms of open sets. This exercise shows that for metric spaces the procedure can be reversed. Unfortunately this is no longer true for arbitrary topological spaces, and so it is impossible to characterise open sets exclusively in terms of convergence of sequences. This difficulty can be overcome if one uses an appropriate generalisation of sequences, called nets. They will be studied systematically in a later chapter.)

Notes

Although the development of the definition of a topological space through properties of open sets in a metric space is fairly standard, it is by no means the only approach to topological spaces. One can as well define such things as closed sets, closure operator, or neighbourhood systems of points etc. and define a topological space by abstraction of any of these concepts. They all lead to equivalent definitions as we shall see later. These other approaches were popular until topologies were defined in terms of abstract open sets.

The property (iv) of Theorem (1.6) is known as the Hausdorff property. Topological spaces satisfying it are known as **Hausdorff spaces.** There are many properties resembling the Hausdorff property in spirit. They are together called separation axioms. We shall study them systematically in a later chapter.

2. Examples of Topological Spaces

As noted in the last section, the metric spaces provide the foremost examples of topological spaces. There are of course many others. Indeed, if metric spaces were the only interesting examples of topological spaces, there would probably be no need to generalise from metric spaces to topological spaces. In this section we present some examples of topological spaces.

1. It may happen that the topology $\mathcal{T}$ on the set X consists only of ϕ and X. It is called the **indiscrete** topology on X. The name is appropriate since the indiscrete topology is induced by the indiscrete pseudometric on X. From the point of view of convergence of sequences this topology is not at all discriminating in that every sequence converges to every point, since there is only one non-empty open set, viz., X.

2. The other extreme is the so-called **discrete** topology on X. Here every set is open; in other words the topology coincides with the power set $P(X)$. It is easy to show that the discrete topology is induced by a discrete metric. In terms of convergence of sequences, it is too strict. The only convergent sequences in a discrete space are those which are eventually constant (Prove!).

3. Somewhere between these two extremes is the so-called cofinite topology. Let X be any set. A subset A of X is said to be **cofinite**, if its complement, $X - A$, is finite. Let $\mathcal{T}$ consist of all cofinite subsets of X and the empty set. (Note that the empty set is cofinite iff X itself is finite.) We leave it to the reader to check that $\mathcal{T}$ is a topology on X. In case X is finite it coincides with the discrete topology but otherwise it is not the same.

Let us study the convergence of sequences w.r.t. the cofinite topology $\mathcal{T}$ on a set X. Suppose $\{x_n\}$ is a sequence in X. We contend that it is convergent iff there is at most one term of it which repeats infinitely often. For suppose first $\{x_n\}$ is convergent. Let, if possible, a and b be two distinct values assumed infinitely often by the sequence. Then there are infinitely many values of n for which x_n is outside $X - \{b\}$. But $X - \{b\}$ is an open set containing a, and if $\{x_n\}$ were to converge to a then $X - \{b\}$ would have to contain x_n for all except finitely many values of n. Thus we see that $\{x_n\}$ cannot converge to a. Similarly it cannot converge to b. But we are assuming $\{x_n\}$ to be convergent. So there exists $c \in X$, different from a and b such that $\{x_n\}$ converges to c. But then, $X - \{a, b\}$ is an open set containing c which fails to contain x_n for infinitely many values of n. We thus get a contradiction. This shows that a convergent sequence can assume at most one value infinitely often.

Conversely suppose $\{x_n\}$ assumes at most one value infinitely often. We have to show that it is convergent. There are two possibilities, either there is a (unique) value, say b, assumed infinitely often by the sequence or there is no such value. We assert that in the first case the sequence converges to b (and to nothing else) while in the second case it converges to every point of X. In the first case, let V be an open set containing b. Then $X - V$ is a finite set and every element in it is assumed at most a finite number of times by the sequence $\{x_n\}$. Hence $X - V$ contains at most finitely many terms of $\{x_n\}$. Consequently $\{x_n\}$ converges to b. Moreover, in this case $\{x_n\}$ cannot converge to anything else since $X - \{b\}$ is an open set which misses infinitely many terms of it. In the case where no value is assumed infinitely often by the sequence $\{x_n\}$, it is clear that it converges to every point of X, for in such a case every non-empty open set contains all except finitely many terms of the sequence.

Thus we have completely characterised convergence of sequences in the cofinite topology. We have given the argument in detail so that the reader may get an idea of the kind of reasoning that is called for in studying convergence. Of course suitable modifications must be made keeping in mind the particular topology involved.

In the same vein one could define the **co-countable topology** on a set by taking the family of all sets whose complements are countable and the empty set. Still more generally, given any cardinal number α, one can form a topology on a set X by taking the empty set and all subsets whose complements in X are of cardinality not exceeding α. Spaces so formed are of little interest except as possible counterexamples.

Before we give further examples, let us pause to compare various topologies on the same set. Suppose X is a set. A topology $\mathcal{T}$ on X is a family of subsets of X; in other words $\mathcal{T}$ itself is a subset of the power set $P(X)$. If $\mathcal{T}_1$, $\mathcal{T}_2$ are both topologies on X then we can compare them by inclusion since they are both subsets of the same set, $P(X)$.

(2.1) Definition: The topology $\mathcal{T}_1$ is said to be **weaker** (or **coarser**) than the topology $\mathcal{T}_2$ (on the same set) if $\mathcal{T}_1 \subset \mathcal{T}_2$ as subsets of the power set. In this case we also say that $\mathcal{T}_2$ is **stronger** (or **finer**) than $\mathcal{T}_1$. The terms **smaller** and **larger** are also used in the sense of inclusion.

It is clear that on any set the indiscrete topology is the smallest or weakest or coarsest of all while the discrete topology is the largest or the finest or the strongest of all topologies on the same set. Any other topology lies somewhere between these two extremes. It may of course happen that two given topologies are not comparable to each other, i.e. neither is stronger than the other. Examples of this will follow later.

The terminology used here deserves some comment, especially, since it is likely to be confusing at times. Suppose $\mathcal{T}_1$, $\mathcal{T}_2$ are topologies on a set X and that $\mathcal{T}_1 \subset \mathcal{T}_2$. This means every open set in $(X, \mathcal{T}_1)$ is also open in $(X; \mathcal{T}_2)$. So if something holds for all open sets in $(X, \mathcal{T}_2)$ then it also holds for all open sets in $(X, \mathcal{T}_1)$. Consequently, the statement that a sequence is convergent in $(X, \mathcal{T}_2)$ is stronger than the statement that the same sequence is convergent in $(X, \mathcal{T}_1)$. Since convergence played a dominant role in the history of topology, this probably explains why the terms 'stronger' and 'weaker' are used. The reader is warned, however, that sometimes they are used exactly the opposite way by analysts. It is therefore necessary to check which convention is used by a particular author.

It is also a little misleading at the outset to call a larger topology as finer and a smaller one as coarser. A justification of this will come naturally after we study connectedness and components. Intuitively, the larger the topology, the more fragmentary a space tends to be. Naturally these fragments (or components as they will technically be called) will be finer in a larger topology and hence the terminology.

Note that the set of all topologies on a set X is partially ordered by the relation 'is weaker than'. The following theorem asserts existence of greatest lower bounds or infima in this partially ordered set.

(2.2) Theorem: Let X be a set and $\{\mathcal{T}_i : i \in I\}$ be an indexed family of topologies on X. Let $\mathcal{T} = \bigcap\limits_{i \in I} \mathcal{T}_i$. Then $\mathcal{T}$ is a topology on X. It is weaker than each $\mathcal{T}_i$, $i \in I$. If $\mathcal{U}$ is any topology on X which is weaker than each $\mathcal{T}_i$, $i \in I$, then $\mathcal{T}$ is stronger than $\mathcal{U}$.

Proof: The proof itself is trivial once we clearly figure out what really is going on. We have a set X. Every topology on X is a subset of $P(X)$ i.e. an element of $P(P(X))$. What we are dealing with is a family of such topologies on X, i.e. with a family of subsets of $P(X)$, or equivalently with a subset of $P(P(X))$, or alternatively with an element of $P(P(P(X)))$! In situations like this it is very important not to confuse among sets of different 'levels'. Intuitively the elements of X (which are of little importance in the present theorem) can be taken to be on the 'ground level'. The set X (as well as all its subsets) are on a higher level. The topologies on X are on a still higher level and the collection of such topologies on the highest level. To say that $\mathcal{T} = \bigcap_{i \in I} \mathcal{T}_i$ means that in $\mathcal{T}$ we put those subsets of X which are present in every topology $\mathcal{T}_i$ of the given family of topologies. It is not the same as the family of sets of the form $\bigcap_{i \in I} A_i$ as A_i ranges over $\mathcal{T}_i$, $i \in I$. Indeed this latter family is of no importance in the present theorem.

Now, coming to the proof itself, let us first verify that $\mathcal{T}$ is a topology on X. Clearly the empty set ϕ belongs to each $\mathcal{T}_i$ since each $\mathcal{T}_i$ is topology on X and so $\phi \in \bigcap_{i \in I} \mathcal{T}_i$, i.e. $\phi \in \mathcal{T}$. Similarly $X \in \mathcal{T}$. Next we show that $\mathcal{T}$ is closed under finite intersections. For this let $A_1, A_2, \ldots, A_n \in \mathcal{T}$ and suppose $A = \bigcap_{j=1}^{n} A_j$. To show $A \in \mathcal{T}$. Now, since $\mathcal{T} = \bigcap_{i \in I} \mathcal{T}_i$, each A_j belongs to each $\mathcal{T}_i$. But $\mathcal{T}_i$, being a topology on X, is closed under finite intersections. So, $A \in \mathcal{T}_i$ for each $i \in I$. But then $A \in \mathcal{T}$ as $\mathcal{T} = \bigcap_{i \in I} \mathcal{T}_i$. The proof that $\mathcal{T}$ is closed under arbitrary unions is similar and left to the reader. The rest of the theorem is just a general property of intersections. ∎

Despite the essential triviality of the argument above, we chose to give it in detail in order to stress upon the reader the importance of clearly understanding a statement before attempting to prove it. It is especially intended for a student who has had no serious encounter with abstract deductive reasonings and for whom mathematics has never meant more than evaluating integrals and solving differential equations. Another healthy habit which pays off immensely in topology (and in all mathematics) is to look for the structural similarity between two arguments. For example, the present proof must have undoubtedly reminded the perceptive reader of the proof of the fact that the intersection of a family of sub-groups of a group is again a subgroup. Such a reader will not only have an easier time in understanding the theorem just proved but will also be in a position to anticipate what is coming.

(2.3) Corollary: Let X be a set and $\mathcal{D}$ a family of subsets of X. Then there exists a unique topology $\mathcal{T}$ on X, such that it is the smallest topology on X containing $\mathcal{D}$.

Proof: Consider the collection of all topologies on X which contain $\mathcal{D}$ (as subsets of $P(X)$). This family is non-empty, for the discrete topology (i.e. the entire power set $P(X)$) surely contains $\mathcal{D}$. Now let $\mathcal{T}$ be the intersection of the members of this collection. Applying the theorem above, $\mathcal{T}$ is a topology on X, it contains $\mathcal{D}$ and clearly it is the smallest topology containing $\mathcal{D}$, for any such topology will be a member of the collection of topologies just considered and hence stronger than its intersection, viz. $\mathcal{T}$. Uniqueness of $\mathcal{T}$ is trivial. ∎

The topology $\mathcal{T}$ so obtained is said to be **generated** by the family $\mathcal{D}$. (Querry: what happens if $\mathcal{D}$ is already a topology on X to begin with?). The corollary asserts, in theory, that we can create a topological space starting with any set X and any family $\mathcal{D}$ of subsets of X. However, in practice the mere existence of the topology $\mathcal{T}$ generated by $\mathcal{D}$ is of little importance unless there is some concrete way of expressing members of $\mathcal{T}$ in terms of members of $\mathcal{D}$. Such a description is actually possible and will be given in the next section. (Querry: What happens in the analogous situation in group theory?).

Returning to examples of topological spaces, let us consider some topologies on $\mathbb{R}$, the set of real numbers. (The numbering continues from the examples given earlier in the section.)

4. **The usual topology** on $\mathbb{R}$ (or any other euclidean space) is defined as the topology induced by the euclidean metric. Note that with this metric the open balls are just bounded open intervals. So all bounded open intervals are indeed open sets in the usual topology on $\mathbb{R}$. Since unbounded open intervals can be expressed as unions of bounded open intervals, they are also open in the usual topology. Thus all open intervals are open sets of $\mathbb{R}$ in the usual topology. The converse is false. However, it is a standard fact from analysis that an open subset of $\mathbb{R}$ can be expressed as the union of countably many, mutually disjoint open intervals. We shall also give a proof of this in Chapter Six. Throughout this book, whenever we speak of $\mathbb{R}$ (or $\mathbb{R}^n$) as a topological space, the usual topology will be understood unless otherwise specified.

5. There is another, and a stronger topology on $\mathbb{R}$, called the **semi-open interval topology**. A subset U is said to be open in this topology if for every $x \in U$, there exists $r > 0$ such that the semi-open interval $[x, x + r)$ is contained in U. The verification that this indeed defines a topology on $\mathbb{R}$ and that the topology so defined is stronger than the usual topology is left to the reader. One can of course consider semi-open intervals of the form $(x - r, x]$ instead of $[x, x + r)$ in the definition and get another topology on $\mathbb{R}$. In a suggestive sense, the two topologies will be the mirror images of each other. However, they are not the same. Indeed they are not even comparable in terms of strength. Their intersection is precisely the usual topology on $\mathbb{R}$ (Prove!). Note that a sequence $\{x_n\}$ converges to y in the semi-open interval topology iff it converges to y from above (or from the right) in the usual topology. For this reason, the semi-open interval topology is also known as the upper limit topology. It has many strange properties and is a source of many counter-examples in general topology.

6. Another weird, although not frequently cited, topology on $\mathbb{R}$ is

called the **scattering topology**. Its open sets are of the form $A \cup B$ where A is an open subset of $\mathbb{R}$ in the usual topology and B is any subset of the irrational numbers. Again, it is trivial to verify that such sets do form a topology. The resulting topological space is called the **scattered** line. Note that every subset of the irrational is open. Consequently no sequence can converge to an irrational number except an eventually constant sequence.

7. Yet another topology on $\mathbb{R}$ can be obtained by taking the empty set ϕ, the whole set and all intervals of the form (a, ∞) where $a \in \mathbb{R}$. This topology is of course much weaker than the usual topology on $\mathbb{R}$. We leave it to the reader to characterise convergence of sequences in this topology. A similar topology can be defined on $\mathbb{Q}$ or $\mathbb{Z}$ as well.

Sometimes some additional structure on a set gives rise to a topology on the set. As we saw, a metric is such a structure. In the next example we show how a linear order on a set determines a certain topology on it.

8. Let a set X be linearly ordered by $\leq$. Declare a subset A of X to be open if for each $x \in A$ there exist $a, b \in X$ such that $a < x < b$ and the 'interval' (a, b) (i.e. the set $\{y \in X : a < y < b\}$) is contained in A. Assume for the moment that X has no smallest and no largest element (this restriction will be removed in the next section). It is easy to show that the collection of open sets is indeed a topology. It is called the **order topology** induced by the order $\leq$. For the real line $\mathbb{R}$, the topology induced by the usual ordering coincides with the usual topology. However by selecting the set and the ordering $\leq$ suitably one can construct many strange spaces. As an example, we invite the reader to compare the usual topology on the plane $\mathbb{R}^2$ with the topology induced by the lexicographic ordering on it (in this ordering $(x_1, x_2) \leq (y_1, y_2)$ iff $x_1 < y_1$ or $x_1 = y_1$ and $x_2 \leq y_2$).

9. A tiny topological space, called the **Sierpinski space** is sometimes useful as a counter-example. Its underlying set is any set with two elements, say, the set $\{a, b\}$, where $a \neq b$. The topology consists of ϕ, $\{a\}$ and $\{a, b\}$. It is probably the simplest example of a topology not induced by a metric (or even a pseudometric).

We now have a fairly long list of examples of topological spaces. It can of course be continued indefinitely. In the next two sections we shall give two general methods for the construction of topological spaces. However, there are a few peculiar spaces which do not result as a consequence of any standard construction. Their main value lies in the counterexamples they provide. We conclude the present section by describing one such space.

10. Let $\mathbb{N}$ be the set of natural numbers, i.e. the set of positive integers. Consider the cartesian product $\mathbb{N} \times \mathbb{N}$. For $k \in \mathbb{N}$, we shall call the subset $\mathbb{N} \times \{k\}$ as the k-th row of $\mathbb{N} \times \mathbb{N}$. The terminology becomes suggestive if points of $\mathbb{N} \times \mathbb{N}$ are pictured in the plane by cartesian coordinates. Now let ∞ be any symbol not in $\mathbb{N} \times \mathbb{N}$ and let $X = (\mathbb{N} \times \mathbb{N}) \cup \{\infty\}$. We shall define a topology on X as follows. Let $\mathcal{T}_1$ be the power set $P(\mathbb{N} \times \mathbb{N})$. Clearly $\mathcal{T}_1 \subset P(X)$. Let $\mathcal{T}_2$ be the collection of those subsets A of X such that $\infty \in A$ and A contains almost all points in almost all rows. Here the term 'almost' means 'with the exception of

finitely many'. Now let $\mathcal{T} = \mathcal{T}_1 \cup \mathcal{T}_2$. Then it is easy to show that $\mathcal{T}$ is a topology on X. The strange properties of this space are developed in an exercise of this section. We shall have many occasions to visit this example again in later chapters.

Exercises

2.1 Prove that in the co-countable topology, the only convergent sequences are those which are eventually constant. In other words, as far as the convergence of sequences is concerned, the co-countable topology behaves the same way as the discrete topology.

2.2 Prove that the Sierpinski space is not obtainable from a pseudo-metric. (Hint: A pseudo-metric on a two point set must be either the indiscrete pseudo-metric or a discrete metric.)

2.3 Prove that the set of all topologies on a set X is a complete lattice. (Hint: The intersection of two topologies is clearly their meet; for finding their join consider the topology generated by their union.)

2.4 Consider the two semi-open interval topologies on $\mathbb{R}$. Prove that their meet is the usual topology while their join is the discrete topology on $\mathbb{R}$.

2.5 Compare the strengths of various pairs of topologies on the real line defined in this section.

2.6 Let (X, d) be a metric space, $\mathcal{T}$ the topology induced by the metric and $\mathcal{U}$ some other topology on X. Prove that $\mathcal{T}$ is weaker than $\mathcal{U}$ iff every open ball is in $\mathcal{U}$. (Hint: Show that every open set is a union of open balls.)

2.7 Prove that the usual topology on the euclidean plane $\mathbb{R}^2$ is strictly weaker than the topology induced on it by the lexicographic ordering.

2.8 For the topological space of Example 10, prove that:

(i) No sequence can converge to a point of $\mathbb{N} \times \mathbb{N}$ unless it is eventually constant.

(ii) No sequence in $\mathbb{N} \times \mathbb{N}$ can converge to ∞. (Hint: If $\{x_n\}$ is a sequence in $\mathbb{N} \times \mathbb{N}$ converging to ∞ then every row will contain at most finitely many terms of it. Excluding these terms from each row, we get an open set containing ∞ which contains no term of this sequence).

(iii) The topology $\mathcal{T}$ is not discrete.

(iv) A sequence converges to a point iff it is eventually constant (in other words, as far as convergence of sequences is concerned the space behaves like a discrete space.)

(v) The space $(X; \mathcal{T})$ is not metrisable even though condition (iv) of Theorem 1.6 of the last section is satisfied. (Hint: In a metrisable space, open sets can be characterised in terms of convergence of sequences.)

(vi) There exists a sequence in $\mathbb{N} \times \mathbb{N}$ which has ∞ as a limit point but no subsequence converging to ∞. (Hint: Enumerate $\mathbb{N} \times \mathbb{N}$.)

2.9 What difficulty would arise in defining the order topology if the set either had a smallest or a largest element?

2.10 Prove that on a finite set with n elements there are at most $2^{(2^n-2)}$ distinct topologies. Verify that this upper bound is attained for $n = 1, 2$.

2.11 Prove that on a set with three elements there are twentynine distinct topologies. Thus the upper bound of the last exercise is far from attained for $n = 3$. Can you find a sharper upper bound?

Notes and Guide to Literature

The general methods of construction of topological spaces are all very standard and will be discussed in this book. But when it comes to peculiar spaces, tailored to serve an ad-hoc purpose such as providing counter-examples, there is no end to their multitude. Willard's book [1] gives quite a few interesting examples of topological spaces. There is also a book entitled 'Counter-examples in Topology' by Steen and Seebach [1].

With an elementary knowledge of ordinal numbers, one can construct many interesting topological spaces. Although we shall not discuss ordinal number *per se*, we shall study these spaces in the last chapter of this book.

See Munkres [1] for an enumeration of the twentynine topologies on a 3-point set. The problem of determining the number of topologies on a finite set is apparently still not completely solved. A recent reference on it is Krishnamurthy [1].

3. Bases and Sub-bases

In the last section we showed that any family of subsets of a set generates a topology on that set. In this section we shall see how the topology so generated can be described intrinsically in terms of the original family of subsets. We begin with an important definition.

(3.1) Definition: Let $(X; \mathcal{T})$ be a topological space. A subfamily $\mathcal{B}$ of $\mathcal{T}$ is said to be a **base** for $\mathcal{T}$ if every member of $\mathcal{T}$ can be expressed as the union of some members of $\mathcal{B}$.

For example, in a metric space every open set can be expressed as a union of open balls and consequently the family of all open balls is a base for the topology induced by the metric. It is not necessary to take all open balls; a moment's reflexion would show that the family of all open balls of rational radii is also a base. Indeed it suffices to take balls of radius $1/n$, $n \in \mathbb{N}$. Before giving further examples let us obtain a simple characterisation of bases.

(3.2) Proposition: Let $(X; \mathcal{T})$ be a topological space and $\mathcal{B} \subset \mathcal{T}$. Then $\mathcal{B}$ is a base for $\mathcal{T}$ iff for any $x \in X$ and any open set G containing x, there exists $B \in \mathcal{B}$ such that $x \in B$ and $B \subset G$.

Proof: First suppose $\mathcal{B}$ is a base for $\mathcal{T}$. Let $x \in X$ and let an open set

G containing x be given. Then G can be written as the union of some members of $\mathcal{B}$, say, $G = \bigcup_{i \in I} B_i$ where I is an index set and $B_i \in \mathcal{B}$ for all $i \in I$. Since $x \in G$, there exists $j \in I$ such that $x \in B_j$. We take this B_j as the set B required in the assertion.

Conversely suppose the given condition holds. Let H be an open set in X, i.e. $H \in \mathcal{T}$. For each $x \in H$, there exists $B_x \in \mathcal{B}$ such that $x \in B_x$ and $B_x \subset H$. (It may happen that for some distinct x, y in H, B_x and B_y are the same sets.) Clearly $H = \bigcup_{x \in H} B_x$. Thus every member of $\mathcal{T}$ can be expressed as the union of some members of $\mathcal{B}$. So $\mathcal{B}$ is a base for $\mathcal{T}$. ∎

Although the preceding proposition is hardly a profound one, it makes it a little easier to test whether a given family is a base for a topology or not. Using it we see at once that the semi-open intervals form a base for the semi-open interval topology on the real line. Also, for the usual topology on the real line, the family of all open intervals with rational end-points is a base. Note that this base is countable although neither the set of real numbers nor the usual topology on it is countable. Spaces for which countable bases exist are quite important and are given a special name.

(3.3) Definition: A space is said to **satisfy the second axiom of countability** or is said to be **second countable** if its topology has a countable base.

The peculiar name 'second countable' suggests that there must also be something called 'first countable' spaces. Such spaces do exist and we shall define them later.

Second countable spaces have many pleasant properties which will be proved later. By way of illustration, we prove only one of them here. But first let us state some definitions. A family $\mathcal{U}$ of sets is said to be a **cover** (or **covering**) of a set A if A is contained in the union of members of $\mathcal{U}$. A **subcover** of $\mathcal{U}$ is subfamily $\mathcal{V}$ of $\mathcal{U}$ which itself is a cover of A. If we are in a topological space then a cover is said to be **open** if all its members are open. With these definitions, we now prove.

(3.4) Theorem: If a space is second countable then every open cover of it has a countable subcover.

Proof: Let $(X; \mathcal{T})$ be a space with a countable base $\mathcal{B}$ and let $\mathcal{U}$ be a given open cover of X. First enumerate $\mathcal{B}$ as $\{B_1, B_2, B_3, \ldots\}$. (In case $\mathcal{B}$ is finite, the same term can be repeated infinitely many times.) Now let $S = \{n \in \mathbb{N} : B_n$ is contained in some member of $\mathcal{U}\}$. For each $n \in S$, fix $U_n \in \mathcal{U}$ s.t. $B_n \subset U_n$. (It may happen that $U_m = U_n$ even though $B_m \neq B_n$). Now let $\mathcal{C} = \{B_n : n \in S\}$ and $\mathcal{V} = \{U_n : n \in S\}$. Clearly $\mathcal{V}$ is a countable sub-family of $\mathcal{U}$ and covers X if $\mathcal{C}$ does. So the theorem will be proved if we show $\mathcal{C}$ is a cover of X. For this let $x \in X$. Then $x \in U$ for some $U \in \mathcal{U}$. By proposition (3.2) above, there is some $k \in \mathbb{N}$ s.t. $x \in B_k$ and $B_k \subset U$. Clearly then, $k \in S$ and so $B_k \in \mathcal{C}$. So $\mathcal{C}$ and consequently $\mathcal{V}$ is a cover of X. ∎

A reader who has studied compactness in real analysis will undoubtedly see the resemblance of the conclusion of this theorem to the definition of compactness. Each of the two conditions as well as second countability implies that the space is small in some sense. We shall systematically study such smallness conditions later on.

Returning to the bases in a general topological space we note that the same topology may have more than one distinct bases but two distinct topologies can never have the same family of subsets as a base for both of them. For, let $\mathcal{T}_1$, $\mathcal{T}_2$ be topologies on a set and each have $\mathcal{B}$ as a base. If $U \in \mathcal{T}_1$ then U can be expressed as a union of some members of $\mathcal{B}$; these are also members of $\mathcal{T}_2$ since $\mathcal{B} \subset \mathcal{T}_2$. But $\mathcal{T}_2$, being a topology, is closed under arbitrary unions. So $U \in \mathcal{T}_2$. Thus $\mathcal{T}_1 \subset \mathcal{T}_2$. Similarly $\mathcal{T}_2 \subset \mathcal{T}_1$ and hence $\mathcal{T}_1 = \mathcal{T}_2$. The argument here can be used to prove the following proposition.

(3.5) Proposition: Let $\mathcal{T}_1$, $\mathcal{T}_2$ be two topologies for a set having bases $\mathcal{B}_1$ and $\mathcal{B}_2$ respectively. Then $\mathcal{T}_1$ is weaker than $\mathcal{T}_2$ iff every member of $\mathcal{B}_1$ can be expressed as a union of some members of $\mathcal{B}_2$. ∎

It is immediate that if $\mathcal{B}$ is a base for a topology $\mathcal{T}$ on a set X, then $\mathcal{B}$ generates $\mathcal{T}$, i.e. $\mathcal{T}$ is the smallest topology containing $\mathcal{B}$; indeed $\mathcal{T}$ consists precisely of those subsets of X which can be expressed as unions of members of subfamilies of $\mathcal{B}$. It is natural to inquire whether we can start with an arbitrary family $\mathcal{B}$ of subsets of a set X and find a topology $\mathcal{T}$ on X for which $\mathcal{B}$ will be a base. Of course in case such a topology exists, it must be unique. But in general no such topology exists as the following example shows:

Consider $\mathbb{R}$, the set of real numbers and let $\mathcal{B}$ consist of all non-degenerate closed bounded intervals, i.e. intervals of the form $[a, b]$ where $a, b \in \mathbb{R}$ and $a < b$. We contend there is no topology on $\mathbb{R}$ for which $\mathcal{B}$ is a base. For suppose $\mathcal{T}$ were such a topology. Since $\mathcal{T}$ is closed under finite intersections, it must in particular contain the intersection of any two members of $\mathcal{B}$. Now any singleton set can be written as an intersection of two non-degenerate closed bounded intervals and so $\mathcal{T}$ contains all singleton sets, from which it follows that $\mathcal{T}$ must be the discrete topology (Prove!). This would mean that every subset of $\mathcal{T}$ can be expressed as a union of non-degenerate closed intervals. This is false; in fact no singleton set can be so expressed. The following proposition tells precisely which families can be bases for topologies.

(3.6) Proposition: Let X be a set and $\mathcal{B}$ a family of its subsets covering X. Then the following statements are equivalent:

(1) There exists a topology on X with $\mathcal{B}$ as a base.

(2) For any $B_1, B_2 \in \mathcal{B}$, $B_1 \cap B_2$ can be expressed as the union of some members of $\mathcal{B}$.

(3) For any $B_1, B_2 \in \mathcal{B}$ and $x \in B_1 \cap B_2$, there exists $B_3 \in \mathcal{B}$ such that $x \in B_3$ and $B_3 \subset B_1 \cap B_2$.

Proof: (1) ⇒ (2). Suppose there exists a topology $\mathcal{T}$ on X for which $\mathcal{B}$ is a base. Let $B_1, B_2 \in \mathcal{B}$. Then $B_1, B_2 \in \mathcal{T}$ and so $B_1 \cap B_2 \in \mathcal{T}$ since $\mathcal{T}$ is closed under finiter intersections. So, by definition of a base, $B_1 \cap B_2$ can be expressed as the union of some members of $\mathcal{B}$.

The proof of the equivalence of (2) and (3) resembles that of Proposition (3.2) and is left as an exercise.

It only remains to show (3) ⇒ (1). Assume that the condition in (3) holds and define $\mathcal{T} = \{G \subset X :$ for all $x \in G$, there exists $B \in \mathcal{B}$ such that $x \in B$ and $B \subset G\}$. We assert that $\mathcal{T}$ is a topology on X. The reader will note that the definition of $\mathcal{T}$ resembles that of the metric topology and so it is not surprising that the proof resembles that of Theorem (1.5) or (1.6). Clearly $\phi \in \mathcal{T}$ while $X \in \mathcal{T}$ since $\mathcal{B}$ is given to be a cover of X. That $\mathcal{T}$ is closed under arbitrary unions is self-evident. It only remains to verify that whenever $G, H \in \mathcal{T}, G \cap H \in \mathcal{T}$. Let $x \in G \cap H$. Then $x \in G$ and $x \in H$. So there exist $B_1, B_2 \in \mathcal{B}$ such that $x \in B_1$, $B_1 \subset G$, $x \in B_2$ and $B_2 \subset H$. Then $x \in B_1 \cap B_2$. So by (3), there exists $B_3 \in \mathcal{B}$ such that $x \in B_3$ and $B_3 \subset B_1 \cap B_2$. But $B_1 \cap B_2 \subset G \cap H$. So $G \cap H \in \mathcal{T}$. Thus $\mathcal{T}$ is a topology on X and in view of its construction and Proposition (3.2), it follows that $\mathcal{B}$ is a base for $\mathcal{T}$. ∎

As an illustration, consider the family $\mathcal{B}$ of all open discs in the plane. (Open triangles or open rectangles would do as well). Then (3) of the last proposition holds true and so there exists a topology $\mathcal{T}$ on the plane with $\mathcal{B}$ as a base for it. We know already that in this case $\mathcal{T}$ is the usual topology on the plane.

The following corollary of the last proposition is immediate.

(3.7) Corollary: If $\mathcal{B}$ is a cover of X and $\mathcal{B}$ is closed under finite intersections then $\mathcal{B}$ is a base for a (unique) topology $\mathcal{T}$ on X. Moreover, $\mathcal{T}$ consists precisely of those subsets of X which can be expressed as unions of subfamilies of $\mathcal{B}$. ∎

Suppose now we start with an arbitrary family $\mathcal{S}$ of subsets of a set X. We let $\mathcal{B}$ be the family of all finite intersections of elements of $\mathcal{S}$. More precisely $\mathcal{B} = \{B \subset X :$ there exist $S_1, S_2, \ldots, S_n \in \mathcal{S}$ $(n \geq 0)$ such that $B = \bigcap_{i=1}^{k} S_i\}$. Note that by taking the intersection of the empty subfamily of $\mathcal{S}$ (i.e. the intersection of zero sets in $\mathcal{S}$) it follows that $X \in \mathcal{B}$. It is also evident that $\mathcal{B}$ is closed under finite intersections and that $\mathcal{S} \subset \mathcal{B}$.

(3.8) Definition: A family $\mathcal{S}$ of subsets of X is said to be a **sub-base** for a topology $\mathcal{T}$ on X if the family of all finite intersections of members of $\mathcal{S}$ is a base for $\mathcal{T}$.

Of course any base for a topology is also a sub-base for the same. In general, however, a subbase can be chosen to be much smaller than a base. For example, for the usual topology on $\mathbb{R}$, the family of all open intervals of the form (a, ∞) or $(-\infty, b)$ for $a, b \in \mathbb{R}$ (or $\mathbb{Q}$) is a sub-base. In the next theorems we characterize sub-bases and show that unlike bases, any family of subsets is a sub-base for some topology.

(3.9) Theorem: Let X be a set, $\mathcal{T}$ a topology on X and $\mathcal{S}$ a family of subsets of X. Then $\mathcal{S}$ is a sub-base for $\mathcal{T}$ iff $\mathcal{S}$ generates $\mathcal{T}$.

Proof: Let $\mathcal{B}$ be the family of finite intersections of members of $\mathcal{S}$. Suppose first that $\mathcal{S}$ is a sub-base for $\mathcal{T}$. We want to show that $\mathcal{T}$ is the smallest topology on X containing $\mathcal{S}$. Now since $\mathcal{S} \subset \mathcal{B}$ and $\mathcal{B} \subset \mathcal{T}$ we at least have that $\mathcal{T}$ contains $\mathcal{S}$. Suppose $\mathcal{U}$ is some other topology on X such that $\mathcal{S} \subset \mathcal{U}$. We have to show that $\mathcal{T} \subset \mathcal{U}$. Now since $\mathcal{U}$ is closed under finite intersections and $\mathcal{S} \subset \mathcal{U}$, $\mathcal{U}$ contains all finite intersections of members of $\mathcal{S}$, i.e. $\mathcal{B} \subset \mathcal{U}$. But again since $\mathcal{U}$ is closed under arbitrary unions and each member of $\mathcal{T}$ can be written as union of some members of $\mathcal{B}$ (by definition of a base), it follows that $\mathcal{T} \subset \mathcal{U}$.

Conversely suppose $\mathcal{T}$ is the smallest topology containing $\mathcal{S}$. We have to show that $\mathcal{S}$ is a sub-base for $\mathcal{T}$, i.e. that $\mathcal{B}$ is a base for $\mathcal{T}$. Clearly $\mathcal{B} \subset \mathcal{T}$ since $\mathcal{T}$ is closed under finite intersections and $\mathcal{S} \subset \mathcal{T}$. Since $\mathcal{B}$ is closed under finite intersections we know by Corollary (3.7) above that there is a topology $\mathcal{U}$ on X such that $\mathcal{B}$ is a base for $\mathcal{U}$. Every member of $\mathcal{U}$ can be expressed as a union of a sub-family of $\mathcal{B}$ and so is in $\mathcal{T}$ since $\mathcal{B} \subset \mathcal{T}$. This means $\mathcal{U} \subset \mathcal{T}$ and consequently $\mathcal{U} = \mathcal{T}$ since $\mathcal{T}$ is the smallest topology containing $\mathcal{S}$. Thus $\mathcal{B}$ is a base for $\mathcal{T}$ and $\mathcal{S}$ is a sub-base for $\mathcal{T}$. ∎

(3.10) Theorem: Given any family $\mathcal{S}$ of subsets of X, there is a unique topology $\mathcal{T}$ on X having $\mathcal{S}$ as a sub-base. Further, every member of $\mathcal{T}$ can be expressed as the union of sets each of which can be expressed as the intersection of finitely many members of $\mathcal{S}$.

Proof: The first assertion follows from the last theorem. To prove that every member of $\mathcal{T}$ has the desired form, let $\mathcal{U}$ consist of all subsets of X which can be expressed as unions of members of $\mathcal{B}$ where $\mathcal{B}$ is the family of finite intersections of members of $\mathcal{S}$. By Corollary (3.7) above, $\mathcal{U}$ is the unique topology having $\mathcal{B}$ as a base. Hence $\mathcal{T} = \mathcal{U}$. ∎

In the last section we observed that any family of subsets of a set X generates a topology on the set X. The preceding two theorems show how the elements of the topology so generated are expressible in terms of members of the original family. Evidently Theorem (3.10) has many applications in the construction of topological spaces. We shall discuss only one such application here. Let n be a positive integer, $\{(X_i, \mathcal{T}_i) : i = 1, 2, 3, \ldots, n\}$ be a family of topological spaces and X be the cartesian product $X_1 \times X_2 \times \ldots \times X_n$. We shall define a certain topology, called the **product topology** on X. By an **open box** in X we mean a set of the form $V_1 \times V_2 \times \ldots V_n$ where $V_i \in \mathcal{T}_i$ for $i = 1, 2, \ldots, n$. The entire set X is clearly an open box. It is also evident that the intersection of two open boxes is again an open box, although their union need not be. By Corollary (3.7) above, the family of open boxes is a base for a unique topology $\mathcal{T}$ on X. This topology is called the **product topology** on X, the space $(X, \mathcal{T})$ is called the **topological product** of the spaces $(X_1, \mathcal{T}_1), (X_2, \mathcal{T}_2), \ldots, (X_n, \mathcal{T}_n)$. The space $(X_i, \mathcal{T}_i)$ is called the i**th co-ordinate space** or the i**th factor** of $(X; \mathcal{T})$.

It may happen that some of the coordinate spaces coincide. In such a case they can be regarded as different copies of the same space. In case each factor X_i equals a given space say Y, for $i = 1, 2, \ldots, n$, then the topological product $X_1 \times \ldots \times X_n$ is called the n**th power** of Y and is often denoted by Y^n. As a foremost example of such powers, let us show that the nth power of the real line (i.e. the topological product of n copies of $\mathbb{R}$ with the usual topology) in fact coincides with $\mathbb{R}^n$ with the usual topology on it. This would remove the ambiguity about the notation $\mathbb{R}^n$, which can be used either for the n-dimensional euclidean space or for the nth power of $\mathbb{R}$.

Let $\mathcal{U}$ denote the usual topology on $\mathbb{R}$. Let $\mathcal{T}_1$ denote the product topology on $\mathbb{R}^n$ while let $\mathcal{T}_2$ denote the usual topology on it (induced by the euclidean metric on $\mathbb{R}^n$). We have to show that $\mathcal{T}_1 = \mathcal{T}_2$. For convenience, we denote a point $(x_1, \ldots, x_n) \in \mathbb{R}^n$ by $\bar{x}$ and so on. Let $d(\bar{x}, \bar{y})$ denote the euclidean distance, $\sqrt{\left(\sum_{i=1}^{n} (x_i - y_i)^2\right)}$ between $\bar{x}$ and $\bar{y}$. $\mathcal{T}_2$ is then the topology induced by the metric d. We first claim $\mathcal{T}_2 \subset \mathcal{T}_1$. It suffices to show that every open ball is a member of $\mathcal{T}_1$. Consider such an open ball $B(\bar{x}, r)$. For any $\bar{y} \in B(\bar{x}, r)$ we shall find $G_y \in \mathcal{T}_1$ such that $\bar{y} \in G_{\bar{y}}$ and $G_{\bar{y}} \subset B(\bar{x}, r)$. It would then follow that $B(\bar{x}, r) = \bigcup_{\bar{y} \in B(\bar{x}, r)} G_{\bar{y}}$ and hence that $B(\bar{x}, r) \in \mathcal{T}_1$. So suppose $\bar{y} \in B(\bar{x}, r)$. Choose ϵ so that $0 < \epsilon < \frac{r - d(\bar{x}, \bar{y})}{\sqrt{n}}$. For $i = 1, 2, \ldots, n$ let V_i be the open interval (in $\mathbb{R}$) of length 2ϵ and centred at y_i and let $G_{\bar{y}} = V_1 \times V_2 \times \ldots \times V_n$. Then each V_i is open in $\mathbb{R}$ and so $G_{\bar{y}}$ is in the product topology $\mathcal{T}_1$. Clearly $\bar{y} \in G_{\bar{y}}$. Further, if $\bar{z} \in G_{\bar{y}}$ then

$$\begin{aligned} d(\bar{x}; \bar{z}) &\leq d(\bar{x}, \bar{y}) + d(\bar{y}, \bar{z}) \\ &= d(\bar{x}, \bar{y}) + \sqrt{\left(\sum_{i=1}^{n} (y_i - z_i)^2\right)} \\ &< d(\bar{x}, \bar{y}) + \sqrt{(n\epsilon^2)} \\ &< d(\bar{x}, \bar{y}) + r - d(\bar{x}, \bar{y}) \\ &= r \end{aligned}$$

Thus we see that $G_{\bar{y}} \subset B(\bar{x}, r)$ and as noted before this shows that $\mathcal{T}_2 \subset \mathcal{T}_1$. For the other way inclusion, suppose $G \in \mathcal{T}_1$. To show that $G \in \mathcal{T}_2$, we proceed from the definition of the metric topology. So let $\bar{x} \in G$. We have to find some $r > 0$ such that $B(\bar{x}, r) \subset G$. Now since the family of open boxes is a base for $\mathcal{T}_1$, there exists an open box say $W = W_1 \times \ldots \times W_n$ containing $\bar{x}$ such that $W \subset G$. Here each W_i is open in $\mathbb{R}$ and contains x_i (since $x \in W$). Find $r_i > 0$ so that the open interval $(x_i - r_i, x_i + r_i)$ is contained in W_i. Let $0 < r < \min\{r_1, \ldots, r_n\}$. Then $r > 0$. Also,

$$\bar{z} \in B(\bar{x}, r) \Rightarrow d(\bar{x}, \bar{z}) < r$$

$$\Rightarrow \sum_{i=1}^{n} (x_i - z_i)^2 < r^2$$

$$\Rightarrow |x_i - z_i| < r < r_i \quad \text{for each } i = 1, 2, \ldots, n$$

$$\Rightarrow z_i \in W_i \quad \text{for each } i = 1, 2, \ldots, n$$

$$\Rightarrow z \in W$$

$$\Rightarrow z \in G.$$

Thus we have $B(\bar{x}, r) \subset G$ as desired. So $\mathcal{T}_1 \subset \mathcal{T}_2$. The geometric idea behind the proof is very simple. It says that any point of an open box can be surrounded by an open ball within the box and vice versa. We illustrate this pictorially for the case $n = 2$, in Figure 1.

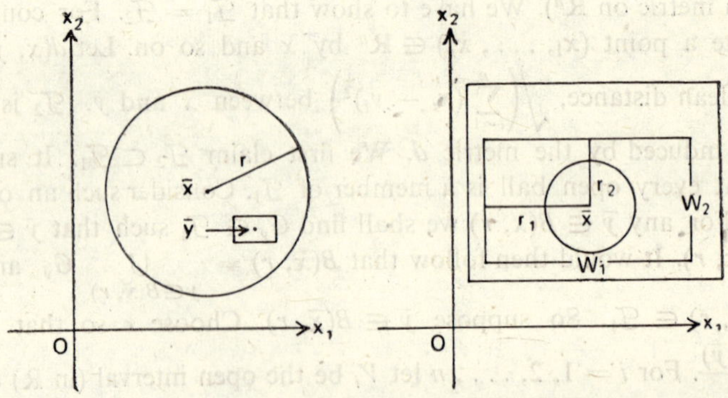

Fig. 1 Metric topology on $\mathbb{R}^2$ is the same as the product topology.

A suitable modification of this argument will establish a more general result, given in the exercises.

We could have as well defined topological products of arbitrary families of topological spaces here. Although nothing radically new is involved, the case of a finite product is easier to deal with since its elements can be 'concretely' described in the form of n-tuples. The definition and properties of arbitrary products will be deferred to a later chapter by which time the reader will have acquired some familiarity with finite products.

Exercises

3.1 Prove that if a space $(X, \mathcal{T})$ has a base $\mathcal{B}$ of cardinality α then the cardinality of $\mathcal{T}$ cannot exceed 2^α. (Hint: Define a surjective function from the set of all sub-families of $\mathcal{B}$ onto $\mathcal{T}$. Note that the cardinality of the set X itself is of little importance here.)

3.2 Prove that the converse of Theorem (3.4) holds for metric spaces. (Hint: For each positive integer n, let $\mathcal{U}_n$ be a countable cover of the space by open balls of radii $1/n$. Show that $\bigcup_{n \in \mathbb{N}} \mathcal{U}_n$ is a countable base for the metric topology.)

3.3 If $(X, \mathcal{T})$ is a second-countable space and $\mathcal{B}$ is a base for $\mathcal{T}$, prove that there exists a countable subfamily $\mathcal{D}$ of $\mathcal{B}$ such that $\mathcal{D}$ is a base for $\mathcal{T}$. (Hint: Let $\mathcal{C}$ be a countable base for $\mathcal{T}$. Using the argument in the proof of Theorem (3.4) express each member of $\mathcal{C}$ as the union of countably many members of $\mathcal{B}$). This result shows that not only has a second countable space a countable base, but a countable base can be extracted from any given base (countable or not) for it.

3.4 Prove that a space is second countable if and only if it has a countable sub-base.

3.5 Prove the equivalence of statements (2) and (3) in Proposition (3.6).

3.6 Let X_1, X_2 be the metric spaces and on $X_1 \times X_2$ put any of the metrics of Exercise (3.4) of Chapter 3. Prove that the metric topology in $X_1 \times X_2$ coincides with the product topology on it. Generalise to the product of n spaces.

3.7 Let $(X_1, \mathcal{T}_1), \ldots, (X_n, \mathcal{T}_n)$ be topological spaces and let $\mathcal{B}_1, \mathcal{B}_2, \ldots, \mathcal{B}_n$ be bases for them respectively. Then prove that the family of 'basic open boxes', i.e. the family $\{B \subset X : B = B_1 \times B_2 \times \ldots \times B_n$ for some $B_i \in \mathcal{B}_i$, $i = 1, 2, \ldots, n\}$ is a base for the product topology on the product $X = X_1 \times X_2 \times \ldots \times X_n$.

3.8 Prove that if each space $(X_i, \mathcal{T}_i)$ is second countable, for $i = 1, 2, \ldots, n$ then so is their topological product.

3.9 In the last section, while defining the order topology on a set X with a linear order $\leq$, we assumed that X had neither a least nor a greatest element. We can remove this restriction by using sub-bases as follows:

For $a \in X$ define $L_a = \{x \in X : x < a\}$ and $R_a = \{x \in X : a < x\}$. Let $\mathcal{S}_1 = \{L_a : a \in X\}$ and $\mathcal{S}_2 = \{R_a : a \in X\}$. Let $\mathcal{S} = \mathcal{S}_1 \cup \mathcal{S}_2$. Then the topology on X having $\mathcal{S}$ as a sub-base is called the **order topology on X.** Prove that in case X has no least and no greatest elements, this definition coincides with the earlier one.

3.10 Let a set X be linearly ordered by $\leq$. Prove that given $a, b \in X$, $a < b$, there exist open sets U, V (in the order topology) such that $a \in U$, $b \in V$ and every member of U is less than every member of V. Prove further that the order topology is the smallest topology on X with this property.

4. Subspaces

So far we considered a set X and various topologies on it. Suppose Y is a subset of X and a topology $\mathcal{T}$ on X is given. It is natural to inquire whether $\mathcal{T}$ induces a topology on Y and if so how the two topologies are related. The most natural thing to do would be to take up various members of $\mathcal{T}$ and to 'truncate' or to 'relativise' them w.r.t. Y, i.e., to take their intersections with Y. More formally let $\mathcal{U} = \{V \subset Y :$ there is some $U \in \mathcal{T}$ such that $V = U \cap Y\}$. It is trivial to verify that $\mathcal{U}$ is a topology on Y; each property required in the definition follows from the corresponding property of $\mathcal{T}$. The topology $\mathcal{U}$ is also denoted by $\mathcal{T}/Y$. It is called the **relative or the subspace topology** on Y induced by $\mathcal{T}$. The space $(Y, \mathcal{T}/Y)$ is called a **subspace** of the space $(X, \mathcal{T})$. $\mathcal{T}/Y$ is said to be obtained by **relativising** $\mathcal{T}$ to Y.

As an example, consider the real line $\mathbb{R}$ as a subset of the plane $\mathbb{R}^2$, by identifying a point $x \in \mathbb{R}$ with the point $(x, 0) \in \mathbb{R}^2$. Let $\mathcal{T}$ be the usual topology, induced by the usual euclidean metric on $\mathbb{R}^2$. We contend that $\mathcal{T}/\mathbb{R}$ is precisely the usual topology on $\mathbb{R}$. For, let $\mathcal{U}$ be the usual topology on $\mathbb{R}$. First we claim $\mathcal{U} \subset \mathcal{T}/\mathbb{R}$, for which it suffices to show that every element of a base for $\mathcal{U}$ is in $\mathcal{T}/\mathbb{R}$. A base for $\mathcal{U}$ can be taken to consist of all bounded open intervals. Each such interval will correspond to the set S of points on the x-axis in $\mathbb{R}^2$ between say $(a, 0)$ to $(b, 0)$, where $a, b \in \mathbb{R}$ and $a < b$. Now let D be the open disc with centre at $\left(\frac{a+b}{2}, 0\right)$ and radius $\frac{b-a}{2}$. D is open in $\mathbb{R}^2$ in usual topology. But clearly $S = D \cap \mathbb{R}$ and so S is open in the relative topology $\mathcal{T}/\mathbb{R}$ on $\mathbb{R}$. Hence $\mathcal{U} \subset \mathcal{T}/\mathbb{R}$. We leave the other-way inclusion to the reader.

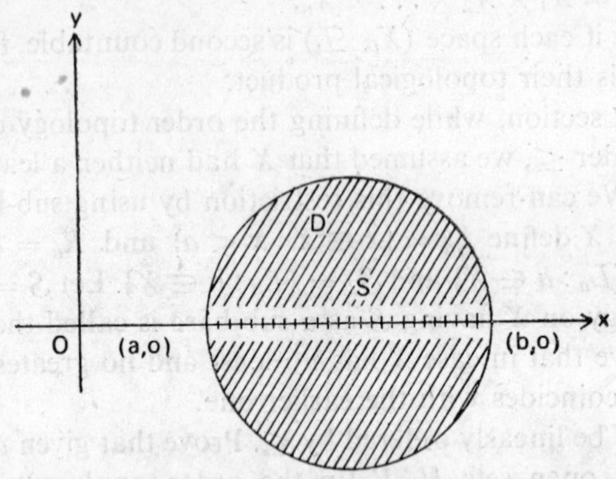

Fig. 2 Relativisation of the topology on $\mathbb{R}^2$ to $\mathbb{R}$

There is of course no restriction as to what the set Y can be. In particular Y itself may or may not be a member of $\mathcal{T}$. Members of $\mathcal{T}/Y$ are said to be open **relative to Y** or open **in** Y. Note that if $V \subset Y$ and V is open in the larger space $(X, \mathcal{T})$ then V is also open relative to Y, for we could write $V = V \cap Y$ and $V \in \mathcal{T}$. The converse is false, it may happen that V is open relative to Y but not open in X. Note for example, that Y is always open relative to itself, although it need not be open in X.

What happens, if we take a subset Z of Y and further relativise $\mathcal{T}/Y$ to Z? Certainly, we would expect that the outcome will be the same as of relativising $\mathcal{T}$ directly to Z in the first place. That this is actually the case is shown by the next proposition. The simple proof is left to the reader.

(4.1) Proposition: Let $Z \subset Y \subset X$, and $\mathcal{T}$ be a topology on X. Then $(\mathcal{T}/Y)/Z = \mathcal{T}/Z$. ∎

A natural question to ask would be how the properties of a space determine those of the subspace $(Y, \mathcal{T}/Y)$. It turns out that some properties of $(X, \mathcal{T})$ carry over smoothly to $(Y, \mathcal{T}/Y)$ while some others break down completely.

(4.2) Definition: A property of topological spaces is said to be **hereditary** if whenever a space has that property, then so does every subspace of it.

A trivial example of a hereditary property is the property of being either an indiscrete or a discrete space. As less trivial examples, metrisability and second countability are hereditary properties as we shall soon prove. We have not yet defined any properties which are not hereditary, but there will be many examples to come, e.g. compactness and connectedness.

(4.3) Proposition: Let $\mathcal{B}$ be a base for a topology $\mathcal{T}$ on a set X and let $Y \subset X$. Let $\mathcal{B}/Y = \{B \subset Y : B \in \mathcal{B}\}$. Then $\mathcal{B}/Y$ is a base for the topology $\mathcal{T}/Y$ on Y.

Proof: We use the characterisation of a base given in proposition (3.2). Let $y \in Y$ and G be an open set in Y containing y. Then $G = H \cap Y$ for some open set H in X. Clearly $y \in H$ and so there exists $B \in \mathcal{B}$ such that $y \in B$ and $B \subset H$. Then $y \in B \cap Y$, $B \cap Y \subset G$ and $B \cap Y \in \mathcal{B}/Y$. This proves that $\mathcal{B}/Y$ is a base for $\mathcal{T}/Y$. ∎

(4.4) Corollary: Second countability is a hereditary property.

Proof: Let $(X, \mathcal{T})$ be a space with a countables base $\mathcal{B}$ and $Y \subset X$. Then certainly $\mathcal{B}/Y$ as defined above is countable and is a base for $\mathcal{T}/Y$. ∎

(4.5) Theorem: Metrisability is a hereditary property.

Proof: Suppose $(X, \mathcal{T})$ is metrisable. Let d be a metric on X which induces the topology $\mathcal{T}$. Let $Y \subset X$ and e be the restriction of d to $Y \times Y$. It is immediate that e is a metric on Y. We claim that e induces the topology $\mathcal{T}/Y$ on Y. To avoid confusion we shall denote the open balls w.r.t. d and e with subscripts, $B_d(x, r)$, $B_e(y, r)$ etc. Note that if $y \in Y$ and $r > 0$ then $B_e(y, r) = B_d(y, r) \cap Y$ and consequently, $B_e(y, r) \in \mathcal{T}/Y$. Now let $\mathcal{U}$ be the metric topology on Y induced by e. We have to show that $\mathcal{U} = \mathcal{T}/Y$. Let $\mathcal{B}$ be the family of open balls (w.r.t. e) in Y. Then $\mathcal{B}$ is a base for $\mathcal{U}$ and as we just saw $\mathcal{B} \subset \mathcal{T}/Y$. So $\mathcal{U} \subset \mathcal{T}/Y$. Conversely let $G \in \mathcal{T}/Y$. Then there is some open set $H \subset X$ such that $G = H \cap Y$. For each $y \in G$, $y \in H$ and there exists $r_y > 0$ such that $B_d(y, r_y) \subset H$. Then $B_d(y, r_y) \cap Y \subset H \cap Y = G$, i.e. $B_e(y, r_y) \subset G$. Clearly then $G = \bigcup_{y \in G} B_e(y, r_y)$ and thus G is a union of open sets w.r.t. the metric topology on Y. Hence $G \in \mathcal{U}$. Thus $\mathcal{T}/Y \subset \mathcal{U}$. Therefore the two topologies $\mathcal{T}/Y$ and $\mathcal{U}$ are the same, or in other words $(Y, \mathcal{T}/Y)$ is metrisable. ∎

Note that the proof above shows more than the theorem claims. It not only shows that a subspace of a metrisable space is metrisable but also gives a construction for a metric on the subspace in terms of a given metric on the original space. This is not an uncommon phenomenon in mathematics, the theorem merely says that something exists but the proof actually gives an explicit construction for it. Sometimes it is not just the theorem but rather its proof that is needed in an application.

Combining Corollary (4.4) above and Theorem (3.4) we get the following theorem due to Lindelöff.

(4.6) Theorem: If $(X; \mathcal{T})$ is second countable and $Y \subset X$ then any cover of Y by members of $\mathcal{T}$ has a countable cover.

Proof: Let $\mathcal{U}$ be a cover of Y by members of $\mathcal{T}$. Let $\mathcal{V} = \{U \cap Y : U \in \mathcal{U}\}$. Then $\mathcal{V}$ is an open cover of Y w.r.t. the relative topology $\mathcal{T}/Y$. Now $(Y, \mathcal{T}/Y)$ is second countable and so by Theorem (3.4) $\mathcal{V}$ has a countable subcover $\mathcal{C}$. For each $C \in \mathcal{C}$ fix $D(C) \in \mathcal{U}$ such that $C = D(C) \cap Y$. Then the family $\{D(C) : C \in \mathcal{C}\}$ is countable sub-family of $\mathcal{U}$ and covers Y. ∎

One of the main applications of subspaces is in the construction of counter-examples. Where hereditary properties are involved, subspaces are not likely to reveal anything unexpected. But when we shall study some non-hereditary properties, especially connectedness, we shall see how strange subspaces can be. Indeed even the most familiar spaces such as the real line or the plane (both with the usual topologies) contain subspaces with extremely weird properties.

Exercises

4.1 Let $\mathcal{T}_1$, $\mathcal{T}_2$ be topologies on a set X and $Y \subset X$. Prove that if $\mathcal{T}_1$ is stronger than $\mathcal{T}_2$ then $\mathcal{T}_1/Y$ is stronger than $\mathcal{T}_2/Y$.

4.2 Give an example of two distinct topologies on a set which relativise to the same topology on some subset. (A trivial example is always possible if the subset is either empty or a singleton. Give a less trivial example.)

4.3 Give an example of two distinct topologies $\mathcal{T}_1$, $\mathcal{T}_2$ on a set X and a subset Y such that $\mathcal{T}_1/Y = \mathcal{T}_2/Y$ and $\mathcal{T}_1/(X - Y) = \mathcal{T}_2/(X - Y)$. (This exercise shows that a topology on a set is not determined uniquely by its relativisations on two complementary subsets. What matters is not only how each subset looks individually in the relative topology but also how the two are fitted together.)

4.4 Suppose $(X, \mathcal{T})$ is a space and $Y \in \mathcal{T}$. Prove that a subset B of Y is open in Y if and only if it is open in X. (One way implication is always true, whether Y is open in X or not. The other way implication can be tersely put as 'open in open is open'.)

4.5 Let a set X be linearly ordered by $\leq$ and $\mathcal{T}$ be the order topology induced by $\leq$ on X. Let $Y \subset X$. Then Y is also linearly ordered by the restriction of $\leq$. Prove that the order topology on Y is weaker than $\mathcal{T}/Y$ and give an example where the two are unequal. (This is in sharp contrast with Theorem (4.5). In general when a topology on a set is induced by some other structure, it is tempting to think that its relativisation to a subset is the same as the topology induced by the restriction of the original structure to that subset. This works well for a metric but fails for an order.)

4.6 Let $\mathcal{T}$ be the semi-open interval topology on $\mathbb{R}$ and let $\mathcal{U}$ be the product topology on $\mathbb{R} \times \mathbb{R}$. Let $L = \{(x, y) \in \mathbb{R}^2 ; x + y = 0\}$. Prove that $\mathcal{U}/L$ is the discrete topology on L. (Hint: Draw a diagram. It is then easy to show that for $(x, y) \in L$, there exists an open box G in $\mathcal{U}$ such that $\{(x, y)\} = L \cap G$.)

4.7 Prove that a discrete space is second countable iff the underlying set is countable.

4.8 Prove that the real line with the semi-open interval topology is not second countable. (Hint: Argue by contradiction, using Exercise (3.8), Corollary (4.4) and the last two exercises.)

4.9 Let $(X_i, \mathcal{T}_i)$, $i = 1, 2, \ldots, n$ be topological spaces and suppose $Y_i \subset X_i$ for each i. Let $X = X_1 \times X_2 \times \ldots \times X_n$ and $Y = Y_1 \times \ldots \times Y_n$. Clearly Y is a subset of X. Prove that the product topology on Y (obtained by putting the relative topology $\mathcal{T}_i/Y_i$ on each Y_i) is the same as the relativisation of the product topology on X. In other words, it does not matter whether we first relativise and then form the product or first take the product and then relativise.

Chapter Five

Basic Concepts

In this chapter we define the building blocks for a systematic study of topological spaces. They include the concepts of closure, which is a refined form of convergence, interior and boundary, which are geometrical in spirit, continuity which is a way of going from one space to another and finally, homeomorphism which is the basic equivalence relation by which objects are classified in topology. We also mention some general problems in topology and define quotient spaces as the solution to one of them.

1. Closed Sets and Closure

(1.1) Definition: Let $(X, \mathfrak{J})$ be a topological space. Then a subset A of X is said to be **closed** in X if its complement $X - A$ is open in X.

The definition is fairly straightforward and one can cite as many examples of closed sets as of open sets. It is fortunate that all closed intervals (bounded or not) of real numbers are indeed closed in the usual topology on the real line. If (X, d) is a metric space, $x \in X$ and $r > 0$, then the **closed ball** with centre x and radius r is defined as the set $\{y \in X : d(x, y) \leq r\}$. We leave it to the reader to verify that each such closed ball is a closed subset in the topology induced by the metric.

A word of warning is perhaps in order. In analogy with everyday usage, a biginner is likely to think that 'closed' is the negation of 'open', that is to say, a set is closed if and only if it is not open. But this is not so. The reason for the misleading terminology is probably that complements of sets are defined in terms of negation. The fact is that the possibilities of a set being open and its being closed are neither mutually exclusive nor exhaustive. Note for example that the empty set and the whole set are always open as well as closed in every space. On the other hand, the set of rationals is neither open nor closed in the usual topology on the real line. A set which is both open and closed is sometimes called a **clopen** set.

It is immediate that a set is open iff its complement is closed. As a result, any statement about open sets can be immediately translated into a corresponding statement about closed sets and vice-versa, as we do in the following theorem.

(1.2) Theorem: Let $\mathcal{C}$ be the family of all closed sets in a topological space $(X, \mathfrak{J})$. Then $\mathcal{C}$ has the following properties:

(i) $\phi \in \mathcal{C}, X \in \mathcal{C}$.

(ii) $\mathcal{C}$ is closed under arbitrary intersections.

(iii) $\mathcal{C}$ is closed under finite unions.

Conversely, given any set X and a family $\mathcal{C}$ of its subsets which satisfies these three properties, there exists a unique topology $\mathfrak{J}$ on X such that $\mathcal{C}$ coincides with the family of closed subsets of $(X, \mathfrak{J})$.

Proof: The first part follows trivially from the definition of a topology and De Morgan's laws. The converse part is equally trivial once it is clearly understood what it says. Here we are given a set X (just a bare set with no topology on it) and some collection $\mathcal{C}$ of its subsets. We are given that properties (i) to (iii) hold for $\mathcal{C}$. We do not know how $\mathcal{C}$ originated, nor do we know whether its members are closed subsets of X. Actually it is meaningless to talk about closed subsets of X unless a topology on X is specified. The theorem says that given such a family $\mathcal{C} \subset P(X)$ we can define a suitable topology $\mathfrak{J}$ on X such that members of $\mathcal{C}$ are precisely the closed subsets of X (w.r.t. the topology $\mathfrak{J}$), and that such a topology is unique.

Having understood what the theorem says, the proof itself is trivial as we have no choice but to let $\mathfrak{J}$ consist of complements (in X) of members of $\mathcal{C}$, i.e. $\mathfrak{J} = \{B \subset X : X - B \in \mathcal{C}\}$. That $\mathfrak{J}$ is a topology on X follows by applying De Morgan's laws. The open subsets of X are precisely the complements of members of $\mathcal{C}$, and hence the closed subsets of X are precisely the members of $\mathcal{C}$ as asserted. Also this condition determines $\mathfrak{J}$ uniquely. ∎

Trivial as the theorem is, its significance is noteworthy. In the definition of a topological space we took 'open set' as a primitive term, that is to say, open sets are not defined (except as members of the topology on the set in question) and nothing is known about their nature save what is implied by the definition of a topology. Everything we do with topological spaces is in terms of open sets. For example, we defined convergence of sequences in a topological space in terms of open sets, and we defined closed sets as complements of open sets. The preceding theorem asserts that this procedure could be reversed. That is, we could as well take 'closed sets' as a primitive concept and then define open sets as complements of closed sets. With this approach our definition of a topological space would be that it is a pair $(X, \mathcal{C})$ where X is a set, $\mathcal{C} \subset P(X)$ and conditions (i), (ii), (iii) above are satisfied. Although nothing is to be gained and nothing is to be lost by adopting this new approach over the usual one, in particular examples of topological spaces it may be more natural to specify the closed sets rather than the open sets. For instance, in the cofinite topology on a set X, it is so easy to tell what the closed subsets are, they are precisely all finite subsets of X and the set X itself.

Any subset of a topological space generates a closed subset called its closure. The definition is as follows:

(1.3) Definition: The **closure** of a subset of a topological space is defined as the intersection of all closed subsets containing it.

In symbols, if A is a subset of a space $(X, \mathfrak{J})$, then its closure is the set

$\bigcap \{C \subset X : C$ closed in X, $C \supset A\}$. It is denoted by $\bar{A}$. Obviously it depends on the topology $\mathfrak{J}$ and when it is important to stress this, it is customary to write $\bar{A}^{\mathfrak{J}}$ or $(\bar{A})_{\mathfrak{J}}$ instead of mere $\bar{A}$. Note further that if $Y \subset X$ and $A \subset Y$ then the closure of A in the space $(X, \mathfrak{J})$ is in general different from its closure in the subspace $(Y, \mathfrak{J}/Y)$. We leave it to the reader to verify that the latter is the intersection of the former with Y. When confusion is likely to arise otherwise, it is usual to write $\bar{A}^Y$ or $(\bar{A})_Y$ to indicate the subspace w.r.t. which the closure is intended. The notations $Cl(A)$ or $C(A)$ or $c(A)$ are also used sometimes to denote the closure. In the next proposition we list down a few properties of closures.

(1.4) Proposition: Let A, B be subsets of a topological space $(X, \mathfrak{J})$.

(i) $\bar{A}$ is a closed subset of X. Moreover it is the smallest closed subset of X containing A i.e. if C is closed in X and $A \subset C$ then $\bar{A} \subset C$.

(ii) $\bar{\phi} = \phi$

(iii) A is closed in X iff $\bar{A} = A$

(iv) $\bar{\bar{A}} = \bar{A}$ or in other words, $c(c(A)) = c(A)$

(v) $\overline{A \cup B} = \bar{A} \cup \bar{B}$.

Proof: (i) and (ii) are immediate consequences of the definition and properties of closed sets. For (iii) we note that if A is closed then it is clearly the smallest closed set containing A and consequently $\bar{A} = A$. Conversely if $\bar{A} = A$ then A is closed since $\bar{A}$ is always a closed set, being the intersection of closed sets. Property (iv) follows by applying (iii) to $\bar{A}$ which is known to be closed. Finally, for (v), note that $\bar{A} \cup \bar{B}$ is first of all a closed set containing $A \cup B$; as $A \subset \bar{A}$ and $B \subset \bar{B}$, and hence $\overline{A \cup B} \subset \bar{A} \cup \bar{B}$. For the other way inclusion, we first observe that whenever $A_1 \subset A_2$, $\bar{A}_1 \subset \bar{A}_2$ (prove !). Now $A \cup B$ contains A as well as B and so $\bar{A}$, $\bar{B}$ are both subsets of $\overline{A \cup B}$. Hence $\bar{A} \cup \bar{B} \subset \overline{A \cup B}$. This completes the proof. ∎

It is instructive to reformulate some of the properties in the last proposition using the terminology of operators. An operator is just another name for a function, except that the term is generally reserved for those functions whose domains are sets of sets or of functions. For example, if $(X, \mathfrak{J})$ is a topological space, then the **closure operator** associated with it is defined as the function $c : P(X) \to P(X)$ such that $c(A) = \bar{A}$ for each $A \in P(X)$. In terms of the closure operator, (ii) to (v) assume the following forms respectively:

(ii)′ ϕ is a **fixed point** of c, i.e. $c(\phi) = \phi$.

(iii)′ The fixed points of c are precisely the closed subsets of X.

(iv)′ c is **idempotent,** i.e. $c \circ c = c$ or in other words for any $A \in P(X)$, $c(c(A)) = c(A)$.

(v)′ c commutes with finite unions.

Note that (ii)′, (iv)′ and (v)′ make no explicit reference to the particular topology $\mathfrak{J}$ on X, while condition (iii)′ shows how one can recover the topology $\mathfrak{J}$ on X if one is given the closure operator induced by it. In particular it follows that two distinct topologies on the same set cannot induce the same closure operator.

Now comes an important question. Suppose we have an abstract operator $\theta : P(X) \to P(X)$. Under what conditions can we find a topology on X whose closure operator will coincide with the given operator θ? Certainly the answer will be in terms of some properties of θ. Worded differently the same question would be 'which properties characterise a closure operator ?' Let us for instance suppose θ satisfies (ii)′, (iv)′ and (v)′. Does it follow that θ is necessarily a closure operator induced by some topology on X? The answer is no as can be seen if we take θ to be the operator which takes every subset of X to the empty set. The only fixed point of θ is the empty set and unless $X = \emptyset$, (iii)′ would be violated. In other words although every closure operator satisfies (ii)′, (iv)′ and (v)′, these three properties do not together characterise a closure operator. It turns out that by adding one more condition one can characterise closure operators as the following theorem shows:

(1.5) Theorem: Let X be a set, $\theta : P(X) \to P(X)$ a function such that

(1) for every $A \in P(X)$, $A \subset \theta(A)$ (this condition is sometimes expressed by saying that θ is an **expansive** operator),
(2) ϕ is a fixed point of θ,
(3) θ is idempotent, and
(4) θ commutes with finite unions.

Then there exists a unique topology $\mathcal{J}$ on X such that θ coincides with the closure operator associated with $\mathcal{J}$. Conversely, any closure operator satisfies these properties.

Proof: The converse part is already established. For the direct implication, suppose $\theta : P(X) \to P(X)$ satisfies (1) to (4). We want to find a topology $\mathcal{J}$ on X such that for every $A \subset X$, $\theta(A) = \bar{A}^{\mathcal{J}}$. If at all such a topology exists then its closed subsets must be precisely the fixed points of θ as we saw above. This gives us a clue to the construction of $\mathcal{J}$. We let $\mathcal{C} = \{A \subset X : \theta(A) = A\}$ and contend that $\mathcal{C}$ has properties (i) to (iii) of Theorem (1.2). Condition (2) shows that $\phi \in \mathcal{C}$ while condition (4) implies that $\mathcal{C}$ is closed under finite unions. To prove that $X \in \mathcal{C}$, we merely note that by (1), $X \subset \theta(X)$ and hence $X = \theta(X)$ since $\theta(X) \subset X$ anyway. It only remains to verify that $\mathcal{C}$ is closed under arbitrary intersections. For this we first note that θ is monotonic, i.e., whenever $A \subset B$, $\theta(A) \subset \theta(B)$, which follows by writing B as $A \cup (B - A)$ and applying (4). Now let $A = \bigcap_{i \in I} A_i$ where I is an index set and $A_i \in \mathcal{C}$ for each $i \in I$. We want to show that $A \in \mathcal{C}$, i.e. $\theta(A) = A$. By (1) we already know $A \subset \theta(A)$. Also $\theta(A) \subset \theta(A_i)$ for each $i \in I$ since θ is monotonic, and so $\theta(A) \subset \bigcap_{i \in I} \theta(A_i)$. But $\theta(A_i) = A_i$ since $A_i \in \mathcal{C}$ for all $i \in I$. Consequently, $\theta(A) \subset A$ and hence $\theta(A) = A$ as desired. So by theorem (1.2), the family $\mathcal{J}$ of complements of members of $\mathcal{C}$ is a topology on X.

It remains to be verified that the closure operator associated with $\mathcal{J}$ coincides with θ. Let $A \subset X$. Then $\bar{A}^{\mathcal{J}}$ (i.e. $\bar{A}$ w.r.t. $\mathcal{J}$) is the intersection of all closed subsets of X containing A. But by very construction, closed subsets of X are precisely the fixed points of θ. Hence $\bar{A} = \bigcap \{B \subset X : A \subset B;$

$\theta(B) = B\}$. Now, whenever $B \supset A$, $\theta(B) \supset \theta(A)$ by monotonicity of θ. So if $B \supset A$ and $\theta(B) = B$ then $B \supset \theta(A)$. But $\bar{A}$ is the intersection of such B's and so $\bar{A} \supset \theta(A)$. For the other way inclusion we note that by condition (3), $\theta(A) \in \mathcal{C}$ while by (1) $A \subset \theta(A)$ whence $\bar{A} \subset \theta(A)$, $\bar{A}$ being the smallest member of $\mathcal{C}$ containing A. Hence for all $A \subset X$, $\theta(A) = \bar{A}$ completing the proof. ∎

The spirit and the significance of the preceding theorem are similar to those of Theorem (1.2). It establishes a one-to-one correspondence between the set of all topologies on a set X and the set of all operators $\theta : P(X) \rightarrow P(X)$ satisfying (1) to (4). It is for this reason that the properties (1) to (4) are called **closure axioms** and the preceding theorem an **axiomatic characterisation of closure operators.**

(1.6) Definition: Let A be a subset of a space X. The A is said to be **dense** in X if $\bar{A} = X$.

Trivially, the entire set X is always dense in itself. Before giving other examples, let us obtain a succinct characterisation of denseness.

(1.7) Proposition: A subset A of a space X is dense in X iff for every non-empty open subset B of X, $A \cap B \neq \emptyset$.

Proof: Suppose A is dense in X and B is a non-empty open set in X. If $A \cap B = \emptyset$ then $A \subset X - B$ whence $\bar{A} \subset X - B$ since $X - B$ is closed. But then $X - B \subsetneq X$ contradicting that $\bar{A} = X$. So $A \cap B \neq \emptyset$. Conversely assume that A meets every non-empty open subset of X. This clearly means that the only closed set containing A is X and consequently $\bar{A} = X$. ∎

It is now easy to show that in the real line with the usual topology the set $\mathbb{Q}$ of all rational numbers, as well as its complement $\mathbb{R} - \mathbb{Q}$ are both dense. Actually, they are also dense subsets w.r.t. the semi-open interval topology, which is a stronger statement. (Why ?)

In this section we defined the closure of a set as the closed set generated by it, i.e. the intersection of all closed sets containing it. This definition is of little value in actually finding the closure of a set as it is often impracticable to consider all closed sets containing a given set. We therefore need a more concrete description of the closure. A similar situation arose in connection with the topology generated by a family of subsets of a set. The mere existence of such a topology is of little value unless we know a way of describing its members in terms of the members of the original family. In the present case, given a subset A of a space X and *a* point $x_0 \in X$ we would like to know precisely when $x_0 \in \bar{A}$. An answer will be provided in the next section.

Exercises

1.1 Let $(X, \mathfrak{J})$ be a topological space and $\mathcal{C}$ be the family of all closed subsets of X. Prove or disprove that $\mathcal{C}$ is the complement of $\mathfrak{J}$ in $P(X)$.

1.2 Let $\mathfrak{J}_1$, $\mathfrak{J}_2$ be topologies on a set X and $A \subset X$. If $\mathfrak{J}_1$ is weaker than $\mathfrak{J}_2$ how are $\bar{A}^{\mathfrak{J}_1}$ and $\bar{A}^{\mathfrak{J}_2}$ related to each other?

1.3 If X is a space and $Y \subset X$ prove that the closed subsets of Y (in the relative topology) are precisely the intersections of closed subsets of X with Y. If $B \subset Y$, prove that whenever B is closed in X it is also closed in Y and that the converse holds if Y is closed in X.

1.4 If X is a space, $Y \subset X$ and $A \subset Y$, prove that $\bar{A}^Y = \bar{A}^X \cap Y$.

1.5 Let X be a set. Define $\theta_1, \theta_2, \theta_3 : P(X) \to P(X)$ by

$$\theta_1(A) = A \text{ for all } A \in P(X)$$

$$\theta_2(A) = \begin{cases} \phi & \text{if } A = \phi \\ X & \text{otherwise} \end{cases}$$

$$\theta_3(A) = \begin{cases} A & \text{if } A \text{ is finite} \\ X & \text{otherwise} \end{cases}$$

Verify that each of $\theta_1, \theta_2, \theta_3$ satisfies the closure axioms. Also in each case find the corresponding topology on X.

1.6 Let $\mathcal{B}$ be a base for a topology $\mathfrak{J}$ on a set X. Prove that a subset $D \subset X$ is dense in X iff D intersects every non-empty member of $\mathcal{B}$. What happens if $\mathcal{B}$ is known to be merely a sub-base of $\mathfrak{J}$?

1.7 Prove that a second countable space always contains a countable dense subset. (Hint: Pick one element each from non-empty members of a countable base.)

1.8 Let D be a dense subset of a space X and let $Y \subset X$. Show by an example that $D \cap Y$ need not be dense in Y (w.r.t. the relative topology on Y). Prove, however, that this is the case if Y is open in X.

1.9 Find out the dense subsets of discrete, indiscrete and cofinite spaces.

1.10 Let $\{(X_i, \mathfrak{J}_i) : i = 1, 2, \ldots, n\}$ be a family of topological spaces and $(X, \mathfrak{J})$ their topological product. For $i = 1, 2, \ldots n$ let D_i be a dense subset of X_i and let $D = D_1 \times D_2 \times \ldots \times D_n$. Prove that D is dense in X.

1.11 Prove that in a metrisable space, the converse of the result in Exercise (1.7) above holds. (Hint: Prove that the family of all open balls of rational radii around points in a dense set is a base for the metric topology.)

1.12 Prove that the real line with the semi-open interval topology is not metrisable. (Hint: Use the last exercise and exercise (4.8) of the last chapter).

Notes and Guide to Literature

The definition of a topological space through closure operators is due to Kuratowski. In the next section we shall relate closure with the intuitive idea of closeness. The fact that the set of rationals is dense in the usual topology on the real line is known (although not exactly in this terminology) for a long time. Intuitively, a dense subset is one whose points are found all over the space. A more appealing formulation will be given in the next section.

2. Neighbourhoods, Interior and Accumulation Points

(2.1) Definition: Let $(X; \mathfrak{J})$ be a topological space, $x_0 \in X$ and $N \subset X$. Then N is said to be a **neighbourhood** of x_0 or x_0 is said to be an **interior point** of N (each w.r.t. $\mathfrak{J}$) if there is an open set V such that $x_0 \in V$ and $V \subset N$.

Note that the set N itself is not required to be open. We warn the reader that some authors do require this to be so. As we shall have occasion to consider neighbourhoods which are not open, we shall put no such restriction. The word 'neighbourhood' is sometimes abbreviated to 'nbd'. In terms of this new concept, we have the following trivial but useful characterisation of open sets.

(2.2) Proposition: A subset of a topological space is open iff it is a neighbourhood of each of its points.

Proof: Let X be a topological space and $G \subset X$. First suppose G is open. Then evidently G is a nbd of each of its points. Conversely suppose G is a nbd of each of its points. Then for each $x \in G$, there is an open set V_x such that $x \in V_x$ and $V_x \subset G$. Clearly then, $G = \bigcup_{x \in G} V_x$. Since each V_x is open so is G. ∎

Trivially if N is a *nbd* of a point x then so is any superset of N. It is also easy to show that the intersection of any two (and hence finitely many) neighbourhoods of a point is again a neighbourhood of that point.

(2.3) Definition: Let $(X, \mathfrak{J})$ be a space and $A \subset X$. Then the **interior** (or more precisely the **$\mathfrak{J}$-interior**) of A is defined to be the set of all interior points of A, i.e. the set $\{x \in A : A \text{ is a nbd of } x\}$. It is denoted by A^0 or int (A), or $\text{int}_{\mathfrak{J}}(A)$ when we want to emphasise its dependence upon $\mathfrak{J}$. As an example, the interior of a closed disc in the plane is the corresponding open disc. The set of rationals has empty interior w.r.t. the usual topology on $\mathbb{R}$, and so does its complement, viz., the set of irrational numbers. The interior of the empty set is empty.

(2.4) Proposition: Let X be a space and $A \subset X$. Then int (A) is the union of all open sets contained in A. It is also the largest open subset of X contained in A.

Proof: Let $\mathcal{U}$ be the family of all open sets contained in A ($\mathcal{U}$ is non empty since $\phi \in \mathcal{U}$). Let $V = \bigcup_{G \in \mathcal{U}} G$. We have to show $V = \text{int}(A)$. Now if $x \in V$ then $x \in G$, for some $G \in \mathcal{U}$. This means A is nbd of x and so $x \in \text{int}\, A$. Conversely, let $x \in \text{int}\,(A)$. Then there is an open set H such that $x \in H$ and $H \subset A$. But then, $H \in \mathcal{U}$ and so $H \subset V$. So $x \in V$. This proves the first assertion of the proposition, and also shows that int (A) is an open set contained in A. To see it is the largest such set, suppose G is an open set contained in A. Then $G \in \mathcal{U}$ and so $G \subset \text{int}(A)$ by the first assertion. ∎

A topology $\mathfrak{J}$ on a set X induces an operator $i : P(X) \to P(X)$ defined by $i(A) = \text{int}(A)$. It is called the **interior operator** associated with $\mathfrak{J}$. Its properties are dual to those of the closure operator. The interior operator determines $\mathfrak{J}$ uniquely, for it is clear that a set is open iff it coincides with its interior. Axiomatic characterisation of interior operators is left to the reader as an exercise.

The **exterior** of a set is defined as the interior of its complement. It is not hard to show that it always coincides with the complement of the closure of the original set. Consequently, it is not an independent concept and does not appear often in topology. It is of some theoretical importance in that, closure could have been defined in terms of it. The sentences 'x is (or lies) inside (or within) A' and 'x is (or lies) outside A' are sometimes used to express that a point x belongs to the interior (respectively exterior) of a set A. They are not particularly recommended as they are likely to be confused with their set-theoretic usage (page 28). The two usages actually coincide if we put the discrete topology on the set in question.

Now let X be a space and $x \in X$. Let $\mathcal{N}_x$ be the set of all neighbourhoods of x in X (w.r.t. the given topology on X). The family $\mathcal{N}_x$ is called the **neighbourhood system** at x. In the following proposition we list some properties of neighbourhood systems.

(2.5) Proposition: Let X be a space and for $x \in X$, let $\mathcal{N}_x$ be the neighbourhood system at x. Then,

(i) If $U \in \mathcal{N}_x$ then $x \in U$.

(ii) For any $U, V \in \mathcal{N}_x$, $U \cap V \in \mathcal{N}_x$.

(iii) If $V \in \mathcal{N}_x$ and $U \supset V$ then $U \in \mathcal{N}_x$.

(iv) A set G is open in X iff $G \in \mathcal{N}_x$ for all $x \in G$.

(v) If $U \in \mathcal{N}_x$ then there exists $V \in \mathcal{N}_x$ such that $V \subset U$ and $V \in \mathcal{N}_y$ for $y \in V$.

Proof: Properties (i) to (iv) are either trivial or already established. In view of (iv), (v) merely says that every nbd of a point contains an open nbd of that point. That this is actually so follows from the definition of a neighbourhood. ∎

The utter triviality of the last proposition suggests that it is probably not worth writing down unless there is some other reason for it. There is indeed a good reason for it, and as the reader might have guessed, it is that the properties above characterize neighbourhood systems as we see in the following.

(2.6) Theorem: Let X be a set and suppose for each $x \in X$, a non-empty family $\mathcal{N}_x$ of subsets of X is given satisfying (i), (ii), (iii) and (v) in the proposition above. Then there is a unique topology $\mathfrak{J}$ on X such that for each $x \in X$, $\mathcal{N}_x$ coincides with the family of all neighbourhoods of x w.r.t. $\mathfrak{J}$.

Proof: If at all such a topology exists, then property (iv) gives us a clue for its construction. We let $\mathfrak{J} = \{U \subset X : U \in \mathcal{N}_x \text{ for all } x \in U\}$ and claim $\mathfrak{J}$ is a topology on X. Clearly $\emptyset \in \mathfrak{J}$. To show that $X \in \mathfrak{J}$ note that for any $x \in X$, $X \in \mathcal{N}_x$ by (iii) as X is a superset of any member of $\mathcal{N}_x$. Property (ii) shows that $\mathfrak{J}$ is closed under finite intersections while using (iii) it follows easily that $\mathfrak{J}$ is closed under arbitrary unions. So $\mathfrak{J}$ is a topology for X. Note that so far we did not use (i) and (v). With our definition of $\mathfrak{J}$, (v) means that for any $x \in X$ and $U \in \mathcal{N}_x$, there exists an open set V (i.e. a member of $\mathfrak{J}$) such that $V \in \mathcal{N}_x$ and $V \subset U$. From (i) we now get that $x \in V$. Thus U contains a member of $\mathfrak{J}$ containing x and is therefore a $\mathfrak{J}$-neighbourhood of the point x. Hence every member of $\mathcal{N}_x$ is neighbourhood of x w.r.t. $\mathfrak{J}$. Conversely let U be a neighbourhood of x w.r.t. $\mathfrak{J}$. Then U contains a member V of $\mathfrak{J}$ such that $x \in V$ and $V \subset U$. But $V \in \mathcal{N}_x$ by the definition of $\mathfrak{J}$ and so by (iii) $U \in \mathcal{N}_x$. Thus neighbourhoods of x w.r.t. $\mathfrak{J}$ are precisely the members of $\mathcal{N}_x$ for each $x \in X$ and this completes the proof. ∎

The significance of this theorem should be obvious, as we have proved several theorems of the same spirit. It asserts that we could have equally well defined a topological space as a pair $(X, \mathcal{N})$ where X is a set, $\mathcal{N}$ is a function from X to the set $P(P(X))$ such that, conditions (i), (ii), (iii) and (v) (along with $\mathcal{N}_x \neq \emptyset$ for all $x \in X$) hold, where $\mathcal{N}_x = \mathcal{N}(x)$. This is yet another approach to topological spaces. It is called 'topology through neighbourhoods'. In this approach, 'neighbourhood' is a primitive term and open sets (and everything else) is defined in terms of neighbourhood. This approach has the advantage that the idea of a neighbourhood is probably more intuitive, inasmuch as it is akin to the concept of nearness. In some examples, it is a little easier to specify the topology by means of neighbourhoods. For example consider the space of Example 10 in Chapter 4, Section 2. Here the set X is $\mathbb{N} \times \mathbb{N} \cup \{\infty\}$. For any $(x, y) \in \mathbb{N} \times \mathbb{N}$, $\{(x, y)\}$ (and hence any subset containing (x, y)) is a nbd of (x, y). The neighbourhoods of ∞ are those subsets of X, which contain ∞ and contain almost all points on almost all rows. This describes the desired topology unambiguously.

Let us now turn to an important concept, that of an accumulation point. Before defining it formally, let us recall what it means in a metric space. Let A be a subset of a metric space $(X; d)$ and let $y \in X$. Then y is said to be an accumulation point (or a limit point) of A if given any $\epsilon > 0$, there exists $x \in A$, $x \neq y$ such that $d(x, y) < \epsilon$. Equivalently every open ball around y must contain at least one point of A, other than y itself (in case y happens to be in A). Since any open set containing y must contain an open ball around y, it follows that y is a limit point of A iff every open set containing y contains at least one element of A other than y. Thus we have shown that the definition of an accumulation point can be rephrased in terms of open sets, without involving the metric d directly. This is exactly what we need in order to extend the definition from metric spaces to arbitrary topological spaces.

(2.7) Definition: Let A be a subset of a topological space X and $y \in X$. Then y is said to be an **accumulation point** of A if every open set containing y contains at least one point of A other than y.

As examples we note that in a discrete space no point is an accumulation point of any set while at the other extreme, in an indiscrete space, a point y is an accumulation point of any set A provided only that A contains at least one point besides y. In the usual topology on the real line, every real number is an accumulation point of the set of rational numbers while the set of integers has no point of accumulation.

For applications ti is convenient to reformulate the definition above slightly by saying that every neighbourhood of y contains at least one point of A other than y. The equivalence of the two versions follows trivially from the definition of a neighbourhood.

(2.8) Definition: Let A be a subset of a space X. Then the **derived set** of A, denoted by A', is the set of all accumulation points of A in X.

Obviously A' depends not only on A but also on the topology under consideration. Properties of derived sets are elementary and left to the reader. With the derived set of a set, we are now in a position to describe the closure of a set more closely.

(2.9) Theorem: For a subset A of a space X, $\bar{A} = A \cup A'$.

Proof: First we claim that $A \cup A'$ is closed or that $X - (A \cup A')$ is open. We do so by showing that $X - (A \cup A')$ is a nbd of each of its points. Let $y \in X - (A \cup A')$. Then since y is not a point of accumulation of A, there exists an open set V containing y such that V contains no point of A except possibly y. But $y \notin A$, so we have $A \cap V = \emptyset$. We claim $A' \cap V$ is also empty. For, let $z \in A' \cap V$. Then V is an open set containing z which is an accumulation point of A. So $V \cap A$ is nonempty, a contradiction. So $A' \cap V = \emptyset$ and hence $V \subset X - (A \cup A')$. This proves that $A \cup A'$ is closed and since it obviously contains A, it also contains $\bar{A}$; i.e. $\bar{A} \subset A \cup A'$.

For the other way inclusion, $A \cup A' \subset \bar{A}$, it suffices to show that $A' \subset \bar{A}$ since we already have $A \subset \bar{A}$. So let $y \in A'$. If $y \notin \bar{A}$ then $y \in X - \bar{A}$ which is an open set since $\bar{A}$ is always a closed set. But y is an accumulation point of A. So $(X - \bar{A}) \cap A \neq \emptyset$ which is a contradiction since $X - \bar{A} \subset X - A$. So $y \in \bar{A}$. This completes the proof. ∎

(2.10) Theorem: For a subset A of a space X, $\bar{A} = \{y \in X :$ every nbd of y meets A non-vacuously$\}$.

Proof: Let $B = \{y \in X : U \in \mathcal{N}_y \Rightarrow U \cap A \neq \emptyset\}$. We have to show $\bar{A} = B$. By the theorem above, this amounts to showing that $A \cup A' = B$. First let $y \in A \cup A'$. If $y \in A$ then certainly every nbd of y meets A at least at the point y and so $y \in B$. If $y \in A'$ then too, by the definition of an accumulation point, every nbd of y contains a point of A and so $y \in B$. Thus $A \cup A' \subset B$. Conversely let $y \in B$. If $y \notin A \cup A'$, then $y \notin \bar{A}$ and so $X - \bar{A}$ is a nbd of y which does not meet A, contradicting that $y \in B$. So

$B \subset A \cup A'$. Thus $B = \bar{A}$. ∎

The preceding theorem tells us how to test whether a particular point is in the closure of a set even though we may not know the full closure. In some sense, it says that $\bar{A}$ consists of all points which are near to A. Indeed, suppose $(X; d)$ is a metric space and A is a non-empty subset of X. For $y \in X$ we define $d(y, A) = \inf \{d(x, y) : x \in A\}$. Intuitively, $d(y, A)$ is the 'shortest distance' of the point y from the set A. It is then easy to show that $\bar{A}$ is precisely the set $\{y \in X : d(y, A) = 0\}$. Even when X is not a metric space, the idea of nearness can be axiomatised as follows:

(2.11) Definition: Let X be a set. *A* **nearness relation** on X is a subset N of $X \times P(X)$ which satisfies the following properties (for $y \in X$ and $A \in P(X)$ we write $y \,\bar{\delta}\, A$ to mean $(y, A) \in N$ and $y \,\bar{\delta}\, A$ to mean $(y, A) \notin N$; '$y \,\delta\, A$' is read as 'y is near A'):

(i) $y \,\bar{\delta}\, \emptyset$ for all $y \in X$

(ii) $y \in A \Rightarrow y \,\delta\, A$ for all $y \in X$, $A \subset X$

(iii) $y \,\delta\, (A \cup B) \Leftrightarrow y \,\delta\, A$ or $y \,\delta\, B$, for all $y \in X$, $A, B \subset X$

(iv) If $y \,\delta\, A$ and $a \,\delta\, B$ for all $a \in A$ then $y \,\delta\, B$, for all $y \in X$, $A, B \subset X$.

An example of a nearness relation is the set $N = \{(y, A) \in X \times P(X) : d(y, A) = 0\}$ where (X, d) is a metric space. In the next theorem we show that the concept of nearness is another approach to a topological space.

(2.12) Theorem: There is a one-to-one correspondence between the set of topologies on a set and the set of all nearness relations on that set.

Proof: Let X be a set. Suppose $\mathfrak{J}$ is a topology on X. For $A \subset X$, $y \in X$ define $y \,\delta\, A$ iff $y \in \bar{A}^{\mathfrak{J}}$. It is easy to check that this gives a nearness relation on X. Conversely suppose a nearness relation on X is given. For $A \subset X$ we let $\theta(A) = \{y \in X : y \,\delta\, A\}$. The conditions (i) to (iv) then easily imply that θ is a closure operator and thus determines a unique topology $\mathfrak{J}$ on X. The proof that these two correspondences are inverses of each other is left to the reader. ∎

Thus we see that the closure of a set consists of all points which are 'near' to it with respect to the corresponding nearness relation. There is one more way of looking at closure intuitively. According to it, a point y is in the closure of a set A iff y is 'approachable' or 'accessible' from A in some sense. It is not hard to show that in a metric space, a point y is in the closure of a set A iff there exists a sequence in A which converges to y. This is not true of all topological spaces. For example, in the space defined in example 10 of Chapter 4, Section 2, the point ∞ clearly belongs to the closure of $\mathbb{N} \times \mathbb{N}$ although no sequence in $\mathbb{N} \times \mathbb{N}$ converges to ∞, as noted in the exercises of that section. The difficulty can be overcome by using an appropriate generalisation of sequences, called 'nets' as we shall see in a later chapter.

We conclude this section by defining the boundary of a subset A of a space X. Our intuition demands that the boundary should consist of those points which can be reached from A as well as from its complement $X - A$. This leads to the following definition:

(2.13) Definition: Let A be a subset of a space X. Then its **boundary** or **frontier** is the set $\bar{A} \cap \overline{(X - A)}$.

It is immediate from the definition that the boundary is always a closed set and that the boundary of a set is the same as the boundary of its complement. Other properties of the boundary are also easy and will be developed through the exercises. To make sure that the definition indeed conforms to our intuition, we invite the reader to check that the boundary of the unit disc (either closed or open) in the complex plane with usual topology is in fact the unit circle. The boundary of a set A is generally denoted by ∂A or $F(A)$.

Exercises

2.1 Prove that the interior of a set is the same as the complement of the closure of the complement of the set, i.e. for a subset A of a space X, int $(A) = X - \overline{(X - A)}$.

2.2 Obtain an axiomatic characterisation of an interior operator (Hint: Either proceed directly or switch from interiors to closures using the last exercise and apply the axiomatic characterisation of a closure operator).

2.3 Prove that in a metric space X, a point y is in the closure of a set A iff there exists a sequence $\{x_n\}$ such that $x_n \in A$ for all n and $\{x_n\}$ converges to y in X.

2.4 Fill in the details of proof of Theorem (2.12).

2.5 For a set A in a space X, prove that $\bar{A}$ is the disjoint union of int (A) with the boundary of A.

2.6 Prove that a set is closed iff it contains its boundary and that it is open iff it is disjoint from its boundary.

2.7 Characterise clopen sets (i.e. those which are both closed and open) in terms of boundaries.

2.8 Let X_1, X_2 be topological spaces and X their topological product. Let $A_1 \subset X_1$, $A_2 \subset X_2$. Prove that

$$\partial(A_1 \times A_2) = (\bar{A}_1 \times \partial A_2) \cup (\partial A_1 \times \bar{A}_2).$$

2.9 Given a topology $\mathfrak{J}$ on a set X, the boundary defines an operator from $P(X)$ into itself. Study properties of this operator, in particular see if it is idempotent, if it commutes with intersections and unions. Also show how $\mathfrak{J}$ can be recovered from the boundary operator.

**2.10 Obtain an axiomatic characterisation of boundary operators.

2.11 Let U be an open set in a space X and let $V = \text{int}(\bar{U})$. Prove that $U \subset V$ and show by an example, that the inclusion may be strict. Prove, however, that U and V always have the same closure.

*2.12 Prove that if A is a subset of a space X then at most 14 distinct sets can result from A by repeated applications of the operators of closure and complementation (w.r.t. X). (Hint: Let α, β respectively denote closure and complementation operators. Since $\alpha^2 = \alpha$ and β^2 is the identity operator, the problem boils down to finding how many distinct 'words' can be formed by alternate use of α and β. Using (2.1) and (2.11) above show that $\alpha\beta\alpha\beta\alpha\beta\alpha\beta = \alpha\beta\alpha\beta$ and hence $\alpha\beta\alpha\beta\alpha\beta\alpha = \alpha\beta\alpha$.)

2.13 Find a subset A of the real line (with usual topology) for which the 14 subsets constructed above are distinct.

2.14 A **proximity** on a set X is a binary relation δ on $P(X)$ such that (writing $A \delta B$ to mean A, B are related and $A \bar{\delta} B$ otherwise):

(i) $\phi \bar{\delta} A$ for any $A \subset X$,
(ii) $A \delta B$ iff $B \delta A$ for $A, B \subset X$,
(iii) $A \delta (B \cup C)$ iff $A \delta B$ or $A \delta C$ for $A, B, C \subset X$,
(iv) $A \cap B \neq \emptyset$ implies $A \delta B$ for $A, B \subset X$,
(v) If $A \delta B$ and $\{b\} \delta C$ for all $b \in B$ then $A \delta C$ where $A, B, C \subset X$.

(a) Prove that every proximity δ on a set X induces a nearness relation on X if we let $y \delta A$ mean $\{y\} \delta A$ for $y \in X$ and $A \subset X$.

*(b) Show by an example that two distinct proximities may induce the same nearness relation on the underlying set.

**(c) Prove that on $\mathbb{R}$, there are 2^c distinct proximities (c being the cardinal number of $\mathbb{R}$) each of which induces the 'usual' nearness relation on $\mathbb{R}$.

2.15 Prove or disprove that in a metric space, a closed ball is the closure of the open ball with the same centre and radius.

2.16 A subset A of a space $(X, \mathcal{J})$ is said to be a **discrete** subset if it has no accumulation points in X. Prove that if A is a discrete subset of X then the relative topology $\mathcal{J}/A$ is discrete. Prove or disprove the converse.

Notes and Guide to Literature

Most of the material in this section is classic. The terms limit point and cluster point are sometimes used for an accumulation point. We do not advocate their use as we shall reserve them for some other concepts to be defined later.

The concepts of nearness relation and proximity have been studied relatively recently although Riesz had hinted at them as early as 1908. It is interesting that although the conditions in the definition of a nearness relation correspond one-by-one to the Kuratowski closure axioms, the equivalence was probably ignored until around 1964 when Lodato [1] and Leader [1] utilised it to define proximity spaces as a generalisation of topological spaces. Exercise (2.12) is a famous problem of Kuratowski. Langford [1] has characterised the sets sought for in Exercise (2.13).

We can also speak of neighbourhoods of sets. If A, N are subsets of a space X, then N is said to be a **neighbourhood** of A if there exists an open set V such that $A \subset V \subset N$. It is clear that this is the case if and only if N is a neighbourhood of each point of A.

3. Continuity and Related Concepts

We already defined continuity of a function from one metric space to another and noted that the definition could be stated in terms of open sets (see Proposition (1.4) of Chapter 4). It follows that the definition can be extended unchanged to topological spaces, as we indeed do.

(3.1) Definition: Let $f : X \to Y$ be a function; $x_0 \in X$ and $\mathfrak{J}$, $\mathcal{U}$ be topologies on X, Y respectively. Then f is said to be **continuous** (or more precisely $\mathfrak{J}$-$\mathcal{U}$ **continuous**) at x_0 if for every $V \in \mathcal{U}$ such that $f(x_0) \in V$, there exists $U \in \mathfrak{J}$ such that $x_0 \in U$ and $f(U) \subset V$.

It is convenient to have some other formulations of continuity at a point. Let us denote the closure w.r.t. $\mathfrak{J}$ and $\mathcal{U}$ by the same symbol. Let us also use δ to denote the nearness relation on X (or Y) induced by $\mathfrak{J}$ (respectively by $\mathcal{U}$). Then we have:

(3.2) Proposition: With the notation above, the following statements are equivalent.

1. f is continuous at x_0.
2. The inverse image (under f) of every neighbourhood of $f(x_0)$ in Y is a neighbourhood of x_0 in X.
3. For every subset $A \subset X$, $x_0 \in \bar{A}$ implies $f(x_0) \in \overline{f(A)}$.
4. For every subset $A \subset X$, $x_0 \, \delta \, A$ implies $f(x_0) \, \delta \, f(A)$.

Proof (1) $\Rightarrow$ (2). Let N be a neighbourhood of $f(x_0)$ in Y. Then there is an open set V in Y such that $f(x_0) \in V$ and $V \subset N$. Since f is continuous at x_0, there is an open set U in X such that $x_0 \in U$ and $f(U) \subset V$. This means $x_0 \in U \subset f^{-1}(V) \subset f^{-1}(N)$ thus showing that $f^{-1}(N)$ is a neighbourhood of x_0.

(2) $\Rightarrow$ (3). Suppose $x_0 \in \bar{A}$ where $A \subset X$. If $f(x_0) \notin \overline{f(A)}$ then by Theorem (2.10) in the last section, there is a neighbourhood N of $f(x_0)$ such that $f(A) \cap N = \emptyset$. This means $f^{-1}(f(A)) \cap f^{-1}(N) = \emptyset$ and hence that $A \cap f^{-1}(N) = \emptyset$ since $A \subset f^{-1}(f(A))$. But by (2), $f^{-1}(N)$ is a neighbourhood of x_0 and so $A \cap f^{-1}(N) \neq \emptyset$, since $x_0 \in \bar{A}$. This is a contradiction.

(3) $\Leftrightarrow$ (4). This is immediate since the nearness relation corresponding to a topology is defined by saying that a point is near a set iff it is in the closure of that set.

(3) $\Rightarrow$ (1). Let V be an open set containing $f(x_0)$. Let $A = X - f^{-1}(V) = f^{-1}(Y - V)$. Then $f(A) \subset Y - V$ and so $\overline{f(A)} \subset Y - V$ as $Y - V$ is closed. So $f(x_0) \notin \overline{f(A)}$ whence $x_0 \notin \bar{A}$ by (3). Hence there is a neighbourhood N of x_0 such that $N \cap A = \emptyset$. Clearly then $f(N) \subset V$ and the proof is completed if we let $U = \text{int}\,(N)$. ∎

Although conditions (1) to (4) are all equivalent, the last two have the advantage that they are probably closer to our intuition of continuity as preservation of nearness. The conditions (1) and (2), on the other hand, go from Y to X although the function f is from X to Y. This point should be especially noted by a beginner, who is most apt to define continuity of f at

x_0 to mean that the image of a nbd of x_0 is a nbd of $f(x_0)$.

Continuity at a point is a local concept. Note that it depends on the particular point, the particular function and also the topologies on the domain and the codomain. In the next definition we globalise the concept.

(3.3) Definition: Let $f : X \to Y$ be a function and $\mathfrak{J}$, $\mathcal{U}$ be topologies on X, Y respectively. Then f is said to be **continuous** (or $\mathfrak{J}$-$\mathcal{U}$ **continuous**) if it is continuous at each point of X.

Again several equivalent formulations can be given. We list some of them in the following theorem.

(3.4) Theorem: Let $(X, \mathfrak{J})$, $(Y, \mathcal{U})$ be spaces and $f : X \to Y$ a function. Then the following statements are equivalent:

1. f is continuous (i.e. $\mathfrak{J}$-$\mathcal{U}$ continuous).
2. For all $V \in \mathcal{U}, f^{-1}(V) \in \mathfrak{J}$.
3. There exists a sub-base $\mathcal{S}$ for $\mathcal{U}$ such that $f^{-1}(V) \in \mathfrak{J}$ for all $V \in \mathcal{S}$.
4. For any closed subset A of $Y, f^{-1}(A)$ is closed in X.
5. For all $A \subset X, f(\bar{A}) \subset \overline{f(A)}$.

Proof: (1) $\Leftrightarrow$ (2). Assume f is continuous at each point of X and V is an open subset of Y. If $x_0 \in f^{-1}(V)$ then $f(x_0) \in V$. But V is then a neighbourhood of $f(x_0)$ and so, by continuity at $x_0, f^{-1}(V)$ is a neighbourhood of x_0. Thus $f^{-1}(V)$ is a nbd of each of its points and so is an open set in X. Conversely suppose (2) holds and let $x_0 \in X$. Given any open set V containing $f(x_0), f^{-1}(V)$ is an open set containing x_0 and moreover $f(f^{-1}(V)) \subset V$. This shows that f is continuous at x_0.

(2) $\Leftrightarrow$ (3): (2) clearly implies (3). For the converse, we note that the inverse image preserves intersections and unions. Let $V \in \mathcal{U}$. Then V can be written as $\bigcup_{i \in I} V_i$ where I is an index set and for each $i \in I$, V_i can be written as the intersection of finitely many members of $\mathcal{S}$ say $S_1^i \cap \ldots \cap S_{r_i}^i$ where $r_i \in \mathbb{N}$, and $S_j^i \in \mathcal{S}$ for $1 \leqslant j \leqslant r_i$. Then $f^{-1}(S_j^i) \in \mathfrak{J}$. But $\mathfrak{J}$ is closed under finite intersections and so $f^{-1}(V_i) \in \mathfrak{J}$. Further $f^{-1}(V) \in \mathfrak{J}$ since $f^{-1}(V) = \bigcup_{i \in I} f^{-1}(V_i)$ and $\mathfrak{J}$ is closed under arbitrary unions. Thus (2) holds.

The equivalence of (2) with (4) follows from the fact that the inverse image preserves complements, i.e. that for any $B \subset Y, f^{-1}(Y - B) = X - f^{-1}(B)$. The equivalence of (5) with (1) is left as an exercise. ∎

As examples of continuous functions, we can cite all known examples from metric spaces. The following theorem, whose proof is trivial and omitted, asserts that certain functions are continuous.

(3.5) Proposition: Compositions of continuous functions are continuous. More specifically, if $f : X \to Y, g : Y \to Z$ are functions, with f continuous at x_0 and g continuous at $f(x_0)$ then $g \circ f$ is continuous at x_0. The identity function on any space is continuous. More generally, if $\mathfrak{J}$, $\mathcal{U}$ are topologies on a set X then the identity function $id_X : X \to X$ is $\mathfrak{J}$-$\mathcal{U}$ continuous iff the topology $\mathfrak{J}$ is stronger than $\mathcal{U}$. Any function from a discrete space and any function into an indiscrete space is continuous. Finally if $(X, \mathfrak{J})$ is a space and $(Y, \mathfrak{J}/Y)$ a subspace then the inclusion function $i : Y \to X$ is $\mathfrak{J}/Y$-$\mathfrak{J}$ continuous. A restriction of a continuous function is continuous. ∎

Continuous functions are also called **maps** or **mappings**. We now prove an important theorem about the product of a finite number of spaces. Although later on we shall prove the same theorem for products of arbitrary number of spaces, using essentially the same techniques, we feel it is worthwhile to do this special case here, for it will familiarise the reader with the essential concepts without entangling him with the technical notations. Let $X_1, X_2, \ldots, X_n$ be sets and let $X = X_1 \times X_2 \times \ldots \times X_n$. For each $i = 1, 2, \ldots, n$ define $\pi_i : X \to X_i$ by $\pi_i(x_1, x_2, \ldots, x_n) = x_i$. Then π_i is called the **projection** on X_i, or the i-th **projection**. It is a surjective function except in the case where some other X_j and hence X is empty. If $x \in X$ then $\pi_i(x)$ is called the **ith coordinate** of x.

(3.6) Theorem: Let $\{(X_i, \mathfrak{J}_i) : i = 1, 2, \ldots, n\}$ be a collection of topological spaces and $(x, \mathfrak{J})$ their topological product. Then each projection π_i is continuous. Moreover, if Z is any space then a function $f : Z \to X$ is continuous if and only if $\pi_i \circ f : Z \to X_i$ is continuous for all $i = 1, 2, \ldots, n$.

Proof: In order to prove that π_i is continuous, let $V_i \in \mathfrak{J}_i$. We have to show that $\pi_i^{-1}(V_i) \in \mathfrak{J}$. But it is easy to check that $\pi_i^{-1}(V_i) = X_1 \times X_2 \times \ldots \times X_{i-1} \times V_i \times X_{i+1} \times \ldots \times X_n$ because to say that $\pi_i(x) \in V_i$ for $x \in X$ puts no restriction on other coordinates of x, only the ith coordinate is required to lie in V_i. Recalling the definition of product topology $\pi_i^{-1}(V_i)$ is a member of the defining base and so is open in X. Hence each projection is a continuous function.

For the second part, suppose $f : Z \to X$ is continuous. Then for each i, $\pi_i \circ f$ is continuous as compositions of continuous functions are continuous. Conversely, suppose each $\pi_i \circ f : Z \to X_i$ is continuous. Denote $\pi_i \circ f$ by f_i. To prove that f is continuous, it suffices to prove, by Theorem (3.4) above, that the inverse image of any member of a base for $(X, \mathfrak{J})$ is an open subset of Z. Now, by definition of the product topology, a base for it consists of all sets V of the form $V_1 \times \ldots \times V_n$ where $V_i \in \mathfrak{J}_1$ for $i = 1, 2, \ldots, n$. It is then immediate that for $z \in Z$, $f(z) \in V$ iff $\pi_i(f(z)) \in V_i$ for all $i = 1, 2, \ldots, n$. In other words, $f^{-1}(V) = \bigcap_{i=1}^{n} f_i^{-1}(V_i)$. But each $f_i^{-1}(V_i)$ is open since f_i is assumed to be continuous. Hence $f^{-1}(V)$, being the intersection of finitely many open sets, is open in Z. This proves that f is continuous and completes the proof. ∎

The preceding theorem is of immense use in checking the continuity of functions into euclidean spaces. As a euclidean space is the topological product of a finite number of copies of the real line (with usual topology), the problem reduces to checking the continuity of real-valued functions. As a simple application, let us prove that the sum of two real-valued continuous functions on a space X is continuous. Let f_1, f_2 be two such functions. Define $f : X \to \mathbb{R} \times \mathbb{R}$ by $f(x) = (f_1(x), f_2(x))$ for $x \in X$. Then by the theorem above f is continuous. Now $f_1 + f_2$ is nothing but the composite of f with the sum function $+ : \mathbb{R} \times \mathbb{R} \to \mathbb{R}$ which is continuous (see Exercise (2.3.11)). So $f_1 + f_2$ is continuous. Similar arguments apply for continuity of the difference, the product and the quotient (whenever defined) of two real-valued continuous functions on the same space.

We now define a few concepts related to continuity.

(3.7) Definition: Let X, Y by spaces. A function $f: X \to Y$ is said to be **open** (respectively **closed**) if whenever A is an open (resp. a closed) subset of X, $f(A)$ is open (resp. a closed) subset of Y.

As the definition shows, openness of a function is in some sense the inverse of continuity. Indeed, if f is a bijection then f is open iff the inverse function f^{-1} is continuous. General properties of open or closed functions are similar to those of continuous functions and easy to prove. However, there are some differences. For example, if $(X, \mathfrak{J})$ is a space, $(Y, \mathfrak{J}/Y)$ is a subspace, then the inclusion map $i : Y \to X$ is in general neither open nor closed. It is open iff Y is open in X and it is closed iff Y is closed in X. Another important point is that the continuity of a function does not really depend upon the topology on the entire codomain but rather on the relative topology on the range. In precise terms let $(X, \mathfrak{J})$ be a space, $f: X \to Y$ a function with range $f(X)$ and $\mathcal{U}_1$, $\mathcal{U}_2$ two topologies on Y. If $\mathcal{U}_1/f(x)$ is the same as $\mathcal{U}_2/f(x)$ then f is $\mathfrak{J}$-$\mathcal{U}_1$ continuous iff it is $\mathfrak{J}$-$\mathcal{U}_2$ continuous. This is no longer true for openness of a function. Consider the function $f: \mathbb{R} \to \mathbb{R}^2$ defined by $f(x) = (x, 0)$ for $x \in \mathbb{R}$. Then f is not open with respect to the usual topologies on $\mathbb{R}$ and $\mathbb{R}^2$. The range of f is the x-axis of $\mathbb{R}^2$. If we regard f as a function from $\mathbb{R}$ to the x-axis (with relative topology) then it is open. Similarly, a restriction of an open function need not be open. These remarks also apply for closed functions.

In order to show that a function is open, it suffices to show that it takes all members of a base for the domain space to open subsets of the co-domain. Using this fact, it is easy to show that the projection functions from a product space to the coordinate spaces are open. It is tempting to think that they are also closed. But this is not the case. Consider the projection $\pi_1 : \mathbb{R}^2 \to \mathbb{R}$, $\pi_1(x, y) = x$; $x, y \in \mathbb{R}$. Let H be the set $\{(x, y) \in \mathbb{R}^2 : xy = 1\}$. H is a closed subset of $\mathbb{R}^2$ as its complement is open. However $\pi_1(H)$ is the set of all non-zero real numbers and it is not a closed subset of $\mathbb{R}$.

Next we come to homeomorphism, which is the fundamental classifying relation in topology. We had already defined this concept in Chapter 3, but we restate the definition.

(3.8) Definition: A **homeomorphsim** from a space X to space Y is a bijection $f: X \to Y$ such that both f and f^{-1} are continuous. When such a homeomorphism exists, X is said to be **homeomorphic** to Y.

As usual, we give some equivalent versions of the definition. The proof of equivalence is left to the reader.

(3.9) Proposition: Let X, Y be topological spaces and $f : X \to Y$ a function. Then the following statements are equivalent:

(1) f is a homeomorphism,

(2) f is a continuous bijection and f is open,

(3) f is a bijection and f^{-1} is an open map,

(4) there exists a function $g : Y \to X$ such that f, g are continuous, $g \circ f = id_X$ and $f \circ g = id_Y$. ∎

We urge the reader to go back to Chapter 3 and read again the material about homeomorphisms and topological properties. Now that these concepts are rigorously defined, he can, for example, prove that second countability is a topological property.

To conclude this section we discuss another important concept, that of an embedding (sometimes called an imbedding). Intuitively, to embed a space X into a space Y should mean that we can identify X with a subspace of Y, where identification is, of course, upto a homeomorphism. This leads to the following definition.

(3.10) Definition: Let $(X, \mathfrak{J})$, $(Y, \mathcal{U})$ be topological spaces. An **embedding** (or **imbedding**) of X into Y is a function $e : X \to Y$ which is a homeomorphism when regarded as a function from $(X, \mathfrak{J})$ onto $(e(X), \mathcal{U}/e(X))$.

(3.11) Proposition: A function $e : X \to Y$ is an embedding iff it is continuous and one-to-one and for every open set V in X, there exists an open subset W of Y such that $e(V) = W \cap Y$.

Proof: The result follows directly from definitions of a homeomorphism and of relative topology. ∎

Inclusion maps are the most immediate examples of embeddings and as the definition implies, these are the only examples upto homeomorphisms. An important problem in topology is to decide when a space X can be embedded in another space Y i.e. when there exists an embedding from X into Y. This is called the **embedding problem**. Theorems asserting the embeddability of a space into some other space which is more manageable than the original space are known as **embedding theorems**. An important embedding theorem will be proved later in this book. On the other hand, theorems which assert that a certain space cannot be embedded into some other space are known as **non-embedding theorems.** Non-embedding theorems are often quite deep and require methods well beyond the general topology. For example, it is by no means trivial to prove that the 2-sphere S^2 cannot be embedded into the euclidean plane.

While we are at it, let us mention two other basic problems in topology. The first one is the so-called extension problem. Suppose we have spaces X and Y. Let $A \subset X$ and suppose that we are given a continuous function f from A (with relative topology) to Y. A function $F : X \to Y$ is said to be an **extension** of f if $F/A = f$, or equivalently if $f = F \circ i$ where i is the inclusion map from A into X. The **extension problem** asks whether f can be continuously extended to X, i.e. whether there exists a continuous function

$F : X \to Y$ which is an extension of f. In terms of diagrams of functions, the question is whether the dotted arrow in Figure 1 can be represented by a function which will make the diagram commute. The problem becomes

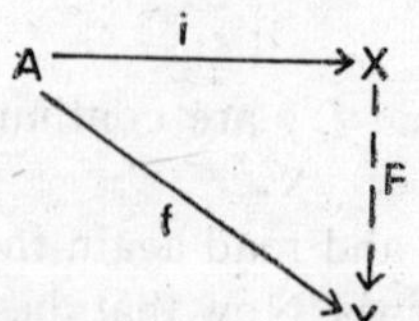

Fig. 1 Extension problem

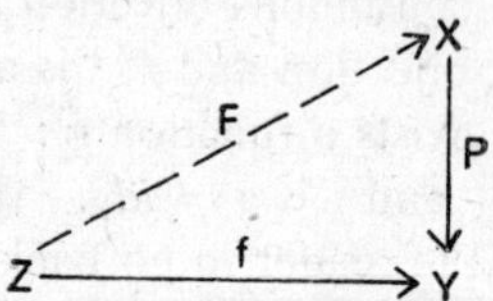

Fig. 2 Lifting problem

trivial if the extension F is not required to be continuous, for then we are free to define F on $X - A$ anyway we like. As an example, of the extension problem we invite the reader to characterise all continuous real-valued functions on the open interval (0, 1) which can be extended continuously to the closed interval [0, 1].

A dual problem is known as the lifting problem. Here we have maps $p : X \to Y$ and $f : Z \to Y$ as shown in Fig. 2. A **lifting** of f is a map $F : Z \to X$ which makes the diagram commute, i.e. F is a lifting of f iff $p \circ F = f$. The **lifting problem** asks whether f has a continuous lifting. A set-theoretic requirement for the existence of a lifting is that the range of f be contained in the range of p. This condition is generally satisfied as in most lifting problems the function p is known to be a surjection. Again, the problem is trivial if F is not required to be continuous. A familiar example of the lifting problem appears in complex analysis. Suppose X and Y are each the complex plane and $p : X \to Y$ is the exponential map, $p(z) = \exp(z) = e^z$. Let Z be a region in the complex plane, and $f : Z \to Y$ a map. Then a lifting of f is nothing but a continuous branch of $\log f$ and a reader familiar with complex analysis would know when such a lifting exists.

It is interesting to view convergence of sequences as an extension problem. Let $\{x_n\}$ be a sequence in a space $(X, \mathfrak{J})$. Let $D = \{1/n : n \in \mathbb{N}\}$ and $D^* = D \cup \{0\}$ each with the subspace topology of $\mathbb{R}$. Define $f : D \to X$ by $f(1/n) = x_n$ for $n \in \mathbb{N}$. Then f is continuous. A continuous extension of f to D^* corresponds to a limit of $\{x_n\}$ in X (cf. Exercise (3.9) below).

The extension and the lifting problems are of a very general nature and a good many problems in topology can be paraphrased either as extension problems or as lifting problems. Obviously then, no general solution can be expected to either of them. There are, however, important theorems which give answers in special cases.

Exercises

3.1 Complete the proof of Theorem (3.4).

3.2 Prove Propositions (3.5) and (3.9).

3.3 Let $(X, \mathfrak{J})$, $(Y, \mathcal{U})$ be topological spaces such that every function from X to Y is $\mathfrak{J} - \mathcal{U}$ continuous. Prove that either $\mathfrak{J}$ is the discrete topology or $\mathcal{U}$ is the indiscrete topology.

3.4 Let $(X_1, \mathfrak{J}_1)$, $(X_2, \mathfrak{J}_2)$ be topological spaces. Prove that $X_1 \times X_2$ is homeomorphic to $X_2 \times X_1$ each being given the product topology. (This shows that the order in which we take products is immaterial).

3.5 More generally, let $\{(X_i, \mathfrak{J}_i) : i = 1, 2, \ldots, n\}$ be a collection of topological spaces and let σ be a permutation of the symbols $1, 2, \ldots, n$. For $i = 1, 2, \ldots, n$ let $Y_i = X_{\sigma(i)}$ and let $Y = Y_1 \times Y_2 \times \ldots \times Y_n$, $X = X_1 \times X_2 \times \ldots \times X_n$ each with product topology. Prove that X and Y are homeomorphic to each other.

3.6 For any three spaces X_1, X_2, X_3 prove that $X_1 \times (X_2 \times X_3)$ is homeomorphic to $(X_1 \times X_2) \times X_3$. Generalise.

3.7 Let X be the topological product of non-empty spaces $X_1, X_2, \ldots, X_n$. Fix points $y_2 \in X_2, y_3 \in X_3, \ldots, y_n \in X_n$. Define $e : X_1 \to X$ by $e(x_1) = (x_1, y_2, y_3, \ldots, y_n)$. Prove that e is an embedding. (This shows that each coordinate space can be looked upon as a subspace of the product space.)

3.8 Let $f : X \to Y$ be a function. The graph of f is defined to be the set $G = \{(x, f(x)) : x \in X\}$. Prove that if f is continuous then G is homeomorphic to X.

3.9 Suppose $f : X \to Y$ is continuous at a point $x_0 \in X$. Prove that whenever a sequence $\{x_n\}$ converges to x_0 in X, the sequence $\{f(x_n)\}$ converges to $f(x_0)$ in Y.

3.10 Let $X = \mathbb{N} \times \mathbb{N} \cup \{\infty\}$ be the space discussed in Example 10 of Chapter 4, Section 2. Define $f : X \to \mathbb{R}$ by $f(x, y) = 0$ for $(x, y) \in \mathbb{N} \times \mathbb{N}$ and $f(\infty) = 1$. Prove that f is not continuous at ∞ although for every sequence $\{x_n\}$ which converges to ∞ in X, $\{f(x_n)\}$ converges to $f(\infty)$.

3.11 Let X be the topological product of spaces $X_1, X_2, \ldots, X_n$. Prove that a sequence $\{y_n\}$ converges to z in X iff for each $i = 1, 2, \ldots, n$ the sequence $\{\pi_i(y_n)\}$ converges to $\pi_i(z)$ in X_i.

3.12 Prove that a continuous real-valued function on the open interval $(0, 1)$ can be extended to the closed interval $[0, 1]$ iff it is uniformly continuous w.r.t. the usual metric. (You may use results from real analysis).

*3.13 Let S^1 be the unit circle in the plane, with the relative topology on it. Define $p : \mathbb{R} \to S^1$ by $p(\theta) = (\cos \theta, \sin \theta)$. Prove that p is continuous. Prove that the identity map $id_{S^1} : S^1 \to S^1$ cannot be lifted to a map from S^1 to $\mathbb{R}$.

3.14 Prove that the set of all homeomorphisms of a space $(X, \mathfrak{J})$ onto itself is a subgroup of the permutation group of X (page 49).

Notes and Guide to Literature

The main idea behind continuity is the preservation of closeness, that is, a continuous function should not throw apart what is held together in the domain space. Although the concept was loosely used for more than a century, it was Heine who first gave a rigorous or 'ϵ-δ' definition of continuity in 1872. This definition became popular and eventually infiltrated even to the undergraduate textbooks on elementary calculus. Its main disadvantage is that a beginner finds it too technical and is not easily convinced that it is consistent with his intuition about continuity. It is probably for this reason that the term 'ϵ-δ mathematics' became a euphemism for the fussy, ritualistic part of mathematics which pays little heed to intuistic appeal. The version of continuity involving nearness of a point to a set remedies the situation and is being tried by Naimpally, Hocking and Cameron [1].

Although a topologist's primary concern is continuity, continuous functions can be quite weird. For example, it can be shown that there exists a continuous function from the unit interval onto the entire closed unit disc in the complex plane. To weed out such examples, one puts additional restrictions such as differentiability, analyticity etc. Of course, in order for these terms to make sense, it is necessary that the domain and the codomain be equipped with some additional structure than mere topology. For example, in case of differentiability both the domain and the codomain must have what is known as a differential structure. The branch of topology which deals with such things is known as differential topology. A very readable book on the subject is by Guillemin and Pollack [1].

On the other hand, recently a trend has emerged to consider functions which satisfy less stringent conditions than continuity. One can for example define 'almost continuous' functions and many other variations of continuity. A reference along this line is Hoyle [1].

4. Making Functions Continuous, Quotient Spaces

In the last section we discussed the continuity of a function w.r.t. the given topologies on its domain and codomain. In the present section we consider the reverse problem. Let $f: X \rightarrow Y$ be a function. Suppose we are given a topology on one of the sets X and Y. How can we topologise the other set so as to make the given function continuous? More generally let us consider families of functions all having either the same domain or the same codomain and ask a similar question. This leads to two problems:

Problem 1: Let $\{(Y_i, \mathfrak{J}_i) : i \in I\}$ be an indexed family of topological spaces, X any set and $\{f_i : i \in I\}$ an indexed collection of functions such that for each $i \in I$, f_i is a function from X to Y_i. What topology $\mathfrak{J}$ on X will make each f_i $\mathfrak{J}$-$\mathfrak{J}_i$ continuous?

Problem 2: Let $\{(Y_i, \mathfrak{J}_i) : i \in I\}$ be an index family of topological spaces, X any set and $\{f_i : i \in I\}$ an indexed family of functions from Y_i into X. What topology $\mathcal{U}$ on X will make each f_i $\mathfrak{J}_i$-$\mathcal{U}$ continuous?

Notice the considerable similarity between the two problems. It can be seen that either problem can be obtained from the other by reversing the directions in which the functions go. For this reason, the two problems are said to be *dual* to each other. Their duality will be reflected through their solutions. By the way, duality is a fairly common occurrence in mathematics. For example, a reader familiar with elementary projective geometry would know the complete duality between points and lines. Or, as we saw earlier in this chapter, the interior and the closure operators are dual to each other. Presence of duality has the advantage that whenever we prove a theorem, we can anticipate its dual theorem and can often prove it by a dual argument. In the present case we shall see ample illustrations of this.

A trivial solution to the first problem would be to let $\mathfrak{J}$ be the discrete topology on X. This is hardly satisfactory. We therefore look for a 'better' solution. Note that if a topology $\mathfrak{J}$ on X renders each f_i continuous, then so will any topology stronger than $\mathfrak{J}$. The idea then is to find as small a topology on X as possible which will make each f_i continuous. Dually, a trivial solution to the second problem would be to let $\mathcal{U}$ be the indiscrete topology on X. If a topology $\mathcal{U}$ on X renders each f_i continuous then so will any topology weaker than $\mathcal{U}$. The idea then is to find as large a topology on X as possible which will make each function f_i continuous. We shall now show that each problem has a best possible solution.

(4.1) Theorem: With the notation of Problem 1, there exists a unique smallest topology $\mathfrak{J}$ on X which makes each f_i continuous.
Proof: First note that if $\mathcal{D}$ is any topology on X for which $f_i : X \to Y_i$ is $\mathcal{D}$-$\mathfrak{J}_i$ continuous, then $\mathcal{D}$ must contain all sets of the form $f_i^{-1}(V_i)$ as V_i ranges over all open subsets of Y_i. This gives us a clue to the construction of the smallest such topology. We let $\mathcal{S} = \{f_i^{-1}(V_i) : V_i \in \mathfrak{J}_i, i \in I\}$. Then $\mathcal{S}$ is a collection of subsets of X. $\mathcal{S}$ itself need not be a topology for X. However we let $\mathfrak{J}$ be the topology generated by $\mathcal{S}$, that is, the topology having $\mathcal{S}$ as a sub-base. Then clearly each f_i is $\mathfrak{J}$-$\mathfrak{J}_i$ continuous. Further if $\mathcal{D}$ is any other topology on X which makes each f_i continuous, then as we just observed $\mathcal{S} \subset \mathcal{D}$ and so $\mathfrak{J} \subset \mathcal{D}$. Thus $\mathfrak{J}$ is the smallest topology on X which makes each f_i continuous. ∎

(4.2) Definition: With the notations above, the weakest topology on X making each f_i continuous is called the **weak topology** determined by the family of functions $\{f_i : i \in I\}$.

The proof of Theorem (4.1) not only shows the existence of the weak topology generated by the given family of functions, but also gives an explicit construction for it. The dual problem is settled in the next theorem. Curiously enough, the proof is simpler than that of the last theorem, as we get the desired topology directly, without resorting to sub-bases.

(4.3) Theorem: With the notation of Problem 2, there exists a unique largest topology $\mathcal{U}$ on X which makes each f_i continuous.
Proof: Let $\mathcal{D}$ be a topology on X which makes each f_i continuous. Then for any $D \in \mathcal{D}$, and $i \in I$, $f_i^{-1}(D)$ is open in X_i. This puts a restriction on how large $\mathcal{D}$ can be and gives us a clue to the construction of the best possible solution. We let $\mathcal{U} = \{A \subset X : f_i^{-1}(A) \in \mathfrak{J}_i \text{ for all } i \in I\}$. Since the inverse

images commute with intersections and unions, it is easy to show that $\mathcal{U}$ is a topology on X and by what we said earlier, it is the strongest topology on X making each f_i continuous. ∎

(4.4) Definition: With the notations above, the strongest topology on X making each f_i continuous is called the **strong topology** determined by the family of functions $\{f_i : i \in I\}$.

We shall now discuss two classic situations in which the problem of finding the weak topology is important. First, let $\{(X_i, \mathfrak{J}_i) : i = 1, 2, \ldots, n\}$ be a collection of topological spaces. Let $X = X_1 \times X_2 \times \ldots \times X_n$, and let $\pi_i : X \to X_i$ be the ith projection function for $i = 1, 2, \ldots, n$. In the last section we saw that each π_i is continuous, provided on X we put the product topology $\mathfrak{J}$. We claim that $\mathfrak{J}$ is in fact the smallest topology which makes each π_i continuous. For suppose $\mathcal{U}$ is any other topology on X such that each π_i is $\mathcal{U}$-$\mathfrak{J}_i$ continuous. Let V be an open box in X, say, $V = V_1 \times V_2 \times \ldots \times V_n$ where $V_i \in \mathfrak{J}_i$ for $i = 1, 2, \ldots, n$. Then clearly $V = \bigcap_{i=1}^{n} \pi_i^{-1}(V_i)$. But each $\pi_i^{-1}(V_i) \in \mathcal{U}$ since π_i is $\mathcal{U}$-$\mathfrak{J}_i$ continuous. So $V \in \mathcal{U}$ since $\mathcal{U}$ is closed under finite intersections. Now the family of all such open boxes is, by definition, a base for the product topology $\mathfrak{J}$ on X. Hence $\mathfrak{J} \subset \mathcal{U}$. We have thus proved the following theorem in the case of finite products.

(4.5) Theorem: The product topology is the weak topology determined by the projection functions.

The second example about weak topologies comes from functional analysis. Let X be a normed linear space. X is then a metric space with the metric induced by the norm. Let X^* be the collection of all continuous, linear functionals on X. Then X^*, regarded as a family of functions from the set X into the space $\mathbb{R}$ (with usual topology) determines the weak topology on X. Note that in general this topology is weaker than the original norm topology.

An important special case of the strong topology is known as the quotient topology and will be discussed shortly.

We now prove two theorems which are of a similar spirit as the theorem about product spaces proved in the last section.

(4.6) Theorem: Let X have the weak topology determined by a family $\{f_i : X \to Y_i \mid i \in I\}$ of functions where each Y_i is a topological space, I being an index set. Then for any space Z, a function $g : Z \to X$ is continuous iff for each $i \in I$, the composite $f_i \circ g : Z \to Y_i$ is continuous.

Proof: If g is continuous then so is each $f_i \circ g$, as the compositions of continuous functions are continuous. It is the converse which is more interesting. We recall that the family $\mathcal{S}$ of all subsets of the form $f_i^{-1}(V_i)$ where V_i is open in Y_i, $i \in I$ is a sub-base for the weak topology on X. So, in order to show that $g : Z \to X$ is continuous, it suffices to show that the inverse image under g of every member of $\mathcal{S}$ is open in Z. Now if $i \in I$ and

V_i is open in Y_i then $g^{-1}(f_i^{-1}(V_i)) = (f_i \circ g)^{-1}(V_i)$ which is open by continuity of $f_i \circ g$. This shows that g is continuous. ∎

The proof of the following theorem, which is dual to the last one is left to the reader.

(4.7) Theorem: Let X have the strong topology determined by a family $\{f_i : Y_i \to X \mid i \in I\}$ of functions where each Y_i is a topological space, I being an index set. Then for any space Z, a function $g : X \to Z$ is continuous iff for each $i \in I$, the composite $g \circ f_i : Y_i \to Z$ is continuous. ∎

The most interesting examples of strong topologies arise when the families determining them are singleton. Let us suppose $f : X \to Y$ is a function and a topology $\mathfrak{I}$ on X is given. Let $\mathcal{U}$ be the strong topology on Y determined by the function f (or more precisely, by the family $\{f\}$). We know that $\mathcal{U}$ consists of those subsets V of Y for which $f^{-1}(V)$ is open in X. But $f^{-1}(V)$ is the same as $f^{-1}(f(X) \cap V)$ where $f(X)$ is the range of the function f. It follows that a subset of Y is open if and only if its intersection with $f(X)$ is open. In other words, the function f and the topology $\mathfrak{I}$ on X directly determine the topology on the range $f(X)$ which is then dragged on to a topology on the entire codomain Y. Our interest therefore lies in $f(X)$ and not in Y. Equivalently, we may suppose that the function f is a surjection to begin with. There is a special name for the strong topology in such a case.

(4.8) Definition: Let $(X; \mathfrak{I})$, $(Y: \mathcal{U})$ be topological spaces and $f : X \to Y$ be an onto function. Then f is said to be a **quotient map** or Y is said to have the **quotient topology** w.r.t. X and f if $\mathcal{U}$ is the strong topology generated by the singleton family $\{f\}$. In' such a case we also say that Y is a **quotient space** of X.

The real justification for the term 'quotient' will be given a little later. For the time being let us note that each factor space is indeed a quotient space of the topological product of spaces. Let $\{(X_i; \mathfrak{I}_i) : i = 1, 2, \ldots, n\}$ be a family of non-empty topological spaces. Let $(X; \mathfrak{I})$ be their topological product. Then for any i, the projection $\pi_i : X \to X_i$ is onto. We claim $\mathfrak{I}_i$ is the quotient topology w.r.t. X and π_i. Let $V \subset X_i$. If V is open in X_i then $\pi_i^{-1}(V)$ is open in X by continuity of π_i. On the other hand, if $\pi_i^{-1}(V)$ is open in X then $V = \pi_i(\pi_i^{-1}(V))$ is open in X_i because the projection functions are open. Thus V is open in X_i iff $\pi_i^{-1}(V)$ is open in X and so $\mathfrak{I}_i$ is the quotient topology. The argument here applies to prove:

(4.9) Proposition: Every open, surjective map is a quotient map. ∎

It is also true that every closed, surjective map is a quotient map.

(4.10) Proposition: Every closed, surjective map is a quotient map.

Proof: Let $f : (X; \mathfrak{I}) \to (Y; \mathcal{U})$ be closed, continuous and onto. Let $V \subset Y$. If V is open in Y, then $f^{-1}(V)$ is open in X by continuity of f. On the other hand suppose $f^{-1}(V)$ is open in X. Then $X - f^{-1}(V)$ is closed in

X and so $f(X - f^{-1}(V))$ is closed in Y, because f is a closed function. Since f is onto, $f(X - f^{-1}(V))$ equals $Y - V$. Thus V is open in Y. This shows that $\mathcal{U}$ is the strong topology determined by $\mathfrak{J}$ and f. Therefore f is a quotient map. ∎

It is not true that every quotient map is either open or closed, as can be shown by simple examples. Actually the class of quotient maps is quite large inasmuch as any surjective function on any topological space becomes a quotient map if we put the quotient topology on its codomain. Consequently quotient spaces can be quite weird as we shall see by examples in the later chapters. Nevertheless there are a few properties which carry over from a space to its quotient space. Such properties have a special name.

(4.11) Definition: A topological property is said to be **divisible** if whenever a space has it, so does every quotient space of it.

Clearly, the property of being a finite space is divisible. As a less trivial example we have:

(4.12) Proposition: The property of being a discrete space is divisible. In other words, every quotient space of a discrete space is discrete.

Proof: Let $f : X \to Y$ be a quotient map where X is a discrete space. Let $\mathcal{U}$ be the quotient topology on Y. We have to show that $(Y, \mathcal{U})$ is discrete. Recall that $\mathcal{U}$ is the strongest topology on Y which makes f continuous. But since the domain X is discrete, any topology on Y renders f continuous. So $\mathcal{U}$ is the strongest topology on Y, that is the discrete topology. ∎

On the other hand, second countability is not a divisible property as will be shown by an example later. In this book, we shall study many topological properties and it is a standing exercise for the reader to check if they are divisible as he comes across them.

Quotient spaces provide an important tool for the construction of new topological spaces from old ones. To see this, it is convenient to look at a quotient space as arising 'intrinsically' out of the domain space and not having to depend on the quotient map. Let $f : X \to Y$ be a surjective function. Then f determines an equivalence relation R on X defined by $x \, R \, y$ iff $f(x) = f(y)$. The equivalence classes of R are precisely the inverse images of singleton subsets of Y. Now let $\mathcal{D}$ be the collection of all equivalence classes under R. $\mathcal{D}$ is called the **quotient set** of X by R and is also denoted by X/R. There is a canonical function $p : X \to X/R$, called the projection which assigns to each $x \in X$, its equivalence class under R. The function $\theta : Y \to X/R$ defined by $\theta(y) = p(x)$ for any $x \in f^{-1}(\{y\})$ is obviously a well-defined bijection and the following diagram is commutative.

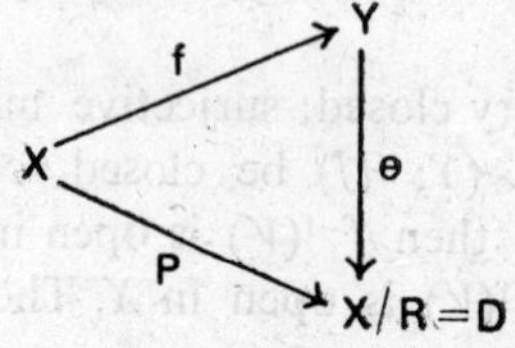

So far, X was merely a set and f a set-theoretic surjection. Suppose now that X, Y are topological spaces and that f is a quotient map. On X/R, we put the strong topology generated by the projection function p. The function θ then becomes continuous as its composite with f, viz., $\theta \circ f$ is continuous. Similarly θ^{-1} is continuous. Thus θ is now not merely a bijection but a homeomorphism. Thus, upto a topological equivalence, we may identify the quotient space Y with the quotient space X/R and the quotient map f with the projection p.

It follows that given a space X, any quotient space of X can be obtained starting from an equivalence relation R on X and putting the strong topology on the set of equivalence classes X/R. This gives us an important tool to generate new topological spaces because there is absolutely no restriction on the relation R on X other than that it be an equivalence relation. In practice it is often convenient to think of the equivalence relation R in terms of the corresponding decomposition $\mathcal{D}$ of X into mutually disjoint subsets. Although the two approaches are equivalent, the latter is more vivid. Thus if $\mathcal{D}$ is an arbitrary decomposition of a space X into mutually disjoint subsets, then the corresponding quotient space is obtained by 'shrinking' or 'identifying' each member of $\mathcal{D}$ to a single point. For this reason, the quotient spaces are sometimes called **identification spaces** and quotient maps as **identification maps.** In common usage it is customary to specify only the non-trivial members of the decomposition, that is, members having more than one point each. For example when we say 'S^1 is obtained by identifying together the two end points of the unit interval', what we mean is that S^1 is homeomorphic to the quotient space of $[0, 1]$ obtained from the decomposition $\mathcal{D}$ whose members are $\{0, 1\}$ and all singleton sets $\{x\}$ for $0 < x < 1$. As in this case, it often happens that only one member of the decomposition is non-trivial. If this member is A, then the quotient space in such a case is denoted by X/A. It is said to be obtained from X by **collapsing or reducing** A to a point. With this notation, S^1 is homeomorphic to $[0, 1]/\{0, 1\}$. More generally we see that if we take the closed unit ball D^n in $\mathbb{R}^n$ and identify its boundary S^{n-1} to a point then we get a quotient space which is homeomorphic to the sphere S^n.

The concept of identification can be used to give a rigorous description of what we intuitively call as 'gluing', 'pasting' or 'sewing'. For example, in Exercise (2.11) of Chapter 3 we described the Mobius band as the space obtained from a square by gluing points on one of its sides with those on the opposite side with the sense reversed. Using quotient spaces we can describe this quite rigorously. Let X be the unit square $[0, 1]\times[0, 1]$ (with the usual topology). Consider the decomposition $\mathcal{D}$ of X all whose members are singleton except that for each $0 \leq x \leq 1$, $(0, x)$ and $(1, 1-x)$ are in the same class. Then the Mobius band is nothing but the quotient space $\mathcal{D}$. Using other identifications of points on the boundary of the unit square we get a variety of interesting spaces, a few of which will be mentioned in the exercises.

Properties of quotient spaces will be developed as we proceed further in subsequent chapters.

Exercises

4.1 Prove theorem (4.7).

4.2 Prove that the only continuous real-valued functions on the Sierpinski space are the constant functions and hence that the weak topology generated by them is indiscrete.

*4.3 Prove that for a metric space, the weak topology determined by the family of all continuous real-valued functions on it coincides with the metric topology. (Hint: Consider functions of the form $f(x) = d(x, A)$ where A is a nonempty subset of the space and d is the metric). Roughly this says that on a metric space there are sufficiently many continuous real-valued functions to recover the original topology.

4.4 A function $f : \mathbb{R} \to \mathbb{R}$ is said to be **continuous from the right or right-continuous** if for every $a \in \mathbb{R}$, $\lim_{x \to a+} f(x) = f(a)$. Find the weak topology on the domain set $\mathbb{R}$, determined by the collection of all right continuous functions on it.

4.5 Let X_1 be the set of rational numbers and X_2 the set of all irrational numbers, each with the usual topology. Let i_1, i_2 be the inclusion functions of X_1, X_2 respectively into $\mathbb{R}$. Find the strong topology on $\mathbb{R}$ which makes i_1, i_2 continuous. (Caution: This topology is much stronger than the usual topology on $\mathbb{R}$).

4.6 Suppose $(X, \mathfrak{J})$ is a topological space and let $\{X_i : i \in I\}$ be a cover of X by its subsets. Assume each X_i has the relative topology $\mathfrak{J}/X_i$ and let $f_i : X_i \to X$ be the inclusion functions. Let $\mathcal{U}$ be the strong topology on X determined by the family $\{f_i : i \in I\}$. Prove that a subset $A \subset X$ is open w.r.t. $\mathcal{U}$ iff for each i, $A \cap X_i$ is open $(X_i, \mathfrak{J}/X_i)$. (Hint: Note that $f_i^{-1}(A) = A \cap X_i$). In general $\mathcal{U}$ is stronger than $\mathfrak{J}$.

4.7 In the last exercise, if each X_i is open (w.r.t. $\mathfrak{J}$) then prove that $\mathcal{U}$ coincides with $\mathfrak{J}$.

4.8 Prove that the conclusion of the last exercise also holds if each X_i is closed in X (w.r.t. $\mathfrak{J}$), provided the index set I is finite. What goes wrong if I is infinite?

4.9 Let X be a space and $\mathcal{U}$ be an open cover of it. Prove that a function f from X into some space Y is continuous iff for each $U \in \mathcal{U}$, the restriction f/U is continuous.

4.10 Prove the same result for a finite closed cover (i.e. a cover each of whose members is closed).

4.11 Prove that the composite of two quotient maps is a quotient map.

4.12 Let R, S be equivalence relations on a space X such that R is finer than S (i.e. for any $x, y \in X$, xRy implies xSy) Prove that the natural function $q : X/R \to X/S$ which associates to each equivalence class under R, the unique equivalence class under S containing it, is a quotient map. (The significance of this result is that in forming a quotient space we may carry out the identification partially at first and then complete it. This often helps in visualising complicated quotient spaces).

4.13 Let $f: X \to Y$ be a homeomorphism and R an equivalence relation on X. For $x, y \in X$ let $f(x) S f(y)$ iff xRy. Prove that this defines an equivalence relation S on Y and that the quotient spaces X/R and Y/S are homeomorphic to each other. What is the significance of this result?

4.14 Let $(X; d)$ be a pseudometric space and let $(Y; e)$ be the metric space obtained from it by the construction in Exercise (3.3.9). Prove that Y is a quotient space of X.

4.15 Define a relation R on $\mathbb{R}$ by xRy iff $x - y \in \mathbb{Q}$. Prove that R is an equivalence relation on $\mathbb{R}$ and that the quotient space $\mathbb{R}/R$ is indiscrete.

4.16 Let X be a space which is the disjoint union of two closed unit discs D_1 and D_2 in the plane. Identify each point on the boundary of D_1 with a corresponding point on the boundary of D_2. Prove that the resulting quotient space is homeomorphic to the unit sphere S^2 in $\mathbb{R}^3$. (Constructions of this type are common. In the present case we say that S^2 is obtained by gluing together two closed discs along their boundaries).

4.17 Obtain the torus surface as a quotient space of the unit square.

4.18 Let X be the unit square $[0, 1] \times [0, 1]$. For each $x \in [0, 1]$ identify $(0, x)$ with $(1, x)$ and $(x, 0)$ with $(1 - x, 1)$. The resulting quotient space is called the *Klein bottle*. On the other hand, if we identify $(0, x)$ with $(1, 1 - x)$ and $(x, 0)$ with $(1 - x, 1)$ for all $x \in [0, 1]$ then the resulting quotient space is called the *projective plane*.

(a) Show these identifications in diagrams and thereby try to visualise these spaces. (Caution: Neither of these spaces can be embedded in $\mathbb{R}^3$; so your attempt to visualise them would require some 'transcendental' imagination.)

(b) Show that the projective plane can also be obtained as a quotient space of the unit sphere S^2 if we identify each point x on S^2 with its antipodal point $-x$. (Hint: Use Exercises (4.12) and (4.13) above). Since every pair of antipodal points on S^2 determines a unique one-dimensional vector subspace of the plane $\mathbb{R}^2$, the projective plane is also called, rather inaccurately but suggestively, the space of all lines in $\mathbb{R}^2$.

Notes and Guide to Literature

Properties of the weak topology on a normed linear space can be found in almost any book on functional analysis, see for example, Dunford and Schwartz [1] or Royden [1]. The proof of non-embeddability of the Klein bottle and the projective plane in $\mathbb{R}^3$ is highly non-trivial and requires elaborate techniques from algebraic topology. A discussion of these spaces may be found in Hocking and Young [1].

Chapter Six

Spaces with Special Properties

In the last two chapters we considered some generalities about topological spaces. In this chapter we consider spaces which satisfy some additional conditions. In the first section we study those conditions which say that a space satisfying them is small in some sense. In the second section we take up the important notion of connectivity and in the third section we localise it. We also briefly discuss paths in a space.

1. Smallness Conditions on a Space

In the fourth chapter we defined second countable spaces and remarked that the existence of a countable base implies that the space is small in the sense that it cannot have too many open sets. We already proved or left as exercises some properties of second countable spaces. These properties themselves are known by some other names and we now define them formally.

(1.1) Definition: A subset A of a space X is said to be a **compact (Lindelöff) subset** of X if every cover of A by open subsets of X has a finite (respectively countable) subcover. A space X is said to be **compact (Lindelöff)** if X is a compact (resp. Lindelöff) subset of itself.

The relationship between a compact subset and a compact space (and similarly between a Lindelöff subset and a Lindelöff space) will be clear from the following simple result.

(1.2) Proposition: Let $(X; \mathfrak{J})$ be a topological space and $A \subset X$. Then A is a compact (Lindelöff) subset of X if and only if the subspace $(A: \mathfrak{J}/A)$ is compact (resp. Lindelöff).

Proof: We give the argument for compactness. Replacing 'finite' by 'countable' in it, we could use it for Lindelöff subsets. Suppose first that A is a compact subset of X. Let $\mathcal{G}$ be an open cover of the space $(A; \mathcal{T}/A)$. Each member G of $\mathcal{G}$ is of the form $H \cap A$ for some $H \in \mathfrak{J}$. For each $G \in \mathcal{G}$, fix $D(G) \in \mathfrak{J}$ such that $G = D(G) \cap A$. Then the family $\{D(G) : G \in \mathcal{G}\}$ is a cover of A by open subsets of X. Since A is a compact subset of X, this cover has a finite subcover, say, $\{D(G_i) : i = 1, 2, \ldots, n\}$ where $G_i \in \mathcal{G}$ for all $i = 1, 2, \ldots, n$. Clearly then, $\{G_1, \ldots, G_n\}$ is a finite subcover of $\mathcal{G}$. This shows that the subspace $(A; \mathfrak{J}/A)$ is compact.

Conversely suppose the subspace $(A; \mathfrak{J}/A)$ is compact. Let $\mathcal{G}$ be a cover of A by open subsets of X. Then $\{G \cap A : G \in \mathcal{G}\}$ is an open cover of the space $(A; \mathfrak{J}/A)$. By compactness of the space $(A; \mathfrak{J}/A)$, this cover has a finite subcover, say, $\{G_i \cap A : i = 1, 2, \ldots, n\}$ where $G_i \in \mathcal{G}$ for $i = 1, 2, \ldots, n$. Clearly then, $\{G_1, \ldots, G_n\}$ is a finite subfamily of $\mathcal{G}$ covering the set A. Thus A is a compact subset of X. ∎

Although this proposition is not very profound, it leads us to an interesting concept. Suppose we have a set X, a subset A and two topologies $\mathfrak{J}_1$, $\mathfrak{J}_2$ on X such that $\mathfrak{J}_1/A = \mathfrak{J}_2/A$. If we know that A is a compact subset of $(X; \mathfrak{J}_1)$ does it follow that A is also a compact subset of $(X; \mathfrak{J}_2)$? In view of the last proposition, the answer is yes. But this is not immediately obvious from the definition. For, knowing that every cover of A by members of $\mathfrak{J}_1$ has a finite subcover is no ground to suppose, *a priori*, that the same would be true about $\mathfrak{J}_2$. The proposition above shows that as long as $\mathfrak{J}_1$ and $\mathfrak{J}_2$ relativise to the same topology on A, A is a compact subset of $(X; \mathfrak{J}_1)$ iff it is a compact subset of $(X; \mathfrak{J}_2)$. In other words, compactness of a subset depends only on the topology induced on it, the topology on the entire space is not directly relevant. To stress this point further we note that a similar statement no longer holds for denseness of a subset. It may very well happen that two distinct topologies $\mathfrak{J}_1$, $\mathfrak{J}_2$ on a set X induce the same topology on A and still A is a dense subset of X w.r.t. $\mathfrak{J}_1$ but not w.r.t. $\mathfrak{J}_2$. A trivial example of this occurs when $\mathfrak{J}_1$ is the indiscrete topology, $\mathfrak{J}_2$ is the discrete topology and A is a singleton subset of X (assuming X has at least two points).

A property of a subset of a topological space is said to be an **absolute property** if it depends only on the relativised topology on that set, otherwise it is called a **relative property**. Thus compactness and the property of being a Lindelöff subset are absolute properties while denseness is a relative property. Similarly being an open set is a relative property. That is why it is meaningless to say that a subset is open unless it is specified (or at least understood) relative to which set.

After this digression we continue the listing of smallness conditions on a topological space.

(1.3) Definition: A space is said to be **separable** if it contains a countable dense subset.

We have already studied some inter-relationship among these conditions, we recapitulate it here.

(1.4) Theorem: Every second countable space is Lindelöff.

Proof: This is merely a restatement of Theorem (4.3.4). ∎

(1.5) Theorem: Every second countable space is separable.

Proof: This is precisely the content of Exercise (5.1.7). ∎

We also noted that the converses of both these theorems hold for metric spaces (See Exercises (4.3.2) and (5.1.11)). In general, however, the converses fail. The real line with the semi-open interval topology is separable but not second countable (see Exercise (4.4.8)). It is a little more difficult to show that the real line with the semi-open interval topology is a Lindelöff space. However, easy examples can be given to show that a Lindelöff space need not be second countable.

As for compactness, it is obvious that every compact space is Lindelöff. The converse is false. The real line with the usual topology is not compact because the cover by open intervals of the form $(-n, n)$ where n varies over all positive integers, has no finite subcover. However, it is second countable and hence Lindelöff.

Trivial examples of compact spaces are all indiscrete spaces and all finite spaces (that is, spaces whose underlying sets are finite). Every cofinite space is compact. For, let X be such a space and let $\mathcal{U}$ be an open cover of X. Take any non-empty member U of $\mathcal{U}$. Then $X - U$ has only finitely many points say $x_1, x_2, \ldots, x_n$. For each such x_i choose $U_i \in \mathcal{U}$ such that $x_i \in U_i$. Then $\{U, U_1, U_2, \ldots, U_n\}$ is a finite subcover of $\mathcal{U}$. Note that a cofinite space is always separable, in fact every infinite subset in it is dense. However it is not second countable unless the underlying set is countable.

A reader familiar with real analysis must have undoubtedly come across the well known Heine-Borel theorem which states that every open cover of a closed bounded interval has a finite subcover. This theorem can be restated as 'every closed and bounded interval is compact'. It turns out that the Heine-Borel theorem (in a form slightly more general than the one stated here) has numerous applications in the proofs of many well-known theorems in analysis. Most of such results can be generalised to compact spaces. To cite one such example, the theorem that every continuous real-valued function on a closed bounded interval has a maximum and minimum can be generalised as follows:

(1.6) Theorem: Every continuous real-valued function on a compact space is bounded and attains its extrema.

Proof: Let X be a compact space and suppose $f : X \to \mathbb{R}$ is continuous. First we show f is bounded. For each $x \in X$, let J_x be the open interval $(f(x) - 1, f(x) + 1)$ and let $V_x = f^{-1}(J_x)$. By continuity of f, V_x is an open set containing x. Note that f is bounded on each V_x. Now the family $\{V_x : x \in X\}$ is an open cover of X and by compactness, admits a finite subcover say $\{V_{x_1}, V_{x_2}, \ldots, V_{x_n}\}$. Let $M = \max\{(f(x_1), f(x_2), \ldots, f(x_n)\} + 1$ and let $m = \min\{f(x_1), f(x_2), \ldots, f(x_n)\} - 1$. Now for any $x \in X$ there is some i such that $x \in V_{x_i}$. Then $f(x_i) - 1 < f(x) < f(x_i) + 1$ and so $m < f(x) < M$ showing that f is bounded. It remains to show that f attains its bounds.

Let L, λ be respectively the supremum and infimum of f over X. If there is no point x in X for which $f(x) = L$, then we define a new function $g : X \to \mathbb{R}$ by $g(x) = \dfrac{1}{L - f(x)}$ for all $x \in X$. Then g is continuous since f is so (see page 120). However g is unbounded, for given any $R > 0$, there exists x such that $f(x) > L - \dfrac{1}{R}$ and hence $g(x) > R$. This contradicts the earlier part of the theorem and shows that f attains the value L. Similarly f attains the infimum λ. ∎

In the proof above, boundedness of f could have been more easily established, had we covered X by sets of the form $f^{-1}(-n, n)$, $n \in \mathbb{N}$ and then taken a finite subcover. We have preferred to present the argument in the form above because it is typical of the way compactness is generally used in topology. For each point of the space, a 'nice' open neighbourhood is constructed (in the present case niceness of a set means that the given function is bounded on it). The space is then covered by all such open neighbourhoods and a finite subcover is extracted. This method is known as the **standard compactness argument** and we shall see many instances of it.

As an application of the last theorem, let us prove a result popularly known as the **Lebesgue covering lemma.** (A direct argument is also possible.)

(1.7) Theorem: Let (X, d) be a compact metric space and let $\mathcal{U}$ be an open cover of X. Then there exists a positive real number r such that for any $x \in X$ there exists $V \in \mathcal{U}$ such that $B(x, r) \subset V$.

Proof: Before we proceed with the proof, we give a word of warning lest the result may seem trivial. For each $x \in X$, we can certainly find $V \in \mathcal{U}$ such that $x \in V$. Then V being open, there exists $r > 0$ such that $B(x, r) \subset V$. This much is trivial and involves nothing more than the definition of an open cover. The crux of the matter is that the number r in this argument may vary as the point x varies, whereas the theorem asserts the existence of a positive r which will work for all $x \in X$. It is tempting to try to prove the result by letting r be $\inf \{r_x : x \in X\}$ where r_x is the radius of some ball around x completely contained in some member of $\mathcal{U}$. The trouble is that even though each r_x is positive, r may be zero. That is where compactness comes into play.

By compactness of the space X, we may suppose that the cover $\mathcal{U}$ is finite say $\mathcal{U} = \{U_1, U_2, \ldots, U_n\}$. We may also assume that each U_i is non-empty and proper (if $U_i = X$ for some i, then any r will work). Let $A_i = X - U_i$. Then each A_i is a non-empty closed set. Define $f_i : X \to \mathbb{R}$ by $f_i(x) = d(x, A_i) = \inf \{d(x, y) : y \in A_i\}$. Then f_i is continuous and is positive on U_i for each $i = 1, 2, \ldots, n$ (see Exercise (3.3.11)). Define $f : X \to \mathbb{R}$ by $f(x) = \max \{f_i(x) : i = 1, 2, \ldots, n\}$. We leave it to the reader to check that f is continuous. Note that f is positive everywhere since for any $x \in X$ there is some i such that $f_i(x) > 0$. So by the last theorem, f has a minimum on X and this minimum is positive. Let r be the minimum. Now let $x \in X$. Then for some $i, f(x) = f_i(x)$. We contend $B(x, r) \subset U_i$. For suppose $d(x, y) < r$. If $y \notin U_i$ then $y \in A_i$ whence, $r > d(x, y) \geq d(x, A_i) = f_i(x) = f(x)$ contradicting that r is the minimum of f. So $B(x, r) \subset U_i$ and we can let U_i be the desired set $V \in \mathcal{U}$. ∎

A number r which satisfies the conclusion of the theorem is called a **Lebesgue number** of the cover $\mathcal{U}$. The Lebesgue covering lemma has numerous applications in analysis. For example, we invite the reader to use it to prove that every continuous real-valued function on a compact metric space is uniformly continuous. Easy examples show that the hypothesis that the space X is compact cannot be dropped from either of the last two theorems. On the other hand, if we replace it by the stronger condition that the space X be finite, then both the theorems reduce to trivialities. This is a common feature about compactness. Some statement is trivially true in the finite case, either false or very difficult to prove in the general case and true but non-trivial in the compact case. It is for this reason that a topologist regards compactness as the next best thing to finiteness and we shall see ample evidence of this.

We shall prove a large number of theorems about compactness throughout this book. For the present we prove a couple of simple propositions.

(1.8) Proposition: Let X be a compact space and suppose $f : X \to Y$ is continuous and onto. Then Y is compact. In other words, every continuous image of a compact space is compact.

Proof: Let $\mathcal{V}$ be any open cover of Y. Let $\mathcal{U} = \{f^{-1}(V) : V \in \mathcal{V}\}$. Then $\mathcal{U}$ is a cover of X and since f is continuous, it is an open cover of X. Since X is compact some finitely many members of $\mathcal{U}$, say $f^{-1}(V_1)$, $f^{-1}(V_2), \ldots, f^{-1}(V_n)$ where $V_1, V_2, \ldots, V_n \in \mathcal{V}$ cover X. But then V_1, $V_2, \ldots, V_n$ cover Y since f is onto. So Y is compact. ∎

(1.9) Definition: A topological property is said to be **preserved under continuous functions** if whenever a space has it so does every continuous image of it, that is if $f : X \to Y$ is continuous and onto and X has the given property, then so does Y. (Note that such a property is necessarily divisible.)

With this terminology, the proposition above reads that compactness is preserved under continuous functions. A similar argument shows that the property of being a Lindelöff space is also preserved under continuous functions. We leave it to the reader to show that separability is also preserved under continuous functions. However, second countability is not preserved under continuous functions. As an example, let $X = \mathbb{N} \times \mathbb{N} \cup \{\infty\}$ with the topology $\mathfrak{J}$ defined in Example 10 of Chapter 4, Section 2. It will be proved later in this section that $(X, \mathfrak{J})$ is not second countable. However since the set X is countable, the space $(X, \mathcal{D})$, $\mathcal{D}$ being the discrete topology, is second countable (see Exercise (4.4.7)). Now the identity function $id_X : (X, \mathcal{D}) \to (X, \mathfrak{J})$ is onto and continuous since $\mathcal{D}$ is stronger than $\mathfrak{J}$. The domain space is second countable but the range is not.

Let us also investigate which of the four smallness conditions considered so far are hereditary. Recall that a topological property is hereditary if whenever a space has it, so does every subspace of it. In Chapter 4 we proved that second countability is a hereditary property. On the other hand separability is not hereditary. As an example consider the product space $\mathbb{R} \times \mathbb{R}$ where each $\mathbb{R}$ is given the semi-open interval topology. Then $\mathbb{R} \times \mathbb{R}$ is separable (Exercise (5.1.10)). But it contains an uncountable subset L which is discrete in the relative topology (Exercise (4.4.6)) and consequently not separable.

The properties of compactness and of being a Lindelöff space are not hereditary. Later on we shall show that any space whatsoever can be embedded into a compact space and that will provide counterexamples. However both of these properties do pass over from the original space to all closed subsets as the following proposition shows:

(1.10) Proposition: If X is a compact (Lindelöff) space and $A \subset X$ is closed in X then A, in its relative topology, is also compact (respectively Lindelöff).

Proof: Suppose X is compact and $A \subset X$ is closed. Let $\mathcal{U}$ be an open cover of A in the relative topology on A. For each $U \in \mathcal{U}$, fix an open set $V(U)$ in X such that $A \cap V(U) = U$. Then the family $\mathcal{V} = \{V(U) : U \in \mathcal{U}\} \cup \{X - A\}$ is an open cover of X and hence admits a finite subcover consisting of, say, $V(U_1)$, $V(U_2), \ldots, V(U_n)$ and possibly $X - A$. But then $\{U_1, U_2, \ldots, U_n\}$ covers A and is a finite subcover of $\mathcal{U}$. This shows that A, in its relative topology, is compact. The proof that A is Lindelöff in case X is, is similar. ∎

(1.11) Definition: A topological property is said to be **weakly hereditary** if whenever a space has it, so does every closed subspace of it.

With this terminology, the last proposition shows that compactness and the Lindelöff property are weakly hereditary. From now on, whenever a new property is introduced, we trust that the reader will make it a habit of checking whether it is preserved under continuous functions, whether it is hereditary or at least weakly hereditary. Note that the proposition above shows that $\mathbb{R} \times \mathbb{R}$ with the product topology induced by the semi-open interval topology on each factor is not Lindelöff, since the closed subspace L is not so. Hence the product of two Lindelöff spaces need not be Lindelöff. Curiously enough the product of any collection of compact spaces (even an infinite one) is compact. This is a highly non-trivial theorem due to Tychonoff and we shall prove it later.

We conclude the section with a study of another smallness condition which is of a local nature in the sense that it deals with what happens near a given point of a space. First we need a definition.

(1.12) Definition: Let X be a space and $x \in X$. Then a local base at x is a collection $\mathcal{L}$ of neighbourhoods of x such that given any neighbourhood N of x there exists $L \in \mathcal{L}$ satisfying $L \subset N$.

For example, the collection of all open neighbourhoods of a point is a local base at that point. If X is a metric space, then the collection of all open balls (closed balls will do as well) centred at a point constitutes a local base at that point. Note that it suffices to take open balls of rational radii, thereby giving a local base which is countable.

(1.13) Definition: A space is said to be **first countable at a point** if there exists a countable local base at that point.

(1.14) Definition: A space is said to be **first countable** or to satisfy the **first axiom of countability** if it is first countable at each point.

As we noted just before giving the definitions, all metric spaces are first countable. The following proposition gives another important class of spaces which are first countable.

(1.15) Proposition: Every second countable space is first countable.

Proof: Let X be a second countable space and $x \in X$. Suppose $\mathcal{B}$ is a countable base for X. Let $\mathcal{L} = \{B \in \mathcal{B} : x \in B\}$. Clearly $\mathcal{L}$ is countable and the proof will be complete if we show that it is a local base at x. For this, let N be a neighbourhood of x. Then there exists an open set V such that $x \in V \subset N$. Since $\mathcal{B}$ is base, there exists $B \in \mathcal{B}$ such that $x \in B \subset V$. But then $B \in \mathcal{L}$, showing that $\mathcal{L}$ is a local base at x. ∎

The converse of the proposition is false. The real line with the semi-open interval topology is not second countable. However it is first countable. For let $x \in \mathbb{R}$. Then the family of all semi-open intervals of the form $\left[x, x + \frac{1}{n}\right)$ where n is a positive integer, is a countable local base at x. Another example would be any uncountable discrete space. Being a metric space, a discrete space is always first countable. As an example of a space which is not first countable, we invite the reader to prove that an uncountable set with the cofinite topology is not first countable at any of its points.

As noted before, metric spaces are the foremost examples of first countable spaces. It turns out that many results about metric spaces involving convergence of sequences generalise to first countable spaces. As examples we cite the results of Exercises (4.1.9) and (5.2.3). By way of illustration we prove one such result.

(1.16) Theorem: Let X, Y be spaces, $x \in X$ and $f: X \to Y$ a function. Suppose X is first countable at x. Then f is continuous at x iff for every sequence $\{x_n\}$ which converges to x in X, the sequence $\{f(x_n)\}$ converges to $f(x)$ in Y.

Proof: The direct implication is easy (see Exercise (5.3.9)) and does not require that X be first countable at x. For the other way implication suppose $\{V_1, V_2, \ldots, V_n \ldots\}$ is a countable local base at X. We let $W_1 = V_1$, $W_2 = V_1 \cap V_2$, $W_3 = V_1 \cap V_2 \cap V_3, \ldots, W_n = V_1 \cap V_2 \cap \ldots \cap V_n, \ldots$ etc. Then the collection $\{W_n : n = 1, 2, 3, \ldots\}$ is also a local base at x. Its advantage over the given base is that it is a nested local base, that is, $W_m \subset W_n$ for $m \geq n$. The reason for switching to W's from V's will be clear in the course of the proof.

In order to show that f is continuous at x, we use one of the characterisations of continuity at a point given in Proposition (5.3.2). Suppose f is not continuous at x. Then there exists a subset A of X such that $x \in \bar{A}$ but $f(x) \notin \overline{f(A)}$. Now for each n, W_n is a neighbourhood of x and since $x \in \bar{A}$, $W_n \cap A \neq \emptyset$. Choose $x_n \in W_n \cap A$, for $n \in \mathbb{N}$. Note that for $m \geq n$,

$x_m \in W_n$ since $W_m \subset W_n$. We claim that the sequence $\{x_n\}$ converges to x in X. For, let G be an open set containing x. Then, since $\{W_1, W_2, \ldots, W_n, \ldots\}$ is a local base at x, there is some n such that $W_n \subset G$. But then for all $m \geq n$, $x_m \in W_n$ and so $x_m \in G$. So $\{x_n\}$ converges to x in X. However, for each n, $f(x_n) \in f(A) \subset \overline{f(A)}$. We are assuming that $f(x) \notin \overline{f(A)}$. Then $Y - \overline{f(A)}$ is an open set which contains $f(x)$ but contains no term of the sequence $\{f(x_n)\}$. Consequently $\{f(x_n)\}$ does not converge to $f(x)$ in Y. This contradicts the hypothesis and shows that f is continuous at X. ∎

Consider now the space $X = \mathbb{N} \times \mathbb{N} \cup \{\infty\}$ discussed in Example 10 of Chapter 4, Section 2. This space is first countable at each $(x, y) \in \mathbb{N} \times \mathbb{N}$. Indeed at any such point the singleton family $\{\{(x, y)\}\}$ is a local base. However, X is not first countable at the point ∞. It is not hard to show this directly, but an indirect argument using the last theorem can be given as follows. Define $f: X \to \mathbb{R}$ by $f(x, y) = 0$ for $(x, y) \in \mathbb{N} \times \mathbb{N}$ and $f(\infty) = 1$. Then f is not continuous at ∞ (see Exercise (5.3.10)) although the condition in the last theorem is satisfied because all the sequences in X which converge to ∞ are eventually constant. Hence X cannot be first countable at ∞, for otherwise f would be continuous at ∞ by the last theorem. We thus have an example of a countable space (i.e. a space whose underlying set is countable) which is not first countable. By Proposition (1.15) X is not second countable, as we had promised to show earlier in this section.

In passing, we mention another smallness condition which is sometimes useful.

(1.17) Definition: A space is said to satisfy the **countable chain condition** if any family of mutually disjoint open sets in it is countable.

(1.18) Proposition: Every separable (and hence every second countable) space satisfies the countable chain condition.

Proof: Suppose D is a countable dense subset of a space X. Let $\mathcal{G}$ be a family of open sets in X such that for $G, H \in \mathcal{G}$, $G \cap H = \emptyset$ unless $G = H$. We have to show $\mathcal{G}$ is countable. Suppose first that $\emptyset \notin \mathcal{G}$. For each $G \in \mathcal{G}$, $G \cap D$ is nonempty since D is dense in X. Choose a point $x_G \in G \cap D$ for $G \in \mathcal{G}$ and define the function $f: \mathcal{G} \to D$ by $f(G) = x_G$. Then f is one-to-one since $x_G = x_H$ would imply $G \cap H \neq \emptyset$, and hence $G = H$. Since the set D is countable, it follows that $\mathcal{G}$ is countable. In case the empty set $\emptyset$ is a member of $\mathcal{G}$ we apply the argument to $\mathcal{G} - \{\emptyset\}$ and find that it is countable. But then $\mathcal{G}$ is also countable. ∎

Exercises

1.1 Prove that the co-countable topology on a set (defined analogously to the cofinite topology, by letting the whole set and all countable subsets be closed) makes it into a Lindelöff space.

1.2 Let X be an uncountable set with the cocountable topology on it. Prove that X is not separable, although it satisfies the countable chain condition.

1.3 Let X be an uncountable set with the cofinite topology on it. Prove that X is separable but not first countable at any point. (Hint: If $x \in X$ and $\{V_1, V_2, \ldots, V_n, \ldots\}$ is a countable local base at x then $\bigcup_{n=1}^{\infty} (X - V_n)$ is countable and so there exists $y \neq x$ such that $y \in V_n$ for all n. Consider $X - \{y\}$).

1.4 Prove that every infinite subset A of a compact space X has at least one accumulation point in X. (Hint: If not, then construct an open cover of X each member of which contains at most one point of A and derive a contradiction.)

1.5 State and prove an analogous result for Lindelöff spaces.

1.6 Let f_1, f_2 be continuous real-valued functions on a space X. Define $f, g : X \to \mathbb{R}$ by $f(x) = \max \{f_1(x), f_2(x)\}$ and $g(x) = \min \{f_1(x), f_2(x)\}$ for $x \in X$. Prove that f, g are continuous. (Hint: First show that whenever a real-valued function h is continuous, so is the function $|h|$. Then note that for any real numbers α, β, $\max \{\alpha, \beta\} = \frac{1}{2}(\alpha + \beta + |\alpha - \beta|)$ and $\min \{\alpha, \beta\} = \frac{1}{2}(\alpha + \beta - |\alpha - \beta|)$. Alternatively note that for any $a \in \mathbb{R}, f^{-1}(a, \infty) = f_1^{-1}(a, \infty) \cup f_2^{-1}(a, \infty)$ and $f^{-1}(-\infty, a) = f_1^{-1}(-\infty, a) \cap f_2^{-1}(-\infty, a)$ etc. and that the open intervals of these types form a sub-base for the usual topology on $\mathbb{R}$.)

1.7 Using the Lebesgue covering lemma, prove that every continuous function from a compact metric space into another metric space is uniformly continuous.

1.8 If $[a, b)$, and $[c, d)$ are two semi-open intervals such that $a \neq c$ prove that either $a \in (c, d)$ or $c \in (a, b)$ or the two are disjoint.

1.9 If $\{[a_i, b_i) : i \in I\}$ is a collection of semi-open intervals, such that $a_i \neq a_j$ for $i \neq j \in I$, prove that there are at most countably many indices $i \in I$ for which a_i is not an interior point (w.r.t. the usual topology on $\mathbb{R}$) of any $[a_j, b_j), j \in I$. (Hint: Use the last exercise, the fact that the real line with the semi-open interval topology is separable and Proposition (1.18).)

*1.10 Prove that the real line with the semi-open interval topology is Lindelöff. (Hint: Use the last exercise along with the fact that the real line with the usual topology is second countable.)

1.11 Prove that separability is preserved under continuous functions. Prove or disprove that first countability is preserved under continuous functions.

1.12 Prove that first countability is a hereditary property.

1.13 Let $\{X_i : i = 1, 2, \ldots, n\}$ be a collection of spaces and X their topological product. For $i = 1, 2, \ldots, n$ let $x_i \in X_i$ and let $\mathcal{L}_i$ be a local base at x_i. Let $x = (x_1, x_2, \ldots, x_n) \in X$. Find a local base at x in terms of $\mathcal{L}_1, \mathcal{L}_2, \ldots, \mathcal{L}_n$. Hence show that if X_i is first countable at x_i for $i = 1, 2, \ldots, n$ then X is first countable at x.

1.14 Prove that the results of exercises (4.1.9) and (5.2.3) hold true if metric spaces are replaced by first countable spaces.

1.15 For a subset of a topological space X, see which of the following properties are absolute and which are relative:

(i) being a finite subset

(ii) being a closed subset

(iii) being a discrete subset (see Exercise (5.2.16) for the definition)

(iv) containing a given point of X.

Notes and Guide to Literature

Although the terminology in this section is by now fairly standard, the reader is warned that some authors use different terms. Second countability is often called complete separability. The term separability was used in old days to mean what we now call as the Hausdorff property (which will be defined later, although we made a mention to it at the end of Chapter 4, Section 1).

Similarly what we call as compact, is called bicompact by some topologists, the term compact being reserved for something weaker (which we shall study as countably compact, in a later chapter). Although we shall consistently follow our terminology throughout this book, the reader should be careful when he refers to other books.

An excellent account of the role of compactness in analysis, amply illustrating how it is the next best thing to finiteness, can be found in the paper by Hewitt [1].

2. Connectedness

In Chapter 3, we mentioned that connectedness is the technical term for the intuitive notion of consisting of a single piece; in other words, to say that a space is not connected means intuitively that it can be split into two or more 'parts'. Let us now formalise this notion. Let X be a topological space. Suppose we write X as the disjoint union of two of its non-empty subsets, say, $X = A \cup B$ where $A \cap B = \emptyset$, $A \neq \emptyset \neq B$. Can we say that A and B are 'parts' of X? If we do so, then every space with two or more points can be decomposed into two parts and no such space will be connected. This is not consistent with our intuition that the familiar spaces like the euclidean spaces, circles, spheres, intervals (all with usual topologies) be connected. So we cannot simply define a connected space as a space which cannot be written as a disjoint union of two of its proper subsets. Some additional restriction on these subsets or 'parts' is needed. Our intuition about splitting requires that the two parts should not only be mutually disjoint but also that they should be far away from each other. As we saw in the last chapter the idea of nearness can be formalised in terms of closures. This leads to the following definition:

(2.1) Definition: A space X is said to be **connected** if it is impossible to find non-empty subsets A and B of it such that $X = A \cup B$ and $\bar{A} \cap \bar{B} = \emptyset$. A space which is not connected is called **disconnected**.

Before giving examples of connected and disconnected spaces it is convenient to slightly reformulate the definition.

(2.2) Proposition: Let X be a space and A, B subsets of X. Then the following statements are equivalent:

1. $A \cup B = X$ and $\bar{A} \cap \bar{B} = \emptyset$.
2. $A \cup B = X$, $A \cap B = \emptyset$ and A, B are both closed in X.
3. $B = X - A$ and A is clopen (i.e. closed as well as open) in X.
4. $B = X - A$ and ∂A (that is, the boundary of A) is empty.
5. $A \cup B = X$, $A \cap B = \emptyset$ and A, B are both open in X.

Proof: (1) $\Rightarrow$ (2). Clearly $\bar{A} \cap \bar{B} = \emptyset$ implies that $A \cap B = \emptyset$ since $A \subset \bar{A}$ and $B \subset \bar{B}$. Also $\bar{A} \subset X - \bar{B} \subset X - B = A$ and so $\bar{A} = A$ showing that A is closed. Similarly B is closed.

(2) $\Rightarrow$ (3) is immediate since the complement of a closed set is open.

(3) $\Rightarrow$ (4). This follows from the fact that the boundary of a clopen set is empty (see Exercise (5.2.7).)

(4) $\Rightarrow$ (5). This requires the converse, viz., that a set with empty boundary is clopen. Also if A is closed, then its complement B is open.

(5) $\Rightarrow$ (1). Assume $X = A \cup B$ where $A \cap B = \emptyset$ and A, B are open. Then $A = X - B$ and $B = X - A$ whence A, B are closed as well and so $\bar{A} = A$, $\bar{B} = B$, showing $\bar{A} \cap \bar{B} = \emptyset$. ∎

(2.3) Proposition: Let X be a space. Then the following are equivalent:

1. X is connected.
2. X cannot be written as the disjoint union of two nonempty closed subsets.
3. The only clopen subsets of X are $\emptyset$ and X.
4. Every nonempty proper subset of X has a nonempty boundary.
5. X cannot be written as the disjoint union of two nonempty open subsets.

Proof: The result is immediate from the definition and the last proposition. ∎

From the definitions we see immediately that every indiscrete space is connected and that the only connected discrete spaces are those which consist of at most one point. The space of rational numbers is disconnected; given any irrational number α the sets $\{x \in \mathbb{Q} : x < \alpha\}$ and $\{x \in \mathbb{Q} : x > \alpha\}$ are both open in the relative topology on $\mathbb{Q}$ and $\mathbb{Q}$ is clearly their disjoint union. Similarly the set of irrational numbers is disconnected. The Sierpinsky space defined in Chapter 4, Section 2 is connected, although it is not indiscrete. It is clear that if a set is connected w.r.t. a topology $\mathfrak{J}$ on it, then it is connected w.r.t. every topology weaker than $\mathfrak{J}$. The following proposition shows that connectedness is preserved under continuous functions.

(2.4) Proposition: Let $f : X \to Y$ be a continuous surjection. Then if X is connected, so is Y.

Proof: Suppose Y is not connected. Then we can write $Y = A \cup B$ where

A, B are disjoint, nonempty and open subsets of Y. But then $X = f^{-1}(A) \cup f^{-1}(B)$. The sets $f^{-1}(A), f^{-1}(B)$ are mutually disjoint, and open since f is continuous. Further each is nonempty since f is onto. This contradicts that X is connected. Hence Y is connected. ∎

The first non-trivial examples of connected spaces come from the real line. The proof uses the completeness of the real numbers. We shall assume it in the order form, that is, every nonempty subset of $\mathbb{R}$ which is bounded above has a least upper bound and every non-empty subset which is bounded below has a greatest lower bound.

(2.5) Theorem: A subset of $\mathbb{R}$ is connected iff it is an interval.

Proof: First note that a subset $X \subset \mathbb{R}$ is an interval iff it has the property that for any $a, b \in X$, $(a, b) \subset X$. (Prove.) Now if X is not an interval then there exist real numbers a, b, c such that $a < c < b$; $a, b \in X$ and $c \notin X$. Let $A = \{x \in X : x < c\}$ and $B = \{x \in X : x > c\}$. Clearly A, B are disjoint, open subsets of X (in the relative topology) since $A = X \cap (-\infty, c)$ and $B = (c, \infty) \cap X$ and $A \cup B = X$. Further $a \in A$, $b \in B$ and hence A, B are nonempty. This shows that X is not connected.

Conversely suppose X is an interval and that $X = A \cup B$ where $\bar{A} \cap \bar{B} = \emptyset$, $A \neq \emptyset$, $B \neq \emptyset$ where the closure is relative to X. Let $a_0 \in A$, $b_0 \in B$. Without loss of generality we may suppose that $a_0 < b_0$. Now let x be the mid-point of the interval from a_0 to b_0, i.e. $x = \dfrac{a_0 + b_0}{2}$. Then $x \in X$ and so x is exactly in one of the sets A and B. If $x \in A$ we rename it as a_1 and rename b_0 as b_1. If $x \in B$, we rename a_0 as a_1 and x as b_1. In any case $[a_1, b_1]$ is an interval with its left end-point in A and the right end-point in B. We can now take the mid-point of $[a_1, b_1]$ and get an interval $[a_2, b_2]$ of half the length with $a_2 \in A$, $b_2 \in B$. Repeating this process ad infinitum, we get a nested sequence of intervals $\{[a_n, b_n] : n = 0, 1, 2, 3, \ldots\}$ such that $a_n \in A$ and $b_n \in B$ for all n. Note that $\{a_n\}$ is a bounded monotonically increasing sequence while $\{b_n\}$ is a bounded monotonically decreasing sequence and that $(b_n - a_n) \to 0$ as $n \to \infty$. By the order completeness of $\mathbb{R}$, both sequences converge to a common limit, say c. Note that $c \in X$ since $a_0 \leqslant c \leqslant b_0$. Also every neighbourhood of c intersects A as well as B. So $c \in \bar{A} \cap \bar{B}$, a contradiction. Hence X is connected. ∎

With minor modifications, the argument above can be applied to any order topology satisfying certain conditions as pointed out in the exercises at the end. The technique employed above is known as the method of successive dissection and the reader must have seen it used in the standard proofs of either the Heine-Borel theorem or the Bolzano-Weierstrass theorem. It is interesting that the Heine-Borel theorem can be deduced from the theorem above. Although this by itself is not very surprising inasmuch as connectedness of $\mathbb{R}$ is equivalent to its completeness, we give below the argument to illustrate another important technique.

(2.6) Theorem: Every closed and bounded interval is compact.

Proof: Since any closed and bounded interval $[a, b]$ (with $a < b$) is homeomorphic to the unit interval $[0, 1]$, it suffices to prove that $[0, 1]$ is compact. Let $\mathcal{U}$ be an open cover of $[0, 1]$. An element U of $\mathcal{U}$ is open relative to $[0, 1]$ and hence is of the form $V \cap [0, 1]$ where V is an open subset of $\mathbb{R}$. Replacing such U's by the corresponding V's, we get a cover $\mathcal{V}$ of $[0, 1]$ by sets which are open in $\mathbb{R}$ and evidently it suffices to show that $\mathcal{V}$ has a finite subcover. (Here, in effect, we are using Proposition (1.2).) Now let $S = \{t \in [0, 1] :$ the interval $[0, t]$ can be covered by finitely many members of $\mathcal{V}\}$. We have to show that $1 \in S$. Evidently $0 \in S$ and so $S \neq \emptyset$. We claim S is both open and closed in $[0, 1]$. First, let $t \in S$. Then $[0, t]$ can be covered by, say, $V_1, V_2, \ldots, V_n \in$ with $t \in V_n$ (say). Since V is open, there exists $\delta > 0$ such that $(t - \delta, t + \delta) \subset V$. Now for any $t' \in (t - \delta, t + \delta) \cap [0, 1]$, the interval $[0, t']$ is also covered by $V_1, V_2, \ldots, V_n$ showing that $t' \in S$. Hence $(t - \delta, t + \delta) \cap [0, 1] \subset S$ and hence S is open, being a neighbourhood of each of its points.

On the other hand suppose $t \in [0, 1] - S$. Choose $V \in \mathcal{V}$ such that $t \in V$. V being open, there is $\delta > 0$ such that $(t - \delta, t + \delta) \subset V$. We claim that $(t - \delta, t + \delta) \cap [0, 1] \subset [0, 1] - S$. For let $t' \in (t - \delta, t + \delta) \cap [0, 1]$. If $t' \in S$ then $[0, t']$ can be covered by, say, $V_1, V_2, \ldots, V_n \in \mathcal{V}$. But then $V_1, V_2, \ldots, V_n$ and V together cover $[0, t]$ contradicting that $t \notin S$. Thus we have shown that $[0, 1] - S$ is also open.

Putting it all together, S is a non-empty clopen subset of $[0, 1]$. But by the theorem above $[0, 1]$ is connected. We are thus forced to conclude that S is the entire interval $[0, 1]$ and hence, in particular, $1 \in S$ as was to be proved. ∎

The way connectedness is used in the proof above is typical. We want to prove something about all points in a connected space. We consider the set of those points for which it is true and show that it is a clopen set. Hence, by connectedness it must be either empty or the whole space. The first possibility is usually easy to rule out.

Now that we know what are the connected subsets of the real line, it follows as a very special case of Proposition (2.4) that a continuous real-valued function on an interval assumes every value between its extrema. This is the well-known intermediate value theorem from elementary calculus. In essence it says that a continuous function from $\mathbb{R}$ to $\mathbb{R}$ maps intervals onto intervals. However this property does not characterise continuity as will be seen from an example in the exercises.

Let us now go back to connected spaces in general. We shall prove some basic results. But first we need a definition.

(2.7) Definition: Two subsets A and B of a space X are said to be **(mutually) separated** if $\bar{A} \cap B = \emptyset$ and $A \cap \bar{B} = \emptyset$.

Here the closure is w.r.t. the space X. The condition is a little stronger than saying that A and B are mutually disjoint. But it is weaker than saying that their closures are disjoint. For example in $\mathbb{R}$, the intervals $(-1, 0)$ and $(0, 1)$ are separated although their closures have 0 as a common point. Note that A and B are separated iff they are disjoint closed subsets of $A \cup B$ with relative topology. Note also that a space is connected iff it is not the union of two non-empty separated subsets.

(2.8) Proposition: Let X be a space and C be a connected subset of X (that is, C with the relative topology is a connected space). Suppose $C \subset A \cup B$ where A, B are mutually separated subsets of X. Then either $C \subset A$ or $C \subset B$.

Proof: Let $G = C \cap A$ and $H = C \cap B$. Then G, H are closed subsets of C since, A, B are closed in $A \cup B$. Also $G \cap H = \emptyset$. But C is connected. So either $G = \emptyset$ or $H = \emptyset$. In the first case $C \subset B$ while in the second, $C \subset A$. ∎

(2.9) Theorem: Let $\mathcal{C}$ be a collection of connected subsets of a space X such that no two members of $\mathcal{C}$ are mutually separated. Then $\bigcup_{C \in \mathcal{C}} C$ is also connected.

Proof: Let $M = \bigcup_{C \in \mathcal{C}} C$. If M is not connected then we could write M as a $A \cup B$ where A, B are nonempty and mutually separated subsets of X. By the proposition above, for each $C \in \mathcal{C}$ either $C \subset A$ or $C \subset B$. We contend that the same possibility holds for all $C \in \mathcal{C}$, i.e. either $C \subset A$ for all $C \in \mathcal{C}$ or $C \subset B$ for all $C \in \mathcal{C}$. If this is not the case, then there exist $C, D \in \mathcal{C}$ such that $C \subset A$ and $D \subset B$. But, A, B are mutually separated and hence their subsets C, D are also mutually separated contradicting the hypothesis. Thus all members of $\mathcal{C}$ are contained in A or all are contained in B. Accordingly $M = A$ or $M = B$, contradicting that A, B are both non-empty. ∎

For applications, it is convenient to have the following corollary of the last theorem.

(2.10) Corollary: Let $\mathcal{C}$ be a collection of connected subsets of a space X and suppose K is a connected subset of X (not necessarily a member of $\mathcal{C}$) such that $C \cap K \neq \emptyset$ for all $C \in \mathcal{C}$. Then $(\bigcup_{C \in \mathcal{C}} C) \cup K$ is connected.

Proof: Let $M = (\bigcup_{C \in \mathcal{C}} C) \cup K$. Let $\mathcal{D} = \{K \cup C : C \in \mathcal{C}\}$. Clearly $M = \bigcup_{D \in \mathcal{D}} D$. By the theorem above, each member of $\mathcal{D}$ is connected since it is a union of two connected sets which intersect (and which are therefore not separated). Now any two members of $\mathcal{D}$ have at least points of K in common and so are not mutually separated. So again by the theorem above, M is connected. ∎

As an application of this corollary we show that the topological product of two connected spaces is connected.

(2.11) Proposition: Let X_1, X_2 be topological spaces and $X = X_1 \times X_2$ with the product topology. Then X is connected.

Proof: If either X_1 or X_2 is empty then so is X and the result holds trivially. So assume X_1, X_2 are both non-empty. Fix a point $y_1 \in X_1$. Then the subset $\{y_1\} \times X_2$ is homeomorphic to X_2 (see Exercise (5.3.7)) and hence is connected. Call it K. For each $x \in X_2$, the set $Y_1 \times \{x\}$ is similarly connected and its intersection with K is non-empty. Also note that $X_1 \times X_2 = (\bigcup_{x \in X_2} X_1 \times \{x\}) \cup K$. So by the corollary above $X_1 \times X_2$ is connected. ∎

It is suggested that the reader draw a picture of $\mathbb{R} \times \mathbb{R}$ to see the construction in the proof above. What we have done is to write the plane as the union of all straight lines parallel to the x-axis. Each such line is connected, but every two distinct parallel lines are mutually separated and so Theorem (2.9) cannot be applied directly. So we let K be the y-axis (or any line parallel to it) and apply the corollary instead. Having proved that the product of two connected spaces is connected we can now extend the result as follows:

(2.12) Corollary: The topological product of any finite number of connected spaces is connected.

Proof: If $X_1, X_2, \ldots, X_{n-1}, X_n$ are spaces (with $n \geq 2$) then $X_1 \times X_2 \times \ldots \times X_n$ is homeomorphic to $(X_1 \times \ldots \times X_{n-1}) \times X_n$ (see Exercise (5.3.6)). The result follows by induction on n and the last proposition. ∎

Another application of the argument in Proposition (2.8) is frequently useful.

(2.13) Proposition: The closure of a connected subset is connected. More generally if C is a connected subset of a space X then any set D such that $C \subset D \subset \bar{C}$ is connected.

Proof: Suppose C is connected and $C \subset D \subset \bar{C}$. If D is not connected then we can write $D = A \cup B$ where A, B are nonempty, disjoint and closed relative to D. Then $C \cap A$, $C \cap B$ are disjoint closed subsets of C whose union is C. But C is connected. So one of them, say, $C \cap B$ is empty. This means $C \subset A$, and hence $\bar{C}^D \subset A$ where the closure is w.r.t. D. But $\bar{C}^D = \bar{C}^X \cap D = D$ since $D \subset \bar{C}^X$. Hence $A = D$ contradicting that B is nonempty. So D is connected. ∎

Proposition (2.4) along with the preceding five results are among the standard ways of showing that a space is connected. Starting from connectedness of intervals in the real line we can now prove that most familiar spaces such as the euclidean spaces $\mathbb{R}^n$, S^n (for $n > 0$), the open or closed balls in euclidean spaces are connected.

If a space is not connected, then intuitively, it splits into two or more pieces or parts. Each piece is called a component. What should be the precise definition? First of all, certainly each piece ought to be a connected subset. Secondly, in order that it be a whole piece it should not be possible to enlarge it while keeping it connected. This suggests the following definition:

(2.14) Definition: A **component** of a space is a maximally connected subset, that is, a connected subset which is not properly contained in any connected subset of that space.

For example, the space $(-1, 0) \cup (0, 1)$ with the usual topology has two components, $(-1, 0)$ and $(0, 1)$. A space is connected iff it has only one component, namely, the whole space itself. On the other hand in a discrete space the only connected subsets are the empty set and the singleton subsets and hence all components are singleton sets. Such a space is said to be **totally disconnected.** Discrete spaces are not the only examples of totally disconnected spaces. The set of rationals, the set of irrationals are also totally disconnected. The real line with semi-open interval topology is another example.

Properties of components are summarised in the following:

(2.15) Theorem: (a) Components are closed sets. (b) Any two distinct components are mutually disjoint. (c) Every nonempty connected subset is contained in a unique component. (d) Every space is the disjoint union of its components.

Proof: (a) Let C be a component of a space X. Then C is connected. Hence by Proposition (2.13) $\overline{C}$ is also connected. Now $C \subset \overline{C}$. But C is maximal w.r.t. the property of being connected, that is, no proper superset of C can be connected. Hence $C = \overline{C}$ and so C is closed. (b) Let C, C' be two components. If $C \cap C'$ is nonempty then by Theorem (2.9), $C \cup C'$ would be connected. But $C \subset C \cup C'$ and $C' \subset C \cup C'$. So again by maximality of C, C' we get $C = C \cup C' = C'$. Thus two distinct components are disjoint. (c) Let A be a nonempty connected subset of X. Let $\mathcal{C}$ be the collection of all connected subsets of X containing A and let $M = \bigcup_{C \in \mathcal{C}} C$. Then any two members of $\mathcal{C}$ intersect and so by theorem (2.9), M is connected. Clearly $A \subset M$. We claim M is a component. For suppose N is a connected subset of X containing M. Then $N \in \mathcal{C}$ and so $N \subset M$, whence $M = N$. In other words, M is a maximally connected subset of X. Thus every nonempty subset is contained in a component. Such a component is unique since two distinct components are disjoint. (d) This assertion follows from the fact that for any $x \in X$, $\{x\}$ is a connected set and hence there is a unique component C of X such that $x \in C$. ∎

The number of components of a space is evidently a topological invariant. A direct or an indirect application of this fact is often useful in showing that certain spaces cannot be homeomorphic to each other. As an example, consider $[0, 1]$ and the unit circle S^1 each with the usual topology. Both have one component each. However in $[0, 1]$ there are points x such that $[0, 1] - \{x\}$ is not connected (such points are called **cut points**). It is clear that if $h : [0, 1] \to S^1$ is a homeomorphism then h would map a cut point of $[0, 1]$ to a cut point of S^1. But it is easy to show that S^1 has no cut points. Hence $[0, 1]$ cannot be homeomorphic to S^1. Clearly the number of cut points is also a topological invariant.

The property of being connected is not hereditary. Also connected sets can be quite weird. For example there exists a connected subset X of the plane and a point x of X such that the set $X - \{x\}$ is totally disconnected! Such a point, when it exists, is called a **dispersion point.**

Exercises

2.1 Prove that a subset X of $\mathbb{R}$ is an interval iff it has the property that for all $a, b \in X$, $(a, b) \subset X$.

2.2 Let $(X, <)$ be linearly ordered set. A *gap* in X is defined as a pair (a, b) where $a, b \in X$ such that $a < b$ and there is no point c in X satisfying $a < c < b$. Prove that X with the order topology is connected iff X is order complete and there are no gaps in X.

2.3 Prove that the complement of $\mathbb{Q} \times \mathbb{Q}$ in the plane $\mathbb{R}^2$ is connected.

2.4 Let S^n be the unit sphere in $\mathbb{R}^{n+1}$. Prove that the space obtained by removing any one point from S^n is homeomorphic to $\mathbb{R}^n$. (Hint: First show that without loss of generality the point removed may be taken to be the 'north pole' $(0, 0, 0, \ldots, 0, 1) \in S^n$. Then use the higher dimensional analogue of the well-known stereographic projection in complex analysis).

2.5 Using the last exercise or otherwise, prove that S^n is connected.

2.6 Give rigorous arguments to show that no two spaces in the following list are mutually homemorphic:
(i) the closed interval $[0, 1]$
(ii) the open interval $(0, 1)$
(iii) the semi-open interval $[0, 1)$
(iv) a triod, that is a space homeomorphic to the figure Y
(v) the unit circle S^1
(vi) a space homeomorphic to the figure X
(vii) a figure eight curve.

2.7 Prove or disprove that the interior and the boundary of a connected set are connected.

2.8 Give an example of a connected closed subset C of $\mathbb{R}^2$ such that $\mathbb{R}^2 - C$ has infinitely many components.

*2.9 Prove that a connected space cannot have more than one dispersion points.

2.10 A metric space (X, d) is said to be **chain connected** if for every $\epsilon > 0$ and for given $x, y \in X$ there exists a finite sequence $\{x_0, x_1, x_2, \ldots, x_n\}$ in X such that $x_0 = x$, $x_n = y$ and $d(x_i, x_{i-1}) < \epsilon$ for all $i = 1, 2, \ldots, n$. Such a sequence is called an ϵ-chain joining x to y.
(a) Prove that a connected metric space is chain connected. (Hint: For each $\epsilon > 0$, apply the trick used in the proof of Theorem (2.6).)
(b) Let $H = \{(x, y) \in \mathbb{R}^2 : x^2y^2 = 1\}$. Prove that with the usual metric, H is chain connected but not connected.

2.11 A space X is said to have the **fixed point property** if every map $f : X \to X$ has a fixed point, i.e. a point $x_0 \in X$ such that $f(x_0) = x_0$.
(a) Prove that a space having the fixed point property must be connected.
(b) Prove that the unit interval $[0, 1]$ has the fixed point property.

2.12 Let $A = \{(x, \sin 1/x) : 0 < x \leqslant 1\} \subset \mathbb{R}^2$ and let $X = \overline{A}$ where the closure is w.r.t. the usual topology on $\mathbb{R}^2$. Prove that X is connected. (Hint: Note that A is the graph of a continuous function on $(0, 1]$ and apply Exercise (5.3.8).)

2.13 Let $p : \mathbb{R} \to \mathbb{R}/R$ be the quotient map in Exercise (5.4.15).

(a) Prove that the set $\mathbb{R}/R$ has cardinal number c.

(b) Let $f = \theta \circ p$ where $\theta : \mathbb{R}/R \to \mathbb{R}$ is some bijection. Prove that f is not continuous at any point of $\mathbb{R}$. Prove also that f maps every non-degenerate interval onto $\mathbb{R}$ and hence that f is open.

2.14 Prove that a continuous bijection from $\mathbb{R}$ to $\mathbb{R}$ must be a homeomorphism. (Hint: First show that such a function is strictly monotonic).

2.15 Let X be the union of semi-open intervals of the form $[2n, 2n + 1)$ for $n \in \mathbb{N}$.

(a) Prove that there exists a continuous bijection from X onto $[0, 1)$. $\left(\text{Hint: Map } [2n, 2n + 1) \text{ onto } \left[1 - \frac{1}{2^{n-1}}, 1 - \frac{1}{2^n}\right) \text{ for } n \in \mathbb{N}\right)$.

(b) Prove that there exists a continuous bijection f from X onto itself which is not a homeomorphism. (Hint: For each $k \in \mathbb{N}$, let $S_k = \{h \in \mathbb{N} : n \text{ is divisible by } 2^{k-1} \text{ but not by } 2^k\}$. Let $X_k = \bigcup_{n \in S_k} [2n, 2n + 1)$. Then X is the disjoint union of X_k's; each X_k is itself homeomorphic to X. Map X_k onto $[2k, 2k + 1)$ by a construction similar to that in (a).)

(c) Let $f : X \to X$ be a bijection as in (b). Prove that f^{-1} is not continuous even though its graph $G(f^{-1})$ is homeomorphic to X. This shows that the converse of the result in Exercise (5.3.8) is false.

2.16 Let X, Y be topological spaces such that each is homeomorphic to a subspace of the other. Must X, Y be homeomorphic to each other? In other words is the topological analogue of the Schröder-Bernstein theorem in set theory true?

Notes and Guide to Literature

The literature on connectedness and related concepts is very vast and it is hard even to give a representative list of references. The book of Hocking and Young [1] gives a fairly extensive account of various concepts and presents a charming wealth of 'pathological' examples. The example in Exercise (2.12) is among the stock counterexamples and we shall return to it in the next section. For a connected subset of $\mathbb{R}^2$ with a dispersion point, see Knaster and Kuratowski [1]. See also the note by J. R. Kline [1].

The fixed point property has been an area of much work. One of the most classic results of topology is the Brouwer fixed point theorem which says that the closed unit ball D^n in $\mathbb{R}^n$ has the fixed point property. An elementary proof of this result can be found in Dugundji [1] although truly elegant proofs require methods beyond general topology. A good expository article on the fixed point property is by Bing [1].

3. Local Connectedness and Paths

In this section we localise the concept of connectedness. In general to say that a topological property holds locally at a point x of a space X is to say that there exists a neighbourhood N of x (in X) which has that property, whether the whole space has that property or not. For example, a space X is said to be **locally metrisable** at a point $x \in X$ if there exists a neighbourhood N of x in X such that the space N (with the subspace topology on it) is metrisable. Sometimes, however, it is not enough to assume the existence of just one neighbourhood having the given property, but it is necessary to require that there are arbitrarily small neighbourhoods of x having that property. Here 'small' is in the sense of inclusion. When the property in question is hereditary (such as metrisability) the problem does not arise because whenever a nbd N has that property so will any other nbd smaller than N. Connectedness is not a hereditary property; so we have to define local connectedness as follows:

(3.1) Definition: A space X is said to be **locally connected at a point** x in it if for every neighbourhood, N of x (in X) there exists a connected neighbourhood M of x such that $M \subset N$. X is said to be **locally connected** if it is locally connected at each of its points.

Put differently, X is locally connected at x iff the family of all connected neighbourhoods of x is a local base at x. Another version is provided by the following proposition.

(3.2) Proposition: A space X is locally connected at a point $x \in X$ iff for every neighbourhood N of x, the component of N containing x is a neighbourhood of x.

Proof: Suppose X is locally connected at x. Let N be a nbd of x and let C be the component of N containing x (caution: C may not be a component of X). We are given that there exists a connected nbd M of x such that $M \subset N$. Then M is contained in a unique component of N and this component must be C since M intersects C. So $M \subset C$ and hence C is a neighbourhood of x. Conversely let N be a nbd of x and let C be the component of N containing x. Then C is a connected neighbourhood of x contained in N and so X is locally connected at x. ∎

It is clear that all indiscrete and all discrete spaces are locally connected. It is also easy to see that the real line with the usual topology is locally connected since we proved in the last section that all intervals are connected. We leave it to the reader that the topological product of a finite number of locally connected spaces is locally connected, from which it follows that all euclidean spaces are locally connected. As examples of spaces which are not locally connected we can cite the spaces of rationals or irrationals with usual relative topologies. Before giving non-trivial examples, let us obtain a simple characterisation of locally connected spaces.

(3.3) Proposition: For a topological space X the following statements are equivalent:

(i) X is locally connected.

(ii) Components of open subsets of X are open (in X).

(iii) X has a base consisting of connected subsets.

(iv) For every $x \in X$ and every neighbourhood N of x there exists a connected open neighbourhood M of x such that $M \subset N$.

Proof: (i) $\Rightarrow$ (ii). Suppose X is locally connected, G is open in X and C is a component of G. (Caution: C need not be a component of X; however it is contained in a component of X.) We have to show that C is open in X. Let $x \in C$. Then G is a neighbourhood of x and C is the component of G containing x. So by local connectedness at x and the last proposition, C is a neighbourhood of x. Thus C is a neighbourhood of each of its points and therefore C is open in X.

(ii) $\Rightarrow$ (iii). Let G be any open subset of X. Write G as the union of its components (in the relative topology on G). All these components are open, connected subsets of X by (ii). This means that every open set can be expressed as the union of some open, connected subsets of X, or in other words that the family of such sets forms a base for X.

(iii) $\Rightarrow$ (iv). Let $x \in X$ and N be a neighbourhood of x. Then there exists an open set V such that $x \in V \subset N$. By (iii) and the characterisation of a base given in Proposition (4.3.2), there exists a connected open set M such that $x \in M$ and $M \subset V$. Hence $M \subset N$.

(iv) $\Rightarrow$ (i). This is immediate. ∎

We are now in a position to give a classic example of a space which is connected but not locally connected. Let X be the set $B \cup A$ where $B = \{(x, 0) \in \mathbb{R}^2 : 0 \leq x \leq 1\}$ and $A = \{(x, y) \in \mathbb{R}^2 : 0 \leq y \leq 1, x = 0$ or $x = 1/n$ for some $n \in \mathbb{N}\}$. The set X is pictured in Figure 1. It consists of infinitely many vertical segments of unit length, including a segment on

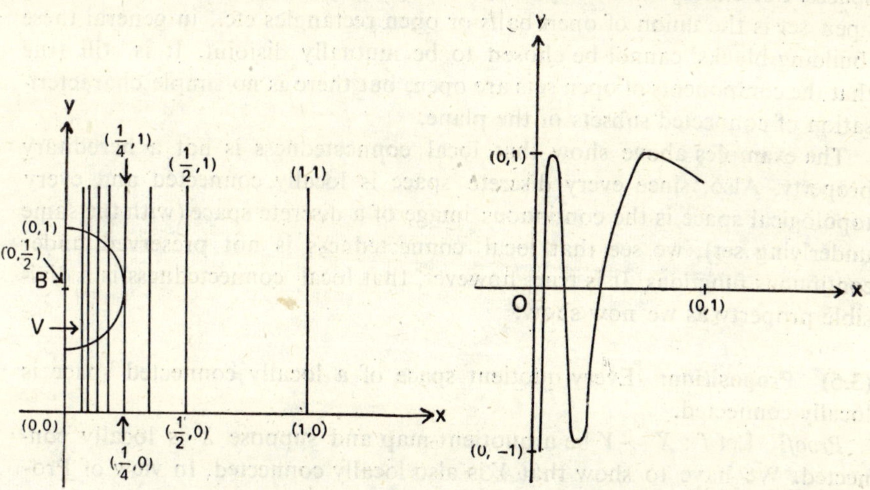

Fig. 1. Comb Space.

Fig. 2. Topologist's sine curve.

the y-axis and a horizontal segment along the x-axis. Give X the relative topology induced by the usual topology on the plane. X is then called a 'comb space'. It is easy to show that X is connected, for each of the vertical segments is connected and meets the horizontal segment B which is also connected. However X is not locally connected. For, let V be the open ball (in X, with usual metric) centred at $(0, \frac{1}{2})$ and with radius $\frac{1}{4}$. The components of V will be portions of the vertical segments. They will all be open (in X) except the one along the y-axis, namely the component $\{(0, y) \in \mathbb{R}^2 : \frac{1}{4} < y < \frac{3}{4}\}$. Another example is the famous 'topologist's sine curve' pictured in Figure 2. This space was analytically defined in Exercise (2.12) of the last section. The proof that it is not locally connected is similar to the proof that the comb space is not locally connected.

As an application of the last proposition we can prove a well-known fact about open subsets of the real line. A proof could have been given earlier, but with the machinery we have at our disposal the result falls out effortlessly.

(3.4) Theorem: Every open subset of the real line (in the usual topology) can be expressed as the union of mutually disjoint open intervals.

Proof: Let V be an open subset of $\mathbb{R}$. Write V as the disjoint union of its components. Each such component is a connected subset of $\mathbb{R}$ and hence an interval by Theorem (2.5). But by the last proposition, each component of V is open in $\mathbb{R}$. So V is the disjoint union of open intervals. ∎

We know that the collection of open intervals is a base for the usual topology on $\mathbb{R}$ and hence that any open subset of $\mathbb{R}$ can be expressed as a union of open intervals. The crux of the preceding theorem is that these intervals could be chosen to be mutually disjoint. There is no analogue of this theorem which holds good for the plane or higher dimensional euclidean spaces. For example, in the plane, although it is easy to show that every open set is the union of open balls or open rectangles etc., in general these 'building blocks' cannot be chosen to be mutually disjoint. It is still true that the components of open sets are open, but there is no simple characterisation of connected subsets of the plane.

The examples above show that local connectedness is not a hereditary property. Also, since every discrete space is locally connected and every topological space is the continuous image of a discrete space (with the same underlying set), we see that local connectedness is not preserved under continuous functions. It is true, however, that local connectedness is a divisible property as we now show:

(3.5) Proposition: Every quotient space of a locally connected space is locally connected.

Proof: Let $f : X \to Y$ be a quotient map and suppose X is locally connected. We have to show that Y is also locally connected. In view of Proposition (3.3), it suffices to prove that components of open subsets of Y are open in Y. So let V be an open subset of Y and let C be a component of V. Then $f^{-1}(V)$ is an open subset of X and we assert that $f^{-1}(C)$ is a union of some components of $f^{-1}(V)$. This amounts to showing that if x is in $f^{-1}(C)$

and D is the component of $f^{-1}(V)$ containing x, then $D \subset f^{-1}(C)$. Since D is connected and f is continuous, we know that $f(D)$ is connected. Also $f(D) \subset V$ and $f(D) \cap C \neq \emptyset$ since $x \in f^{-1}(C) \cap D$. So $f(D) \cup C$ is connected by Theorem (2.9). But C is a maximally connected subset of $f^{-1}(V)$. So $f(D) \cup C = C$ or $f(D) \subset C$. Hence $D \subset f^{-1}(C)$ and we have thus shown that $f^{-1}(C)$ is the union of some components of V. But since X is locally connected and $f^{-1}(V)$ is open, each of its components is open in X. So $f^{-1}(C)$ is open in X. This shows that C is open in Y as Y has the strong topology generated by $\{f\}$. Thus we have shown that components of open subsets of Y are open and so Y is locally connected. ∎

In particular it follows that every quotient space of the unit interval is locally connected. In the next chapter we shall see that any surjective map from the unit interval is a quotient map provided the range space satisfies a certain condition. The last proposition would then be useful to show the non-existence of certain maps.

We conclude this chapter with a brief discussion of paths. Although they are of no great importance in general topology, we include them because the idea behind them appeals to our intuition very easily.

In practice we regard a path as a track or a route. Thus a path in a space would be some subset of it satisfying some condition to the effect that it should have a length but no breadth. In topology, however, we incorporate the concept of a motion in the definition of a path. Thus in order to specify the path traversed by a particle in motion, not only must we give the various positions assumed by the particle, but we must also tell at which moments of time the particle attained these positions. As time is usually measured starting from the initial position as zero, we are led to the following definition.

(3.6) Definition: A **path** in a topological space X is a continuous function α from the unit interval $[0, 1]$ into X. The points $\alpha(0)$ and $\alpha(1)$ are called respectively the **initial** and the **terminal points** or the **beginning** and the **end** of α. We say α is a path from $\alpha(0)$ to $\alpha(1)$ or that it joins $\alpha(0)$ to $\alpha(1)$.

We could of course replace the unit interval by any closed, bounded interval $[a, b]$ where $a < b$. However, topologically the generality gained is illusory since $[a, b]$ is homeomorphic to $[0, 1]$. A path is sometimes also called a **curve**; however, this term is generally reserved for paths in a euclidean space. A path α is said to be **simple** if the function α is injective and it is said to be **closed** if $\alpha(0) = \alpha(1)$. A **simple closed** path is a closed path α which is injective except for $\alpha(0) = \alpha(1)$ (that is, 0 and 1 are the only distinct points which are taken by α to the same point). The essential point about a path α is that it is a function and is not to be confused with its range, i.e. the set of points traversed by it. Consider the paths $\alpha, \beta : [0, 1] \to \mathbb{R}$ by $\alpha(t) = t$ and $\beta(t) = t^2$ for $t \in [0, 1]$. Then α, β are distinct paths although they cover the same set of points in $\mathbb{R}$.

(3.7) Definition: A space X is said to be **path-connected** if for every two points $x, y \in X$, there exists a path α in X such that $\alpha(0) = x$ and $\alpha(1) = y$.

Clearly the real line, all euclidean spaces and any convex subsets of them are all path-connected. The unit sphere S^n is also path-connected for $n \geqslant 1$ since given any two distinct points on S^n there is an arc of a great circle joining them and it can be parametrised as a path (in fact as a simple path). It is easy to show that the topological product of path-connected spaces is path-connected.

(3.8) Proposition: Every path-connected space is connected.

Proof. Let X be a path-connected space. If X is empty then certainly it is connected. Suppose X is non-empty. Fix some point $x_0 \in X$. For each $x \in X$, we are given a path α_x in X such that $\alpha_x(0) = x_0$ and $\alpha_x(1) = x$. Let C_x be the range of α_x. Then C_x is connected since it is a continuous image of the unit interval which is connected. Clearly $X = \bigcup_{x \in X} C_x$. For any two $x, y \in X$, $x_0 \in C_x \cap C_y$ and so C_x, C_y are not mutually separated. So by Theorem (2.9) X is connected. ∎

The proposition is sometimes useful in showing that a space is connected, when a direct proof of connectedness tends to be complicated. For example, suppose F is a countable subset of the plane $\mathbb{R}^2$. We show that its complement $\mathbb{R}^2 - F$ is connected by showing that it is in fact path-connected. Let $x, y \in R^2 - F$. If $x = y$ then certainly a constant path joins x and y. Suppose $x \neq y$. Let L be the perpendicular bisector of the line segment joining x and y.

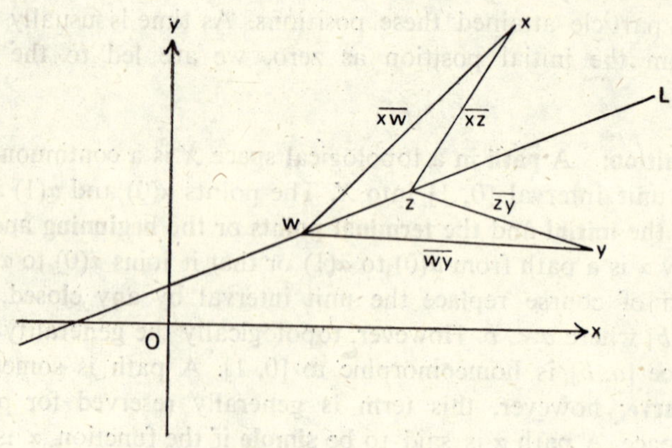

Fig. 3. Path-connectedness of $\mathbb{R}^2$ minus a countable set.

For each $z \in L$ consider the path $\alpha_z : [0, 1] \to \mathbb{R}^2$ defined by

$$\alpha_z(t) = \begin{cases} 2tz + (1 - 2t)x & \text{if } 0 \leqslant t \leqslant 1/2 \\ (2t - 1)y - 2tz + 2z & \text{if } \frac{1}{2} \leqslant t \leqslant 1 \end{cases}$$

Then α_z is a path joining x and y. The range of α_z is the union of the line segments from x to z and from z to y. If neither of these segments intersects F then α_z can be considered as a path in $\mathbb{R}^2 - F$. Note that for distinct points z and w in L the segments $\overline{xz}$ and $\overline{xw}$ have only one point (viz. x) in common and so $\overline{xz} \cap F$ is disjoint from $\overline{xw} \cap F$. Similarly $\overline{zy} \cap F$ is disjoint from $\overline{wy} \cap F$. But F is countable. It follows that there are at most countably many points z on L for which the path α_z attains values in F. But the line L is uncountable and so there exists a point $z \in L$ such that the path α_z can be taken to be a path from x to y in $\mathbb{R}^2 - F$. Hence $\mathbb{R}^2 - F$ is path-connected.

The converse of Proposition (4.3) is false. A classic counterexample is the topologists's sine curve X, pictured in Figure 2. Recall that $X = A \cup B$ where $A = \{(x, \sin 1/x) \in R^2 : 0 < x \leqslant 1\}$ and $B = \{(0, y) \in \mathbb{R}^2, -1 \leqslant y \leqslant 1\}$. Then X is connected (see Exercise (2.12)). It is clear that A and B are each path-connected, since they are respectively homeomorphic to the semi-open interval $(0, 1]$ and the unit interval $[0, 1]$, both of which are path-connected. A rigorous proof that X itself is not path-connected will be given in the next chapter and it will be based on the fact that X is not locally connected. As a less appealing example let X be a set of cardinality greater than c, the cardinality of $\mathbb{R}$. For example, we may let X be the power set of $\mathbb{R}$. Let $\mathfrak{J}$ consist of the empty set $\emptyset$ and all subsets of X whose complements have cardinality c or less. It is easy to show that $\mathfrak{J}$ is a topology on X (see the remark following the definition of the cofinite topology in Chapter 4, Section 2). With this topology X is connected because every two open sets have a nonempty intersection unless one of them is empty. For the same reason, every open subset of X is connected and hence X is locally connected. However, X is not path-connected. If α is a path in X, then its range is a connected subset of X of cardinality at most c. The only such sets are the empty set and singleton subsets. It follows that the only paths in X are the constant ones and hence that X is not path-connected.

When a space is not path-connected, it can be decomposed into path-connected subspaces as we now show.

3.9 Proposition: In any space X, the binary relation defined by letting $x \sim y$ for $x, y \in X$ iff there exists a path in X from x to y, is an equivalence relation.

Proof: Reflexivity of the relation follows trivially by considering constant paths. For symmetry, suppose α is a path in X from x to y. Define $\beta : [0, 1] \to X$ by $\beta(t) = \alpha(1 - t)$ for $t[0, 1]$. Now β is continuous since it is the composition $\alpha \circ f$ where $f : [0, 1] \to [0, 1]$ is the map $f(t) = (1 - t)$, $t \in [0, 1]$. Then β is a path from y to x. For transitivity, suppose α is a path from x to y and β is a path from y to z. Define $\gamma : [0, 1] \to X$ by

$$\gamma(t) = \begin{cases} \alpha(2t) & \text{if } 0 \leqslant t \leqslant 1/2 \\ \beta(2t - 1) & \text{if } 1/2 \leqslant t \leqslant 1. \end{cases}$$

Then γ is well-defined since $\alpha(1) = y = \beta(0)$. Also γ is continuous since its restriction to each of the two closed subsets $[0, 1/2]$ and $[1/2, 1]$ of $[0, 1]$ is continuous (see Exercise (5.4.10)). Clearly γ is a path in X from x to z and so $x \sim z$. Hence $\sim$ is an equivalence relation on X. ∎

The path γ defined in the proof above is called the **composition** or **concatenation** of the paths α and β. It is defined iff the end of α is the beginning of β.

(3.10) Definition: The equivalence classes under the equivalence relation $\sim$ defined above are called the **path-components** of the space X.

It is clear from the definition that every space is the disjoint union of its path components. The following proposition establishes the analogy between components and path-components.

(3.11) Proposition: A subset C is a path-component of a space X iff C is a maximal subset of X w.r.t. the property of being path-connected.

Proof: Suppose C is a path-component of a space X. We claim C is path-connected. Let $x, y \in C$. We certainly know that there is a path α in X from x to y. We assert that the range of α is contained in C. For otherwise, there exists $s \in (0, 1)$ such that $\alpha(s) \notin C$. Now define $\beta : [0, 1] \to X$ by $\beta(t) = \alpha(st)$. Then β is a path in X from x to $\alpha(s)$. This means that x and $\alpha(s)$ are in the same equivalence class under $\sim$, contradicting that $\alpha(s) \notin C$. Hence α can be considered as a path in C, showing that C is path-connected. Moreover if D is a proper superset of C, then a point of $D - C$ cannot be joined to a point in C by a path in X, and certainly not by a path in D, showing that D is not path connected. Hence C is a maximally path-connected subset of X.

Conversely suppose C is path-connected and is not a proper subset of any path-connected subset of X. Then any two points of C can be joined by a path in C and hence, *a fortiori*, by a path in X. Hence C is contained in some equivalence class under $\sim$, say $C \subset C'$. But as we just saw, every path-component of X is path-connected. In particular C' is path-connected and so $C = C'$ by maximality of C. Thus C is a path-component of X. ∎

The concept of path-connectedness can be localised the same way as connectedness. The localisation as well as other results on path-connectedness are left as exercises.

Exercises

3.1 Let $X_1, X_2, \ldots, X_n$ be topological spaces and X their topological product. Suppose X_i is locally connected at a point x_i for $i = 1, 2, \ldots, n$. Let $x = (x_1, x_2, \ldots, x_n) \in X$. Prove that X is locally connected at x.

3.2 Prove that an open subspace of a locally connected space is locally connected.

3.3 A topological space X is said to be **homogeneous** if given any two points x and y in X, there exists a homeomorphism h of X onto itself such that $h(x) = y$ (in terms of group actions this means that the group of automorphisms of X acts transitively on X).

(a) Prove that this is a topological property, that the euclidean spaces, all unit spheres, all discrete spaces and all indiscrete spaces are homogeneous.

(b) Prove that the unit interval is not homogeneous.

(c) Prove that a homogeneous space is locally connected if and only if it is locally connected at any one point (more generally this is true of any topological property which is local in character).

3.4 Find precisely at which points the spaces in Figures 1 and 2 are not locally connected.

3.5 A **broom space** (or a pencil) is defined as follows: For an angle α, let R_α denote the portion inside the closed unit disc of the ray which makes an angle α with the x-axis. In polar coordinates, $R_\alpha = \{(r, \theta) \in \mathbb{R}^2 : 0 \leqslant r \leqslant 1, \theta = \alpha\}$. Let $B = \bigcup \{R_\alpha : \alpha = 0 \text{ or } \alpha = 1/n \text{ for } n \in \mathbb{N}\}$. The space B is pictured in Figure 4.

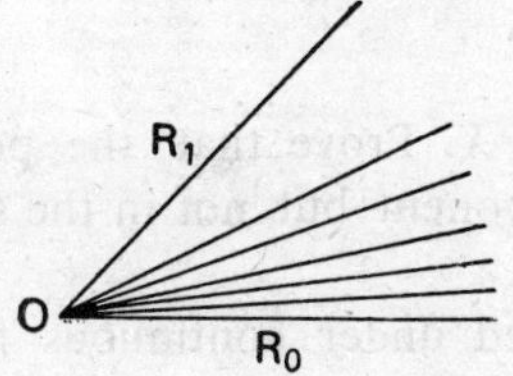

Fig. 4 Broom space.

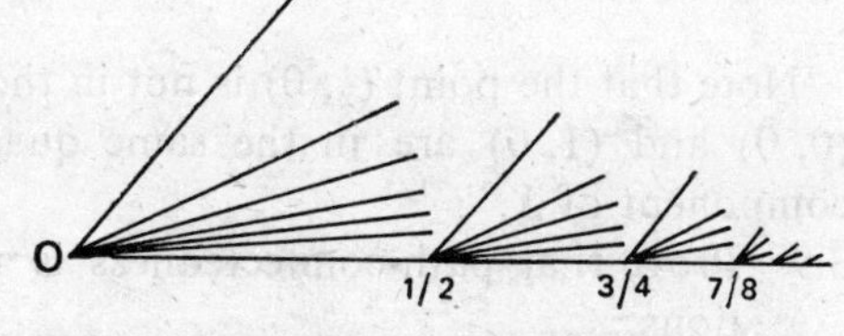

Fig. 5 String of Brooms.

Prove that the broom space B is connected and that it is not locally connected at any point of the ray R_0 except the origin. (Hint: Note that if a connected subset of B contains points from distinct rays then it must contain the origin.)

3.6 The string of brooms pictured in Figure 5 is obtained by dividing the unit interval [0, 1] into subintervals of length $\frac{1}{2^n}$, $n \in \mathbb{N}$ and placing copies of the broom space on each of them.

*(a) Express points in this space analytically.

(b) Prove that this space is locally connected at the point 1 although the only connected open neighbourhood of 1 is the whole space.

3.7 Does part (b) of the last exercise contradict the equivalence of statements (1) and (4) in Proposition (3.3) ? (cf. Exercise (1.3.9) (v)).

3.8 Let X be a topological space. Define a binary relation on X by $x \sim y$ for $x, y \in X$ iff there exists no clopen subset of X containing x but not y. In other words, $x \sim y$ iff whenever X is written as the disjoint union of two open sets, x and y are always in the same set.

(a) Prove that $\sim$ is an equivalence relation on X. The equivalence classes are called **quasi-components** of X.

(b) Prove that each component of X is contained in a quasi-component of X. (Hint: If there exists a connected set containing x and y then $x \sim y$ by Proposition (2.8).)

(c) Prove that if X is locally connected then every component of X is also a quasi-component.

(d) Let $A_n = \left\{\left(x, \frac{1}{n}\right) \in \mathbb{R}^2, 0 \leqslant x \leqslant 1\right\}$ for $n \in \mathbb{N}$ and let A_0 be

$\{(x, 0) \in \mathbb{R}^2 : 0 \leqslant x < \frac{1}{2} \text{ or } \frac{1}{2} < x \leqslant 1\}$. Let $X = \bigcup_{n=0}^{\infty} A_n$ with the relative topology of the plane. X is pictured in Figure 6.

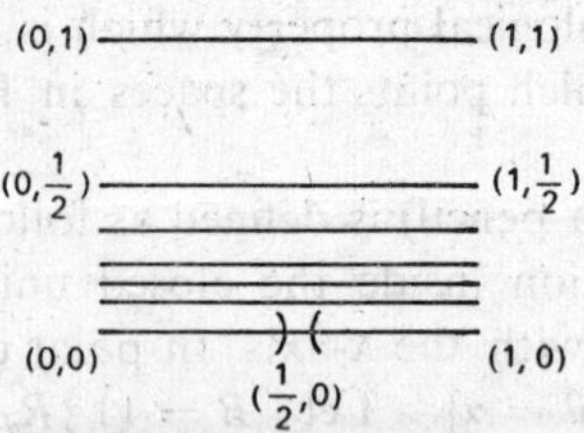

Fig. 6 Distinction between quasi-components and components.

Note that the point $(\frac{1}{2}, 0)$ is not in the space X. Prove that the points $(0, 0)$ and $(1, 0)$ are in the same quasi-component but not in the same component of X.

3.9 Prove that path-connectedness is preserved under continuous functions.

3.10 Prove that the topological product of a finite number of non-empty spaces is path-connected iff each coordinate space is so.

3.11 Define local path-connectedness. Characterise it the way local connectedness is characterised. Prove also that it is a divisible property.

3.12 Prove that a connected, locally path-connected space is path-connected.

3.13 Prove that a closed path determines a map from S^1. (Hint: obtain S^1 as a quotient space of $[0, 1]$.)

Notes and Guide to Literature

The definition of local connectedness at a point is not quite standard. Some authors, for example, Hocking and Young [1] require that the points under consideration possess arbitrarily small open connected neighbourhoods. Exercise (3.6) above shows that our definition is strictly weaker than this. However, as Proposition (3.3) shows, the two notions yield equivalent definitions of local connectedness of the whole space.

The results proved and the counterexamples given in this section give only the glimpse of the richness of the theory of local connectedness. Once again we refer the reader to Hocking and Young [1] for further information.

Chapter Seven

Separation Axioms

Although we have studied a good deal about topological spaces so far, we have not yet studied any conditions which guarantee that a sufficient number of open sets exists. Indeed, the only open sets whose existence was absolutely required so far were the empty set and the entire set. In the case of an indiscrete space, these are the only open sets. Indiscrete spaces are not very interesting topologically as most theorems reduce to trivialities when specialised to indiscrete spaces. Every sequence converges to every point and every point is near every non-empty set. The reason for this extreme behaviour is that there is no open set which will distinguish between points of the space, in the sense that it will contain some points of the space but will exclude some others. In other words given two distinct points, there are no open sets which will separate them from each other.

Since every concept in topology is defined in terms of open sets, in order to make non-trivial and interesting statements about a space, it is necessary that the space possess a fairly rich collection of open sets. In this chapter we shall study various degrees of such richness. We shall define a number of related conditions all of which assert the existence of open sets which will contain something but which will also exclude something else. For this reason, the conditions are known as separation axioms. We already noted in Chapter 4, Section 1 that all metric spaces satisfy one such condition, namely the Hausdorff property. There are other axioms of a similar spirit.

1. Hierarchy of Separation Axioms

The separation axioms are of various degrees of strengths and they are called T_0, T_1, T_2, T_3 and T_4 axioms in ascending order of strength, T_0 being the weakest separation axiom.

(1.1) Definition: A topological space X is said to **satisfy the T_0-axiom,** or is said to be a **T_0 space** if given any two distinct points in X, there exists an open set which contains one of them but not the other.

It is easy to see why the T_0 axiom is the weakest separation axiom. For if a space X is not T_0, then there would exist two distinct points x, y in X such that every open set in X either contains both x and y or else contains neither of them. In such a case, x and y may as well be regarded as topologically identical, inasmuch as any topological statement about one of them will imply a corresponding statement about the other and vice versa. For example, a sequence in X will converge to x iff it converges to y. Thus T_0 condition is the minimum requirement if we want to distinguish between x and y topologically.

Examples of T_0 spaces include, first of all, all metric spaces. Note that a pseudo-metric space is a metric space iff it is a T_0 space (Prove!). As another example, let $\mathcal{T}$ be the topology on $\mathbb{R}$ whose members are $\emptyset$, $\mathbb{R}$ and all sets of the form (a, ∞) for $a \in \mathbb{R}$ (see Example 7 of Chapter 4, Section 2). Note that in this space, for $x, y \in \mathbb{R}$ with $x < y$, there exists an open set containing y but not x (for example the interval (x, ∞)) although there exists no open set which contains x but not y. In the next separation axiom we remove this asymmetry.

(1.2) Definition: A space X is said to **satisfy the T_1-axiom** or is said to be a **T_1-space** if for every two distinct points x and $y \in X$, there exists an open set containing x but not y (and hence also another open set containing y but not x).

Again, all metric spaces are T_1. It is obvious that every T_1 space is also T_0 and the space $(\mathbb{R}, \mathcal{T})$ above shows that the converse is false. Thus the T_1-axiom is strictly stronger than T_0. (Sometimes a beginner fails to see any difference between the two conditions. The essential point is that given two distinct points, the T_0-axiom merely requires that *at least one* of them can be separated from the other by an open set whereas the T_1-axiom requires that *each one* of them can be separated from the other.) The following proposition characterises T_1-spaces.

(1.3) Proposition: For a topological space $(X, \mathcal{T})$ the following are equivalent:

(1) The space X is a T_1-space.

(2) For any $x \in X$, the singleton set $\{x\}$ is closed.

(3) Every finite subset of X is closed.

(4) The topology $\mathcal{T}$ is stronger than the cofinite topology on X.

Proof: Only the equivalence of (1) with (2) needs to be established. The rest follow from properties of closed sets and the definition of the cofinite topology. Assume (1) holds and let $x \in X$. We claim that $X - \{x\}$ is a neighbourhood of each of its points. For, let $y \in X - \{x\}$. Then x, y are distinct points of X and so by the T_1-axiom, there exists an open set, say, V such that $y \in V$ and $x \notin V$. But this means $V \subset X - \{x\}$ and hence $X - \{x\}$ is a neighbourhood of y. So $X - \{x\}$ is open and therefore $\{x\}$ is closed in X. Conversely assume (2). Let x, y be distinct points of X. Then $X - \{y\}$ is an open set which contains x but not y. Thus the T_1-axiom holds in X. ∎

In view of the proposition above, the cofinite topology on a set X is the weakest topology which makes X into a T_1-space. T_1 spaces share some of the pleasant properties of metric spaces. As an illustration, we prove the following simple proposition whose truth in the case of metric spaces must already be known to the reader.

(1.4) Proposition: Suppose y is an accumulation point of a subset A of a T_1 space X. Then every neighbourhood of y contains infinitely many points of A.

Proof: Let N be a neighbourhood of y and let $F = A \cap (N - \{y\})$. We claim that the set F is infinite. For, if not, $X - F$ will be an open set containing y and so $N \cap (X - F)$ will be a neighbourhood of y. Evidently this neighbourhood contains no point of A, except possibly y, contradicting that y is an accumulation point of A. So F is infinite, showing that every neighbourhood of y contains infinitely many points of A. ∎

Despite their nice properties, T_1 spaces are not totally free of anomalous behaviour, the most serious of which is probably in connection with convergence of sequences. In a T_1-space a sequence may converge to more than one point, in fact, it may even converge to every point of the space as in the case of the cofinite topology on an infinite set. This is certainly inconsistent with our expectation that limits of sequences be unique, a natural requirement since so many things in mathematics are defined as limits of something. Uniqueness of limits holds in spaces which have the Hausdorff property, which we now formally state as our next separation axiom.

(1.5) Definition: A space X is said to **satisfy the T_2 axiom** (or the **Hausdorff property**) or is said to be a T_2 (or **Hausdorff**) **space** if for every distinct point $x, y \in X$ there exist disjoint open sets U, V in X such that $x \in U$ and $y \in V$.

Evidently, every T_2 space is T_1 since the condition in the definition implies that U contains x but not y and that V contains y but not x. The converse is false. An infinite set with the cofinite topology is T_1 but not T_2, in fact no two open sets in it are disjoint unless one of them is empty. All metric spaces are T_2 as proved in Theorem (4.1.5). However, there are spaces which are T_2 but which are not metrisable, the real line with the semi-open interval topology is such a space (see Exercise (5.1.12)).

(1.6) Proposition: In a Hausdorff space, limits of sequences are unique.

Proof: Let $\{x_n\}$ be a sequence in a Hausdorff space X and suppose $x_n \to x$ and $x_n \to y$ as $n \to \infty$. We have to show that $x = y$. If not, then there exist open sets, U, V in X such that $x \in U$, $y \in V$ and $U \cap V = \emptyset$. Then there exist $N_1, N_2 \in \mathbb{N}$ such that $x_n \in U$ for all $n \geq N_1$ and $x_n \in V$ for all $n \geq N_2$. Let m be an integer greater than both N_1 and N_2. Then $x_m \in U \cap V$ contradicting that $U \cap V = \emptyset$. So $x = y$ and thus limits of sequences in X, when they exist, are unique. ∎

Note that the proposition does not say that in a T_2 space, every sequence has a unique limit. Indeed, it makes no assertion about the existence of a limit. It simply says that the limit, in case it exists, is unique.

The three separation axioms T_0, T_1 and T_2 considered so far deal with the

11(45-90/1980)

separation of two distinct points, say, x and y of a space. Suppose we replace one or both these points by closed sets. We then get two new separation axioms.

(1.7) Definition: A space X is said to be **regular at a point** $x \in X$ if for every closed subset C of X not containing x, there exist disjoint open sets U, V such that $x \in U$ and $C \subset V$. X is said to be **regular** if it is regular at each of its points.

(1.8) Definition: A space X is said to be **normal** if for every two disjoint closed subsets C and D of X there exist two disjoint open sets U and V such that $C \subset U$ and $D \subset V$.

Informally, a space is regular if every point can be separated from every closed subset (not containing it) and it is normal if every two mutually disjoint closed subsets can be separated from each other. These conditions can be satisfied in a vacous manner if the space fails to have very many closed sets. For example, an indiscrete space is regular because if x is a point not in a closed set C then C must be the empty set and hence we can take (actually we *have to* take) U and V to be respectively the whole space and the empty set. Similarly an indiscrete space is normal because the only way two closed subsets of it can be mutually disjoint is when one of them is empty. This argument also shows that the space $(\mathbb{R}, \mathcal{T})$ considered above is normal. Note however that it is not regular.

Thus the conditions of regularity and normality would impose real restrictions on a space only if it had a fair number of closed subsets. The simplest sets one would like to be closed are the singleton subsets. This is precisely the T_1-axiom. When this condition is combined with regularity and normality respectively, we get the following conditions:

(1.9) Definition: A topological space is said to **satisfy the T_3-axiom** or is said to be a **T_3-space** if it is regular and T_1.

(1.10) Definition: A topological space is said to **satisfy the T_4-axiom** or is said to be a **T_4-space** if it is normal and T_1.

The consistency of this terminology with the earlier notations T_0, T_1, T_2 is established by the following theorem.

(1.11) Theorem: The axioms T_0, T_1, T_2, T_3 and T_4 form a hierarchy of progressively stronger conditions.

Proof: We have to show that each axiom from T_1 onwards implies the preceding one. We have already seen that T_1 implies T_0 and T_2 implies T_1. The implication that T_3 implies T_2 follows easily from the definition of regularity by taking the closed set to be a sigleton set. Note that normality does not, by itself, imply regularity as we saw in an example above. However, in presence of the T_1-axiom, we can apply the condition of normality to a singleton set and a given closed set and see that regularity holds.

Hence T_4 implies T_3. ▮

None of the implications in the theorem above is reversible. We already gave examples of spaces which are T_0 but not T_1 and of spaces which are T_1 but not T_2. An example of a space which is Hausdorff but not regular is provided by the following topology on the real line.

Let $\mathcal{U}$ be the usual topology on $\mathbb{R}$ and let C be the set $\{1/n : n \in \mathbb{N}\}$. Let $\mathcal{S}$ be the smallest topology on $\mathbb{R}$ containing $\mathcal{U} \cup \{\mathbb{R} - C\}$. Then $\mathcal{S}$ makes $\mathbb{R}$ a T_2 space since $\mathcal{S}$ is stronger than $\mathcal{U}$ and $(\mathbb{R}, \mathcal{U})$ is a T_2-space. Note that C is closed in $\mathbb{R}$ w.r.t. $\mathcal{S}$ (although not with respect to $\mathcal{U}$) and that $0 \notin C$. The proof that 0 cannot be separated from C by open sets in $\mathcal{S}$ is left to the reader. Thus $(\mathbb{R}, \mathcal{S})$ is not a regular space. An example of a space which is T_3 but not T_4 is a little complicated and will be given later.

Spaces satisfying the T_4-axiom are quite important because on one hand the T_4-axiom is strong enough to have many interesting consequences (as we shall see later in this chapter) and on the other hand the class of T_4 spaces is large enough to include two important classes, the class of all compact Hausdorff spaces and that of all metric spaces. That all compact T_2 spaces are normal will be proved in the next section. Here we prove that all metric spaces are T_4.

(1.12) Theorem: All metric spaces are T_4 (and hence T_3 as well).

Proof: We already know that metric spaces are Hausdorff and hence T_1. It only remains to show that they are normal. Let (X, d) be a metric space and let C, D be disjoint, closed subsets of X. If either C or D is empty then we could separate them by the empty set and the set X. Suppose then that both C, D are non-empty. Define $f : X \to \mathbb{R}$ by $f(x) = d(x, C) - d(x, D)$ for $x \in X$. Then f is continuous, being the difference of two continuous real valued functions (see Exercise (3.3.11)). Note that $d(x, C) = 0$ iff $x \in C$ and $d(x, D) = 0$ iff $x \in D$ (see Chapter 5, Section 2). So f is positive on D and negative on C. Now let $U = f^{-1}(-\infty, 0)$ and $V = f^{-}(0, \infty)$. Then U, V are open subsets of X since they are the inverse images of open subsets of $\mathbb{R}$. Also $C \subset U$ and $D \subset V$ as noted above. Finally $U \cap V = \emptyset$ completing the proof that X is normal. ▮

There is a condition which is between the axioms T_3 and T_4. It is called the Tychonoff condition and we define it here.

(1.13) Definition: A space X is said to be **completely regular** if for any point $x \in X$ and closed set C not containing x, there exists a continuous function $f : X \to [0, 1]$ such that $f(x) = 0$ and $f(y) = 1$ for all $y \in C$, where the continuity is w.r.t. the usual topology on the unit interval $[0, 1]$. A space is said to be a **Tychonoff** space if it is completely regular and T_1.

There is of course nothing special about the unit interval and it could be replaced by any other closed bounded interval $[a, b]$ or even by the real line. What matters is that the function f takes distinct values at x and on the set C. Put in a different language, we see that in a completely regular space, a point can be separated (or distinguished) from a closed set by means of a continuous real-valued function. This condition is significantly different from the earlier separation axioms which assert separation by means of mutually disjoint open sets. However it is related to them, as we now show.

(1.14) Proposition: Every completely regular space is regular. Every Tychonoff space is T_3.

Proof: Suppose X is a completely regular space. Let $x \in X$ and C be a closed subset of X not containing x. We are given a continuous function $f: X \to [0, 1]$ which assumes the value 0 at x and the value 1 at all points of C. Now $[0, 1]$ is a Hausdorff space (in fact it is a metric space). Let G, H be disjoint open sets in $[0, 1]$ containing 0 and 1 respectively. Let $U = f^{-1}(G)$ and $V = f^{-1}(H)$. Then U, V are mutually disjoint, open sets in X and clearly $x \in U$, $C \subset V$. This shows that X is a regular space whenever it is completely regular. If in addition X is also T_1 then X is T_3 whenever it is a Tychonoff space. ∎

The converse of this proposition is false. A nontrivial counter-example is not easy.

Normality does not imply complete regularity, in fact, it does not even imply regularity as we saw above. However, in presence of the T_1 condition, a normal space is a Tychonoff space. This is a non-trivial result which we shall prove later. It shows that T_4 is stronger than Tychonoff while the proposition above shows that Tychonoff itself is stronger than T_3. For this reason, Tychonoff spaces are sometimes called **$T_{3\frac{1}{2}}$-spaces**! This clumsy notation is inevitable since the use of T_3 and T_4 is now fairly standard. There are also stronger separation axioms than T_4. One such will be pointed out in the exercises at the end.

For applications it is convenient to have the following reformulations of the definitions of regularity and normality.

(1.15) Proposition: For a topological space X the following statements are equivalent.

(1) X is regular.

(2) For any $x \in X$ and any open set G containing x there exists an open set H containing x such that $\overline{H} \subset G$.

(3) The family of all closed neighbourhoods of any point of X forms a local base at that point.

Proof: (1) $\Rightarrow$ (2). Suppose X is regular, $x \in X$ and G is an open set containing x. Then $X - G$ is a closed set not containing x. So by regularity of X there exist open sets U, V such that $x \in U$, $(X - G) \subset V$ and $U \cap V = \emptyset$. Then $U \subset X - V$ and hence $\overline{U} \subset X - V$ since $X - V$ is a closed set. But $X - V \subset G$ and thus $\overline{U} \subset G$. So we can let $H = U$ and (2) holds.

(2) $\Rightarrow$ (3). Let $x \in X$ and N be a neighbourhood of x in X. Let G be the interior of N in X. Then G is an open set containing x and so by (2) there exists an open set H such that $x \in H$ and $\overline{H} \subset G$. Then $\overline{H}$ is a closed neighbourhood of x contained in N. Hence the family of all closed neighbourhoods of x is a local base at x.

(3) ⇒ (1). Suppose $x \in X$ and C is a closed subset not containing x. Then $X - C$ is a neighbourhood of x. So by (3), there exists a closed neighbourhood M of x such that $M \subset X - C$. Let $U = \text{int}(M)$ and $V = X - M$. Then U, V are mutually disjoint open sets and clearly $x \in U$. Also $C \subset V$. This shows that X is a regular space. ∎

(1.16) Proposition: For a topological space X the following are equivalent:

(1) X is normal.

(2) For any closed set C and any open set G containing C, there exists an open set H such that $C \subset H$ and $\overline{H} \subset G$.

(3) For any closed set C and any open set G containing C, there exists an open set H and a closed set K such that $C \subset H \subset K \subset G$.

Proof: The argument is similar to that of the last proposition and is left to the reader. Note that (3) can be informally stated as, in any closed-open inclusion we can insert an open-closed inclusion. ∎

Perhaps in no other section so far, there are so many definitions as in this section. We conclude this section by commenting whether the properties defined in this section are preserved under continuous functions, whether they are hereditary, etc. Note that every discrete space satisfies all the separation axioms mentioned so far. Since any space whatsoever can be expressed as the continuous image of a discrete space, it is clear that none of the properties defined in this section is preserved under continuous functions. However, most of them are hereditary, with the exception of normality. By way of illustration, we prove that regularity is a hereditary property. It will be instructive for the reader to find out precisely where the argument would break down in attempting to prove that normality is hereditary. An actual counterexample will be given later.

(1.17) Proposition: Regularity is a hereditary property.

Proof: Suppose X is a regular space and Y is a subspace of X. Let $y \in Y$ and D be a closed subset of Y not containing y. Then D is of the form $C \cap Y$ where C is a closed subset of X. Note that $y \notin C$ for otherwise $y \in D$. Hence by regularity of X, there exist open sets U, V (in X) such that $y \in U$, $C \subset V$ and $U \cap V = \emptyset$. Let $G = U \cap Y$, $H = V \cap Y$. Then G, H are open in the relative topology on Y. Also $y \in G$, $D \subset H$ and $G \cap H = \emptyset$. Thus the space Y (with the relative topology) is regular. ∎

The question whether the topological product of spaces satisfying a separation axiom also satisfies it will be taken up in the next chapter where we shall deal with products formally. Here again, the only abnormal behaviour is exhibited by normality. The product of two normal spaces need not be normal. A counter-example will be given later.

Finally, as regards divisibility, it turns out that none of the separation axioms is divisible. If on $\mathbb{R}$ we define an equivalence relation R by letting xRy iff $x - y \in \mathbb{Q}$ then the quotient space X/R is indiscrete (see Exercise (5.4.15)). Thus we see that none of the T_0, T_1, T_2, T_3, T_4 and Tychonoff conditions is divisible. In this example the quotient space is regular and normal. That these properties are not divisible can be shown by another example. Let X be the set $\mathbb{R}x\{0, 1\}$ in the plane $\mathbb{R}^2$ with the usual topology. X is the union of two straight lines. Identify $(x, 0)$ with $(x, 1)$ for all $x \neq 0 \in \mathbb{R}$. The resulting quotient space Y is T_1 but not Hausdorff since the points $\{(0, 0)\}$ and $\{(0, 1)\}$ are distinct and cannot be separated from each other. It follows that Y is neither regular nor normal even though the space X has both these properties and the quotient map is open. The situation is somewhat better if the quotient map is closed and the members of the decomposition are compact, as we shall see in the next section. It is interesting to note that in the present example every point of Y has a neighbourhood which is homeomorphic to the real line. For this reason, Y is an example of what are known as locally euclidean spaces. Note in particular that every point of Y has a neighbourhood which is regular (in its relative topology), even though Y is not regular at the points $\{(0, 0)\}$ and $\{(0, 1)\}$. This means that although we have defined regularity at a point of a space, it is not a local concept like first countability or local connectedness!

Exercises

1.1 Prove that the properties T_0, T_1, T_2 and complete regularity are all hereditary.

1.2 Prove that normality is a weakly hereditary property.

1.3 Prove that the product of a finite number of non-empty spaces is T_0, T_1, T_2 regular or completely regular iff each of the coordinate spaces has the respective property.

1.4 Prove that in the space $(\mathbb{R}, \mathcal{S})$ considered above, if U, V are open sets containing 0 and the set C respectively then $U \cap V \neq \emptyset$. (Hint: Show that U must contain a set of the form $(-1/n, 1/n) \cap (\mathbb{R} - C)$ for some $n \in \mathbb{N}$. Then for $m > n$, $1/m \in \bar{U}$).

1.5 Prove that the co-countable topology on an uncountable set does not make it a Hausdorff space although limits of sequences in it are unique (See Exercise (4.2.1).) This example shows that the converse of Proposition (1.6) is false.

*1.6 Prove that the converse of Proposition (1.6) does hold if the space is assumed to be first countable. (Even without first countability it holds if one uses the appropriate generalisation of sequences, called nets, which we shall study later.)

1.7 Prove Proposition (1.16).

*1.8 For a set Y, the **diagonal** ΔY is defined to be the set $\{(y, y) \in Y \times Y : y \in y\}$. Prove that a space Y is T_2 iff the diagonal ΔY is a closed subset of $Y \times Y$ in the product topology.

1.9 Let Y be a Hausdorff space. Prove that for any space X and any two maps $f, g : X \to Y$ the set $\{x \in X : f(x) = g(x)\}$ is closed in X. (Hint: If U, V are neighbourhoods of $f(x)$ and $g(x)$ respectively then $f^{-1}(U) \cap g^{-1}(V)$ is a neighbourhood of x).

1.10 Prove that the property in the last exercise actually characterises

Hausdorff spaces. (Hint: Apply the property to the two projections from the product $Y \times Y$ onto Y and use Exercise (1.8).)

1.11 A space X is said to be **completely normal** if for every two mutually separated subsets C and D of it, there exist open sets, U, V such that $C \subset U$, $D \subset V$ and $U \cap V = \emptyset$. A completely normal and T_1 space is called T_5 **space.**

(a) Prove that every completely normal space is normal.

(b) Prove that all metric spaces are T_5.

(An example of a space which is T_4 but not T_5 will be given later.)

1.12 Let R be an equivalence relation on a space X. Prove that the quotient space X/R is T_1 if and only if the equivalence classes of R are closed in X.

1.13 Let A be a non-empty closed subset of a space X. Prove that the quotient space X/A, obtained by collapsing A to a point is T_2 whenever X is T_3.

1.14 Let A, B be mutually disjoint, non-empty, closed subsets of a space X. Let Y be the quotient space obtained by identifying A to a point and B to another point (in other words, Y is obtained from X by the decomposition whose members are A, B and $\{x\}$ for $x \in X - (A \cup B)$). Prove that if X is T_4 then Y is T_2.

1.15 Let X be a T_3 space which contains two closed subsets A, B which cannot be separated from each other by open sets. (In other words, X is a T_3 space which is not normal; an example of such a space will be given later.) Prove that the quotient space X/A is T_2 but not T_3.

1.16 Let X be any space. Define a binary relation R on X by letting xRy iff $x \in \overline{\{y\}}$ and $y \in \overline{\{x\}}$.

(a) Prove that R is an equivalence relation on X.

(b) Prove that the quotient space X/R is T_1 and that the quotient map $p : X \to X/R$ is closed. (Actually if a subset A of X is closed then $p^{-1}(p(A)) = A$.)

(c) Prove that any map from X into a T_1 space factors uniquely through p. In other words, given any map $f : X \to Y$ where Y is a T_1 space, there exists a unique map $g : X/R \to Y$ such that $f = g \circ p$.

Notes and Guide to Literature

The hierarchy of separation axioms may be thought of as a set of progressively stronger conditions imposed upon a topological space to make it resemble a metric space. As observed by Kelley [1], the terms 'regular', 'completely regular', 'normal' etc. are excellent examples of the time-honoured custom of referring to a problem we cannot handle as abnormal, irregular, improper, degenerate, inadmissible and otherwise undesirable. The terminology using increasing subscripts for increasingly stronger conditions, is due to Alexandroff and Hopf [1].

The T_2-axiom, introduced by Hausdorff under the name of separability is probably the most useful separation axiom especially where convergence is concerned. Indeed many topologists do not bother to consider spaces which are not Hausdorff. This attitude is typified in the Bourbaki school where 'regular' means what we call as T_3, 'normal' means what we call as T_4 and even the definition of compactness includes the Hausdorff property! It was probably the book of Kelley that prominently pointed out that life is not all that hopeless without T_2. An interesting article on this issue, titled 'Life without T_2' is due to Wilansky [1]. The characterisations of T_2 property given in the exercises are frequently useful.

D. S. Scott [1] has constructed a large class of T_0 spaces with interesting applications to logic.

For examples of T_3 spaces which are not completely regular, see Novak [1] or Thomas [1].

Two extremely important and non-trivial characterisations of normality will be proved later in this chapter. Stronger notions such as 'full normality' and 'perfect normality' have also been studied (see Tukey [1], A. H. Stone [1]). Dieudonné, in 1944, defined paracompactness, which is a much stronger condition than normality. Paracompact spaces will be briefly studied later in this book.

2. Compactness and Separation Axioms

The main theme of this section is that as far as separation axioms are concerned, compact sets behave very much like finite sets, which is of course a corollary of the maxim that compactness is the next best thing to finiteness. We already alluded to it in the first section of the last chapter. We shall illustrate it more fully in this section.

To begin with, suppose we have a Hausdorff space X. This means that any two distinct singleton subsets say $\{x\}$, $\{y\}$ of X can be separated from each other by disjoint open sets. Suppose we replace the singleton set $\{y\}$ by a finite subset F. It is then easy to show that $\{x\}$ and F can be separated from each other by disjoint open sets. Indeed, suppose the distinct elements of F are $y_1, y_2, \ldots, y_n$. For $i = 1, 2, \ldots, n$ let U_i, V_i be open sets such that $x \in U_i$, $y_i \in V_i$ and $U_i \cap V_i = \emptyset$. Now let $U = \bigcap_{i=1}^{n} U_i$ and $V = \bigcup_{i=1}^{n} V_i$. Then clearly U, V are open sets, $x \in U$, $F \subset V$ and $U \cap V = \emptyset$. We have used here that the intersection of finitely many open sets is open. The argument will break down precisely at this point in case the set F is infinite. If, however, F is compact, things are not so bad, for we could apply a standard compactness argument.

(2.1) Proposition: Let X be a Hausdorff space, $x \in X$ and F a compact subset of X not containing x. Then there exist open sets U, V such that $x \in U$, $F \subset V$ and $U \cap V = \emptyset$.

Proof: For each $y \in F$, there exist open sets U_y, V_y such that $x \in U_y$, $y \in V_y$ and $U_y \cap V_y = \emptyset$. The family $\{V_y : y \in F\}$ is an open cover of F.

Since F is compact, there is a finite subcover, say $\{V_{y_1}, V_{y_2}, \ldots, V_{y_n}\}$. Let $U = \bigcap_{i=1}^{n} U_{y_i}$ and $V = \bigcup_{i=1}^{n} V_{y_i}$. Then U, V are disjoint open subsets, $x \in U$ and $F \subset V$. ∎

Before proving further results, let us draw some important corollaries from this proposition. The first one is a direct consequence of it.

(2.2) Corollary: A compact subset in a Hausdorff space is closed.

Proof: Suppose X is a T_2 space and F is a compact subset of X. Then by the proposition above, for any $x \in X - F$ there exist open sets U, V such that $x \in U$, $F \subset V$ and $U \cap V = \emptyset$. In particular, $U \cap F = \emptyset$ and hence $U \subset X - F$. Thus $X - F$ is a neighbourhood of each of its points. So $X - F$ is open and F closed. ∎

(2.3) Corollary: Every map from a compact space into a T_2 space is closed. The range of such a map is a quotient space of the domain.

Proof: Suppose $f : X \to Y$ is continuous where X is compact and Y is Hausdorff. Let C be a closed subset of X. Then C is compact by Proposition (6.1.10) and so $f(C)$ is compact by Proposition (6.1.8). But then $f(C)$ is closed in Y by the corollary above. Hence images of closed sets in X are closed in Y, i.e. the map f is closed. Let Z be the range of f. Then f, regarded as a map from X onto Z is a quotient map by Proposition (5.4.10). Consequently Z is a quotient space of X. ∎

(2.4) Corollary: A continuous bijection from a compact space onto a Hausdorff space is a homeomorphism.

Proof: Let $f : X \to Y$ be a continuous bijection where X is compact and Y is Hausdorff. We claim f is open. Let G be an open subset of X. Then $X - G$ is closed and hence $f(X - G)$ is closed in Y by the corollary above. But $f(X - G) = Y - f(G)$ because f is a bijection. So $f(G)$ is open in Y. Thus f is a continuous, open bijection and hence a homeomorphism. ∎

(2.5) Corollary: Every continuous, one-to-one function from a compact space into a Hausdorff space is an embedding.

Proof: This is immediate from the last corollary. ∎

These corollaries are of frequent use. As a typical application, we can prove that there can be no continuous one-to-one map from the unit circle S^1, into the real line. For if f were such a map then $f(S^1)$ would be homeomorphic to S^1 by the corollary above since S^1 is a compact space. (Prove !) But then $f(S^1)$ would be a compact, connected subspace of $\mathbb{R}$, whence $f(S^1)$ must be a closed, bounded interval. But such an interval cannot be homeomorphic to S^1 since it has a cut point whereas S^1 has no cut points.

Another interesting consequence is that if X is any set then a compact topology on X (i.e. topology on X which makes it a compact space) cannot be properly stronger than a Hausdorff topology. For, suppose $\mathcal{T}_1$, $\mathcal{T}_2$ are topologies on X such that $(X, \mathcal{T}_1)$ is compact and $(X, \mathcal{T}_2)$ is Hausdorff. If

$\mathcal{T}_1 \supset \mathcal{T}_2$ then the identity function $id_X : X \to X$ is $\mathcal{T}_1 - \mathcal{T}_2$ continuous and hence a homeomorphism by Corollary (2.4). So $\mathcal{T}_1 = \mathcal{T}_2$. It follows that in the class of all compact topologies on a set, every Hausdorff topology is maximal and that in the class of all Hausdorff topologies on a set, every compact topology is a minimal one.

As yet another example of the use of these corollaries, we make good a promise given in the last chapter, namely to show that the topologist's sine curve X is not path-connected. Recall that $X = A \cup B$ where $A = \{(x, \sin 1/x) \in \mathbb{R}^2 : 0 < x \leqslant 1\}$ and $B = \{(0, y) \in \mathbb{R}^2 : -1 \leqslant y \leqslant 1\}$. We claim there is no path in X from $(1, \sin 1)$ to $(0, 0)$. For, if possible, suppose α is such a path. Let Y be the range of α. Then Y is a compact, connected subset of X, and contains the points $(1, \sin 1)$ and $(0, 0)$. Then Y must contain the entire set A; for if a point $(b, \sin 1/b)$ is not in Y then the sets $\{(x, y) \in Y : x > b\}$ and $\{(x, y) \in Y : x < b\}$ give a disconnection of Y. Thus $A \subset Y$. But Y is compact and hence closed in X by Corollary (2.2). So $\bar{A} \subset Y$. Since A is dense in X, it follows that $Y = X$, i.e. that α is onto. Hence by Corollary (2.3), X is a quotient space of the unit interval $[0, 1]$. This contradicts the result of Proposition (6.3.5) since $[0, 1]$ is locally connected while X is not so. Thus there is no path in X joining $(1, \sin 1)$ to $(0, 0)$. It now follows that X has two path components, A and B, although it has only one component, namely, X itself.

Returning to the consequences of compactness in separation we get the following:

(2.6) Theorem: Every compact Hausdorff space is a T_3 space.

Proof: Let X be a compact, Hausdorff space. Then every closed subset of X is compact and so the space X is regular by Proposition (2.1). Since X is also T_1 (being T_2), the result follows. ∎

Using arguments analogous to those of Proposition (2.1), we get:

(2.7) Proposition: Let X be a regular space, C a closed subset of X and F a compact subset of X, such that $C \cap F = \emptyset$. Then there exist open sets U, V such that $C \subset U$, $F \subset V$ and $U \cap V = \emptyset$. ∎

As a corollary of this proposition, it follows that every compact, regular space is normal. However, a stronger result is true and since we shall need it later, we prove it here. It is due to Tychonoff.

(2.8) Theorem: Every regular, Lindelöff space is normal.

Proof: Let X be regular, Lindelöff space and let C, D be disjoint, closed subsets of X. By the characterisation of regularity proved in Proposition (1.15), we get for each $x \in C$, an open set U_x containing x such that $\bar{U}_x \subset X - D$ and similarly for each $y \in D$, an open set V_y containing y such that $\bar{V}_y \subset X - C$. The sets C, D are closed subsets of a Lindelöff space and hence they are themselves Lindelöff spaces by Proposition (6.1.10). So the open covers $\{U_x : x \in C\}$ and $\{V_y : y \in D\}$ of C, D respectively have countable subcovers say $\{U_n : n = 1, 2, \ldots\}$ and $\{V_n : n = 1, 2, \ldots\}$. It is now tempting to let $U = \bigcup_{n=1}^{\infty} U_n$ and $V = \bigcup_{n=1}^{\infty} V_n$. Unfortunately although $\bar{U}_n \subset X - D$ for each n, we cannot deduce from it that $\bar{U} \subset X - D$ because the closure of U may be larger than $\bigcup_{n=1}^{\infty} \bar{U}_n$ (however, this argument would be valid in case X were compact, for then the subcover could be chosen to be finite and the closure operator does commute with finite unions).

So we resort to a simple trick. For each $n \in \mathbb{N}$, let $G_n = U_n - \bigcup_{i=1}^{n} \bar{V}_i$ and $H_n = V_n - \bigcup_{i=1}^{n} \bar{U}_i$. Note that G_n, H_n are open sets for all n. Let $G = \bigcup_{n=1}^{\infty} G_n$ and $H = \bigcup_{n=1}^{\infty} H_n$. We contend $C \subset G$. For, let $x \in C$. Then $x \in U_n$ for some $n \in \mathbb{N}$. Also $x \notin \bar{V}_m$ for all m since $\bar{V}_m \subset X - C$ for all m. Hence $x \in G_n$ and so $x \in G$. Similarly $D \subset H$. Thus G, H are open sets in X containing C and D respectively and to complete the proof we need only show that $G \cap H = \emptyset$. If this is not so, then there exist $m, n \in \mathbb{N}$ such that $G_m \cap H_n \neq \emptyset$. Without loss of generality we may suppose $m \leq n$. Let $x \in G_m \cap H_n$. Then $x \in U_m \subset \bar{U}_m$ which contradicts that $x \in H_n$. So $G \cap H = \emptyset$ and X is normal. ∎

(2.9) Corollary: Every regular, second countable space is normal.

Proof: This is immediate since every second countable space is a Lindelöff space (see Theorem (6.1.4)). ∎

(2.10) Corollary: Every compact Hausdorff space is T_4.

Proof: By Theorem (2.6) above every compact Hausdorff space is regular and hence by Theorem (2.8) it is also normal. Moreover, it is a T_1 space since it is T_2 space. Putting together, every compact Hausdorff space is T_4. ∎

Proposition (2.11) and Corollary (2.10) can also be deduced from the following theorem due to Wallace.

(2.11) Theorem: Let A, B be compact subsets of topological spaces X, Y respectively. Let W be an open subset of $X \times Y$ containing the rectangle $A \times B$. Then there exist open sets U, V in X, Y respectively such that $A \subset U$, $B \subset V$ and $U \times V \subset W$.

Proof: The result is trivial if either A or B is empty. So assume A, B are both nonempty. Fix $b \in B$. For each $a \in A$, W is an open neighbourhood of the point $(a, b) \in X \times Y$. So by the definition of the product topology, there exist open sets G_a, H_a in X, Y respectively such that $a \in G_a$, $b \in H_a$ and $G_a \times H_a \subset W$. (Actually, G_a, H_a depend not only on a but on b as well. But as we are dealing with a fixed $b \in B$, we suppress it from the notation). The family $\{G_a : a \in A\}$ is an open cover of the compact set A. Let $\{G_{a_1}, \ldots, G_{a_n}\}$ be a finite subcover. Let G_b: $\bigcup_{i=1}^{n} G_{a_i}$ and $H_b = \bigcap_{i=1}^{n} H_{a_i}$. Then G_b, H_b are open sets in X, Y respectively such that $A \subset G_b$, $b \in H_b$ and

$G_b \times H_b \subset W$. These sets depend on the point $b \in B$. Note that so far we used only compactness of A and not of B.

We are about half done. We now let b vary over B. For each $b \in B$ we find open sets G_b, H_b as above. The family $\{H_b : b \in B\}$ is an open cover of the compact set B. Let $\{H_{b_1}, H_{b_2}, \ldots, H_{b_m}\}$ be a finite subcover. Let $U = \bigcap_{i=1}^{m} G_{b_i}$ and $V = \bigcup \{H_{b_i} : i = 1, \ldots, m\}$. Then U, V clearly have the desired properties. ∎

This theorem is of great importance in applications. Its content can be expressed informally by saying that any neighbourhood of a compact rectangle contains an open rectangular neighbourhood. Lest the theorem sound trivial, we give here an example to show that compactness of both A and B is essential. Let X and Y each be the real line with the usual topology. Let $A = \mathbb{R}$ and $B = \{0\}$. Then $A \times B$ is just the x-axis in the plane $\mathbb{R} \times \mathbb{R}$. Let W be the open set, pictured in Figure 1, i.e. $W = \left\{(x, y) \in \mathbb{R}^2 : x = 0, \text{ or } x > 0 \text{ and } -\frac{1}{x} < y < \frac{1}{x}, \text{ or } x < 0 \text{ and } \frac{1}{x} < y < -\frac{1}{x}\right\}$. Then W is an open set containing $A \times B$. But there is no open set V containing B (i.e. containing 0) such that $A \times V \subset W$. Here Wallace's theorem fails because A is not compact.

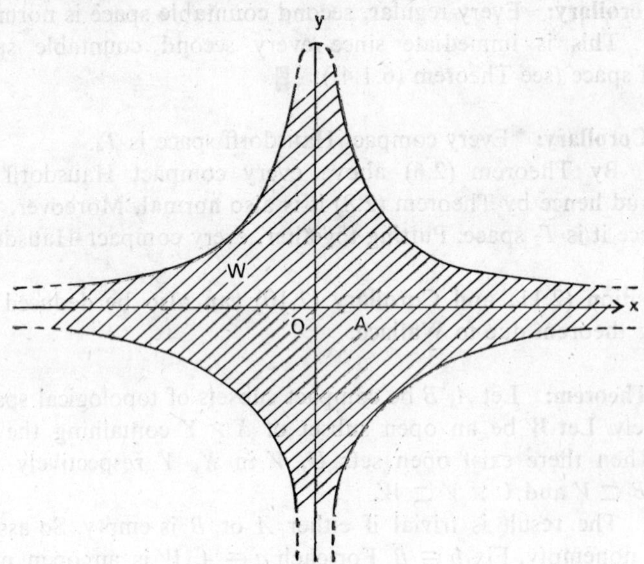

Fig. 1. How Wallace's theorem fails.

It is not immediately obvious how Wallace's theorem implies the results of Proposition (2.1) and Corollary (2.10). By way of illustration we show how Proposition (2.1) follows from Wallace's theorem. Let X be a T_2-space, $x \in X$ and F a compact set not containing x. Then $\{x\}$ and F are compact subsets of X and $\{x\} \times F$ does not meet the diagonal Δx in $X \times X$. Now since X is a Hausdorff space, ΔX is a closed subset of $X \times X$

(see Exercise (1.8)). So $X \times X - \Delta X$ is an open set containing the rectangle $\{x\} \times F$ and hence by Wallace's theorem there exists an open rectangle $U \times V$ containing $\{x\} \times F$ such that $U \times V \subset X \times X - \Delta X$. This means $x \in U$, $F \subset V$ and $U \cap V = \emptyset$ as desired.

Let us now see how compact sets behave w.r.t. complete regularity.

(2.12) Proposition: Let X be a completely regular space. Suppose F is a compact subset of X, C is a closed subset of X and $F \cap C = \emptyset$. Then there exists a continuous function from X into the unit interval which takes the value 0 at all points of F and the value 1 at all points of C.

Proof: If F is the empty set, the function which is identically 1 will work. Similarly if C is empty, the identically zero function will work. Let us assume then that C and F are both non-empty. For each $x \in F$, there exists a map $f_x : X \to [0, 1]$ such that $f_x(x) = 0$ and $f_x(y) = 1$ for all $y \in C$. If the set F were finite we could easily get the result by taking the minimum (or the product) of the finite family of functions $\{f_x : x \in F\}$. Since F is compact (and not necessarily finite), we do the next best thing, namely, to apply a standard compactness argument. For each $x \in F$, let U_x be the set $f_x^{-1}([0, \frac{1}{2}))$. Then U_x is an open set containing x and so the family $\{U_x : x \in F\}$ is an open cover of the set F. By compactness of F, there exists a finite subcover, say, $\{U_{x_1}, U_{x_2}, \ldots, U_{x_n}\}$. Now define $f : X \to [0, 1]$ by $f(x) = \min \{f_{x_1}(x), f_{x_2}(x), \ldots, f_{x_n}(x)\}$ for $x \in X$. Then f is continuous (see Exercise (6.1.6)). Also f assumes the value 1 at all points of C since each f_{x_i} does so for $i = 1, 2, \ldots, n$. However, f may not vanish identically on the set F. This difficulty can be corrected as follows. We certainly know that $f(F)$ is a subset of the semi-open interval $[0, \frac{1}{2})$ since for $x \in F$, there exists i such that $0 \leq f_{x_i}(x) < \frac{1}{2}$ and $f(x) \leq f_{x_i}(x)$. Now let g be a continuous function from the unit interval $[0, 1]$ into itself such that $g([0, \frac{1}{2})) = \{0\}$ and $g(1) = 1$. For example we could let $g(t) = 0$ for $0 \leq t \leq \frac{1}{2}$ and $g(t) = 2t - 1$ for $\frac{1}{2} \leq t \leq 1$. The composite $g \circ f$ vanishes identically on F and assumes the value 1 at all points of C. ∎

In the last section we remarked that when the members of a decomposition are compact and the projection map is closed, the quotient space shares many of the nice properties of the original space. We now study a few results of this type. Throughout the remainder of this section, X will be a space and $\mathscr{D}$ some decomposition of X. We shall regard $\mathscr{D}$ as a quotient space of X and $p : X \to \mathscr{D}$ will denote the projection map. Note that if $D \in \mathscr{D}$ then D is a point of the space $\mathscr{D}$ while it is a subset of the space X.

Before we proceed, it is convenient to have a characterisation of the requirement that p be closed, in terms of $\mathscr{D}$ and the topology on X. Note that for a subset S of X, $p^{-1}(p(S))$ is in general large than S. If R denotes the equivalence relation on X corresponding to the decomposition $\mathscr{D}$ then $p^{-1}(p(S))$ is the set $\{x \in X : x \, R \, y \text{ for some } y \in S\}$. It is clear that $p^{-1}(p(S)) = S$ if and only if S is the union of some members of $\mathscr{D}$ and that this is the case iff every member of $\mathscr{D}$ is either completely contained in S or completely contained in $X - S$. There is a name for such sets.

(2.13) Definition: A subset S of X is said to **saturated** (w.r.t. the decomposition $\mathcal{D}$) if $p^{-1}(p(S)) = S$; equivalently S is **saturated** if there exists a subfamily $\mathcal{C}$ of $\mathcal{D}$ such that $S = \bigcup_{C \in \mathcal{C}} C$.

Note that for any $A \subset X$, the set $p^{-1}(p(A))$ is always saturated. Also the complement of a saturated subset is saturated. Note also that if A, B are mutually disjoint and at least one of them is saturated then $p(A)$ and $p(B)$ are mutually disjoint. In terms of saturated sets we can now tell when the projection map is closed.

(2.14) Proposition: With the notation above, the quotient map $p : X \to \mathcal{D}$ is closed if and only if for any $D \in \mathcal{D}$ and any open set G (in X) containing D, there exists a saturated open set H such that $D \subset H \subset G$.

Proof: Assume first that p is closed. Let $D \in \mathcal{D}$ and an open subset G (of X) containing D be given. Let $K = p^{-1}(p(X - G))$ and $H = X - K$. Then K is closed since p is closed and continuous. So H is open in X and clearly it is saturated since K is so. Also $H \subset G$ since $X - G \subset K$. It remains to show that $D \subset H$. For this, let $x \in D$. Since $D \subset G$, it follows that $p(x) = D \notin p(X - G)$. So $x \notin K$ and hence $x \in H$ as desired.

Conversely assume that the given condition holds and suppose C is a closed subset of X. We have to show that $p(C)$ is closed in $\mathcal{D}$. In view of the fact that $\mathcal{D}$ has the quotient topology on it, this amounts to showing that $p^{-1}(p(C))$ is closed in X. So let $V = X - p^{-1}(p(C))$. We claim that V is a neighbourhood of each of its points and hence is open. Note first that V is the union of those members of $\mathcal{D}$ which are disjoint from C. Hence V is saturated and does not intersect C. Now let $x \in V$. Let D be the unique member of $\mathcal{D}$ containing x. Then $D \subset X - C$ which is open. By the given condition there exists a saturated open set H such that $D \subset H \subset X - C$. Then H is the union of some members of $\mathcal{D}$. None of these members intersects C. So $H \subset V$. Thus $x \in D \subset H \subset V$, showing that V is a neighbourhood of x. ∎

The advantage of this proposition is that we can now tell whether a projection map is closed intrinsically, in terms of the topology of the domain and the corresponding decomposition, without directly involving the quotient topology. We put it to immediate use in the following theorem.

(2.15) Theorem: Suppose $\mathcal{D}$ is a decomposition of a space X each of whose members is compact and suppose the projection $p : X \to \mathcal{D}$ is closed. Then the quotient space $\mathcal{D}$ is Hausdorff or regular according as X is Hausdorff or regular.

Proof: Assume first that X is Hausdorff. Let C, D be distinct elements of $\mathcal{D}$. Then C, D are compact subsets of X and since X is T_2, we can apply a standard compactness argument along with Proposition (2.1) to get open subsets U, V of X such that $C \subset U$, $D \subset V$ and $U \cap V = \emptyset$. By the last proposition, there exist saturated open sets G, H such that $C \subset G \subset U$ and $D \subset H \subset V$. Clearly $p(G)$, $p(H)$ are mutually disjoint subsets of $\mathcal{D}$, containing C, D respectively. Also $p^{-1}(p(G)) = G$ is open in X and so $p(G)$ is open in $\mathscr{D}$ by definition of the quotient topology. Similarly $p(H)$ is open. It thus follows that $\mathscr{D}$ is a Hausdorff space.

Next, suppose X is regular. Let $A \in \mathscr{D}$ and suppose C is a closed subset of $\mathscr{D}$ not containing A. Then $p^{-1}(C)$ is a closed subset of X which is disjoint from A. By Proposition (2.7), there exist open subsets U, V containing A and $p^{-1}(C)$ respectively such that $U \cap V = \emptyset$. Note that $p^{-1}(C)$ is the union of some members of $\mathscr{D}$. For each of these we apply the last proposition and get a saturated open subset contained in V. The union of all such open saturated sets gives an open, saturated subset H such that $p^{-1}(C) \subset H \subset V$. Also there exists a saturated open subset G such that $A \subset G \subset U$. The rest of the argument is now similar to that given in the last paragraph. ∎

It is interesting that under the hypothesis of the last theorem, second countability of X implies that of $\mathscr{D}$. Before proving it, we introduce some notation. For a subset V of X let $K(V)$ denote the union of those members of $\mathscr{D}$ which are contained in V. Evidently $K(V) \subset V$ and $X - K(V) = p^{-1}(p(X - V))$. It thus follows that if p is closed and V is open then $K(V)$ is open. Moreover, $p(K(V))$ is open since $p^{-1}(p(K(V))) = K(V)$.

(2.16) Theorem: With the hypothesis of the last theorem, if X is second countable, so is $\mathscr{D}$.

Proof: Let $\mathscr{B}$ be a countable base for X. Let $\mathscr{U}$ be the family of all finite unions of members of $\mathscr{B}$. Then $\mathscr{U}$ is also countable. Let $\mathscr{L} = \{p(K(U)) : U \in \mathscr{U}\}$. Then $\mathscr{L}$ is a countable family of open sets in $\mathscr{D}$. We contend that $\mathscr{L}$ is a base for the quotient topology on $\mathscr{D}$. For this, let $A \in \mathscr{D}$ and G be an open subset of $\mathscr{D}$ containing A. Then $p^{-1}(G)$ is an open subset of X containing A. Since A is compact, we can find a finite number of members of $\mathscr{B}$ covering A whose union is contained in $p^{-1}(G)$. This means, there exists $U \in \mathscr{U}$ such that $A \subset U \subset p^{-1}(G)$. Then $p(K(U))$ is an open subset of $\mathscr{D}$ containing A and contained in G. Since $p(K(U)) \in \mathscr{L}$ by definition, it follows that $\mathscr{L}$ is a base for $\mathscr{D}$. ∎

The hypothesis that all the members of $\mathscr{D}$ are compact cannot be dropped. A counterexample will be given in the exercises.

Exercises

2.1 Prove that the unit circle S^1 is compact. (Hint: Either use the higher dimensional analogue of the Heine-Borel theorem or else express S^1 as a continuous image of the unit interval).

2.2 For any map $f: S^1 \to \mathbb{R}$ prove that there exists a point $x_0 \in S^1$ such that $f(x_0) = f(-x_0)$. (Hint: Consider the sets $\{x \in S^1 : f(x) > f(-x)\}$ and $\{x \in S^1 : f(x) < f(-x)\}$ and use connectedness of S^1. Note that this result is stronger than saying that S^1 cannot be embedded in $\mathbb{R}$.)

*2.3 Let A, B be closed subsets of S^1 such that $S^1 = A \cup B$. Prove that at least one of A and B contains a pair of mutually antipodal points. (Hint: Apply the preceding exercise to a suitably defined real-valued map on S^1.)

2.4 Let X be any infinite set with a distinguished element $*$. Let $\mathcal{T}$ be the topology on X consisting of the empty set and all subsets of X containing $*$. Prove that X has a compact subset whose closure is not compact.

2.5 Prove that the closure of a compact subset of a regular space is compact.

2.6 Prove that the real line with the semi-open interval topology is normal. (Hint: Show that it is a regular space and then use Exercise (6.1.10).)

2.7 Consider $\mathbb{R} \times \mathbb{R}$ with the product topology where $\mathbb{R}$ is given the semi-open interval topology. Let A, B be the sets $\{(x, -x) : x \in \mathbb{Q}\}$ and $\{(x, -x) : x \in \mathbb{R} - \mathbb{Q}\}$ respectively. Let L be the line $\{(x, -x) : x \in \mathbb{R}\}$.

(a) Prove that A and B are both closed subsets of $\mathbb{R} \times \mathbb{R}$. In fact show that every subset of L is closed in $\mathbb{R} \times \mathbb{R}$ (see Exercise (4.4.6)).

**(b) Prove that if U is any open set in $\mathbb{R} \times \mathbb{R}$ containing A, then $\bar{U} \cap B \neq \emptyset$.

(c) Prove that the space $\mathbb{R} \times \mathbb{R}$ is not normal.

(This would of course follow from (b). However, an argument not based on (b) can be given as follows. If $\mathbb{R} \times \mathbb{R}$ were normal then for every subset C of L there would exist open sets $G(C)$ and $H(C)$ in $\mathbb{R} \times \mathbb{R}$ such that $C \subset G(C)$, $L - C \subset H(C)$ and $G(C) \cap H(C) = \emptyset$. Let D be the set $\mathbb{Q} \times \mathbb{Q}$. Then D is dense in $\mathbb{R} \times \mathbb{R}$. Define $\theta : P(L) \to P(D)$ by $\theta(C) = G(C) \cap D$. Prove that θ is one-to-one and obtain a contradiction, comparing the cardinalities of the power sets $P(L)$, $P(D)$.)

2.8 Give an example of a space in which every compact subset is closed but which is not Hausdorff.

2.9 Prove that a subset of $\mathbb{R}$ with usual topology is compact iff it is closed and bounded.

*2.10 Characterise all compact subsets of $\mathbb{R}$ with the semi-open interval topology.

2.11 Deduce corollary (2.10) from the theorem of Wallace.

2.12 Prove that if $\mathcal{D}$ is a decomposition of a normal space X and the quotient map $p : X \to \mathcal{D}$ is closed then $\mathcal{D}$ is normal (without any hypothesis about compactness of members of $\mathcal{D}$).

2.13 Let $A = \mathbb{R} \times \{0\} \subset \mathbb{R}^2$. Then A is a closed subset of $\mathbb{R}^2$.

(a) If $\{V_n : n = 1, 2, \ldots\}$ is a sequence of open sets in $\mathbb{R}^2$ containing A, prove that there exists an open set V containing A such that V does not contain any V_n.

(Hint: For each $n \in \mathbb{N}$ find a non-zero real number a_n such that $(n, a_n) \in V_n$. Construct V so that it does not contain (n, a_n) for all $n \in \mathbb{N}$).

(b) Prove that the quotient space R^2/A is not first countable even though the projection map is closed.

Notes and Guide to Literature

The results in this section are all very standard. The class of all Hausdorff topologies on a space has been an area of much interest. See for example, Ramanathan [1].

The result of Exercise (2.1) (and also of (2.2)) generalises to maps from S^n to $\mathbb{R}^n$ for all $n \geq 1$. This is a classic result known as the Borsuk-Ulam theorem. However the proof in higher dimensions is not easy. A fairly elementary proof, using induction on n, can be found in Dugundji [1].

The result of Exercise (2.7) (b) was first observed by Sorgenfrey [1]. Later on we shall prove it using completeness of $\mathbb{R}$. The argument given for part (c) of that exercise is due to F. B. Jones.

3. The Urysohn Characterisation of Normality

In showing that complete regularity implies regularity we saw that separation of a point from a closed set by means of a continuous function implies separation by open sets. This also holds for separation of two closed sets as the following proposition, which is analogous to Proposition (1.14), shows.

(3.1) Proposition: Let A, B be subsets of a space X and suppose there exists a continuous function $f: X \to [0, 1]$, such that $f(x) = 0$ for all $x \in A$ and $f(x) = 1$ for all $x \in B$. Then there exist disjoint open sets U, V such that $A \subset U$ and $B \subset V$.

Proof: We simply choose any two disjoint open sets G, H in $[0, 1]$ containing 0 and 1 respectively (for example we could let $G = [0, \frac{1}{2})$ and $H = (\frac{1}{2}, 1]$ and set $U = f^{-1}(G)$ and $V = f^{-1}(H)$. The assertion follows from the given properties of f. ∎

(3.2) Corollary: If a space X has the property that for any two mutually disjoint closed subsets A, B of it, there exists a continuous function $f: X \to [0, 1]$ taking the value 0 at all points of A and the value 1 at all points of B, then X is normal.

Proof: This follows immediately from the last proposition and the definition of normality. ∎

The interesting thing is that the converse of the corollary above is true. This is a non-trivial result due to Urysohn and is known as the Urysohn characterisation of normality. The present section is aimed at proving this result. But before doing so, let us comment why it is such a remarkable theorem. Firstly, its analogue does not hold for regular spaces, that is to say given a point x and a closed subset C ($x \notin C$) of a regular space X we may not always be able to find a continuous function $f: X \to [0, 1]$ such that $f(x) = 0$ and $f(y) = 1$ for all $y \in C$. In fact there do exist regular, Hausdorff spaces (i.e. T_3 spaces) on which every continuous real valued function is constant. Another novel feature of the converse of corollary (3.2) is that its hypothesis deals purely with subsets of X but the conclusion guarantees the existence of a real-valued function with certain properties. A similar theorem about metric spaces will not be so remarkable because in the very definition of a metric space we are given a real-valued function with certain nice properties using which the desired function can be constructed rather easily. In the definition of a normal topological space, no function is given. The desired function is to be constructed from subsets of the space and the construction is nothing short of ingenious.

Without further ado we now state the theorem. Despite its non-triviality, it is called **'Urysohn's lemma'** because Urysohn used it as a lemma to prove something else (which we shall study later as the Urysohn metrisation theorem).

(3.3) Theorem: A topological space X is normal if and only if it has the property that for every two mutually disjoint, closed subsets A, B of X, there exists a continuous function $f: X \to [0, 1]$ such that $f(x) = 0$ for all $x \in A$ and $f(x) = 1$ for all $x \in B$.

Proof: Sufficiency of the condition is established in Corollary (3.2). The proof of necessity will be very long and we present it through a series of lemmas.

We are given some information about the space X and our task is to find a continuous function $f: X \to [0, 1]$ satisfying certain conditions. Here we follow an important technique in mathematics. When you want to find something, simply assume that it exists and analyse it. This analysis often provides important clues to the construction or the location of the thing you are looking for. Of course, it is then to be supplemented by a rigorous proof that the thing you have constructed or located is in fact what you wanted. It is somewhat like this. Suppose an inspector of police is investigating a mysterious death. He tentatively supposes that it is a case of a murder. He then visits the site of the death. Based on the observations and the inquiries he makes, he figures that the murderer (if at all it was a murder) must be, say, a strong, left-handed man living in the vicinity and a close associate of the deceased. He then searches and finds a man M who answers this description. This does not conclusively prove that M is the murderer, nor even that the death was due to a murder but it at least gives a hope for the answer. To actually prove that M committed the murder would require further proof.

So, in the present case we have to construct a continuous function f. We are given some information about the subsets of X. Let us, then, see how the existence of a continuous function $f: X \to [0, 1]$ implies the existence of a certain family of subsets of X.

(3.4) Lemma: Let $f: X \to [0, 1]$ be continuous. For each $t \in \mathbb{R}$ let $F_t = \{x \in X : f(x) < t\}$. Then the indexed family $\{F_t : t \in \mathbb{R}\}$ has the following properties:

(i) F_t is an open subset of X for each $t \in \mathbb{R}$

(ii) $F_t = \emptyset$ for $t < 0$ (Actually F_0 is also $\emptyset$ but this is not very important.)

(iii) $F_t = X$ for $t > 1$

(iv) For any $s, t \in \mathbb{R}$, $s < t \Rightarrow \overline{F}_s \subset F_t$.

Moreover, for each $x \in X, f(x) = \inf \{t \in \mathbb{Q} : x \in F_t\}$.

Proof: Note that F_t is the inverse image (unfer f) of the set $(-\infty, t)$ which is open in $\mathbb{R}$. So, by continuity of f, each F_t is open, showing (i), (ii) and (iii) follow easily from the fact that f takes values in the unit interval. For (iv) let r be any number between s and t and let $C = \{x \in X : f(x) \leq r\}$. Then $F_s \subset C \subset F_t$. But C is closed in X by continuity of f. So $\overline{F}_s \subset C$ and $\overline{F}_s \subset F_t$.

For the remainder of the lemma, let $x \in X$ and let G_x be the set $\{t \in \mathbb{Q} : x \in F_t\}$. We have to show $f(x) = \inf G_x$. G_x is nonempty because $t \in G_x$ for all rational $t > f(x)$. Also $t \geq 0$ for all $t \in G_x$ and so G_x is bounded below. Let $y = \inf G_x$. Now, for any $t \in G_x$, $x \in F_t$ and so $f(x) < t$. Hence $f(x) \leq \inf G_x$, i.e. $f(x) \leq y$. Suppose $f(x) < y$. Let q be a rational number between $f(x)$ and y. Then $q \notin G_x$ since $q < y$. Hence $x \notin F_q$. So $f(x) \geq q$, which is a contradiction. Hence we get $f(x) = y$. ∎

The preceding lemma will not really be used in the proof, but its significance is profound. It shows that a continuous function $f : X \to [0, 1]$ induces a certain indexed family of open subsets of X and moreover that we can recover the function f if we knew some of these sets, namely the sets F_t for $t \in \mathbb{Q}$. Note that the fact that the set of rationals is dense was used above. Of course it is also true that for $x \in X, f(x) = \inf \{t \in \mathbb{R} : x \in F_t\}$. The advantage of $\mathbb{Q}$ over $\mathbb{R}$ is that the former is countable while the latter is not.

Can we now reverse the procedure? That is, if we are given an indexed family of open sets $\{F_t : t \in \mathbb{Q}\}$ in X and if we define f by $f(x) = \inf \{t \in \mathbb{Q} : x \in F_t\}$, is f continuous? The answer is provided in the following lemma.

(3.5) Lemma: Let X be a topological space and suppose $\{F_t : t \in \mathbb{Q}\}$ is a family of sets in X such that

(1) F_t is open in X for each $t \in \mathbb{Q}$

(2) $F_t = \emptyset$ for $t \in \mathbb{Q}, t < 0$

(3) $F_t = X$ for $t \in \mathbb{Q}, t > 1$

(4) $\overline{F}_s \subset F_t$ for $s, t \in \mathbb{Q}, s < t$.

For $x \in X$, let $f(x) = \inf \{t \in \mathbb{Q} : x \in F_t\}$. Then f is a continuous real-valued function on X and it takes values in the unit interval [0, 1].

Proof: For $x \in X$, let $G_x = \{t \in \mathbb{Q} : x \in F_t\}$. Condition (3) shows that G_x is non-empty and condition (2) shows that it is bounded below (by 0) for all $x \in X$. So the function $f(x) = \inf G_x$ is certainly a well-defined, real-valued function on X. Also conditions (2) and (3) easily imply that for each $x \in X$, $0 \leq f(x) \leq 1$ and hence f takes values in the unit interval. It only remains to prove that f is continuous. For this, note that the family of all intervals of the form $(-\infty, a)$ or (b, ∞) for $a, b \in \mathbb{R}$ is a sub-base for the usual topology on $\mathbb{R}$. Hence, in view of Theorem (5.3.4) continuity of f will be established if we can prove that for any $s \in \mathbb{R}$, the sets $\{x \in X : f(x) < s\}$ and $\{x \in X : f(x) > s\}$ are open in X. We do this separately.

Let $s \in \mathbb{R}$. Let H be the set $\{x \in X : f(x) < s\}$. It is tempting to think that H is precisely the set F_s and hence is trivially open. Unfortunately this need not be so even when s is rational. Nevertheless, we claim $H = \bigcup\{F_t : t \in \mathbb{Q}, t < s\}$. For, suppose first, $x \in H$. Then $f(x) < s$. Since $f(x) = \inf G_x$ and $f(x) < s$, there exists $q \in G_x$ such that $q < s$. This means that $x \in F_q$ and hence $x \in \bigcup\{F_t : t \in \mathbb{Q}; t < s\}$. Conversely, we have to show that if $t \in \mathbb{Q}$ and $t < s$ then $F_t \subset H$. Let $x \in F_t$. Then clearly $f(x) \leq t$ and so $f(x) < s$, showing $x \in H$. Thus we have shown that H is the union of F_t's, for $t \in \mathbb{Q}$, $t < s$. But each F_t is an open subset of X by (1) and so H is open in X.

Next, let $K = \{x \in X : f(x) > s\}$. We show that K is open in X by showing that its complement $X - K$ is closed. To do this, we claim that $X - K = \bigcap\{\bar{F}_t : t \in \mathbb{Q}, t > s\}$. For, suppose first that $x \in X - K$. Then $f(x) \leq s$. Suppose $t \in \mathbb{Q}$ and $t > s$. Then $f(x) < t$. Since $f(x) = \inf G_x$, there exists $q \in G_x$ such that $q < t$. But then $x \in F_q$ and by (4), $\bar{F}_q \subset \bar{F}_t$. So $x \in \bar{F}_t$ for all $t \in \mathbb{Q}$ for which $t > s$. Conversely suppose $x \in \bigcap\{\bar{F}_t : t \in \mathbb{Q}, t > s\}$. We must show that $f(x) \leq s$. If not, then $s < f(x)$. Let q, t be rational numbers such that $s < q < t < f(x)$. Then clearly $x \notin \bar{F}_q$ for otherwise $x \in F_t$ by (4), and so $t \in G_x$ violating that $f(x) = \inf G_x$. Thus $q \in \mathbb{Q}$, $q > s$ and $x \notin \bar{F}_q$, a contradiction. This establishes that $X - K$ is an intersection of closed sets and therefore is closed in X. As noted before, this completes the proof of the continuity of f and of the lemma. (A simpler argument is also possible and will be given as an exercise.) ∎

In view of the preceding lemmas, the problem of finding a continuous function on a space reduces to the problem of constructing a family $\{F_t : t \in \mathbb{Q}\}$ of subsets with certain conditions. Countability of $\mathbb{Q}$ allows us to apply an inductive method in the construction.

The proof of Theorem (3.4) can now be easily completed. We are given disjoint closed subsets A and B of a normal space X. We want a continuous real-valued function f on X such that $f(x) = 0$ for all $x \in A$ and $f(x) = 1$ for all $x \in B$. We define a family of sets $\{F_t : t \in \mathbb{Q}\}$ satisfying the conditions of the last lemma. For $t < 0$, and $t > 1$ we have no choice but to let F_t be respectively the empty set and the set X. Let $F_1 = X - B$. Define F_0 to be any open set containing A such that $\bar{F}_0 \subset X - B$ (such a set exists by the characterisation of normality given in Proposition (1.16)). For rational numbers between 0 and 1 we proceed as follows.

Enumerate the set of rationals in [0, 1] as $\{q_0, q_1, q_2, q_3, \ldots, q_n, \ldots\}$ with $q_0 = 0$ and $q_1 = 1$. (An explicit formula for such an enumeration could be given but is not important for our purpose; for us what matters is that each rational in [0, 1] occurs exactly once in this enumeration). Now F_{q_0}, F_{q_1} are already defined. Consider q_2. Clearly $q_0 < q_2 < q_1$. Define F_{q_2} to be any open set such that $\bar{F}_{q_0} \subset F_{q_2} \subset \bar{F}_{q_2} \subset F_{q_1}$. Such a set exists by normality of X. Now suppose $n \geq 3$ and that the open sets $F_{q_1}, F_{q_2}, \ldots, F_{q_{n-1}}$ have already been defined so as to satisfy condition (4) of the lemma. Consider q_n. Let q_i be the largest among those of q_0, q_1, q_2,

$\ldots, q_{n-1}$, which are less than q_n, i.e. $q_i = \max\{q_r : 0 \leq r \leq n-1, q_r < q_n\}$. Similarly let q_j be the smallest among those of $q_0, q_1, \ldots, q_{n-1}$ which are greater than q_n. Then $q_i < q_n < q_j$. By the inductive hypothesis $\bar{F}_{q_i} \subset F_{q_j}$. By normality of X, there exists an open set F_{q_n} such that $\bar{F}_{q_i} \subset F_{q_n}$ and $\bar{F}_{q_n} \subset F_{q_j}$. Then condition (4) continues to be satisfied with this F_{q_n} included in the set of the F's defined so far. This completes the inductive step in the definition and also shows that the family of sets $\{F_t : t \in \mathbb{Q}\}$ satisfies all the conditions in the last lemma. So the function defined by $f(x) = \inf\{t \in \mathbb{Q} : x \in F_t\}$ is a continuous function from X into $[0, 1]$. Now, if $x \in A$ then $x \in F_0$ and $f(x) = 0$. Similarly if $x \in B$ then $x \notin F_t$ for any $t \in \mathbb{Q}$, $t \leq 1$ and so $f(x) = 1$. Thus the proof of Theorem (3.4) is complete. ∎

We urge the reader to read the proof again and again until he can do justice to this beautiful theorem of Urysohn (who, incidentally, died at the age of twentyfour!). As an immediate consequence of it we get:

(3.6) Corollary: All T_4 spaces are completely regular and hence Tychonoff.

Proof: Simply apply the theorem to the case where one of the closed sets is a singleton. ∎

Note that in the proof above nowhere is it claimed, nor is it true, that f is 0 only on A and 1 only on B. There may be points outside $A \cup B$ at which f is 0 or 1. In other words we merely claim that $A \subset f^{-1}(\{0\})$ and $B \subset f^{-1}(\{1\})$ and not that $A = f^{-1}(\{0\})$ or $B = f^{-1}(\{1\})$. We invite the reader to prove that in case X happens to be a metric space, then f could indeed be so chosen as to take the value 0 precisely on A and the value 1 precisely on B. Another sufficient condition (weaker than metrisability) which will ensure this will be given in the next section as an exercise.

A function whose existence is asserted by the Urysohn's lemma is called a **Urysohn function.**

Exercises

3.1 Prove that both Lemma (3.4) and (3.5) continue to hold if the set $\mathbb{Q}$ is replaced by any dense subset of $\mathbb{R}$.

3.2 A rational number p/q (in reduced form) is said to be **dyadic** if q is a power of 2. Prove that a real number is a dyadic rational iff it has a dyadic expansion which terminates. (More generally, one can define triadic or p-adic rationals for any prime p and prove a similar result.)

3.3 Obtain an explicit enumeration of the set of all dyadic rational numbers in the interval $(0, 1)$. (Hint: Consider the sequence .1, .01, .11, .001, .101, .011, .111, ... where the numbers are in dyadic expansion. This result shows that had we used dyadic rationals in the proof of Urysohn's lemma, we could have given the proof a little more explicitly.)

3.4 Let $f : X \to [0, 1]$ be a continuous function where X is a topological space. For each $t \in \mathbb{R}$ let $G_t = \{x \in X : f(x) \leq t\}$. Obtain properties of the family $\{G_t : t \in \mathbb{Q}\}$ of subsets of X similar to those in Lemma

(3.4) and show how f can be recovered from this family. Then prove the corresponding analogue of Lemma (3.5) and use it to give an alternate (although essentially similar) proof of Theorem (3.3).

3.5 Suppose X is a metric space and A, B are non-empty disjoint, closed subsets of X. Prove that there exists a continuous function $f: X \to [0, 1]$ such that $A = f^{-1}(\{0\})$ and $B = f^{-1}(\{1\})$. (Hint: Adapt the argument used in proving that all metric spaces are normal and get a map $g: X \to \mathbb{R}^2$ which takes A into the positive 'x-axis' and B into the positive 'y-axis'. Now compose with a suitable inverse trigonometric function.)

3.6 Prove that every continuous real-valued function on an indiscrete space is constant, even though such a space is always normal. Why does this not contradict Theorem (3.4)?

3.7 Prove that a connected, T_4 space with at least two points must be uncountable (i.e. the underlying set must be uncountable). (Hint: A Urysohn function on such a space must be onto).

3.8 Prove that there exists no countable, connected, T_3 space. (Hint: Such a space is Lindelöff and hence normal by Theorem (2.8).).

3.9 Give an example of a proof (from whatever branch of Mathematics) where a desired thing is found by assuming tentatively its existence and thereby obtaining clue to its construction.

3.10 In Lemma (3.5), if $x_0 \in X$, establish continuity of f at x_0 directly. (Hint: If V is a neighbourhood of $f(x_0)$, find $p, q \in \mathbb{Q}$ such that $f(x_0) \in (p, q)$ and $[p, q] \subset V$. Now prove, $F_q - \bar{F}_p \subset f^{-1}(V)$.)

Notes and Guide to Literature

Urysohn's lemma (and the metrisation theorem in whose proof this 'lemma' was used) is among the most classic theorems in general topology.

The technique of defining a real-valued function by means of a family of subsets indexed by $\mathbb{Q}$ (or other suitable subset of $\mathbb{R}$) is an important one and is used, for example, to prove that every T_0 topological group is a Tychonoff space, see Hewitt and Ross [1]. We shall also prove it later using uniform spaces.

An example of a T_3 space (with two or more points) on which every real-valued continuous function is constant was given by Novak [1]. This example is quite complicated. Recently, Thomas [1] has given a simple example of a T_3-space which is not completely regular.

In contrast with the result of Exercise (3.8), there do exist countable, connected, T_2 spaces, see Gustin [1] or Bing [2].

4. Tietze Characterisation of Normality

We mentioned in Chapter 5, Section 3 that one of the most important problems in topology is the extension problem. In the present section we consider an instance of an extension problem.

Suppose X is a topological space, A is a subset of X and $f: A \to \mathbb{R}$ is a continuous function (w.r.t. the subspace topology on A and the usual topology on $\mathbb{R}$). We want to find a continuous extension of f to the space X, that is we want a continuous function $F: X \to \mathbb{R}$ such that for any $x \in A$, $F(x) = f(x)$. Of course such an extension may not always exist. For example, let $X = [0, 1]$, $A = (0, 1]$ and define $f: A \to \mathbb{R}$ by $f(x) = \sin 1/x$. Then f cannot be continuously extended to $[0, 1]$ because there exist sequences $\{x_n\}$, $\{y_n\}$ in A which converge to 0 in $[0, 1]$ such that $\{f(x_n)\}$, $\{f(y_n)\}$ converge to distinct limits in $\mathbb{R}$ and this would violate the continuity of any extension of f to $[0, 1]$ (see Exercise (5.3.9)).

Note that in this example the set A is dense in X and the crux of the argument is that if at all an extension F exists then $F(0)$ is uniquely determined by the values of f on A. More generally, we have:

(4.1) Proposition: Let A be a subset of a space X and let $f: A \to \mathbb{R}$ be continuous. Then any two extensions of f to X agree on $\bar{A}$. In other words, if at all an extension of f exists its values on $\bar{A}$ are uniquely determined by values of f on A.

Proof: Suppose $F, G: X \to \mathbb{R}$ are both extensions of f. Let $C = \{x \in X: F(x) = G(x)\}$. Clearly $A \subset C$. But since $\mathbb{R}$ is a Hausdorff space, C is closed in X by Exercises (1.9). Hence $\bar{A} \subset C$ which implies the result. ∎

The general problem of extending f from A to X can be broken into two steps: (i) to extend f from A to $\bar{A}$, and (ii) to extend it further from $\bar{A}$ to X. The proposition just proved says that the first part has at most one solution and the example given above shows that there may actually be no solution. An important special case in which a solution exists will be proved much later in this book. For the time being it is the second part of the extension problem that interests us. Let us then suppose that f has already been extended somehow to $\bar{A}$ and inquire whether it can further be extended to X. This means that without loss of generality, A may be regarded as a closed subset of X. Even then, an extension of f may not always exist. In fact the existence of such extensions puts a strong condition on the space X as we see in the following proposition.

(4.2) Proposition: Suppose a topological space X has the property that for every closed subset A of X, every continuous real valued function on A has a continuous extension to X. Then X is normal.

Proof: Let B and C be disjoint closed subsets of X. Let $A = B \cup C$ and define $f: A \to \mathbb{R}$ by $f(x) = 0$ for $x \in B$ and $f(x) = 1$ for $x \in C$. Then A is a closed subset of X. Also the function f is well defined and continuous (see Exercise (5.4.10)). By hypothesis, there exists a continuous function $F: X \to \mathbb{R}$ which extends f. Then $F(x) = 0$ for $x \in B$ and $F(x) = 1$ for $x \in C$. From this, as we have seen many times, it follows that there are disjoint open sets containing B and C respectively. Hence the space X is normal. ∎

This proposition is by no means profound. The interesting point is that its converse is true, and like the Urysohn's lemma it is extremely non-trivial to prove. Proposition (4.2) along with its converse is called the **Tietze characterisation of normality.** The proof of the converse requires the use of Urysohn's lemma and the notion of uniform convergence of sequences of functions. Although a reader familiar with analysis has undoubtedly encountered this concept, we repeat it here for the sake of completeness.

(4.3) Definition: Let X be a topological space and (Y, d) a metric space. Then a sequence of functions $\{f_n\}$ from X to Y is said to converge **uniformly** on X to a function $f: X \to Y$ if for every $\epsilon > 0$, there exists $N \in \mathbb{N}$ such that for all $n \geq N$, and for all $x \in X$, $d(f_n(x), f(x)) < \epsilon$. The sequence $\{f_n\}$ is said to converse **pointwise** to f if for every $x \in X$ the sequence $\{f_n(x)\}$ converges to $f(x)$ in Y.

It is immediate that uniform convergence implies pointwise convergence. That the converse is false is shown by taking $X = Y = [0, 1]$ and defining $f_n(x) = x^n$ for $n \in \mathbb{N}$. In this case the limit function f is 0 on $[0, 1)$ and 1 at 1. Note that $\{f_n\}$ converges to f pointwise and although each f_n is continuous, f is not. With uniform convergence things are better.

(4.4) Proposition: Let X, (Y, d), $\{f_n\}$ and f be as above and suppose $\{f_n\}$ converges to f uniformly. Then if each f_n is continuous, so is f.

Proof: Let $x_0 \in X$ and let V be an open neighbourhood of $f(x_0)$ in Y. Choose $\epsilon > 0$ so that $B(f(x_0), \epsilon) \subset V$. Let $N \in \mathbb{N}$ be such that for all $n \geq N$ and for all $x \in X$, $d(f_n(x), f(x)) < \epsilon/3$. Since f_N is continuous at x_0 there exists an open neighbourhood W of x_0 such that for all $x \in W$, $d(f_N(x), f_N(x_0)) < \epsilon/3$. Now for any $x \in W$ we have

$$d(f(x), f(x_0)) \leq d(f(x), f_N(x)) + d(f_N(x), f_N(x_0)) + d(f_N(x_0), f(x_0))$$
$$< \epsilon/3 + \epsilon/3 + \epsilon/3 = \epsilon.$$

Hence $f(W) \subset B(f(x_0), \epsilon) \subset V$ showing that f is continuous at x_0. Since $x_0 \in X$ was arbitrary, f is continuous. ∎

We can also define uniform and pointwise convergence of a series of functions which are real valued. Nothing new is really involved because all the definitions are in terms of the sequence of partial sums. We shall need the following result:

(4.5) Proposition: Let $\sum_{n=1}^{\infty} M_n$ be a convergent series of non-negative real numbers. Suppose $\{f_n\}$ is a sequence of real valued functions on a space X such that for each $x \in X$ and $n \in \mathbb{N}$, $|f_n(x)| \leq M_n$. Then the series $\sum_{n=1}^{\infty} f_n$ converges uniformly to a real valued function on X.

Proof: By the comparison test for series, for each $x \in X$, the series $\sum_{n=1}^{\infty} f_n(x)$ is absolutely convergent. Denote its sum by $f(x)$. Then f is a real valued function on X. Let $\{s_n\}$ be the sequence of the partial sums of the series $\sum_{n=1}^{\infty} f_n$. We claim that $\{s_n\}$ converges to f uniformly on X. Note that for each $x \in X$ and $n \in \mathbb{N}$, $|s_n(x) - f(x)| \leq \sum_{k=n+1}^{\infty} M_k$. Since the series $\sum_{k=1}^{\infty} M_k$ is given to be convergent, given $\epsilon > 0$, we can find $N \in \mathbb{N}$ such that for all $n \geq N$, $\sum_{k=n+1}^{\infty} M_k < \epsilon$. But then for any $x \in X$ and any $n \geq N$, $|s_n(x) - f(x)| < \epsilon$. Hence $\sum_{n=1}^{\infty} f_n$ converges uniformly to f on X. ∎

We now have all the machinery to prove the Tietze extension theorem. First we establish it for functions into the interval $[-1, 1]$. The reason for this technical assumption will be clear in the course of the proof. The idea of the proof is to approximate the given function uniformly by a series of Urysohnlike functions.

(4.6) Theorem: Let A be a closed subset of a normal space X and suppose $f: A \to [-1, 1]$ is a continuous function. Then there exists a continuous function $F: X \to [-1, 1]$ such that $F(x) = f(x)$ for all $x \in A$.

Proof: Let $B_1 = \{x \in A : f(x) \leq -\frac{1}{3}\}$ and $C_1 = \{x \in A : f(x) \geq \frac{1}{3}\}$. By continuity of f, B_1 and C_1 are closed subsets of A and hence of X since A is closed in X. Clearly $B_1 \cap C_1 = \emptyset$. By a variation of Urysohn's lemma there exists a continuous function $f_1 : X \to [-\frac{1}{3}, \frac{1}{3}]$ such that $f_1(x) = -\frac{1}{3}$ for all $x \in B_1$ and $f_1(x) = \frac{1}{3}$ for $x \in C_1$. (Urysohn's lemma deals with the interval $[0, 1]$. However, given any $a, b \in \mathbb{R}$ with $a < b$, there exists a homeomorphism of $[a, b]$ onto $[0, 1]$ carrying a to 0 and b to 1; see the remark made after the definition of complete regularity in Section 1). Note that for any $x \in A$, $|f(x) - f_1(x)| \leq \frac{2}{3}$. Define $g_1 : A \to [-\frac{2}{3}, \frac{2}{3}]$ by $g_1(x) = f(x) - f_1(x)$. We now apply the earlier argument to g_1 instead of f. Thus let $B_2 = \{x \in A : g_1(x) \leq -\frac{2}{9}\}$ and $C_2 = \{x \in A : g_1(x) \geq \frac{2}{9}\}$. Using the appropriate variation of the Urysohn's lemma, we get a continuous function $f_2 : X \to [-\frac{2}{9}, \frac{2}{9}]$ which equals $-\frac{2}{9}$ on B_2 and $\frac{2}{9}$ on C_2. Note again that for any $x \in A$, $|f(x) - f_1(x) - f_2(x)| = |g_1(x) - f_2(x)| \leq \frac{4}{9}$. Hence we get a continuous function $g_2 : A \to [-\frac{4}{9}, \frac{4}{9}]$ defined by $g_2(x) = f(x) - f_1(x) - f_2(x)$ for $x \in A$. Then we repeat this argument for g_2 to get a map $f_3 : X \to [-\frac{4}{27}, \frac{4}{27}]$ such that for all $x \in A$, $g_3(x) \in [-\frac{8}{27}, \frac{8}{27}]$ where $g_3(x) = f(x) - f_1(x) - f_2(x) - f_3(x)$. Continuing in this manner, we get a sequence $\{f_n\}$ of continuous functions on X satisfying:

(i) f_n is continuous for each $n \in \mathbb{N}$.

(ii) $|f_n(x)| \leq \dfrac{2^{n-1}}{3^n}$ for all $n \in \mathbb{N}$ and all $x \in X$.

(iii) $|f(x) - \sum_{i=1}^{n} f_i(x)| \leq \dfrac{2^n}{3^n}$ for all $n \in \mathbb{N}$ and $x \in A$.

Now, the series $\sum_{n=1}^{\infty} \dfrac{2^{n-1}}{3^n}$ is convergent (being a geometric series with common ratio less than 1) and so by the proposition above, $\sum_{n=1}^{\infty} f_n$ converges uniformly, say to, F on X. But each f_i and hence each partial sum of the series $\sum_{n=1}^{\infty} f_n$ is continuous and so by Proposition (4.4), F is continuous. Finally, statement (iii) above shows that for each $x \in A$, $\sum_{n=1}^{\infty} f_n(x)$ converges to $f(x)$. But $\sum_{n=1}^{\infty} f_n(x)$ converges to $F(x)$ for all $x \in X$. Hence $f(x) = F(x)$ for all $x \in A$, completing the proof. (Querry: Which property of the real line is being tacitly used here?) ∎

There is no difficulty in showing that the theorem still holds if the interval $[-1, 1]$ is replaced by any other closed bounded interval $[a, b]$ where $a < b \in \mathbb{R}$, since any such interval is homeomorphic to $[-1, 1]$. However when one tries to prove it for an open interval, there is a little difficulty. Consider for example that we had a continuous function $f: A \to (-1, 1)$ where A is a closed subset of a normal space X. We can apply the argument of the proof above to get an extension F of f. The trouble is that for some $x \in X$, it may well happen that $f_n(x) = \frac{2^{n+1}}{3^n}$ for all n $\left(\text{or that } f_n(x) = -\frac{2^{n-1}}{3^n} \text{ for all } n\right)$ and hence that $F(x) = 1$ (or $F(x) = -1$). Thus although f takes values in $(-1, 1)$, in general we can only ensure that F takes values in $[-1, 1]$ and not necessarily in $(-1, 1)$. This difficulty can be overcome by a neat trick.

(4.7) Theorem: Let A be a closed subset of a normal space X and suppose $f: A \to (-1, 1)$ is continuous. Then there exists a continuous function $F: X \to (-1, 1)$ such that $F(x) = f(x)$ for ali $x \in A$.

Proof: Define $g(x) = \frac{f(x)}{1 + |f(x)|}$ for $x \in A$. Then g is continuous and takes values in $(-\frac{1}{2}, \frac{1}{2})$. So by the theorem above, it has an extension, $G: X \to [-\frac{1}{2}, \frac{1}{2}]$. Now let $B = \{x \in X : G(x) = \frac{1}{2} \text{ or } G(x) = -\frac{1}{2}\}$. Then B is a closed subset of X since $B = G^{-1}(\{-\frac{1}{2}, \frac{1}{2}\})$. Note that $A \cap B = \emptyset$, for if $x \in A$ then $G(x) = g(x) = \frac{f(x)}{1 + |f(x)|}$ and $G(x) = \pm\frac{1}{2}$ would mean $2|f(x)| = 1 + |f(x)|$, a contradiction. So by Urysohn's lemma, there exists a continuous function $h: X \to [0, 1]$ which equals 0 on B and 1 on A. Define $H(x) = h(x)G(x)$ for $x \in X$. Then H is continuous and takes values only in $(-\frac{1}{2}, \frac{1}{2})$. Now define $F: X \to \mathbb{R}$ by $F(x) = \frac{H(x)}{1 - H(x)}$ for $x \in X$. F is a well-defined continuous function on X and it takes values in $(-1, 1)$ as H takes values in $(-\frac{1}{2}, \frac{1}{2})$. So we regard F as a continuous function from X into $(-1, 1)$. It remains to verify that F is an extension of f. Let $x \in A$. If $f(x) \geq 0$ then $H(x) = h(x) \cdot G(x) = G(x) = g(x) = \frac{f(x)}{1 + f(x)} \geq 0$ and so $F(x) = f(x)$ by direct calculation of $F(x)$. Similarly $f(x) = F(x)$ when $f(x) \leq 0$. This shows that f has F as an extension. ∎

Since any open interval and the real line are homeomorphic to $(-1, 1)$, we get the following corollary.

(4.8) Corollary: Any continuous real-valued function on a closed subset of a normal space can be extended continuously to the whole space. ∎

Exercises

4.1 Prove that there exists a map $r: \mathbb{R} \to [0, \infty)$ such that $r(x) = x$ for all $x \in [0, \infty)$. (In other words r is a right inverse to the inclusion map of $[0, \infty)$ into $\mathbb{R}$. Such a map is called a **retraction** and in the present case we say that $[0, \infty)$ is a **retract** of $\mathbb{R}$.)

4.2 Prove that if A is a closed subset of a normal space X then any map $f: A \to [0, \infty)$ can be continuously extended to a map from X to $[0, \infty)$. (Hint: First obtain an extension regarding f as a map from A to $\mathbb{R}$ and then follow it by a retraction of $\mathbb{R}$ onto $[0, \infty)$.)

4.3 Does the Tietze extension theorem hold for maps into $\mathbb{R}$ if on $\mathbb{R}$ we put the semi-open interval topology?

4.4 Prove that the Tietze extension theorem holds for maps into euclidean spaces with the usual topology. (Hint: Extend each coordinate function separately.)

4.5 A subset A of a space X is said to be a **G_δ-set** (or simply a G_δ) in X if it is the intersection of a countable number of open sets in X.
   (a) Prove that if $f: X \to Y$ is a map and B is a G_δ in Y then $f^{-1}(B)$ is a G_δ in X.
   (b) If X is a space, $A \subset X$ and $f: X \to \mathbb{R}$ is a map which vanishes precisely on A then A is a closed G_δ in X. (Hint: Note that $\{0\}$ is a G_δ in $\mathbb{R}$).
   (c) Prove that in a normal space, the converse of (b) holds. (Hint: Construct a uniformly convergent sequence of Urysohn functions.)

4.6 A topological space is said to be **perfectly normal** if it is normal and every closed set in it is a G_δ set in it.
   (a) Prove that all metric spaces are perfectly normal.
   (b) Prove that if A, B are mutually disjoint, closed subsets of a perfectly normal space X then there exists a map $f: X \to [0, 1]$ such that $f = 0$ precisely on A and $f = 1$ precisely on B.

4.7 Let $X = \mathbb{R} \cup \{\infty\}$ where ∞ is any symbol not in $\mathbb{R}$. On X we put a topology as follows: let $\mathcal{T}_1 = P(\mathbb{R})$ and let $\mathcal{T}_2 = \{A \subset X : \infty \in A$ and $X - A$ is finite$\}$. Then it is easily seen that $\mathcal{T} = \mathcal{T}_1 \cup \mathcal{T}_2$ is a topology on X.
   (a) Prove that X is compact, T_2 with this topology.
   (b) Prove that the set $\{\infty\}$ is not *a* G_δ set in X. Thus X is an example of a space which is normal but not perfectly normal. Another example will be given in the next chapter.
   (c) Characterise all continuous functions from X to $\mathbb{R}$.

Notes and Guide to Literature

Tietze's extension theorem, like Urysohn's lemma, is a classic theorem in topology and analysis. Many extensions of it have been obtained. See, for example, Dugundji [2].

Retracts were first defined by Borsuk [1], in 1931. This notion has since played an important role not only in topology but in other branches of mathematics as well. The theory of retracts is closely related to the theory of extension of maps. Two excellent references on this topic are Hu [1] and Borsuk [1], the former more readable, the latter more exhaustive.

The space $(X, \mathcal{T})$ constructed in Exercise (4.7) may seem a little artificial, but it is a powerful counterexample. It is called the one-point or Alexandroff compactification of $\mathbb{R}$ with discrete topology. This is a general construction and we shall undertake a formal study later. Johnson [1] has given an example of a non-metrisable, perfectly normal space.

Chapter Eight

Products and Coproducts

So far we defined the topological product of a finite number of topological spaces and proved a number of facts about them. In this chapter, we extend the study to the products of arbitrary families of topological spaces. Nothing substantially new is involved except perhaps the definition of the product. A reader who has grasped the finite products well will find the present chapter fairly easy. We also discuss briefly the coproducts or sums of topological spaces. Although they are of little importance as compared to products, we include them chiefly to stress their duality with products.

1. Cartesian Products of Families of Sets

Suppose we have a finite number of sets say $X_1, X_2, \ldots, X_n$. Then their Cartesian product, usually denoted by $\prod_{i=1}^{n} X_i$ is defined, as we know, as the set of all ordered n-tuples $(x_1, x_2, \ldots, x_n)$ with $x_i \in X_i$ for $i = 1, 2, \ldots, n$. It is not hard to generalise this definition when the number of sets is infinite but countable. Let us say that the sets are indexed as $X_1, X_2, X_3, \ldots X_n, \ldots$. Then $\prod_{n=1}^{\infty} X_n$ is defined as the set of all sequences $(x_1, x_2, x_3, \ldots, x_n, \ldots)$ with $x_i \in X_i$ for all $i = 1, 2, 3, \ldots, n, \ldots$. But what if we had an uncountable index set, say, I and a family of sets indexed by I say $\{X_i : i \in I\}$? In this case, the elements of the product (which is conveniently denoted by $\prod_{i \in I} X_i$) cannot be written down by merely 'putting together' entries from the X_i's, $i \in I$. How can then we generalise the concept of a cartesian product from a finite collection of sets to an arbitrary collection of sets?

To overcome this difficulty, we adopt a trick which is commonly used in mathematics. Whenever we want to generalise a certain concept from one situation to another, and the generalisation cannot be carried out directly, we paraphrase the concept to be generalised into a new form. In this new form, the generalisation becomes almost self-evident. We have already seen instances of this trick. For example, the usual definitions of convergence and continuity for metric spaces are given in terms of the metric concerned and therefore they cannot be carried over to arbitrary topological spaces in that form. So we reformulated these concepts in terms of open sets without involving the metric explicitly. In this new form, the definitions apply unchanged to all topological spaces.

Let us try a similar thing for the appropriate generalisation of product. The product of finitely many sets, $X_1, X_2, \ldots, X_n$ consists of all ordered n-tuples $x = (x_1, x_2, \ldots, x_n)$ with $x_i \in X_i$ for $i = 1, 2, \ldots, n$. The n-tuple x is uniquely determined by its entries and the order in which they occur. It follows that we can regard x as a function defined on the set $\{1, 2, \ldots, n\}$. If we denote x_i by $x(i)$ then the function x belongs to $\prod_{i=1}^{n} X_i$ iff $x(i) \in X_i$ for all $i \in \{1, 2, \ldots, n\}$. Note that $\{1, 2, \ldots, n\}$ is the index set here. Thus we can regard an element of the product $\prod_{i=1}^{n} X_i$ as a function x defined on the index set $\{1, 2, \ldots, n\}$ and satisfying $x(i) \in X_i$ for all i in the index set. Conversely any such function determines an element of the product $\prod_{i=1}^{n} X_i$, for given such a function f, we associate to it the n-tuple $(f(1), f(2), \ldots, f(n)) \in \prod_{i=1}^{n} X_i$. Thus we have succeeded in describing a finite product as the collection of certain functions whose domain is the index set. There is no difficulty in specifying their codomain. One can always suppose that these functions take values in the set $\bigcup_{i=1}^{n} X_i$.

It is now obvious how to define the cartesian product of an arbitrary indexed collection $\{X_i : i \in I\}$ of sets.

(1.1) Definition: Let $\{X_i : i \in I\}$ be an indexed family of sets. Then its **cartesian product**, denoted by $\prod_{i=I} X_i$ is defined as the set of all functions x from the index set I into $\bigcup_{i \in I} X_i$ such that $x(i) \in X_i$ for all $i \in I$, that is, $\prod_{i \in I} X_i = \{x : I \to \bigcup_{i \in I} X_i \mid x(i) \in X_i, \text{ for all } i \in I\}$. When I is understood, we can write $\prod X_i$ for $\prod_{i \in I} X_i$. The notations $\times_{i \in I} X_i$ and $\times X_i$ are also common.

(1.2) Definition: Let $x \in \prod_{i \in I} X_i$ and $j \in J$, then the **jth coordinate of** x is defined $x(j)$, that is, the value of x at j. The set X_j is called the ***j*th factor** of the product $\prod_{i \in I} X_i$.

In practice it is customary to denote the jth coordinate of x by x_j rather than by $x(j)$. Accordingly, the element x is often written as $\{x_i\}_{i \in I}$. In other words, we continue to think of elements of $\prod_{i \in I} X_i$ as ordered collections of elements from the X_i's. The idea of a function is merely a device which is used for the sake of making precise what is intuitively appealing. Thus, when we formally work with products, we treat their elements as functions from the corresponding index sets. But a 'feeling' for products comes only when one regards them as consisting of all ordered collections of entries from the corresponding sets. The reader is urged to master both the approaches and the ability to switch from one to the other when necessary. Disregard of the formal approach will make the study of products mathematically unsound while disregard of the intuitive approach will make it dry.

As a typical illustration of the difference between the two approaches, let us define the projection functions from the product onto the coordinate sets. Let $X = \prod_{i \in I} X_i$ where $\{X_i : i \in I\}$ is an indexed family of sets and let $j \in I$. We want to define the jth projection π_j from X onto X_j. Let $x \in X$. If we write x as $\{x_i\}_{i \in I}$ then it is obvious that $\pi_j(x)$ ought to be x_j, the jth coordinate of x. A formal definition would, however, take the following form:

(1.3) Definition: Let $\{X_i : i \in I\}$ be an indexed family of sets and let $X = \prod_{i \in I} X_i$. For $j \in I$, the j**th projection** is the function (usually denoted by) $\pi_j : X \to X_j$ defined by $\pi_j(x) = x(j)$ for $x \in X$.

The definition does look a little awkward to a beginner, but it is the only precise way of expressing the underlying idea. Note that the domain of the function π_j itself consists of functions on the index set I (of which j is a member).

The properties of projections established in the case of finite products (see Chapter 6, Section 3) continue to hold for arbitrary products as well and will not be repeated. However, a few simple observations and a little terminology will considerably simplify our task of defining the product topology in the next section. We shall throughout use the notation that X is the cartesian product $\prod_{i \in I} X_i$ where $\{X_i : i \in I\}$ is an indexed family of sets. Note that if B_i is a subset of X_i for $i \in I$ then $\prod_{i \in I} B_i$ is, strictly speaking, not a subset of $\prod_{i \in I} X_i$ because elements of the former are functions from I into $\bigcup_{i \in I} B_i$ whereas those of the latter are functions from I into $\bigcup_{i \in I} X_i$. However, since $\bigcup_{i \in I} B_i$ is a subset of $\bigcup_{i \in I} X_i$, any function into $\bigcup_{i \in I} B_i$ can also be regarded as a function into $\bigcup_{i \in I} X_i$ and, modulo this understanding, we regard $\prod_{i \in I} B_i$ as a subset of $\prod_{i \in I} X_i$. (The technical difficulty involved here can be circumvented by making a flat convention that all sets under consideration are contained in some universal set M. In such a case, the products $\prod_{i \in I} X_i$ and $\prod_{i \in I} B_i$ can both be defined as the sets of functions from I into M satisfying the appropriate conditions.)

(1.4) Definition: A **box** in X is a subset B of X of the form $\prod_{i \in I} B_i$ for $B_i \subset X_i$, $i \in I$. For $j \in I$, B_j is called the j**th side** of the box B. A box B is said to be **large** if all except finitely many of its sides are equal to the respective sets X_i's, that is to say, if there exist $j_1, j_2, \ldots, j_n \in I$ such that $B_i = X_i$ for all $i \in I - \{j_1, j_2, \ldots, j_n\}$.

Thus a large box is a box which has only finitely many 'short' sides.

It is easy to justify the term 'box' geometrically if one considers the cartesian product of copies of the set of real numbers. In case the index set I is finite, every box is a large box. Let us also introduce another term whose justification is again left to the reader.

(1.5) Definition: A **wall** in X is a set of the form $\pi_j^{-1}(B_j)$ for some $j \in I$ and some $B_j \subset X_j$. We also say this set is a **wall on $\mathbf{B}_j$**.

Note that a non-empty wall is a box at most one of whose sides is 'short'. In particular every such wall is a large box. The intersection of two walls $\pi_j^{-1}(B_j)$ and $\pi_k^{-1}(B_k')$ is not a wall if $j \neq k$. But if $j = k$, then it is a wall on $B_j \cap B_k'$. These simple observations lie behind the proof of the following proposition.

(1.6) Proposition: A subset of X is a box iff it is the intersection of a family of walls. A subset of X is a large box iff it is the intersection of finitely many walls.

Proof: Suppose $B = \prod_{i \in I} B_i$ where $B_i \subset X_i$ for all $i \in I$ is a box in X. For $i \in I$, let $W_i = \pi_i^{-1}(B_i)$. Then each W_i is a wall in X. We claim $B = \bigcap_{i \in I} W_i$. For $x \in B$ iff the ith coordinate of x belongs to B_i for all $i \in I$; or in other words, $x \in B$ iff $\pi_i(x) \in B_i$ for all $i \in I$. Hence $x \in B$ iff $x \in W_i$ for all $i \in I$. Thus we see that B can be written as an intersection of walls. Conversely suppose $\{W_\alpha : \alpha \in A\}$ is a family of walls in X. Then for each $\alpha \in A$, there exists $j(\alpha) \in I$ such that $W_\alpha = \pi_{j(\alpha)}^{-1}(B_{j(\alpha)})$ for some subset $B_{j(\alpha)}$ of $X_{j(\alpha)}$. For each $\alpha \in A$ fix such $j_{(\alpha)} \in I$ and $B_{j(\alpha)} \subset X_{j(\alpha)}$. Now for $j \in I$, let $C_j = \bigcap \{B_{j(\alpha)} : \alpha \in A, j(\alpha) = j\}$. In case there are no α's in A for which $j(\alpha) = j$, C_j is to be the set X_j. Let B be the box $\prod_{i \in I} C_i$. We claim that $B = \bigcap_{\alpha \in A} W_\alpha$. First suppose $x \in B$ and $\alpha \in A$. Let $j = j(\alpha)$. Then $\pi_{j(\alpha)}(x) = \pi_j(x) \in C_j \subset B_{j(\alpha)}$ and so $x \in \pi_{j(\alpha)}^{-1}(B_{j(\alpha)})$, i.e. $x \in W_\alpha$. Hence $B \subset \bigcap_{\alpha \in A} W_\alpha$. Conversely suppose $x \in W_\alpha$ for all $\alpha \in A$. Let $j \in I$. Then $\pi_{j(\alpha)}(x) \in B_{j(\alpha)}$ for all $\alpha \in A$ for which $j(\alpha) = j$. Hence $\pi_j(x) \in C_j$ for all $j \in I$. Hence $x(j) \in C_j$ for all $j \in I$. So $x \in B = \prod_{i \in I} C_i$. Hence $\bigcap_{\alpha \in A} W_\alpha \subset B$. Combining the two together, we have proved that every intersection of walls is a box. This completes the proof of the first assertion of the proposition. The proof of the second assertion is similar except that we take into account only those indices $j \in I$ for which the jth side is possibly not equal to the entire set X_j. The details are left to the reader. ∎

As a corollary of this proposition, we get

(1.7) Proposition: The intersection of any family of boxes is a box. The intersection of a finite number of large boxes is a large box.

Proof: Every box is an intersection of walls. Therefore an intersection of boxes is an intersection of intersections of walls and hence an intersection of walls. But an intersection of walls is a box. So the intersection of a family of boxes is a box. For the second part we note that a large box is the intersection of finitely many walls. Hence the intersection of finitely many large boxes will again be the intersection of finitely many walls and hence a large box. ∎

We have given the proof above verbally in order to avoid clumsy notations. The reader should, however, know how to say these types of things in precise terms if need be.

We now turn to an aspect of cartesian products which is important from the point of view of axiomatic set theory. Let $X = \prod_{i \in I} X_i$ where $\{X_i : i \in I\}$ is an indexed family of sets. Then elements of X are functions defined on I with a certain condition, namely, that the value at $i \in I$ be a member of the set X_i. Such a function then amounts to making a simultaneous choice of one element from each X_i for $i \in I$. Conversely any such choice determines a unique element of the product $\prod_{i \in I} X_i$. For this reason elements of X are also called *choice functions*.

An important question is 'Do choice functions necessarily exist?' The answer is obviously no if for some $j \in I$, X_j is the empty set. But what if it is given that each member of the indexed family of sets $\{X_i : i \in I\}$ is non-empty? In this case an affirmative answer requires the use of the axiom of choice. Indeed, it is easy to see that the axiom of choice is equivalent to the statement that the cartesian product of nonempty sets is always non-empty.

We note that the definition of a cartesian product is set-theoretic in the sense that it is compatible with bijections, which are the fundamental equivalence relations in set theory. To be precise suppose I is an index set, $\{X_i : i \in I\}$ and $\{Y_i : i \in I\}$ are two families of sets indexed over I. If there is a bijection $f_i : X_i \to Y_i$ for each $i \in I$ then there is a bijection between $\prod_{i \in I} X_i$ and $\prod_{i \in I} Y_i$. Indeed, we simply define $f: \prod_{i \in I} X_i \to \prod_{i \in I} Y_i$ by $(f(x))(i) = f_i(x(i))$ for $i \in I$, for $x \in X$. In simpler terms, the ith coordinate of $f(x)$ is the image, under f_i, of the ith coordinate of x. It is easy to show that f is a bijection by constructing its inverse function in terms of the inverse function $f_i^{-1} : Y_i \to X_i$ for $i \in I$.

The observation made here allows us to define the products of arbitrary indexed families of cardinal numbers.

(1.8) Definition: Let $\{\alpha_i : i \in I\}$ be an indexed family of cardinal numbers. Then its **product** $\prod_{i \in I} \alpha_i$ is defined to be the cardinal number of the product set $\prod_{i \in I} X_i$ where for each $i \in I$, X_i is a set of cardinality α_i.

The comment made above shows that this definition is unambiguous in that it does not depend on the choice of the sets X_i as long as they have the right cardinality. Elementary results about products of cardinal numbers are easy to establish and are left to the reader.

An important special case of cartesian products occurs when all sets in the family are equal. Suppose for each i in an index set I, X_i is some fixed set Y. In this case the product $\prod_{i \in I} X_i$ is clearly the set of all functions from I into Y. In case I and Y are both finite sets, with say m and n elements respectively, then it is easy to see that there are n^m distinct functions from I into Y. For this reason, the set of all such functions is denoted by Y^I and this notation is retained even when I and Y are possibly infinite. It is called the **Ith power of Y, or the product of I-many copies of Y or Y raised to I.** All these terms are very suggestive and frequently used. They also have the advantage that some of the properties suggested by their resemblance with the exponential functions are indeed true. We prove one such property here.

(1.9) Theorem: For any sets Y, I and J, $(Y^I)^J = Y^{I \times J}$ upto a set-theoretic equivalence (that is upto a bijection).

Proof: Let S and T respectively denote the sets on the left and right hand side. S is the set of all functions from J into Y^I (which itself is the set of all functions from I into Y) and T is the set of all functions from $I \times J$ into Y. We have to show that S and T are set-theoretically equivalent. Define $\lambda : S \to T$ as follows. Let $f \in S$. Then f is a function from J into Y^I and we want $\lambda(f)$ to be a function from $I \times J$ into Y. Let $(i, j) \in I \times J$. We have to define $\lambda(f)((i, j))$ as an element of Y. Now $f(j) \in Y^I$, that is, $f(j)$ itself is a function from I into Y. If we evaluate this function at $i \in I$, we get an element of Y, namely, $f(j)(i)$. We let $\lambda(f)(i, j))$ be $f(j)(i)$. This defines λ. To show it is one-to-one suppose $f, g \in S$ and $\lambda(f) = \lambda(g)$. Then for any $(i, j) \in I \times J$, $f(j)(i) = g(j)(i)$, which means that the function $f(j)$ identically equals the function $g(j)$. As this holds for all $j \in J$, $f = g$. Hence λ is one-to-one. To show it is onto, suppose $h \in T$. Then h is a function from $I \times J$ into Y. Define $f : J \to Y^I$ as follows.

Let $j \in J$. We let $f(j)$ be the function from I into Y defined by $f(j)(i) = h(i, j)$ for $i \in I$. Then $f \in S$ and from the very definition, $\lambda(f) = h$. So λ is onto. This λ is a bijection from S onto T or, in other words, S and T are equal upto a bijection. ∎

This theorem is not really a very profound result. The only reason to give it here is its proof. The construction in it is a special case of what is known as an adjunction. Usually there is something aesthetically appealing in such constructions.

We conclude the section with a brief discussion of the dual notion of the coproduct of an indexed family of sets.

(1.10) Definition: Let $\{X_i : i \in I\}$ be an indexed family of sets. Then its **coproduct** or **direct sum** is defined to be the set $\{(x, i) \in (\bigcup_{i \in I} X_i) \times I : x \in X_i\}$.

There are two standard notations for the coproducts. One is $\sum_{i \in I} X_i$, where the 'sigma' signifies its close relationship with the union of X_i's (as we shall see soon). The other one is $\coprod_{i \in I} X_i$ where the inverted 'pi' stresses the duality between coproducts and products. We shall prefer the first notation and as with products we shall sometimes write merely Σ when the index set I is understood.

It is obvious that ΣX_i is the union of the family of sets $\{X_i \times \{i\} : i \in I\}$. For $i \in I$, $X_i \times \{i\}$ is of course just a copy of X_i. For $i \neq j$ in I, the sets $X_i \times \{i\}$ and $X_j \times \{j\}$ are always mutually disjoint even though X_i and X_j may overlap. Thus $\underset{i \in I}{\Sigma} X_i$ is nothing but the union of mutually disjoint copies of the X_i's and is for this reason called the **disjoint union or the disjunctified union** of the X_i's. In case the X_i's are mutually disjoint to start with, $\underset{i \in I}{\Sigma} X_i$ may even be identified with $\underset{i \in I}{\bigcup} X_i$.

For each $j \in I$, there is a function $\lambda_j : X_j \to \underset{i \in I}{\Sigma} X_i$ defined by $\lambda_j(x) = (x, j)$ for all $x \in X_j$. Obviously each λ_j is one-to-one and it is called the jth **injection.** X_j is often called the jth-**summand.** The properties of the injections are dual to those of the projections.

Exercises

1.1 Justify the terms 'box' and 'wall' geometrically for products of copies of the real-line.

1.2 Prove the second assertion of the statement of Proposition (1.6).

1.3 Prove proposition (1.7) rigorously, using appropriate notations.

1.4 Let $\{X_i : i \in I\}$ be an indexed family of sets and let θ be a permutation of I (that is a bijection of I onto itself). For $i \in I$, let $Y_i = X_{\theta(i)}$. Prove that there is a bijection between $\underset{i \in I}{\Pi} X_i$ and $\underset{i \in I}{\Pi} Y_i$ (cf. Exercise (5.3.5)).

1.5 Let $\{X_i : i \in I\}$ be an indexed family of sets and suppose the index set I is the union of mutually disjoint subsets say $I = \underset{\alpha \in A}{\bigcup} I_\alpha$. For each $\alpha \in A$ let $Y_\alpha = \underset{i \in I_\alpha}{\Pi} X_i$. Prove that there is a bijection between $\underset{i \in I}{\Pi} X_i$ and $\underset{\alpha \in A}{\Pi} Y_\alpha$. Comment on the significance of this result.

1.6 Prove that if a product is non-empty, then each projection function is onto.

1.7 For cardinal numbers α, β define α^β to be the cardinal number of the set A^B where A, B are sets with cardinality α, β respectively. Prove that this is well-defined.

1.8 Define the sum of an indexed family of cardinal numbers by means of coproducts of sets.

1.9 For any cardinal numbers α, β, γ prove that
(i) $(\alpha)^{\beta \cdot \gamma} = (\alpha^\beta)^\gamma$
(ii) $\alpha^{(\beta+\gamma)} = \alpha^\beta \cdot \alpha^\gamma$

*1.10 Let $\{\alpha_i : i \in I\}$, $\{\beta_i : i \in I\}$ be collections of cardinal numbers. Assume for each $i \in I$, $\alpha_i < \beta_i$. Prove that $\underset{i \in I}{\Sigma} \alpha_i < \underset{i \in I}{\Pi} \beta_i$. (This result is known as **König's theorem.)**

1.11 Deduce from König's theorem that for any cardinal number α, $\alpha < 2^\alpha$ where 2 is the cardinal number of any set with two distinct elements.

1.12 Prove that König's theorem is equivalent to the axiom of choice.

Notes and Guide to Literature

This section is preparatory to the next one. The terms 'box', 'large box' and 'wall' are not standard and are introduced here as they will prove very convenient in the next section.

The statement of König's theorem is taken from Goffmann [1].

2. The Product Topology

In the last section we defined the Cartesian product of an indexed family of sets, say, $\{X_i : i \in I\}$. Suppose now each X_i is endowed with some topology $\mathcal{T}_i$. Then we want to topologise the product set $\prod_{i \in I} X_i$ by putting a 'suitable' topology on it. This topology would naturally depend on the topologies $\mathcal{T}_i$, $i \in I$. It will be called the product topology on the product set $\prod_{i \in I} X_i$. Its construction is motivated by the requirement that all projection functions be continuous.

(2.1) Definition: Let $\{(X_i; \mathcal{T}_i) : i \in I\}$ be an indexed collection of topological spaces and let X be the cartesian product $\prod_{i \in I} X_i$ and for each $i \in I$ let $\pi_i : X \to X_i$ be the projection function. Then the **product topology** on X is the smallest topology on X which makes each π_i continuous; in other words, it is the weak topology determined by the family of projection functions $\{\pi_i : i \in I\}$. The set $\prod_{i \in I} X_i$ with the product topology is called the **(topological) product** of the family $\{(X_i, \mathcal{T}_i) : i \in I\}$. The spaces X_i are called the **co-ordinate spaces** or the **factor spaces** of X.

The terminology as well as the justification for the existence and uniqueness of the product topology come from Chapter 5, Section 4. In Theorem (5.4.1) an explicit construction was given for the weak topology determined by a family of functions. In view of the special importance of the product topology it is helpful to recall the construction and record it as a theorem.

(2.2) Theorem: Let $\mathcal{T}$ be the product topology on the set $\prod_{i \in I} X_i$ where $\{(X_i, \mathcal{T}_i) : i \in I\}$ is an indexed collection of topological spaces. Then the family of all subsets of the form $\pi_i^{-1}(V_i)$ for $V_i \in \mathcal{T}_i$, $i \in I$ is a sub-base for $\mathcal{T}$. Also the family of all large boxes all of whose sides are open in the respective spaces is a base for $\mathcal{T}$.

Proof: The first assertion follows from the proof of Theorem (5.4.1). For the second, let V be a box say $V = \prod_{i \in I} V_i$ with $V_i \subset X_i$ for each $i \in I$. We are given each V_i is open in X_i and $V_i = X_i$ for $i \in I$ except for $i = i_1$, $i_2, \ldots, i_n$ (say). Then $V = \bigcap_{j=1}^{n} \pi_{i_j}^{-1}(V_{i_j})$ as we saw in the proof of Proposition (1.6). Hence V is the intersection of finitely many members of the sub-base considered above. On the other hand any such intersection is clearly a large box all whose sides are open in respective spaces. So the second assertion follows from the first assertion in view of the definition of a sub-base. ∎

(2.3) Definition: The sub-base and the base given by the theorem above are called respectively the **standard sub-base and the standard base** for the product topology.

With the terminology introduced in the last section, the standard sub-base consists of all walls on open sets while the standard base consists of all large boxes with open sides. Since every non-empty open set must contain a non-empty, basic open set, it follows that a non-empty open set G in a product space is large in the sense that $\pi_i(G) = X_i$ for almost all $i \in I$. Note in particular that if H_i is an open subset of X_i for each $i \in I$, then the product $\prod_{i \in I} H_i$ is in general not an open subset of $\prod_{i \in I} X_i$. It will be so iff $H_i = X_i$ for all except finitely many indices in I (when I itself is finite, this restriction is vacuous). Thus the product of open sets need not be an open subset. In sharp contrast, however, the product of closed sets is always closed as we show.

(2.4) Proposition: Let C_i be a closed subset of a space X_i for $i \in I$. Then $\prod_{i \in I} C_i$ is a closed subset of $\prod_{i \in I} X_i$ w.r.t. the product topology.

Proof: Let $X = \prod_{i \in I} X_i$ and $C = \prod_{i \in I} C_i$. We claim $X - C$ is an open set in the product topology on X. Let $x \in X - C$. Note that $C = \bigcap_{i \in I} \pi_i^{-1}(C_i)$ and so $x \notin C$ implies that there exists $j \in I$ such that $\pi_j(x) \notin C_j$. Let $V_j = X_j - C_j$ and let $V = \pi_j^{-1}(V_j)$. Then V_j is an open subset of X_j and so V is an open subset (in fact a member of the standard sub-base) in the product topology on X. Evidently, $\pi_j(x) \in V_j$ and so $x \in V$. Moreover $C \cap V = \emptyset$ since $\pi_j(C) \cap \pi_j(V) = \emptyset$. So $V \subset X - C$. Thus $X - C$ is a neighbourhood of each of its point. So $X - C$ is open and C closed in X. ∎

Before proving further results about topological products, we remark that the usual approach to them is to define the product topology in terms of the standard sub-base and then to prove the statement of Definition (2.1) as a theorem. We have reversed this procedure as we have already defined weak topologies generated by families of functions. From the property of the weak topology proved in Theorem (5.4.6) we get the following theorem as an immediate consequence.

(2.5) Theorem: Let $(X, \mathcal{T})$ be the topological product of an indexed family of topological spaces, $\{(X_i, \mathcal{T}_i) : i \in I\}$ and let Y be any topological space. Then a function $f : Y \to X$ is continuous (w.r.t. the product topology on X) iff for each $i \in I$, the composition $\pi_i \circ f : Y \to X_i$ is continuous. ∎

Many results which were proved earlier (often through exercises) for the topological products of finitely many spaces continue to hold for the topological products of arbitrary families of topological spaces. As examples we prove a few results.

(2.6) Theorem: The projection functions are open.

Proof: Let $(X, \mathcal{T})$ be the topological product of $\{(X_i, \mathcal{T}_i) : i \in I\}$. Fix $j \in I$. We have to show that the projection $\pi_j : X \to X_j$ is an open function. Let G be an open set in X (in the product topology $\mathcal{T}$). We claim $\pi_j(G)$ is a neighbourhood of each of its points. A typical point of $\pi_j(G)$ is of the form $\pi_j(x)$ for $x \in G$. Now since G is open, there exists a member V of the standard base for the product topology such that $x \in V \subset G$. Let $V = \prod_{i \in I} V_i$. Then clearly $\pi_j(V) = V_j \subset \pi_j(G)$ and $\pi_j(x) \in V_j$. But V_j is an open subset of X_j. Thus $\pi_j(G)$ is a neighbourhood of each of its points and hence an open subset in X_j. So the function π_j is open. ∎

(2.7) Theorem: If the product is non-empty, then each coordinate space is embeddable in it.

Proof: Let X be the topological product of an indexed family of spaces $\{(X_i, \mathcal{T}_i) : i \in I\}$. Since X is nonempty so is each X_i for $i \in I$. Now fix $j \in I$. We want to show that X_j can be embedded into X. For each $i \neq j$ in I, fix some $y_i \in X_i$. Now define $e : X_j \to X$ by

$$e(x)(i) = \begin{cases} y_i \text{ for } i \neq j \in I \\ x \text{ for } i = j \end{cases}$$

In other words, $e(x)$ is that element of $\prod_{i \in I} X_i$ whose jth coordinate is x and all other coordinates are equal to the respective chosen y_i's. Evidently e is one-to-one, since $e(x) = e(x')$ would in particular imply, $e(x)(j) = e(x')(j)$ whence $x = x'$. Also the composite $\pi_j \circ e$ is the identity map on X_j while for any $i \neq j$, $\pi_i \circ e$ is the constant map taking all points of X_j to y_i. In either case, the composite $\pi_i \circ e$ is continuous for all $i \in I$. So by Theorem (2.5) above e is continuous. To show that e is an embedding it only remains to show that e is open when regarded as a function from X_j onto $e(X_j)$. Note that $e(X_j)$ is the box $\prod_{i \in I} Y_i$ where $Y_j = X_j$ and $Y_i = \{y_i\}$ for $i \neq j \in I$. Now let V be an open subset of X_j. It is then easy to show that $e(V) = \pi_j^{-1}(V) \cap e(X_j)$ as each side equals the box $\prod_{i \in I} Z_i$ where $Z_j = V$ and $Z_i = \{y_i\}$ for $i \neq j \in I$. Now the set $\pi_j^{-1}(V)$ is open in X and so $e(V)$ is an open set in $e(X_j)$ in the relative topology. As noted above, this completes the proof that e is an embedding. ∎

Note that in case each X_i is a T_1-space then the sets $\{y_i\}$ are closed in the respective sets X_i for $i \neq j$ while X_j is certainly closed in X_j. So by Proposition (2.4), $e(X_j)$ is a closed subset of X. Thus in such a case, any weakly hereditary property which is possessed by the product is also enjoyed by each coordinate space.

In view of Theorem (2.6) and Proposition (5.4.9) it follows that each factor space is a quotient space of the product provided the latter is non-empty. It is natural to ask whether the converse is true, that is, given that Y is a quotient space of X, does there exist a space Z such that X is homeomorphic to $Y \times Z$? This is seen to be false from simple examples. Let X be the unit interval $[0, 1]$ and Y the unit circle S^1. Then Y is a quotient space of X. However there is no space Z such that X is homeomorphic to $Y \times Z$. For if such a space Z exists then the preceding theorem would imply that S^1 can be embedded into $[0, 1]$, which is not the case as we saw in the last chapter.

An important special case of topological product occurs when all spaces in $\{(X_i, \mathcal{T}_i) : i \in I\}$ are equal to some space Y. In this case the topological product (like the Cartesian product) is denoted by Y^I and is called the ***I*th power** of the space Y. The cases where Y is the real line or the unit interval $[0, 1]$ are especially important.

(2.8) Definition: A **cube** is a space of the form $[0, 1]^I$ where I is some set. If the set I is denumerable the cube is called a Hilbert cube.

Evidently if I and J are sets of the same cardinality then the cube $[0, 1]^I$ is homeomorphic to the cube $[0, 1]^J$. Hence upto homeomorphism there is only one Hilbert cube and it is called **the Hilbert cube.** The importance of the Hilbert cube in general topology is due not only to its nice intrinsic properties (a few of which will be proved in this book) but also to the fact that every second countable metric space can be embedded into it (as we shall prove in the next chapter where this point will be further elaborated).

The powers $\mathbb{R}^I$ of the real line with the usual topology on it are of importance as counterexamples when the set I is uncountable. When I is finite, we of course get the euclidean spaces. The space $\mathbb{R}^{\mathbb{N}}$ is a separable metric space as we shall prove later in this chapter. The infinite dimensional euclidean space $\mathbb{R}^\infty$ is a subspace of $\mathbb{R}^{\mathbb{N}}$.

Another very important case of a power Y^I arises when Y is a discrete space with two points. Such a space is commonly denoted by $\mathbb{Z}_2$ and the power $\mathbb{Z}_2^I$ is called a **Cantor discontinuum,** the terminology coming from the fact that $\mathbb{Z}_2^{\mathbb{N}}$ is homeomorphic to the well-known Cantor ternary set in $\mathbb{R}$ (the proof is left as an exercise). Their special importance stems from the fact that the set $\mathbb{Z}_2$ can be given an algebraic structure as well; it is a field with two elements. Consequently the powers $\mathbb{Z}_2^I$ are what are known as Boolean algebras.

It is very easy to characterise convergence of a sequence in a product space in terms of convergence in each coordinate space.

(2.9) Proposition: Let $X = \prod_{i \in I} X_i$, each X_i being a topological space. Suppose $\{x_n\}$ is a sequence in X and that $x \in X$. Then $\{x_n\}$ converges to x in X iff for each $i \in I$, the sequence $\{\pi_i(x_n)\}$ converges to $\pi_i(x)$ in X_i.

Proof: First suppose that $\{x_n\}$ converges to x in X. Let $i \in I$. Since π_i is continuous it follows that $\{\pi_i(x_n)\}$ converges to $\pi_i(x)$ by Exercise (5.3.9). Conversely suppose that for each $i \in I$, $\{\pi_i(x_n)\}$ converges to $\pi_i(x)$ in X_i. Let G be an open set in X containing x. Then G contains an element of the standard base for the product topology, say, V such that $x \in V \subset G$. Let $V = \prod_{i \in I} V_i$ where V_i is open in X_i for all i and $V_i = X_i$ for all $i \in I$ except possibly for $i = i_1, i_2, \ldots, i_r$ (say). Now for each $k = 1, 2, \ldots, r$, V_{i_k} is an open subset containing $\pi_{i_k}(x)$ and since $\pi_{i_k}(x_n)$ converges to $\pi_{i_k}(x)$, there exists $N_k \in \mathbb{N}$ such that for all $n \geq N_k$, $\pi_{i_k}(x_n) \in V_{i_k}$. Let $N = \max\{N_1, \ldots, N_k\}$. Then evidently for all $n \geq N$, $\pi_i(x_n) \in V_i$ for all $i \in I$ and so $x_n \in \prod_{i \in I} V_i$. Thus $x_n \in G$ for all $n \geq N$. Since G was an arbitrary open set containing x, it follows that $\{x_n\}$ converges to x in X. ∎

We note that the essential property of the product topology used in this proof is that it is the weak topology generated by the family of projections $\{\pi_i : i \in I\}$ and that an analogous result would hold for any weak topology.

It is instructive to reformulate the proposition above slightly. Recall that x as well as each x_n is a choice function, all on the same domain I. If we rewrite $\pi_i(x)$ as $x(i)$ (which is, in fact, the definition of π_i) and $\pi_i(x_n)$ as $x_n(i)$ then the proposition above reads that $x_n \to x$ in X iff $x_n(i) \to x(i)$ for each $i \in I$. This is precisely the definition of pointwise convergence of the sequence $\{x_n\}$ of functions. For this reason, the theorem above is lucidly paraphrased as 'convergence in the product topology is pointwise (or co-ordinate-wise)' and the product topology is often called the **topology of pointwise convergence or of coordinate-wise convergence.**

As in the last section, we conclude with a brief discussion of coproducts. Let $\{(X_i, \mathcal{T}_i) : i \in I\}$ be an indexed family of topological spaces and let X be the coproduct $\sum_{i \in I} X_i$. We put on X the strong topology $\mathcal{T}$ determined by the family $\{\lambda_i : i \in I\}$ of all injections into X. The space $(X; \mathcal{T})$ is called the **topological coproduct** or the **topological sum** of the given family of topological spaces. It is easy to see that each λ_i is an embedding. Recall that $\lambda_i(X_i)$ is the subset $X_i \times \{i\}$ of X. Clearly a subset G of X is open if and only if its intersection with each $X_i \times \{i\}$ is open. A similar statement holds for closed subsets. The spaces $(X_i, \mathcal{T}_i)$, $i \in I$ are called the **constituent spaces** or **summands** of the coproduct $(\sum_{i \in I} X_i, \mathcal{T})$.

The properties of coproducts are dual to those of products. For example, a function from a coproduct into a topological space is continuous if and only if its composite with each injection is continuous. A few results about coproducts will be given as exercises. However, topologically, the coproducts are not as interesting as the products. This is so because in a coproduct, each constituent space is 'isolated' from the others in the sense that it is both a closed and an open subset. Consequently any information about the coproduct is obtained by merely putting together information about constituent spaces. Nevertheless, coproducts are useful for a convenient description of certain topological spaces.

Exercises

2.1 Prove that if X_i is a T_1-space for each $i \in I$ then $\prod_{i \in I} X_i$ is also a T_1-space in the product topology. (Hint: Use Proposition (2.4)).

2.2 Prove that the converse also holds, assuming that X is nonempty. What may go wrong if X is empty?

2.3 Let $\{(X_i, \mathcal{T}_i) : i \in I\}$ be a family of spaces and suppose $J \subset I$. Prove that $\prod_{i \in J} X_i$ can be embedded in $\prod_{i \in I} X_i$, provided $X_i \neq \emptyset$ for $i \in I - J$.

2.4 Let Y be any topological space and let I, J be any sets. Prove that the powers $(Y^I)^J$ and $Y^{I \times J}$ are homeomorphic to each other. (Hint: A bijection between the two was already defined in Theorem (1.9). The problem is to show that it and its inverse are continuous. For this, apply Theorem (2.5).)

2.5 Let $X_i : i \in I$ be an indexed family of spaces and suppose θ is a permutation of the index set I. For $i \in I$, let $Y_i = X_{\theta(i)}$. Prove that the topological products $\prod_{i \in I} X_i$ and $\prod_{i \in I} Y_i$ are homeomorphic to each other.

2.6 Let $\{X_i : i \in I\}$ be an indexed family of spaces and let I be the disjoint union of some subsets of it, say $I = \bigcup_{\alpha \in A} I_\alpha$ where A is some index set, $I_\alpha \subset I$ for each $\alpha \in A$ and $I_\alpha \cap I_\beta = \emptyset$ for $\alpha \neq \beta$ in A. For each $\alpha \in A$, let $Y_\alpha = \prod_{i \in I} X_i$. Prove that the topological products $\prod_{i \in I} X_i$ and $\prod_{\alpha \in A} Y_\alpha$ are homeomorphic to each other.

2.7 Let X be a set, $\{Y_i : i \in I\}$ a family of spaces. Suppose for each $i \in I$, $f_i : X \to Y_i$ is a function and let $\mathcal{T}$ be the weak topology on X determined by the family $\{f_i : i \in I\}$. Prove that a sequence $\{x_n\}$ converges to x in X iff for each $i \in I$, the sequence $\{f_i(x_n)\}$ converges to $f_i(x)$ in Y_i.

2.8 Let Y_i be a subspace of X_i for $i \in I$. Prove that the product topology on $\prod_{i \in I} Y_i$ coincides with the relative topology on it as a subset of $\prod_{i \in I} X_i$.

2.9 The Cantor ternary set C is a subset of $[0, 1]$ defined as follows:

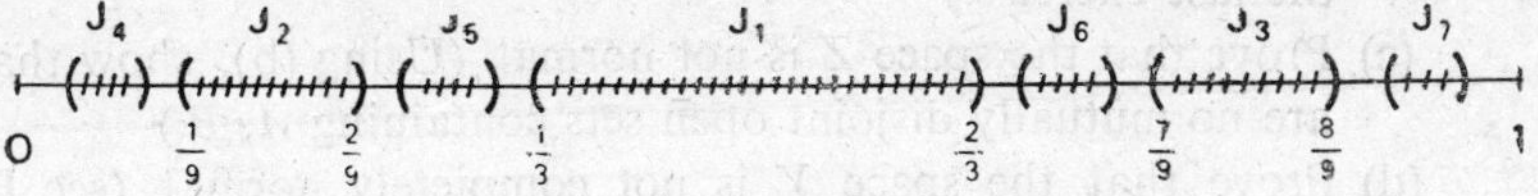

Fig. 1 Definition of the Cantor ternary set.

Let J_1 be the open interval $(\frac{1}{3}, \frac{2}{3})$, that is J_1 is the middle third of $[0, 1]$. Then the complement of J_1 in $[0, 1]$ consists of the disjoint union of the intervals, $[0, \frac{1}{3}]$ and $[\frac{2}{3}, 1]$. Let J_2, J_3 be respectively their open middle thirds, that is, $J_2 = (\frac{2}{9}, \frac{8}{9})$ and $J_3 = (\frac{7}{9}, \frac{8}{9})$. Remove these also from $[0, 1]$. The remainder then consists of four disjoint closed intervals. Let J_4, J_5, J_6, J_7 be their open middle thirds. Remove them and repeat the process *ad infinitum* to get a sequence of open intervals $J_1, J_2, J_3, \ldots, J_n, \ldots$. C is then defined as $[0, 1] - \bigcup_{n=1}^{\infty} J_n$.

(a) Prove that C is compact. (Hint: It is a closed subset of $[0, 1]$.)
(b) Prove that C is totally disconnected.
(c) Prove that the Cantor set C consists precisely of those real numbers in $[0, 1]$ that have a ternary expansion which does not involve 1.
*(d) Prove that C is homeomorphic to the power space $\mathbb{Z}_2^{\mathbb{N}}$. (Hint: Using (c) define a bijection from C onto $\mathbb{Z}_2^{\mathbb{N}}$.)
(e) Prove that $C \times C$ is homeomorphic to C. (Hint: Apply a very special case of Exercise (2.6) to (d) above.)

*2.10 Let $X = [0, 1]^I$ where the set I is uncountable. Let $\theta \in X$ be the point each of whose coordinates is 0 (i.e. $\theta(i) = 0$ for all $i \in I$). Prove that the set $\{\theta\}$ is a closed subset of X but that it is not a G_δ set. (Hint: That $\{\theta\}$ is closed follows from exercise (2.1). For the other part, suppose if possible that $\{\theta\} = \bigcap_{n=1}^{\infty} G_n$ where each G_n is open in X. Then each G_n contains a large box, say V_n, containing θ. All sides of V_n are equal to $[0, 1]$ except those with indices coming from a finite subset, say, I_n of I. Let $J = \bigcup_{n=1}^{\infty} I_n$. Then J is countable and so there exists $k \in I - J$. Define $x \in X$ by $x(k) = 1$ and $x(i) = 0$ for all $i \neq k$. Then show that $x \in \bigcap_{n=1}^{\infty} V_n$ thus getting a contradiction.)

*2.11 Let X, θ be as above and let $Y = X \times [0, 1]$ (actually Y is homeomorphic to X but this fact will not be needed). Let $Z = Y - \{(\theta, 0)\}$. Let $A = X \times \{0\} - \{(\theta, 0)\}$ and $B = \{\theta\} \times [0, 1] - \{(\theta, 0)\}$.
(a) Prove that A, B are closed subsets of Z (although not of Y).
(b) Let V be an open subset of Z (and hence of Y) containing B. Prove that there exists $x \in X$, $x \neq \theta$ such that $(x, 1/n) \in V$ for all $n \in \mathbb{N}$. (Hint: Note that V is an open neighbourhood of $(\theta, 1/n)$ for all $n \in \mathbb{N}$. The crux of the argument then is similar to that of the last exercise.)
(c) Prove that the space Z is not normal. (Using (b), show that there are no mutually disjoint open sets containing A, B.)
(d) Prove that the space Y is not completely normal (see Exercise (7.1.11) for definition).

2.12 Verify that the injections into a coproduct are embeddings, that they are closed as well as open.

2.13 Let $\{(x_i; \mathcal{T}_i) : i \in I\}$ and $\{Y_i, \mathcal{U}_j) : j \in J\}$ be families of topological spaces and X, Y their coproducts respectively. Prove that $X \times Y$ is homeomorphic to the coproduct of a certain family of spaces.

2.14 Prove that a locally connected space is the topological sum of the family of its components. (Hint: Recall that in such a space the components are open as well as closed.)

2.15 Prove that $\mathbb{R} \times \mathbb{R}$ with the order topology induced by the lexicographic ordering on it is homeomorphic to the topological sum of copies of $\mathbb{R}$ with the usual topology.

2.16 Generalise Proposition (2.4) by showing that if $A_i \subset X_i$ for $i \in I$, then $\prod_{i \in I} \bar{A}_i$ is the closure of the box $\prod_{i \in I} A_i$ in the product topology.

Notes and Guide to Literature

The definition of the product topology, as well as many of its standard properties are due to Tychonoff ([1] and [2]).

The Cantor ternary set is an invaluable tool of analysts to provide counter-examples. Although its measure-theoretic properties are of little interest to us, it has been a very fertile source of research even for the topologists. The properties given in Exercise (2.9) are but a glimpse of the wealth of topological results that are known about the Cantor set.

Later on we shall show that all cubes are compact and T_2 and hence normal. In view of this, the space X in Exercise (2.10) provides an example of a space which is normal but not perfectly normal, while the spaces Y and Z in Exercise (2.11) show that normality is not a hereditary property. The Hilbert cube will be of particular interest to us in the next chapter. It has a few strange properties which will not be proved in this book. For example, it is homogeneous as shown in Fort [1]. This is certainly startling because neither the unit interval [0, 1] nor any of its finite powers is homogeneous. Homogeneity of [0, 1] means intuitively that it has no 'corners'!

3. Productive Properties

In this and the next section we shall be concerned with finding out whether a certain topological property carries over from coordinate spaces to their topological products. In other words suppose X is the topological product of an indexed family $\{(X_i, \mathcal{T}_i) : i \in I\}$ of topological spaces. If each X_i has a topological property can we say that X also has it? The answer will evidently depend upon the property itself and also upon how large the index set I is. Depending on it we make the following definitions.

3.1 Definition: A topological property is said to be **productive** if whenever $\{(X_i, \mathcal{T}_i) : i \in I\}$ is an indexed family of spaces having that property, the topological product $\prod_{i \in I} X_i$ also has it. The property is said to be **countably productive** if this condition holds for all countable index sets I and it is said to be **finitely productive** if the condition holds whenever the index set I is finite.

As an example, Exercise (2.1) shows that the T_1 property is a productive property. Obviously every productive property is countably productive and every countably productive property is finitely productive. In the last few chapters we dealt with only finite products (that is, with products of a finite number of spaces) and by way of exercises we pointed out that quite a few topological properties are finitely productive. It turns out that many of these are actually productive while a few others (especially those involving countability) are not productive but are nevertheless countably productive. In this section we shall consider the former. Sometimes the converse is also true; that is, whenever a non-empty product has a certain property so does every coordinate space. It is necessary to assume that the product space is non-empty. For, if the product is empty then one of the coordinate spaces is empty and the other coordinate spaces can be any arbitrary spaces. So it is impossible in such a case to say anything about their properties just by knowing that the product (which is the empty set) has a certain property. Thus throughout this and the next section it will be assumed that we are dealing with products of non-empty spaces.

As typtical results, we prove that most of the separation properties are productive.

(3.2) Theorem: A topological product is T_0, T_1, T_2 or regular iff each coordinate space has the corresponding property.

Proof: First note that all the properties in the theorem are hereditary. Now if $X = \prod_{i \in I} X_i$ then each coordinate space X_j is homeomorphic to a subspace of X by Theorem (2.7). So whenever X is T_0, T_1, T_2 or regular, so will be each X_j. It is the converse that really needs some arguments. Exercise (2.1) shows that the T_1 property is productive. We show next that the T_2 property is productive. Suppose X_i is a T_2 space for each $i \in I$ and let $X = \prod_{i \in I} X_i$. Let x, y be distinct points of X. Note that x, y are both functions on the index set I and so, to say that they are distinct means that they assume distinct values at some index. So there exists $j \in I$ such that $x(j) \neq y(j)$, or in a different notation, $\pi_j(x) \neq \pi_j(y)$ in X_j. Since X_j is a Hausdorff space, there exist disjoint open sets, U, V in X_j such that $x \in U$, $y \in V$. Now let $G = \pi_j^{-1}(U)$, $H = \pi_j^{-1}(V)$. Then G, H are open subsets of X. Also $x \in G$, $y \in H$ and $G \cap H = \Phi$ showing that X is a Hausdorff space.

An essentially similar argument applies to show that T_0 is a productive property. However when we try to adapt it for regularity, we are in trouble. For suppose $X = \prod_{i \in I} X_i$, $x \in X$ and C is a closed subset of X not containing x. In this case it may happen that even though $x \notin C$, still $\pi_i(x) \in \pi_i(C)$ for all $i \in I$. Secondly, even if we could find some $j \in I$ such that $\pi_j(x) \notin \pi_j(C)$, there is no guarantee that $\pi_j(C)$ is closed in X_j even though C is closed in X (for projection maps need not be closed, although they are always open). We therefore modify the argument as follows. Suppose each X_i is regular. Let $X = \prod_{i \in I} X_i$, $x \in X$ and C be a closed subset of X not containing x. Then $X - C$ is an open set containing x. So there exists a member V of the standard base for the product topology such that $x \in V$

and $V \subset X - C$. Let $V = \prod_{i \in I} V_i$ where each V_i is open in X_i and $V_i = X_i$ for all $i \in I$ except possibly for $i = i_1, i_2, \ldots, i_n$ (say). Let $x_{i_r} = \pi_{i_r}(x)$ for $r = 1, 2, \ldots, n$. Then the open set V_{i_r} contains x_{i_r} in X_{i_r} for $r = 1, 2, \ldots, n$. So by regularity of each coordinate space (see Proposition (7.1.15)) there exists an open set U_{i_r} in X_{i_r} such that $x_{i_r} \in U_{i_r}$ and $\bar{U}_{i_r} \subset V_{i_r}$. For $i \neq i_1, i_2, \ldots, i_n$, let $U_i = X_i$ and consider the box $U = \prod_{i \in I} U_i$. Clearly U is an open set in X and $x \in U$. Also U is contained in the box $\prod_{i \in I} \bar{U}_i$ which itself is contained in V. But by Proposition (2.4), the set $\prod_{i \in I} \bar{U}_i$ is closed in X and hence contains $\bar{U}$ (actually it equals $\bar{U}$ by Exercise (2.16)). Thus $x \in U$ and $\bar{U} \subset X - C$. Therefore U and $X - \bar{U}$ are mutually disjoint open subsets containing x and C respectively. This shows that X is regular. ∎

We already noted that normality is not even finitely productive (see Exercise (7.2.7)), and hence it is not a productive property. Complete regularity, however, is a productive property, but the proof is not so routine as that of the last theorem. To begin with, a slight reformulation of the definition shows that a space X is completely regular iff for any $x \in X$ and for any open set V containing x, there exists a map $f: X \to [0, 1]$ such that $f(x) = 0$ and $f(y) = 1$ for all $y \notin V$. The following lemma shows that it suffices if this condition is satisfied for members of a sub-base.

(3.3) Lemma: Let $\mathcal{S}$ be a sub-base for a topological space X. Then X is completely regular iff for each $V \in \mathcal{S}$ and for each $x \in V$, there exists a continuous function $f: X \to [0, 1]$ such that $f(x) = 0$ and $f(y) = 1$ for all $y \notin V$.

Proof: If X is completely regular, then certainly the condition holds for all open sets and so, in particular, for members of $\mathcal{S}$. Conversely suppose each $V \in \mathcal{S}$ satisfies the given condition. Let G be an open set in X and let $x \in X$. By definition of a sub-base there exist $V_1, V_2, \ldots, V_n \in \mathcal{S}$ such that $x \in \bigcap_{i=1}^{n} V_i \subset G$. For each $i = 1, 2, \ldots, n$ we have a map $f_i : X \to [0, 1]$ which vanishes at x and takes the value 1 outside V_i. Define $f: X \to [0, 1]$ by $f(y) = 1 - (1 - f_1(y))(1 - f_2(y)) \ldots (1 - f_n(y))$. Clearly f is continuous and $f(x) = 0$ and f takes the value 1 outside $\bigcap_{i=1}^{n} V_i$ and hence outside G. This proves that X is completely regular. ∎

(3.4) Theorem: A product of topological spaces is completely regular iff each coordinate space is so.

Proof: Once again, the direct implication follows from Theorem (2.7) and the fact that complete regularity is a hereditary property. For the converse suppose $X = \prod_{i \in I} X_i$ where each X_i is completely regular. Let $\mathcal{S} = \{\pi_j^{-1}(V_j) : j \in I, V_j \text{ open in } X_j\}$ be the standard sub-base for the product topology and suppose $x \in \pi_j^{-1}(V_j)$. Then $\pi_j(x) \in V_j$ and so by complete regularity of X_j, there exists a map $f: X_j \to [0, 1]$ such that $f(\pi_j(x)) = 0$

and f takes value 1 on $X_j - V_j$. Then the composite $f \circ \pi_j : X \to [0, 1]$ vanishes at x and takes value 1 on $\pi_j^{-1}(X_j - V_j)$ i.e. on $X - \pi_j^{-1}(V_j)$. Hence the condition of the last lemma is satisfied by every member of the subbase $\mathcal{S}$. So X is completely regular. ∎

(3.5) Corollary: A topological product of spaces is Tychonoff iff each coordinate space is so.

Proof: Since by definition Tychonoff property means the combination of complete regularity with T_1, the result follows by merely putting together Theorems (3.2) and (3.4). ∎

After separation axioms let us turn to properties involving connectedness. We begin with the

(3.6) Theorem: A product of spaces is connected iff each coordinate space is connected.

Proof: Let $X = \prod_{i \in I} X_i$. Note that for each $j \in I$, X_j is the image of X under the projection map π_j. So if X is connected so is X_j by Proposition (6.2.4). Conversely suppose each X_i for $i \in I$ is a connected space. We want to show that X is connected. In Corollary (6.2.12) we proved that connectedness is a finitely productive property and we shall use it here. Fix some point say $x = \{x_i\}_{i \in I}$ in X and let C be the component of X containing x. We claim that C is dense in X. For let G be any nonempty open set in X. Then G contains a nonempty basic open set, say, $V = \prod_{i \in I} V_i$ where V_i is open in X_i for all $i \in I$ and $V_i = X_i$ for all $i \in I$ except possibly for $i = i_1, i_2, \ldots, i_n$ (say). Let $Z = \prod_{i \in I} Z_i$ where $Z_i = \{x_i\}$ for $i \neq i_1, i_2, \ldots, i_n$ and $Z_{i_r} = X_{i_r}$ for $r = 1, 2, \ldots, n$. Then Z is homeomorphic to the finite product $X_{i_1} \times X_{i_2} \times \ldots \times X_{i_n}$ (see the proof of Theorem (2.7), also see Exercise (2.3)). Since each X_{i_r} is connected, Z is connected by Corollary (6.2.12). Note that $x \in Z$. So $Z \cup C$ is connected by Theorem (6.2.9). This implies that $Z \subset C$ as C is a maximal connected subset of X. On the other hand $Z \cap V$ is non-empty because it contains all points of the form $y = \{y_i\}_{i \in I}$ where $y_i = x_i$ for $i \neq i_1, i_2, \ldots, i_n$ and y_{i_r} is any arbitrary element of V_{i_r} for $r = 1, 2, \ldots, n$ (note that each V_{i_r} is non-empty because V is non-empty). So $C \cap V \neq \Phi$ and finally $C \cap G \neq \Phi$. Thus C intersects every non-empty open subset of X. Therefore C is dense by Proposition (5.1.7). Hence $\overline{C} = X$. But on the other hand, components are always closed subsets by Theorem (6.2.15). So $C = \overline{C}$. Putting it all together, $C = X$ and hence X is connected because C is so. ∎

This proof is peculiar for two reasons. First of all connectedness for X is not established directly from definition, but rather by showing that X equals its own component. Secondly, the proof that connectedness is productive is done in two steps. First it is shown that it is finitely productive and this fact is then crucially used in proving that is it productive. The arguments for the two steps are quite different. This is in sharp contrast with Theorem (3.2) where the proof of productivity is independent of the cardinality of the index set I.

Local connectedness is not productive in nature. For example, the discrete space $\mathbb{Z}_2$ with two points is locally connected. But the product $\mathbb{Z}_2^{\mathbb{N}}$, which is homeomorphic to the Cantor set C as noted in Exercise (2.9 (d)) is not so. However, it is possible to tell precisely when a topological product is locally connected.

(3.7) Theorem: A product of spaces is locally connected iff each coordinate space is locally connected and all except finitely many of them are connected.

Proof: Let $X = \prod_{i \in I} X_i$. First suppose X is locally connected. For each $j \in I$, the projection function $\pi_j : X \to X_j$ is continuous, onto and open by Theorem (2.6). So each X_j is a quotient space of X by Proposition (5.4.9). Applying Proposition (6.3.5) we see that X_j is locally connected. Now fix any $x \in X$. Then there exists a connected neighbourhood M of x in X (as a matter of fact there exist lots of such neighbourhoods). Let V be a basic open set containing x such that $x \in V$ and $V \subset M$. We know then that $\pi_i(V) = X_i$ for all except finitely many $i \in I$. Hence $\pi_i(M) = X_i$ for all except finitely many i's. But since M is connected and π_i is continuous, this means that all except finitely may X_i's are connected.

Conversely suppose each X_i is locally connected and also that X_i is connected for all i except possibly for $i = i_1, i_2, \ldots, i_n$ (say). We have to show that X is locally connected at each of its points. Let $x \in X$ and suppose G is an open subset of X containing x. As usual G contains a large open box V containing x. Let $V = \prod_{i \in I} V_i$ where V_i is open in X_i for all $i \in I$ and $V_i = X_i$ for all $i \in I$ except possibly $i = j_1, j_2, \ldots, j_m$ (say). Some of these indices may overlap with the indices $i_1, i_2, \ldots, i_n$, Let $S = \{i_1, i_2, \ldots, i_n, j_1, j_2, \ldots, j_m\}$. Certainly S is a finite subset of I. Now for each $i \in S$, let W_i be a connected neighbourhood of $\pi_i(x)$ in X_i such that $W_i \subset V_i$ while for $i \in I - S$ let $W_i = X_i$. Let W be the box $\prod_{i \in I} W_i$. Then W is connected by Theorem (3.6) above. Also $x \in W$. Moreover if we let $U_i = \text{int } W_i$ in X_i for $i \in I$ then the box $U = \prod_{i \in I} U_i$ is an open set in X containing x and contained in W. So W is a connected neighbourhood of x in X. Finally, by its very construction, $W \subset V \subset G$. Since G was an arbitrary open set in X containing x, we have shown that X is locally connected at x. But since x is also arbitrary, X is locally connected. ∎

In almost all the proofs given so far, a crucial and direct use was made of the standard base for the product topology. We trust that by now the reader has grasped the technique fairly well. By contrast, in proving the productivity of a few other properties, a somewhat different technique is used. We illustrate it below for path-connectedness.

(3.8) Theorem: A product of topological spaces is path-connected iff each coordinate space is path-connected.

Proof: As usual, let $X = \underset{i \in I}{\Pi} X_i$. The direct implication follows from Exercise (6.3.9) in view of the fact that each projection function π_i is continuous and onto. Conversely suppose each X_i is path-connected. Let x, y be points in X. For each $i \in I$, there exists a path α_i in X_i from $\pi_i(x)$ to $\pi_i(y)$. Define $\alpha : [0, 1] \to X$ by $\alpha(t)(i) = \alpha_i(t)$ for all $i \in I$ and $t \in [0, 1]$. In other words, $\alpha(t)$ is the point whose i-th coordinate is $\alpha_i(t)$, that is, $\alpha(t) = \{\alpha_i(t)\}_{i \in I}$. We claim α is continuous. In view of Theorem (2.5), it suffices to show that for each $i \in I$, the composite $\pi_i \circ \alpha$ is continuous. But for $t \in [0, 1]$, $\pi_i(\alpha(t))$ is, by the very definition of α, equal to $\alpha_i(t)$. Hence $\pi_i \circ \alpha = \alpha_i$ which is given to be continuous. So α is a path in X. Clearly $\alpha(0) = x$ since $\alpha(0)(i) = \alpha_i(0) = \pi_i(x) = x(i)$ for all $i \in I$. Similarly $\alpha(1) = y$. Thus any two points of X can be joined by a path and so X is path-connected. ∎

The proof is simple but noteworthy because of an important trick adopted in it. The problem was to construct a certain function from [0, 1] into the product ΠX_i and we did it by 'putting together' the family of functions $\{\alpha_i : i \in I\}$ from [0, 1] into the coordinate spaces X_i, $i \in I$. In the next chapter we shall exploit this technique fully.

The dual notion of productive property is additive property. Thus a topological property is said to be **additive** if whenever each summand in a topological coproduct has it, so does the coproduct. Finitely additive and countably additive properties are defined by putting appropriate restrictions on the number of summands. Inasmuch as each constituent space is an open subspace of the coproduct, it is easy to see that any local property is additive in character. In the exercises we shall mention a few results about additive properties.

Exercises

3.1 Prove that a product of spaces is totally disconnected iff each coordinate space is so. (See Chapter 6, Section 2, for the definition).

3.2 Let $\{(X_i, \mathcal{T}_i) : i \in I\}$ be a family of spaces and X the set $\underset{i \in I}{\Pi} X_i$. Prove that the family of all boxes with open sides is a base for a topology on X. This topology is called the **box topology** and X with this topology is called **the box product** of the spaces $(X_i, \mathcal{T}_i)$, $i \in I$.

3.3 (a) Prove that in general the box topology is stronger than the product topology and that the two coincide when the index set is finite.

(b) Examine which of the topological properties considered so far carry over from a family of topological spaces to their box product.

3.4 A space Y is said to be an **absolute extensor** if for every closed subset A of a normal space X, every map from A to Y has a continuous extension to a map from X to Y. The Tietze extension theorem amounts to saying that the unit interval [0, 1] and $\mathbb{R}$ are absolute extensors.

(a) Prove that a retract of an absolute extensor is an absolute extensor (cf. Exercise (7.4.1)).

(b) Prove that the topological product of a family of spaces is an absolute extensor iff each coordinate space is so (cf. Exercise (7.4.4)).

3.5 Give an example of a Tychonoff space which is not normal. (Hint: Show that the space Z of Exercise (2.11) is a Tychonoff space).

3.6 A space X is said to be **binormal** if the product $X \times [0, 1]$ is normal. Prove that every binormal space is normal.

3.7 Find under what conditions a topological product is locally path-connected.

3.8 Prove that metrisability, first countability and second countability are not productive properties. (Hint: Prove that the space X in Exercise (2.10) is not first countable and hence neither metrisable nor second countable, even though the unit interval $[0, 1]$ has all these properties.)

3.9 To what extent is the property of being a discrete space productive?

3.10 Same question for indiscreteness.

3.11 Prove that all the separation properties, first countability and local connectedness are additive and that compactness is a finitely additive property.

Notes and Guide to Literature

The results of this section are very standard. The reader must have noted the total absence of compactness in this section. It is true that compactness is a productive property. But the proof is far from trivial and will be given later.

The box topology coincides with the product topology in the case of products of finitely many spaces. For arbitrary products it would appear at first sight that the box topology is somehow more natural than the product topology. However, the box topology has some serious disadvantages. Convergence in it cannot be described in terms of convergence in coordinate spaces; in other words, Proposition (2.9) fails for the box product. Moreover, the box product of compact spaces is in general not compact. Nevertheless box products provide some interesting counterexamples, see Munkres [1].

See Willard [1] for interesting comments on binormal spaces.

4. Countably Productive Properties

In Exercise (3.8) we pointed out that the properties of metrisability and first or second countability are not productive. In Exercises (4.3.6), (6.1.13), and (4.3.8) respectively we asked the reader to show that these properties are finitely productive. It turns out that they are in fact countably productive. We shall prove this in the present section.

We begin with metrisability. We need a technical device which is of interest in its own rights.

(4.1) Proposition: Let (X, d) be a metric space and let λ be any positive real number. Then there exists a metric e on X such that $e(x, y) \leq \lambda$ for all $x, y \in X$ and e induces the same topology on X as d does.

Proof: Define $e : X \times X \to \mathbb{R}$ by $e(x, y) = \min \{\lambda, d(x, y)\}$. We claim that e is a metric on X. The properties of positivity and symmetry are immediate from the definition. For triangle inequality, suppose $x, y, z \in X$. Then $e(x, z) \leq \lambda$ and so $e(x, z) \leq e(x, y) + e(y, z)$ holds when either $e(x, y) = \lambda$ or $e(y, z) = \lambda$. The only remaining case is when $e(x, y) = d(x, y) < \lambda$ and $e(y, z) = d(y, z) < \lambda$. In this case, $d(x, z) \leq d(x, y) + d(y, z)$. But $e(x, z) \leq d(x, z)$ and so, $e(x, z) \leq e(x, y) + e(y, z)$. Thus the triangle inequality holds for $x, y, z \in X$ and so e is a metric on X. That $e(x, y) \leq \lambda$ for all $x, y \in X$ is immediate from the very definition of e. It only remains to show that the topology induced by e is the same as that induced by d. For this, note that the family of all open balls whose radii do not exceed λ is a base for the respective metric topology. But these balls are identical whether w.r.t. d or w.r.t. e. Thus the two metric topologies have a common base and therefore coincide with each other. ∎

The significance of this proposition is that as far as the metric topology is concerned, there is no loss of generality in assuming that the metric is bounded. Moreover, the bound can be chosen to be any positive number at will. We immediately put this proposition to task in the following theorem.

(4.2) Theorem: Metrisability is a countably productive property.

Proof: For convenience we give the proof assuming that the index set is the set $\mathbb{N}$. In case the index set is finite the argument can be easily modified and is actually much simpler. So let $\{(X_n, d_n) : n \in \mathbb{N}\}$ be a collection of metric spaces. Let $\mathcal{T}_n$ be the metric topology on X_n for $n \in \mathbb{N}$ and let $(X, \mathcal{T})$ be the topological product of $\{(X_n, \mathcal{T}_n) : n \in \mathbb{N}\}$. We have to prove that there is a metric d on X which induces the topology $\mathcal{T}$. We shall actually do better in that we shall give an explicit formula for d in terms of the metrics d_n's.

First of all we use the proposition above and suppose that d_n is bounded by 2^{-n} for all $n \in \mathbb{N}$, for otherwise we replace d_n by another metric which also induces the same topology and which is bounded by 2^{-n}. Denote points of $X = \prod_{n \in \mathbb{N}} X_n$ by sequences $x = \{x_n\}$ with $x_n \in X_n$ for all $n \in \mathbb{N}$. Now define $d : X \times X \to \mathbb{R}$ by $d(x, y) = \sum_{n=1}^{\infty} d_n(x_n, y_n)$. This sum is well-defined since its n-th term is bounded by 2^{-n} and the series $\sum_{n=1}^{\infty} 2^{-n}$ is convergent. The positivity, symmetry and triangle inequality of d follow easily from the corresponding properties of the metrics $\{d_n : n \in \mathbb{N}\}$. So d is a metric on X. Let $\mathcal{U}$ be the metric topology induced by d. We assert that $\mathcal{U}$ coincides with $\mathcal{T}$. This would prove that $(X: \mathcal{T})$ is metrisable, as claimed.

First suppose $G \in \mathcal{U}$, i.e. G is open in the metric topology induced by d.

We show G is open in the product topology too. Let $x = \{x_n\} \in G$. Then there exists $r > 0$ such that $B(x, r) \subset G$. Now choose $N \in \mathbb{N}$ such that $\sum_{n=N+1}^{\infty} 2^{-n} < \frac{r}{2}$. For each $n = 1, 2, \ldots, N$, let $V_n = B\left(x_n, \frac{r}{2N}\right)$ where the ball is w.r.t. the metric d_n on X_n. Let $V_n = X_n$ for $n > N$. Let $V = \prod_{n \in \mathbb{N}} V_n$. Then evidently $x \in V$ and V is an open set in the product topology $\mathcal{T}$ on X. Further $V \subset B(x, r)$. For let $y \in V$. Now

$$\begin{aligned} d(x, y) &= \sum_{n=1}^{\infty} d_n(x_n, y_n) = \sum_{n=1}^{N} d_n(x_n, y_n) + \sum_{n=N+1}^{\infty} d_n(x_n, y_n) \\ &\leq N \cdot \frac{r}{2N} + \sum_{n=N+1}^{\infty} 2^{-n} \left(\text{since } y_n \in B\left(x_n, \frac{r}{2N}\right) \text{ for } n = 1, 2, \ldots N\right) \\ &\leq \frac{r}{2} + \frac{r}{2} \quad \text{(by choice of } N\text{)} \\ &= r. \end{aligned}$$

Hence $V \subset B(x, r) \subset G$. Thus G is a neighbourhood of each of its points and therefore G is open in the product topology. So $\mathcal{U} \subset \mathcal{T}$.

Conversely suppose G is open in the product topology. Let $x = \{x_n\} \in G$. Let V be a standard basic open set such that $x \in V$ and $V \subset G$. Let $V = \prod_{n \in \mathbb{N}} V_n$ where each V_n is open in X_n and $V_n = X_n$ for all $n > N$ (say) for some $N \in \mathbb{N}$. For $n = 1, 2, \ldots, N$ let $r_n = d_n(x_n, X_n - V_n)$ in case $V_n \neq X_n$, otherwise let $r_n = 2^{-n}$. Let $r = \min\{r_1, r_2, \ldots. r_N\}$. Then $r > 0$ since each r_n is positive (why?). We claim that $B(x, r) \subset V$. For if $y = \{y_n\} \in B(x, r)$ then $\sum_{n=1}^{\infty} d_n(x_n, y_n) < r$ and so $d_n(x_n, y_n) < r \leq r_n$ for each $n = 1, 2, \ldots, N$. But this means that $y_n \in V_n$ for $n = 1, 2, \ldots, N$. Also for $n > N$, $y_n \in V_n$ since $V_n = X_n$. So $y_n \in V_n$ for all $n \in \mathbb{N}$, showing that $y \in V$. Hence $B(x, r) \subset V \subset G$. Thus G contains an open ball around each of its points. So G is open w.r.t. the metric topology. This shows that $\mathcal{T} \subset \mathcal{U}$. More elegantly, note that for each n, $\pi_n : X \to X_n$ is continuous (in fact uniformly continuous) w.r.t. the metrics d and d_n and hence $\mathcal{U} - \mathcal{T}_n$ continuous. Since $\mathcal{T}$ is the weak topology generated by $\{\pi_n : n \in \mathbb{N}\}$, it follows that $\mathcal{T} \subset \mathcal{U}$. So $\mathcal{T}$ and $\mathcal{U}$ coincide. ∎

The reader is urged not to get carried by the computational aspect of the proof. The underlying idea is simple. In the case of finitely many metric spaces the proof means geometrically that an open rectangular box around any point contains an open ball around that point and vice versa. In the infinite case, the distance between two points $\{x_n\}$ and $\{y_n\}$ is approximately equal to the sum $\sum_{n=1}^{N} d_n(x_n, y_n)$ for large N. Effectively, we are approximating the general case by the finite case.

To settle the case of first and second countability we need a couple of simple results whose proofs are left as exercises to the reader.

(4.3) Proposition: Let $f : X \to Y$ be continuous and open and $x_0 \in X$.

Then if X is first countable at x_0, Y is first countable at $f(x_0)$. Consequently first countability is preserved under continuous open functions. ∎

(4.4) Proposition: Second countability is preserved under continuous open functions. ∎

In the next theorems we answer precisely when topological products are first/second countable. We recall that the products are assumed to be non-empty. In case they are empty, the theorems are either trivially true or trivially false.

(4.5) Proposition: Let $X = \prod_{i \in I} X_i$ and $x \in X$. Then X is first countable at x iff for each $i \in I$, X_i is first countable at $\pi_i(x)$ and for all except countably many i's, X_i is the only neighbourhood of $\pi_i(x)$ in X_i.

Proof: Suppose first that X is first countable at x. Then applying proposition (4.3) above to the projection maps π_i, we see that X_i is first countable at $\pi_i(x)$ for all $i \in I$. Let $V^1, V^2, V^3, \ldots$, be a local base at x in X. We use superscripts instead of subscripts since the subscripts will be needed to denote elements of I. Without loss of generality, we may suppose each V^n is open and in fact a member of the standard base for the product topology. Then for each n, $V^n = \prod_{i \in I} V_i^n$ where $V_i^n = X_i$ for all $i \in I - J_n$ for some finite subset J_n of I. Let $J = \bigcup_{n=1}^{\infty} J_n$. Then J is a countable subset of I. We claim that for all $i \in I - J$, X_i is the only open set (in X_i) containing $\pi_i(x)$. For if this were not the case then there exists $j \in I - J$ and an open set G_j in X_j containing $\pi_j(x)$ such that $G_j \underset{\neq}{\subset} X_j$. Now let $G = \pi_j^{-1}(G_j)$. Then G is an open set containing x. So by the definition of a local base, there exists $n \in \mathbb{N}$ such that $V^n \subset G$. Now $j \notin J_n$ and so $X_j = V_j^n = \pi_j(V^n) \subset \pi_j(G) = G_j$, a contradiction. This completes the proof of the direct implication.

For the converse suppose each X_i is first countable at $\pi_i(x)$ for all $i \in I$ and that there exists a countable subset J of I such that for all $i \in I - J$, X_i is the only open set in X_i containing $\pi_i(x)$. For each $j \in J$, let $\mathcal{L}_j$ be a countable open local base at $\pi_j(x)$. Let $\mathcal{F}$ be the collection of all finite subsets of J. Since J is countable, so is $\mathcal{F}$. For each $F \in \mathcal{F}$, let $\mathcal{L}_F = \{ \prod_{i \in I} Y_i : Y_i = X_i$ for $i \notin F$ and $Y_j \in \mathcal{L}_j$ for $j \in F\}$. In other words, $\mathcal{L}_F$ consists of those large open boxes whose 'short' sides belong to $\mathcal{L}_j$ for $j \in F$. Since each $\mathcal{L}_j$ is countable and F is finite, $\mathcal{L}_F$ is countable. Note also that each member of $\mathcal{L}_F$ is an open neighbourhood of x in X. Now let $\mathcal{L} = \bigcup_{F \in \mathcal{F}} \mathcal{L}_F$. Then $\mathcal{L}$ is countable since each $\mathcal{L}_F$ is countable and $\mathcal{F}$ is countable. The proof will be complete if we show that $\mathcal{L}$ is a local base at x. To this end, let G be an open set containing x. Then G contains a standard basic open set V containing x. Let $V = \prod_{i \in I} V_i$. Note that for $i \notin J$, $V_i = X_i$ since X_i is the only open set containing $\pi_i(x)$. Let $F = \{j \in I : V_j \neq X_j\}$. Then F is a finite subset of J. For each $j \in F$ there exists $U_j \in \mathcal{L}_j$ such that $U_j \subset V_j$. Let $U_j = X_j$ for all other $j \in I$. Let $U = \prod_{i \in I} U_i$. Then U is an open nbd of x and also $U \in \mathcal{L}_F$ and hence $U \in \mathcal{L}$. Clearly $U \subset V \subset G$. Thus $\mathcal{L}$ is a countable local base at x and so X is first countable at x. ∎

(4.6) Theorem: A product is first countable iff each coordinate space is first countable and all except finitely many coordinate spaces are indiscrete.

Proof: Let $X = \prod_{i \in I} X_i$. Suppose X is first countable. Then by Proposition (4.3) above, each X_i is first countable. Let J be the set $\{i \in I : X_i$ is not indiscrete$\}$. We have to show that J is countable. Suppose this is not so. Then for each $j \in J$ there exists a nonempty, proper open subset G_j of X_j. Let $G_j = X_j$ for $j \in I - J$ and let x be any point in $\prod_{i \in I} G_i$. Then X is first countable at x. But there are uncountably many indices j for which X_j is not the only neighbourhood of $\pi_j(x)$ in X_j. This contradicts the last proposition and establishes that all except countably many coordinate spaces are indiscrete. The converse implication follows immediately from the last proposition. ∎

The situation with second countability is similar and comes next.

(4.7) Theorem: A topological product is second countable iff all coordinate spaces are so and all except countably many are indiscrete spaces.

Proof: Let $X = \prod_{i \in I} X_i$. If X is second countable so is each X_i as seen by applying Proposition (4.3). Moreover, in this case, X is also first countable and so by the last theorem, all except countably many X_j's are indiscrete. Conversely suppose each X_i is second countable and the set $J = \{i \in I : X_i$ is not indiscrete$\}$ is countable. Let $\mathcal{B}_j$ be a countable base for X_j for $j \in J$. For a finite subset F of J let $\mathcal{B}_F$ be the collection of all large boxes whose 'short' sides come from $\mathcal{B}_j$ for $j \in F$, that is, $\mathcal{B}_F = \{\prod_{i \in I} Y_i : Y_i = X_i$ for $i \notin F$ and $Y_j \in \mathcal{B}_j$ for $j \in F\}$. Since each $\mathcal{B}_j$ is countable and F is finite, $\mathcal{B}_F$ is countable. Let $\mathcal{F}$ be the collection of all finite subsets of J and let $\mathcal{B} = \bigcup_{F \in \mathcal{F}} \mathcal{B}_F$. Then $\mathcal{F}$ and hence $\mathcal{B}$ are countable. Note also that each member of $\mathcal{B}$ belongs to the standard base for the product topology and hence is an open set. It only remains to show that $\mathcal{B}$ itself is a base for the product topology. The argument here resembles closely that in the latter half of the proof of Proposition (4.5). We leave it to the reader. ∎

Countable productivity of first and second countability follows from the last two theorems respectively. But actually they say a little more. An indiscrete space is not significantly different from a singleton space and singleton spaces are the neutral elements for products, that is to say, topologically you do not get anything new by taking products with singleton spaces. The theorems above then say in essence that the properties of first and second countability are not productive beyond countable index sets except in the trivial case where the other factors are the neutral elements.

An important consequence of Theorems (4.2) and (4.7) is,

(4.8) Theorem: Every subspace of the Hilbert cube is a second countable metric space.

Proof: The unit interval [0, 1] is a second countable metric space. So by Theorems (4.2) and (4.7), the product of countably many copies of [0, 1] is also a second countable metric space. Thus the Hilbert cube $[0, 1]^{\mathbb{N}}$ is second countable and metrisable. But both these properties are hereditary. The assertion follows from this. ∎

The converse of this theorem is also true in the sense that every second countable metric space is homeomorphic to a subspace of the Hilbert cube. This will be proved in the next chapter.

As regards other conditions involving countability, it is easy to show that the Lindelöff property is not even finitely productive. However the case of separability is interesting. It is easy to show that it is a countably productive property but that it is not productive in general. However, the power Y^I of a separable space continues to be separable even when the index set I is uncountable, provided the cardinality of I does not exceed c, the cardinality of the unit interval. We establish this result first for the case where the index set I is the semi-open interval [0, 1). The reason for this peculiar choice will be clear in the course of the proof.

(4.9) Theorem: Let Y be a separable space and let $I = [0, 1)$. Then the product space Y^I is separable.

Proof: Let $X = Y^I$. Note that elements of X are functions from I into Y. Let S be the set of all step functions in X, that is, functions for which there exist $0 = a_0 < a_1 < a_2 < \ldots < a_n = 1$ such that f is constant on each $[a_{i-1}, a_i)$ for $i = 1, 2, \ldots, n$. Now suppose D is a countable dense subset of Y. Let E be the set of those step functions for which the numbers $a_1, a_2, \ldots, a_{n-1}$ above are all rational and the values of the functions are in D. Since the number of partitions of [0, 1) in intervals with rational end points is countable and D is countable, it is clear that E is a countable subset of X. We show that E is dense in X. For this, it suffices to show that E intersects every non-empty member of the standard base for the product topology on X. Let V be such a member. Then V is of the form $\prod_{i \in I} V_i$ where each V_i is a nonempty open subset of Y and $V_i = Y$ for all $i \in I$ except possibly for $i = i_0, i_1, \ldots, i_n$ (say). Without loss of generality we may suppose that $i_0 < i_1 < \ldots < i_n$. Let $a_1, \ldots, a_n$ be rational numbers in [0, 1) such that $i_{k-1} < a_k < i_k$ for $k = 1, 2, \ldots, n$. Let $a_0 = 0$, $a_{n+1} = 1$. For each $k = 1, 2, \ldots, n$, $D \cap V_{i_k} \neq \Phi$. Fix some $d_{i_k} \in D \cap V_{i_k}$. Now define $e : I \to Y$ by $e(t) = d_{i_k}$ for $t \in [a_k, a_{k+1})$, $k = 0, 1, \ldots, n$. Then $e \in E$. Moreover, $e(i_k) \in V_{i_k}$ for $k = 0, 1, \ldots, n$ while for other i's, $e(i) \in V_i = Y$. So $e \in \prod_{i \in I} V_i = V$. Hence $E \cap V \neq \Phi$. Thus E is a countable, dense subset of X and so X is separable. ∎

Note that in the proof we used the fact that the set of rationals is countable and order-dense in $\mathbb{R}$. However, the dependence upon the particular index set $[0, 1)$ can be removed now.

(4.10) Theorem: Let J be any set of cardinality not exceeding c. Then for any separable space Y, the power Y^I is separable.

Proof: If J is the empty set then Y^J is a singleton space which is separable. Suppose J is nonempty. Then since the cardinal number of J is at most that of $[0, 1)$ there exists a bijection $\theta : J \to K$ for some subset K of $[0, 1)$. The bijection θ clearly establishes a homeomorphism between the power Y^J and Y^K (cf. Exercise (2.5)) and so the problem reduces to showing that Y^K is separable. Now let $L = [0, 1) - K$. Then $Y^{[0, 1)}$ is homeomorphic to the product $Y^K \times Y^L$ (see Exercise (2.6)). Then there exists a continuous function from $Y^{[0, 1)}$ onto Y^K (namely, the first projection associated with the product $Y^K \times Y^L$). By the theorem above, $Y^{[0, 1)}$ is separable and by Exercise (6.1.11)), Y^K is separable. As noted before this means that Y^J is separable. ∎

The theorem does not hold when the cardinality of J exceeds c. The advantage in considering powers instead of arbitrary products $\prod_{i \in I} X_i$ of separable spaces is that the same dense subset D of X_i was needed in the proof of Theorem (4.9). However, by suitably modifying the argument, it can be shown that a product $\prod_{i \in I} X_i$ is separable when each X_i is separable and the index set I has cardinality at most c (Exercise (4.6) below). In particular it follows that separability is a countably productive property.

Exercises

4.1 Justify why the numbers r_n defined in the proof of Theorem (4.2) are positive.

4.2 Prove propositions (4.3) and (4.4).

4.3 Complete the proof of Theorem (4.7) by showing that the family $\mathscr{B}$ constructed in it is a base for the product topology.

4.4 Give proofs of Theorems (4.6) and (4.7) without appealing to Propositions (4.3) and (4.4). (Hint: These propositions were used only to deduce that each coordinate space is first or second countable. This part could as well have been done using Theorem (2.6)).

4.5 Prove that a topological product is metrisable iff all coordinate spaces are metrisable and all except countably many are singleton spaces. (Hint: Note that a metric space is first countable.)

4.6 Generalise Theorems (4.9) and (4.10) to the case of a product of at most c many separable spaces. (Hint: Assume first that the index set I is $[0, 1)$ and for each i, $g_i : D_i \to \mathbb{N}$ is a bijection where D_i is a dense subset of X_i. Use functions in $\prod_{i \in I} X_i$ which are step functions modulo these bijections.)

4.7 Suppose a product $\prod_{i \in I} X_i$ is separable and T_2. Prove that all X_i's are separable and that all except c many of them are singleton, that is, show that the set $J = \{j \in I : X_j$ is not singleton$\}$ has cardinality at most c. (Hint: For $j \in J$, let U_j, V_j be a pair of mutually disjoint non-empty open subsets of X_j. Suppose D is a countable dense subset of X. For $j \in J$ let $\theta(j) = D \cap \pi_j^{-1}(V_j)$. Show that θ is a one-to-one function from J into the power set $P(D)$. The Hausdorff property is needed only to ensure the existence of two mutually disjoint, non-empty sets, which is needed to show that θ is one-to-one.)

4.8 Give an example of a separable, Tychonoff space which is not metrisable. (Hint: Consider a suitable cube.)

4.9 Prove that if $\mathcal{T}$ is the semi-open interval topology on $\mathbb{R}$ then the product $\mathbb{R} \times \mathbb{R}$ is not Lindelöff even though $(\mathbb{R}, \mathcal{T})$ is.

4.10 Prove that second countability, separability and the Lindelöff property are countably additive.

Notes and Guide to Literature

All results in this section are well-known except possibly those involving separability. These can be found in Dugundji [1]. There are many other ways of metrising a countable product of metric spaces than the one given in the proof of Theorem (4.2). One could use, for example, the Pythagorean metric or any of the L^p metrics for $1 < p < \infty$. The metric we have used is the L^1 metric in this terminology.

Chapter Nine

Embedding and Metrisation

In this chapter we shall prove two most classic theorems in general topology, one due to Tychonoff and the other to Urysohn. They obtain respectively characterisations of Tychonoff spaces and of second countable metric spaces. Apart from their intrinsic values, they serve an instructive purpose for us as they provide us with an opportunity to apply almost everything we studied in the last three chapters—except the material dealing with compactness and connectedness. It will be fair to say that with the end of the present chapter the reader will have acquired a substantial and coherent piece of general topology. Moreover, the technique used is also important elsewhere in mathematics.

1. Evaluation Functions into Products

In Section 2 of the last chapter, while proving that path-connectedness is a productive property we defined a function into the product space, being given a function into each coordinate space. The construction of such a function is purely set-theoretic and in view of its importance we define it formally.

(1.1) Definition: Let $\{Y_i : i \in I\}$ be an indexed family of sets. Suppose X is a set and let for each $i \in I$, $f_i : X \rightarrow Y_i$ be a function. Then the function $e : X \rightarrow \prod_{i \in I} Y_i$ defined by $e(x)(i) = f_i(x)$ for $i \in I$, $x \in X$ is called the **evaluation function** of the indexed family $\{f_i : i \in I\}$ of functions.

In other words, for each $x \in X$, the i-th coordinate of $e(x)$ is obtained by evaluating the i-th function f_i at x. This justifies the term evaluation function. Intuitively evaluation function is obtained by 'listing together the information given by various f_i's'. To illustrate this, suppose X is the set of all students in a class and $f_1, f_2, f_3, \ldots,$ etc. are functions specifying respectively, say, the age, the sex, the height etc. of members of X. Then the evaluation function is like a catalogue which lists against each student all the information available about the student. For example, a typical entry in this catalogue might be e (Mr. X.Y.Z.) = (21, Male, 5 ft. 6 in., . . .). The reader may find this example helpful in developing a feeling for evaluation functions, which may otherwise appear rather formal.

The following proposition characterises evaluation functions.

(1.2) Proposition: Let $\{Y_i : i \in I\}$ be a family of sets, X a set and for each $i \in I, f_i : X \to Y_i$ a function. Then the evaluation function is the only function from X into ΠY_i whose composition with the projection $\pi_i : \Pi Y_i \to Y_i$ equals f_i for all $i \in I$.

Proof: Let $e : X \to \Pi Y_i$ be the evaluation function of the family $\{f_i : i \in I\}$. Then for any $i \in I$ by very definition of e, $\pi_i(e(x)) = e(x)(i) = f_i(x)$ and so $\pi_i \circ e = f_i$. Conversely suppose $e' : X \to \Pi Y_i$ satisfies that $\pi_i \circ e' = f_i$ for all $i \in I$. Let $x \in X$. Then for any $i \in I$, $e'(x)(i) = \pi_i(e'(x)) = f_i(x) = e(x)(i)$ and so $e(x) = e'(x)$. But since $x \in X$ was arbitrary this means that $e' = e$. Thus e is the only function from X into ΠY_i having the given property. ∎

In terms of diagrams, the proposition above says that given any family $\{f_i : i \in I\}$ there is a unique function from X to ΠY_i which makes the following diagram commute for each $i \in I$.

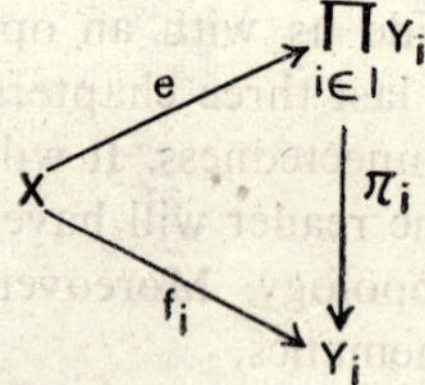

Although we shall not need it in future, it is perhaps worthwhile at this stage to note that the property proved here of ΠY_i and of the projections π_i characterises products. We state this precisely in the following theorem.

(1.3) Theorem: Suppose $\{Y_i : i \in I\}$ is an indexed family of sets, Z is a set and $\{\theta_i : Z \to Y_i \mid i \in I\}$ is a family of functions such that for any set X and any family $\{f_i : X \to Y_i \mid i \in I\}$ of functions, there exists a unique function $e : X \to Z$ satisfying $\theta_i \circ e = f_i$ for all $i \in I$. Then there exists a bijection h from Z to ΠY_i such that for each $i \in I$, $\theta_i = \pi_i \circ h$. Moreover this bijection is unique. In other words, upto a bijection, Z is the product of Y_i's and θ_i's are the projection functions.

Proof: The proof is really simple, but to really appreciate it the reader is asked to draw appropriate diagrams of functions. Let $h : Z \to \Pi Y_i$ be the evaluation function of the family $\{\theta_i : i \in I\}$. Then by definition, for each $i \in I$, $\pi_i \circ h = \theta_i$. It only remains to prove that h is a bijection. We do so by finding an inverse for h. Take $X = \Pi Y_i$ and for $i \in I$ take f_i to be the projection $\pi_i : \Pi Y_i \to Y_i$. Then by the hypothesis of the theorem, there exists a unique function $k : \Pi Y_i \to Z$ such that for each $i \in I$, $\theta_i \circ k = \pi_i$. Then for each $i \in I$, we have $\pi_i \circ (h \circ k) = \pi_i$. But on the other hand $\pi_i \circ (id) = \pi_i$ for all $i \in I$ where id is the identity function on ΠY_i. Thus, applying the uniqueness part of proposition (1.2), taking $X = \Pi Y_i$, and $f_i = \pi_i$ for $i \in I$ we see that $h \circ k$ is the identity function on ΠY_i. Similarly $\theta_i \circ (id_Z) = \theta_i$ for all $i \in I$ where $id_Z =$ identity function on Z. But from $\theta_i \circ k = \pi_i$ and $\pi_i \circ h = \theta_i$ we also get $\theta_i \circ (k \circ h) = \theta_i$ for all $i \in I$. Then again, by the uniqueness part of the hypothesis, taking $X = Z$ and $f_i = \theta_i$ for $i \in I$, we get that $k \circ h = id_Z$. So h and k are inverses of each other.

Thus h is a bijection. Uniqueness of h follows from the preceding proposition. ∎

Let us now obtain a condition under which the evaluation function is one-to-one. First we need a definition.

(1.4) Definition: An indexed family $\{f_i : i \in I\}$ of functions all defined on the same domain X is said to **distinguish points** if for any distinct x, y in X there exists $j \in I$ (which may depend upon both x and y) such that $f_j(x) \neq f_j(y)$.

The following proposition is now almost self-evident.

(1.5) Proposition: The evaluation function of a family of functions is one-to-one if and only if that family distinguishes points.

Proof: For $i \in I$, let $f_i : X \to Y_i$ be a function and let $e : X \to \Pi Y_i$ be the evaluation function. Let x, y be distinct points of X. Then $e(x) \neq e(y)$ iff there exists $j \in I$, such that $e(x)(j) \neq e(y)(j)$. But $e(x)(j) = f_j(x)$ by definition of e. Similarly $e(y)(j) = f_j(y)$. So the condition that $e(x) \neq e(y)$ is equivalent to saying that there exists $j \in I$ (which may depend upon both x and y) such that $f_j(x) \neq f_j(y)$. Since x, y are arbitrary the result follows. ∎

It is obvious from the definition that the smaller the family $\{f_i : i \in I\}$, the less likely it is to distinguish points. If the set X is small, then a small family of functions on X may ensure that the evaluation function is one-to-one. But as the set X becomes larger, the same family may not be adequate for this, because it may fail to distinguish between points of the larger set. In such a case it is necessary to enlarge the family of functions. This observation should be obvious from common sense too. Let us recall the example where X was the set of all students in a class and the functions $f_1, f_2, f_3, \ldots$ describe various characteristics such as age, sex, height etc. of students. In a small class, it is very unlikely that two distinct students will have the same age, the same sex and the same height. That is to say, the family consisting of the corresponding three functions is most likely to distinguish points in a small class. In a big class, however, it is possible to find two distinct students with the same age, sex and height. In such a class some other characteristics such as weight must be used in order to distinguish between students.

The discussion so far was purely set-theoretic in that the sets we dealt with were not assumed to be equipped with any additional structure such as topology. Let us now suppose that we are dealing with topological spaces X and Y_i for $i \in I$. Let $f_i : X \to Y_i$ be a function for $i \in I$. We can then ask when the corresponding evaluation function $e : X \to \Pi Y_i$ where ΠY_i carries the product topology is continuous. The answer is very simple.

(1.6) Proposition: With the notations above, the evaluation function is continuous iff each f_i is continuous.

Proof: By proposition (1.2) we have $\pi_i \circ e = f_i$ for all $i \in I$. The result now follows from Theorem (8.2.5). ∎

The evaluation, when continuous, is also called the **evaluation map.** Proposition (1.5) still applies and shows that the evaluation map is one-to-one iff the family $\{f_i : i \in I\}$ of maps distinguishes points. It does not follow, however, under these conditions that e is a topological embedding. A sufficient condition which will ensure this will be studied in the next section.

Exercises

1.1 Let $f_1, f_2, f_3 : \mathbb{R} \to \mathbb{R}$ be defined by $f_1(x) = \cos x$, $f_2(x) = \sin x$, $f_3(x) = x$ for $x \in \mathbb{R}$. Describe the evaluation maps of the families $\{f_1, f_2\}$, $\{f_1, f_3\}$, $\{f_1, f_2, f_3\}$. Which of these families distinguish points?

1.2 Prove the topological analogue of Theorem (1.3).

1.3 Let $\{G_i : i \in I\}$ be a collection of groups. Let $G = \prod_{i \in I} G_i$. A binary operation can then be defined on G by 'coordinate-wise multiplication'.
(a) Make this definition precise.
(b) Verify that with this binary operation G is in fact a group and that each projection function π_i, $i \in I$ is a group homomorphism.

1.4 Let H be a group and $\{f_i : H \to G_i \mid i \in I\}$ an indexed family of group homomorphisms. Prove that the evaluation function $e : H \to \prod_{i \in I} G_i$ is a group homomorphism.

1.5 State and prove the group-theoretic analogue of Theorem (1.3).

1.6 If X is a Tychonoff space, prove that the family of all continuous real-valued functions on X distinguishes points.

1.7 Let $\mathcal{F}$ be a family of continuous real-valued functions on S^1. If $\mathcal{F}$ distinguishes points, prove that $\mathcal{F}$ must have at least two elements. Give an example of such a family having only two elements. More generally, prove that on S^n there exists a family of $n + 1$ real-valued continuous functions which distinguishes points.

1.8 Let $\{X_i : i \in I\}$ be an indexed family of sets, ΣX_i its coproduct and $\lambda_i : X \to \Sigma X_i$ the ith injection for $i \in I$. Prove that for any set Y and any family $\{f_i : i \in I\}$ where f_i is a function from X_i to Y for $i \in I$, there exists a unique function $p : \sum_{i \in I} X_i \to Y$ such that $p \circ \lambda_i$ for all $i \in I$.

1.9 Prove the topological analogue of the result in the last exercise.

1.10 Suppose $\{X_i : i \in I\}$ is an indexed family of sets; Z is a set and $\{\mu_i : X_i \to Z \mid i \in I\}$ is a family of functions with the property that given any set Y and any family $\{f_i : X_i \to Y \mid i \in I\}$ of functions, there exists a unique function $p : Z \to Y$ satisfying $p \circ \mu_i = f_i$ for all $i \in I$. Then prove that there is a unique bijection h from ΣX_i to Z such that for each $i \in I$, $h \circ \lambda_i = \mu_i$.

1.11 Prove the topological analogue of the result in the last exercise.

Notes and Guide to Literature

The results of this section are simple, almost to the point of being trivial. But the construction of the evaluation function is important. Theorem (1.3) and its analogues in Exercises (1.2) and (1.5) show that products could be characterised upto the respective equivalence relation (a bijection for sets, a homeomorphism for topological spaces and an isomorphism for groups) by the existence and uniqueness requirements of certain functions. This theorem has many analogues in various branches of mathematics in which products are defined. In fact, the hypothesis of Theorem (1.3) (or rather, something similar to it) is taken as the definition of products in category theory which serves to unify the constructions in various branches of mathematics (see the remarks at the end of chapter 3, section 1).

The last four exercises are obviously dual to the properties of products proved in this section.

2. Embedding Lemma and Tychonoff Embedding

We continue with the notations of the last section, except that from now, on we shall be dealing exclusively with topological spaces and continuous functions. In the last section we saw under what conditions the evaluation map is one-to-one. We now obtain a condition that will ensure that the evaluation map is an embedding.

(2.1) Definition: An indexed family of functions $\{f_i : X \to Y_i \mid i \in I\}$, where X, Y_i are topological spaces, is said to **distinguish points from closed sets** in X if for any $x \in X$ and any closed subset C of X not containing x, there exists $j \in I$ such that $f_j(x) \notin \overline{f_j(C)}$ in Y_j.

The definition must have reminded the alert reader of complete regularity, because there too, a point was separated from a closed set by means of a real-valued map. The resemblance is brought out in the next proposition.

(2.2) Proposition: A topological space is completely regular iff the family of all continuous real-valued functions on it distinguishes points from closed sets.

Proof: Let X be a topological space and let $\mathcal{F}$ be the family of all continuous real-valued functions on X. Suppose first that X is completely regular. Let a point $x \in X$ and a closed subset C of X, not containing x be given. Then there exists a continuous function $f : X \to [0, 1]$ such that $f(x) = 0$ and $f(C) \subset \{1\}$ (actually $f(C) = \{1\}$ except when $C = \Phi$). Then we can regard f as a function from X to $\mathbb{R}$ (or more precisely take the composite of f with the inclusion function of $[0, 1]$ into $\mathbb{R}$). Then $f \in \mathcal{F}$ and evidently $f(x) \notin \overline{f(C)}$ since $\{1\}$ is a closed subset of $\mathbb{R}$. So $\mathcal{F}$ distinguishes points from closed sets in X.

Conversely suppose that $\mathcal{F}$ distinguishes points from closed sets. Let $x \in X$ and C be a closed subset of X not containing x. Then there exists a map $f : X \to \mathbb{R}$ such that $f(x) \notin \overline{f(C)}$. Now $\{f(x)\}$ and $\overline{f(C)}$ are disjoint closed sub-sets of $\mathbb{R}$ which is normal space. So there exists a continuous function $g : \mathbb{R} \to [0, 1]$ which takes the values 0 and 1 respectively on them. Let $h : X \to [0, 1]$ be the composite $g \circ f$. Then clearly $h(x) = 0$ and $h(y) = 1$ for all $y \in C$. Thus we see that X is a completely regular space. ∎

Our interest in families of functions which distinguish points from closed sets comes from the following result.

(2.3) Proposition: Let $\{f_i : X \to Y_i \mid i \in I\}$ be a family of functions which distinguishes points from closed sets in X. Then the corresponding evaluation function $e : X \to \prod_{i \in I} Y_i$ is open when regarded as a function from X onto $e(X)$.

Proof: Let V be an open subset of X. We have to show that $e(V)$ is an open subset of $e(X)$. A typical point of $e(V)$ is of the form $e(x)$ for some $x \in V$. Now $X - V$ is a closed subset of X not containing x. So by the hypothesis, there exists $j \in I$ such that $f_j(x) \notin \overline{f_j(X - V)}$. Let $G = Y_j - \overline{f_j(X - V)}$. Then G is an open subset of Y_j and so $\pi_j^{-1}(G)$ is an open subset of $\prod_{i \in I} Y_i$. We claim that $\pi_j^{-1}(G) \cap e(X) \subset e(V)$. For suppose $y \in \pi_j^{-1}(G) \cap e(X)$. Then $y = e(z)$ for some $z \in X$. Also $\pi_j(y) \in G$ and so $\pi_j(e(z)) \in G$ whence $f_j(z) \in G$ since $\pi_j \circ e = f_j$. From this it follows that $z \in V$, since otherwise $f_j(z) \in f_j(X - V) \subset \overline{f_j(X - V)} = Y_j - G$. Thus $y = e(z) \in e(V)$ as was to be shown. Now, the set $\pi_j^{-1}(G) \cap e(X)$ is open in the relative topology on $e(X)$ and clearly it contains $e(x)$, showing that $e(V)$ is a neighbourhood of $e(x)$ in $e(X)$. But $e(x)$ was a typical point of $e(V)$. Thus $e(V)$ is a neighbourhood of each of its points in the relative topology on $e(X)$. So $e(V)$ is an open subset of $e(X)$. Since this holds for all open sets V in X, we see that e, regarded as a function from X onto $e(X)$ is open. ∎

The converse of this proposition is false. Let $X = \mathbb{R}^2$ and f_1, f_2 be the two projections on $\mathbb{R}$. Then the evaluation function is simply the identity map, which is open. But the family $\{f_1, f_2\}$ does not distinguish points from closed sets in X.

Putting together what we have proved so far, we get an important result called the **embedding lemma.**

(2.4) Theorem: Let $\{f_i : X \to Y_i \mid i \in I\}$ be a family of continuous functions which distinguishes points and also distinguishes points from closed sets. Then the corresponding evaluation map is an embedding of X into the product space $\prod_{i \in I} Y_i$.

Proof: Let $e : X \to \prod Y_i$ be the evaluation function. Continuity of e follows from Proposition (1.6). That it is one-to-one follows from Proposition (1.5) while the last proposition shows that e is an open map when regarded as a function from X to $e(X)$. These are precisely the requirements of an embedding. ∎

The word 'lemma' belies the importance of the last theorem. It appears in the name because Tychonoff originally used this result to prove his famous embedding theorem which we are about to state. Before doing so we remark that the embedding lemma is of value only when the spaces Y_i, $i \in I$ are more tangible and manageable than the space X. There is, for example, little point in applying the embedding lemma to the singleton family $\{id_X : X \to X\}$. This family certainly distinguishes points and also distinguishes points from closed sets. But the embedding that we get from the last theorem is simply the embedding $id_X : X \to X$ which is absolutely useless in giving us any new information about X.

Often the embedding lemma is applied to the case where each $Y_i, i \in I$ is the same space, say, Y. In this case the index set I itself can be chosen to be a family $\mathcal{F}$ of maps from X to Y and the embedding that we get is into the power space $Y^{\mathcal{F}}$. In applications, on one hand the family $\mathcal{F}$ must be large enough to satisfy the hypothesis of the embedding lemma, while on the other hand, it should not be too large for otherwise the power space $Y^{\mathcal{F}}$ will be too unwieldy.

We now prove the celebrated embedding theorem due to Tychonoff.

(2.5) Theorem: A topological space is a Tychonoff space iff it is embeddable into a cube.

Proof: By Corollary (8.3.5) we know that every cube is a Tychonoff space. But the Tychonoff property is hereditary by Exercise (7.1.1). So every subspace of a cube and hence every space homeomorphic to a subspace of a cube is a Tychonoff space. Conversely suppose a space X is a Tychonoff space. Let $\mathcal{F}$ be the family of all continuous functions from X into $[0, 1]$. Then since X is completely regular, $\mathcal{F}$ distinguishes points from closed sets in X. But since X is also T_1, all singleton sets are closed and so it follows that $\mathcal{F}$ distinguishes points as well. So by the last theorem the evaluation map $e : X \to [0, 1]^{\mathcal{F}}$ is an embedding of X into the cube $[0, 1]^{\mathcal{F}}$. ∎

The significance of this theorem is profound. There can be so many abstract Tychonoff spaces. The theorem says that if we look at all possible cubes and their subspaces then we need not go beyond in order to study Tychonoff spaces. It is like having a library in which you can always find at least one copy of every book on a particular subject. Although such a library may not literally contain all the books on that subject, effectively it does so. Theorems of this type are always desirable for they reduce the study of abstract objects to the study of something concrete.

Exercises

2.1 Let Y be a space and I any index set. Then the **diagonal** ΔY in Y^I is defined to be the set of all constant functions in Y^I, with the relative topology.

(a) Prove that in case I has two elements, this definition is consistent with the one given in Exercise (7.1.8).

(b) Use embedding lemma to prove that ΔY is homeomorphic to Y.

2.2 It was proved in Theorem (8.2.7) that in the case of a non-empty product, every coordinate space is embeddable into it. Prove the same result using the embedding lemma.

2.3 Prove that every pseudo-metric space is completely regular. (Hint: A desired continuous function can be defined in terms of the pseudo-metric).

2.4 Let $(X, \mathcal{T})$ be a space. Let $\mathcal{D}$ be the indiscrete topology on X. Prove that the function $id_X : X \to X$ is $\mathcal{T}$-$\mathcal{D}$ continuous and that the family $\{id_X\}$ distinguishes points of X.

2.5 Prove that a space is completely regular iff it can be embedded into a product of pseudo-metric spaces. (Hint: For the direct implication use Proposition (2.2) along with the last exercise and apply the embedding lemma).

In the remaining exercises, you may assume as known the theorem that in a compact, Hausdorff space the components coincide with the quasi-components. (For the definition of a quasi-component see Exercise (6.3.8).)

2.6 Let X be a compact, Hausdorff, totally disconnected space. Prove that X has a base consisting of clopen sets.

2.7 Let X be as above and let $\mathbb{Z}_2$ be a discrete space with two points. Prove that the family of all maps from X into $\mathbb{Z}_2$ distinguishes points from closed sets.

2.8 Prove that every compact, Hausdorff, totally disconnected space can be embedded into a Cantor discontinuum (that is, a space of the form $\mathbb{Z}_2^I$ for some index set I).

2.9 Let G be a group, $\{H_i : i \in I\}$ a family of groups and $\{h_i : G \to H_i \mid i \in I\}$ a family of group homomorphisms which distinguishes points. Prove that G is isomorphic to a subgroup of the product group $\prod_{i \in I} H_i$.

Notes and Guide to Literature

The Tychonoff embedding theorem was indeed proved by Tychonoff. The technique has been extended to other embedding problems. It is known, for example, that every Boolean algebra is isomorphic to a sub-algebra of a power $\mathbb{Z}_2^I$. This is an understatement of the famous Stone representation theorem (M. H. Stone [1]).

A proof of the fact that in a compact, Hausdorff space, components coincide with quasi-components can be found in Hocking and Young [1].

3. The Urysohn Metrisation Theorem

In the last section we proved that every Tychonoff space is embeddable in a cube. All cubes share those topological properties of the unit interval [0, 1] which are productive in character. There are however a few properties such as metrisability, second countability etc. which are only countably productive as we saw in Section 4 of the last chapter. The Hilbert cube, therefore, enjoys a special importance among all cubes. In this section we shall topologically characterise the subspaces of Hilbert cube, in other words, we shall find necessary and sufficient conditions under which a space can be embedded into the Hilbert cube. To this end we prove the following theorem of Urysohn.

(3.1) Theorem: A space is embeddable in the Hilbert cube if and only if it is second countable and T_3.

Proof: By Theorem (8.4.8), every subspace of the Hilbert cube is second countable and metrisable. But every metrisable space is T_3 by Theorem (7.1.12). This proves the necessity of the condition. For sufficiency, assume X is a second countable and T_3 space. Then first of all X is normal by Corollary (7.2.9). Now let $\mathcal{B}$ be a countable base for X. Enumerate $\mathcal{B}$ as $\{B_1, B_2, \ldots, B_n, \ldots\}$ where in case $\mathcal{B}$ is actually finite we repeat some member of $\mathcal{B}$ infinitely often. Now let $I = \{(m, n) \in \mathbb{N} \times \mathbb{N} : \overline{B}_m \subset B_n\}$. I is then a countable set. For each $i \in I$, say $i = (m, n)$, we apply the Urysohn's lemma to the disjoint closed sets $\overline{B}_m$ and $X - B_n$ of X and get a map $f_i : X \to [0, 1]$ which takes the value 0 on $\overline{B}_m$ and the value 1 on $X - B_n$. We claim that the family $\{f_i : i \in I\}$ distinguishes points from closed sets in X. For let $x \in X$ and let C be a closed subset of X not containing x. Then $x \in X - C$ which is open. So for some $n \in \mathbb{N}$, $x \in B_n$ and $B_n \subset X - C$. By regularity of X, there exists an open set G in X such that $x \in G$ and $\overline{G} \subset B_n$. But again because $\mathcal{B}$ is a base, there exists $m \in \mathbb{N}$ such that $x \in B_m$ and $B_m \subset G$. Clearly then $(m, n) \in I$. Let $i = (m, n)$. Then the corresponding function f_i vanishes at 0 and takes the value 1 on $X - B_n$ and in particular on C as $C \subset X - B_n$. So $f_i(x) \notin \overline{f_i(C)}$ in $[0, 1]$.

Thus the family of maps $\{f_i : i \in I\}$ distinguishes points of X from closed sets. But since X is also a T_1 space, all singleton sets are closed and so the family $\{f_i : i \in I\}$ distinguishes points as well. All the conditions of the embedding lemma are now satisfied and so the corresponding evaluation map $e : X \to [0, 1]^I$ is an embedding. In case I is countably infinite, this completes the proof because $[0, 1]^I$ is then homeomorphic to $[0, 1]^{\mathbb{N}}$ which is the Hilbert cube. If, however, I is finite we have proved a little more than the theorem asserts, because a finite-dimensional cube (that is, one with the index set I finite) is certainly embeddable into the Hilbert cube. ∎

In the sufficiency part of the proof above, second countability of the space X was used twice. First it was used to conclude that X is normal (for which it would have of course sufficed if X were merely Lindelöff instead of second countable). Later on we used it to extract a countable family of real-valued maps on X which distinguished points from closed sets. From the fact that X is a Tychonoff space, we already know that the family $\mathcal{F}$ of all maps from X to $[0, 1]$ distinguishes points from closed sets. The problem was to see if some smaller subfamily of $\mathcal{F}$ would also do the job. The smaller the space X, the better are the chances of finding such a small family. This is where second countability, which is a smallness condition, helps.

The following corollary is known as **Urysohn's metrisation theorem.**

(3.2) Corollary: A second countable space is metrisable iff it is T_3.

Proof: Every metrisable space, whether second countable or not, is T_3. For the converse, we note that every second countable T_3 space is homeomorphic to a subspace of the Hilbert cube. Since the Hilbert cube is metrisable and metrisability is a hereditary property, the result follows. ∎

The theorem above characterises metric spaces in the class of all second countable spaces. However it does not characterise them in the class of all topological spaces. This means that it does not answer precisely under what conditions an abstract topological space is metrisable. This question was completely answered by Nagata and Smirnov. A full discussion is beyond the scope of this book. We content ourselves by merely stating the result.

(3.3) Definition: Let X be a topological space. Then a family $\mathcal{U}$ of subsets of X is said to be **locally finite** if for each $x \in X$, there exists a neighbourhood N of x which intersects only finitely many members of $\mathcal{U}$.

(3.4) Definition: Let X be a topological space. Then a family $\mathcal{V}$ of subsets of X is said to be **σ-locally finite** if it can be written as the union of countably many subfamilies each of which is locally finite.

It is obvious that every finite family of subsets is locally finite and every countable family of subsets is σ-locally finite. In the real-line, the family of all intervals of the form $[n, \infty)$, $n \in \mathbb{N}$ is locally finite.

The general metrisation theorem states that a topological space is metrisable iff it is regular, T_1 and has a σ-locally finite base. Since every second countable space has a σ-locally finite base, Theorem (3.2) follows immediately from the general metrisation theorem.

Exercises

3.1 Give an example of a space which is second countable and T_2 but not metrisable. (Hint: Consider the space $(\mathbb{R}, \mathcal{S})$ defined in Chapter 7, Section 1).

3.2 Give an example of a metric space which is not second countable.

3.3 Prove that if $\mathcal{C}$ is a locally finite family of closed sets then $\bigcup_{C \in \mathcal{C}} C$ is a closed set.

3.4 Prove that if $\mathcal{C}$ is a locally finite family and $D = \bigcup_{C \in \mathcal{C}} C$ then $\bar{D} = \bigcup_{C \in \mathcal{C}} \bar{C}$. Also prove that $\{\bar{C} : C \in \mathcal{C}\}$ is locally finite.

3.5 Let $\mathcal{C}$ be a locally finite family of closed subsets of a space X. Suppose $\mathcal{C}$ covers X and $f : X \to Y$ is a function where Y is some topological space. Prove that f is continuous iff for each $C \in \mathcal{C}$, $f/C : C \to Y$ is continuous.

3.6 Let X be a second countable metric space. Suppose $\mathcal{D}$ is a decomposition of X each of whose members is compact and suppose the projection $p : X \to \mathcal{D}$ is closed. Prove that $\mathcal{D}$ is metrisable. (Hint: Use Theorems (7.2.15) and (7.2.16).)

Notes and Guide to Literature

A proof of the general metrisation theorem can be found in Kelley [1]. Many other metrisation theorems have been proved recently; see for example, Martin [1] or Hung [1].

Although the results proved in this chapter are of great theoretical importance, for a particular space they may not necessarily be the best ones. There is considerable extravagance in the embedding lemma. Let X be a space and $\{f_i : i \in I\}$ a family of maps on X. In order to make the evaluation map e an open map, we assumed that this family distinguishes points from closed sets in X. This is only a sufficient and not a necessary requirement as we noted after Proposition (2.3). As a result, the family $\{f_i : i \in I\}$ is often far larger than is absolutely necessary to make e an embedding. The situation is somewhat better when X is second countable. Even then, Theorem (3.1) gives an embedding only into the Hilbert cube, whereas, it may sometimes be possible to embed a space into a much smaller cube. A most classic result to this effect is that every n-dimensional, second countable, T_3 space can be embedded into the $(2n + 1)$-dimensional cube. See Hurewicz and Wallman [1].

Chapter Ten

Nets and Filters

We remarked earlier that convergence is one of the most important concepts in topology, especially in its aspects dealing with analysis. We have so far been discussing convergence of sequences from time to time. In many theorems involving convergence of sequences in a topological space, we found that one way implication was true in any topological space but the converse required the assumption of first countability. For example, in a Hausdorff space the limits of convergent sequences are unique by Proposition (7.1.6). But this property does not characterise Hausdorff spaces (see Exercise (7.1.5)). This 'inadequacy of sequences' is due to the fact that the convergence of sequences in a space does not uniquely determine its topology (see Exercise (4.2.8) (iii) and (iv)). A question then arises as to whether we can somehow extend the concept of convergence to something more general and thereby ensure that all important notions in topology can be characterised in terms of convergence of this new type of objects.

There are two ways of answering this question. The objects they deal with are respectively called nets and filters. In this chapter we shall study both these concepts. Although they lead to essentially equivalent theories, the nets have the advantage that they are a very natural and direct generalization of sequences with which we are all too familiar. On the other hand, there are some things about filters that have no natural analogues for nets. Moreover, filters can be defined and studied in other contexts such as Boolean algebras. As a result, recently filters are more in vogue than nets. As an interesting (and non-trivial) application of filters, it will be proved in this chapter that compactness is a productive property.

1. Definition and Convergence of Nets

We want to define a net as a generalization of a sequence. Let us therefore look at the definition of a sequence a little closely. A sequence in a set X is a function $f : \mathbb{N} \to X$. A natural attempt to generalise this definition would be to replace the domain $\mathbb{N}$ by an abstract set, say, D. But we cannot simply say that a net is a function $f : D \to X$. If we do so, we shall get nothing more than the general theory of functions from one set to another. In order to give meaning to the concept of convergence of a net it is necessary that the set D be equipped with some additional structure which will satisfy some of the properties enjoyed by the set $\mathbb{N}$. Now, there are so many additional structures on the set of positive integers. For example, there are the binary operations of addition and multiplication. However, they rarely come into picture when we are dealing with convergence of sequences. What really matters in convergence is the order structure on $\mathbb{N}$. Let us say a **follows** b if $a \geq b$, for $a, b \in \mathbb{N}$. A few properties of this binary relation are:

(i) For any $a, b, c \in \mathbb{N}$, if $a \geq b$ and $b \geq c$, then $a \geq c$.
(ii) For any $a \in \mathbb{N}$, $a \geq a$.
(iii) For any $a, b \in \mathbb{N}$, $a \geq b$ and $b \geq a$ imply $a = b$.
(iv) For any $a, b \in \mathbb{N}$, either $a \geq b$ or $b \geq a$.
(v) Given any non-empty subset S of $\mathbb{N}$, there exists $x \in X$ such that $a \geq x$ for all $a \in S$.

The reader must have noted that these properties are precisely the respective formulations of transitivity, reflexivity, anti-symmetry, law of dichotomy and the well-ordering property. Now, when we want to put an additional structure on the domain D of a net, we have to consider a binary relation $\geq$ on D which satisfies a few of these (or other) properties. Exactly which ones should we select? Once again, we are confronted with a choice between generality and non-triviality. Experience shows that the transitivity and reflexivity are too vital not to be included while well-ordering is irrelevant because convergence deals with what happens as the argument of the function becomes larger and larger. Anti-symmetry is useful but not indispensible most of the time. The remaining property, namely, the law of dichotomy, is a bit too strong. What is really needed is a simple consequence of it. The idea is that in defining convergence, the elements of the domain set should be getting larger and larger.

With these observations we make the following definition.

(1.1) Definition: A **directed set** is a pair $(D, \geq)$ where D is a non-empty set and $\geq$ a binary relation on D satisfying:

(i) For all $m, n, p \in D$, $m \geq n$ and $n \geq p$ imply $m \geq p$.
(ii) For all $n \in D$, $n \geq n$.
(iii) For all $m, n \in D$, there exists $p \in D$ such that $p \geq m$ and $p \geq n$.

We also say that the relation $\geq$ **directs** the set D. It is clear that the law of dichotomy implies that condition (iii) holds. As a foremost example of a directed set we have $(\mathbb{N}, \geq)$ where $\geq$ is the usual ordering on $\mathbb{N}$. Any lattice is also a directed set. Before giving other examples we note that condition (iii) can be replaced by the apparently stronger one that for any finite number of elements $n_1, n_2, \ldots, n_k \in D$, there exists p such that $p \geq n_i$ for all $i = 1, \ldots, k$. The stronger version follows from the version in the definition by induction, using transitivity of $\geq$.

Another important example of a directed set is the neighbourhood system $\mathcal{N}_x$ of a point x in a topological space X. Here for $U, V \in \mathcal{N}_x$, we define $U \geq V$ to mean $U \subset V$. Note that this is in sharp contrast with our intuition that 'U is greater than V' should mean that U contains V. Perhaps it is better to read $\geq$ not as 'is greater than' but rather as 'follows'. Conditions (i) and (ii) hold trivially while (iii) follows from the fact the that intersection of two neighbourhoods is again a neighbourhood. The family of all open neighbourhoods also gives a directed set.

Yet another example is worth mentioning here for its historical significance in the definition of Riemann integrals. A **partition** of the unit interval [0, 1] is a finite sequence $P = \{a_0, a_1, \ldots, a_n\}$ such that $0 = a_0 < a_1 < \ldots < a_n = 1$. The interval $[a_{i-1}, a_i]$, $i = 1, 2, \ldots, n$ is called the i-th **subinterval** of P. Such a partition is said to be **refinement** of another partition $Q = \{b_0, b_1, \ldots, b_m\}$ if each subinterval of P is contained in some subinterval of Q. Let us write $P \geq Q$ to mean that P is a refinement of Q. It is easy to see that $\geq$ directs the set of all partitions of [0, 1]. Conditions (i) and (ii) of the definition are immediate. For (iii) note that if P and Q are partitions, then the partition obtained by superimposing them together is a common refinement of P and Q.

We now formally state the definition of a net.

(1.2) Definition: A **net** in a set X is a function $S : D \to X$ where D is a directed set.

Clearly a sequence is a very special case of a net. As with sequences, given a net $S : D \to X$ and $n \in D$, it is customary to write S_n instead of $S(n)$. Examples of nets other than sequences will come up as we proceed. For its historical significance, let us define the **Riemann net** corresponding to a bounded real-valued function f on the unit interval [0, 1]. Let D be the set of all pairs (P, ξ) where P is a partition of [0, 1] say $P = \{a_0, a_1, \ldots, a_n\}$ and $\xi = \{\xi_1, \ldots, \xi_n\}$ is a finite sequence such that $\xi_i \in [a_{i-1}, a_i]$ for $i = 1, \ldots, n$. Given two elements (P, ξ) and (Q, η) of D, let us say $(P, \xi) \geq (Q, \eta)$ iff P is a refinement of Q and for each j, $\eta_j = \xi_i$ where i is so defined that the j-th sub-interval of Q is contained in the i-th sub-interval of the partition P. It is easy to show that $\geq$ directs D. Now define the net $S : D \to \mathbb{R}$ by $S(P, \xi) = \sum_{i=1}^{n} f(\xi_i)(a_i - a_{i-1})$. This is, of course, nothing but what is called in integral calculus as the Riemann sum of the function f for the partition P and the choice of points ξ_i in the i-th sub-interval of P. The Riemann integral, as the reader recalls, is defined as the limit of such Riemann sums as the partitions get more and more refined. More generally, the notion of a limit of a net can be formalised as follows.

(1.3) Definition: Let $(X; \mathcal{T})$ be a topological space and let $S : D \to X$ be a net. Then S is said to **converge** to a point $x \in X$ if given any open set U containing x, there exists $m \in D$ such that for all $n \in D$, $n \geq m$ implies that $S_n \in U$. When this happens we also say that x is a **limit** of the net S in X.

Note that the convergence depends as much on the topology $\mathcal{T}$ as on the net S. The striking similarity between the definitions of convergence of a sequence and that of a net leads us to expect that whatever is true about sequences should also be true about nets. This is more or less correct, the natural exceptions being those cases where some special properties of the set $\mathbb{N}$ (which is the domain of all sequences), such as its countability, are involved. As a matter of fact, where sequences fail, nets can do the job. As a typical result in this vein, we have the following theorem (cf. Exercises (7.1.5) and (7.1.6)).

(1.4) Theorem: A topological space is Hausdorff iff limits of all nets in it are unique.

Proof: Suppose X is a Hausdorff space, $S : D \to X$ is a net in X and S converges to x and y in X. We have to show that $x = y$. If this is not so, then there exist open sets, U, V such that $x \in U$, $y \in V$ and $U \cap V = \Phi$. By definition of convergence, there exist m_1, $m_2 \in D$ such that for all $n \in D$, $n \geq m_1$ implies $S_n \in U$ and $n \geq m_2$ implies $S_n \in V$. Now because D is a directed set, there exists $n \in D$ such that $n \geq m_1$ and $n \geq m_2$. But then $S_n \in U \cap V$, a contradiction. So $x = y$, establishing the necessity of the condition. Conversely suppose that the limits of all nets in a space X are unique. If X is not Hausdorff then there exist two distinct points x, y in X which do not have mutually disjoint neighbourhoods in X. Let $\mathcal{N}_x$, $\mathcal{N}_y$ be the neighbourhood systems in X at x and y respectively. Let $D = \mathcal{N}_x \times \mathcal{N}_y$ and for (U_1, V_1), $(U_2, V_2) \in D$, define $(U_1, V_1) \geq (U_2, V_2)$ iff $U_1 \subset U_2$ and $V_1 \subset V_2$. This makes D a directed set and we define a net $S : D \to X$ as follows. For any $U \in \mathcal{N}_x$ and $V \in \mathcal{N}_y$ we know that $U \cap V \neq \Phi$. Define $S(U, V)$ to be any point in $U \cap V$; (in essence, we are using the axiom of choice here). We claim that the net S so defined converges to x. For let G be an open neighbourhood of x. Then $(G, X) \in D$. Now if $(U, V) \geq (G, X)$ in D then $U \subset G$ and so $S(U, V) \in U \cap V \subset U \subset G$. Thus S converges to x. Similarly S converges to y also, contradicting the hypothesis. So X is T_2. ∎

Note that condition (iii) in the definition of a directed set was used crucially in the proof of the direct implication. The proof of the converse implication illustrates the advantage nets have over sequences. In a sequence, the domain is always the set of positive integers, while in defining nets, we have considerable freedom in the choice of the directed set.

There is, however, a corresponding disadvantage. When we want to find limits of nets, the domain sets can often be too large and such techniques as induction which work smoothly for sequences no longer apply for arbitrary nets. For example, if want to evaluate the Riemann integral of a function f on $[0, 1]$ directly from its definition as the limit of the Riemann net, it is hardly feasible to do so by computing all possible Riemann sums corresponding to all possible partitions of $[0, 1]$ and all possible choices of the points ξ_i! It is then customary to consider only a few of these partitions, for example, those in which all sub-intervals are of equal length, and the choice of ξ_i is also restricted to be either one of the end points or the mid-point of the corresponding sub-interval. Of course it must be justified that taking limits of these 'nice' Riemann sums does give the same value as taking limits of all possible Riemann sums. We carry out this justification in the general context of all nets (not just Riemann nets).

(1.5) Definition: A subset E of a directed set D is said to be **eventual** if there exists $m \in D$ such that for all $n \in D$, $n \geq m$ implies that $n \in E$. A net $S : D \to X$ is said to be **eventually** in a subset A of X if the set $S^{-1}(A)$ is an eventual subset of D.

In informal but suggestive terms, a subset E is eventual iff it contains all elements of D 'after a certain stage' and a net $S : D \to X$ is eventually in A iff A contains all its terms after a certain stage. For sequences, 'after a certain stage' is synonymous with 'for all except finitely many' but in general this is not so for arbitrary nets. It is obvious that a net converges to a point iff it is eventually in every neighbourhood of it.

(1.6) Proposition: If $(D, \geq)$ is a directed set and E is an eventual subset of D, then E, with the restriction of $\geq$ is a directed set. Moreover a net $S : D \to X$ where X is a topological space, converges to x in X iff the restriction $S/E : E \to X$ converges to x in X.

Proof: Let $\geq'$ denote the relation $\geq$ restricted to E. Transitivity and reflexivity of $\geq'$ follow from the corresponding properties of $\geq$. For the remaining property let $m_1, m_2 \in E$. By definition, there exists $m \in D$ such that $n \geq m$ implies $n \in E$ for all $n \in D$. Now there exists $n \in D$ such that $n \geq m_1$, $n \geq m_2$ and $n \geq m$ (see the extension of condition (iii) in the definition of a directed set). Clearly $n \in E$ and $n \geq' m_1$, $n \geq' m_2$. This proves that $\geq'$ directs E. Now suppose $S : D \to X$ is a net in a topological space X and $x \in X$. Let T denote the restriction S/E. Suppose S converges to x. Let U be an open set in X containing x. There exists $m_1 \in D$ such that for all $n \in D$, $n \geq m_1$ implies $S(n) \in U$. Now let m be as above. Find $m_2 \in D$ such that $m_2 \geq m$ and $m_2 \geq m_1$. Then $m_2 \in E$. Also for any $n \in E$, $n \geq' m_2$ implies that $n \geq m_2$ in D and hence that $n \geq m_1$ by transitivity. So $T(n) = S(n) \in U$. This shows that T converges to x in X. The proof that convergence of T to x implies that of S to x is even simpler and left to the reader. ∎

There is a weaker condition than eventuality which is also useful.

(1.7) Definition: Let $(D, \geq)$ be a directed set. A subset F of D is said to be a **cofinal** subset of D if for every $m \in D$, there exists $n \in F$ such that $n \geq m$. A net $S : D \to X$ is said to be **frequently** in a subset A of X if $S^{-1}(A)$ is a cofinal subset of D.

It is obvious that every eventual subset is a cofinal subset. The converse is false. In $\mathbb{N}$, any infinite subset is cofinal but not necessarily eventual. In informal terms, a subset is confinal if it contains arbitrarily large elements of the directed set. As an important example of a cofinal set, note that a cofinal set in the neighbourhood system of a point in a topological space is nothing but a local base at that point. In the set of all partitions of the unit interval [0, 1], the set of partitions of the form $\left\{0, \frac{1}{n}, \frac{2}{n}, \ldots, \frac{n-1}{n}, 1\right\}$

for $n \in \mathbb{N}$ is a cofinal subset.

The usefulness of cofinal sets in the evaluation of limits comes from the following proposition, whose proof is left to the reader.

(1.8) Proposition: Let F be a cofinal subset of a directed set $(D, \geq)$. Then the restriction of $\geq$ to F makes it a directed set. If a net $S : D \to X$ converges to a point $x \in X$, so does its restriction $S/F : F \to X$. ∎

Thus, if it is known beforehand that S is convergent then its limits can be found by restricting it to any cofinal subset. It may happen that the restriction of a net to a cofinal subset converges to something although the original net does not. However, the original net and the limit of its restriction to a cofinal subset are not entirely unrelated.

(1.9) Definition: Let $S : D \to X$ be a net. A point $x \in X$ is said to be a **cluster point** of S if for every neighbourhood U of x in X, and $m \in D$, there exists in $n \in D$ such that $n \geq m$ and $S_n \in U$. Equivalently, x is a cluster point of S iff for every neighbourhood U of x, S is frequently in U.

The definition is a generalization of the notion of a limit point of a sequence. It is obvious that if a net S converges to x then x is a cluster point of X. Actually a stronger result holds.

(1.10) Proposition: Suppose $S : D \to X$ is a net and F is a cofinal subset of S. If $S/F : F \to X$ converges to a point x in X, then x is a cluster point of S.

Proof: Let U be a neighbourhood of x in X. Then there exists $m_1 \in F$ such that for any $n \in F$, $n \geq m_1$ implies that $S_n \in U$. Now let $m \in D$ be given. Choose $n_1 \in D$ such that $n_1 \geq m$ and $n_1 \geq m_1$. Next choose $n \in F$ such that $n \geq n_1$. Then $S_n \in U$. So S is frequently in U and since U was arbitrary, x is a cluster point of S. ∎

In analysis, it is well-known that a point is a limit point of a sequence of real numbers iff there exists a subsequence converging to it. A similar assertion is also true for nets. But first we must define what a subnet is. It is natural to think that a subnet should be the restriction of a net to a cofinal subset of its domain. Unfortunately, this simple-minded definition does not turn out to be adequate for certain purposes, where it is convenient to allow subnets to have domains other than subsets of the domain of the original net.

(1.11) Definition: Let $S : D \to X$ and $T : E \to X$ be nets. Then T is said to be a **subnet** of S if there exists a function $N : E \to D$ such that (i) $T = S \circ N$, and (ii), for any $n \in D$, there exists $p \in E$ such that for all $m \in E$, $m \geq p$ implies $N(m) \geq n$ in D, where the same notation $\geq$ is used to denote the binary relations directing D and E.

Note that the function N is not required to be order preserving. The first condition ensures that every value assumed by T is also assumed by S (hence the name 'subnet') while the second condition informally means that as m becomes sufficiently large in E, $N(m)$ becomes and stays as large as desired (or 'tends to ∞') in D. It is a little stronger than merely saying that the range of N is a cofinal subset of D. Note that a sub-sequence of a sequence is also a subnet of it but a sequence may have subnets which are not subsequences.

We now prove the expected relationship between cluster points and subnets.

(1.12) Theorem: Let $S : D \to X$ be a net in a topological space and let $x \in X$. Then x is a cluster point of S iff there exists a subnet of S which converges to x in X.

Proof: First suppose $T : E \to X$ is a subnet of S converging to x in X. Let $N : E \to D$ be the function given in the definition of a subnet. Let U be any neighbourhood of x in X, and let $m_1 \in D$ be given. Then there exists $p \in E$ such that for all $m \in E$, $m \geq p$ implies $N(m) \geq m_1$. Also because T converges to x, there exists $q \in E$ such that for all $m \in E$, $m \geq q$ implies $T(m) \in U$, i.e. $S(N(m)) \in U$. Now choose $m \in E$ such that $m \geq p$ and $m \geq q$ and let $n = N(m)$. Then $n \geq m_1$ and $S_n \in U$. Since m_1 and U were arbitrary, it follows that x is a cluster point of S.

Conversely suppose x is a cluster point of S. We construct a subnet T of S as follows. Let $\mathcal{N}_x$ be the neighbourhood system of the point x in X (any cofinal subset of $\mathcal{N}_x$ would do as well). Let $\geq$ denote the given binary relation on the directed set D. We define E to be the set $\{(n, U) \in D \times \mathcal{N}_x : S_n \in U\}$. For $(n, U), (m, V) \in E$ we let $(n, U) \geq (m, V)$ mean $n \geq m$ in D and $U \subset V$. It is easy to show that the binary relation $\geq$ directs the set E. Now define $T : E \to X$ by $T(n, U) = S(n)$ for $(n, U) \in E$. Then T is a net in X and actually it is a subnet of S, because if we define $N : E \to D$ by $N(n, U) = n$, we see that both the conditions of the definition are satisfied. It only remains to verify that T converges to x in X. For this, let G be a given neighbourhood of x in X. Since x is a cluster point of S, S is frequently in G. In particular fix any $n \in D$ such that $S_n \in G$. Then $(n, G) \in E$. Now for any $(m, U) \in E$, $(m, U) \geq (n, G)$ implies that $T(m, U) = S_m \in U \subset G$. Thus T converges to x in X as desired. ∎

The proof of the converse uses the freedom allowed in the definition of a subnet as regards the choice of its domain. If the domain is required to be a subset of the domain of the original net, the theorem is false. In Example 10, of Chapter 4, Section 2, we considered the space $\mathbb{N} \times \mathbb{N} \cup \{\infty\}$. Any enumeration of $\mathbb{N} \times \mathbb{N}$ gives a sequence in X having ∞ as a cluster point, and so by the theorem above, there exists a subnet converging to ∞. However, no net whose domain is a subset of $\mathbb{N}$ can converge to ∞ (cf. Exercise (4.2.9)).

Exercises

1.1 Let X be any set and $\mathcal{F}$ the set of all finite subsets of X. For F, $G \in \mathcal{F}$, define $F \geq G$ to mean $F \supset G$. Prove that $\geq$ directs $\mathcal{F}$.

1.2 The notion of refinement can be extended to covers of an arbitrary

set X. A cover $\mathcal{U}$ is said to be a **refinement** of another cover $\mathcal{V}$ if every member of $\mathcal{U}$ is contained in some member of $\mathcal{V}$, that is for each $U \in \mathcal{U}$ there exists $V \in \mathcal{V}$ such that $U \subset V$. If $\mathcal{U}$ is a refinement of $\mathcal{V}$, we write $\mathcal{U} \geq \mathcal{V}$.
(a) Prove that $\geq$ directs the set of all covers of the set X.
(b) Prove that in general $\geq$ is not anti-symmetric.

1.3 Let X be a topological space with a sub-base $\mathcal{S}$ and $x \in X$. Prove that a net in X converges to x in X iff the condition in the definition of convergence holds for all neighbourhoods of x which are members of $\mathcal{S}$.

1.4 Let X be the topological product of a family of spaces $\{X_i : i \in I\}$. Prove that a net $S : D \to X$ converges to a point $x \in X$ iff for each $i \in I$ the net $\pi_i \circ S$ converges to $\pi_i(x)$ in X_i.

1.5 State and prove a similar result for the weak topology generated by a family of functions having a common domain set.

1.6 Let $\mathcal{T}_1$, $\mathcal{T}_2$ be topologies on a set X and suppose $\mathcal{T}_1 \subset \mathcal{T}_2$. How are the convergences of nets in $(X; \mathcal{T}_1)$ and $(X; \mathcal{T}_2)$ related to each other?

1.7 Prove Proposition (1.8).

1.8 Prove that in an indiscrete space, every net converges to every point and that this property characterises indiscrete spaces.

1.9 Obtain a characterisation of discrete spaces in terms of convergence of nets in it. Can a discrete space be characterised in terms of convergence of sequences in it? (Cf. Exercise (4.2.9)).

1.10 Let $S : D \to X$ be a net in a space X and for each $n \in D$, let $A_n = \{S_m : m \in D, m \geq n\}$. Prove that a point $x \in X$ is a cluster point of S iff $x \bigcap\limits_{n \in D} \bar{A}_n$.

1.11 Let $\{x_n\}$ be a sequence in a space X, having a cluster point x in X. Prove that if X is first countable at x then there exists a subsequence (and not just a subnet) of $\{x_n\}$ which converges to x.

Notes and Guide to Literature

The convergence of nets is called the **Moore-Smith convergence** after Moore and Smith [1]. McShane [1] gives an expository account. Inadequacy of sequences for arbitrary topological spaces is long known, see a recent article by Conway [1].

The origin of the term 'net' (and also 'filter') is obscure. The key word is probably 'mesh'. In classical Riemann integration one lets the mesh of the partition approach zero. Mesh is a measure of fineness or refinement and in everyday practical life the purpose of nets and filters is to refine materials. The limit is what one gets as the degree of refining becomes larger and larger.

2. Topology and Convergence of Nets

In the last section we proved that the Hausdorff property can be characterised in terms of a certain property involving convergence of nets. Also in Exercises (1.8) and (1.9) we asked the reader to characterise indiscrete and discrete topologies in terms of convergence of nets. In this section we go even deeper and show that every topology and every topological property can be characterised in terms of convergence of nets. In order to do this it suffices to obtain a characterisation of open sets because everything in topology depends ultimately on open sets. Since we have already characterised open sets in terms of closed sets, closures etc., it will suffice to characterise any one of these in terms of convergence of nets. We begin with the closure points.

(2.1) Proposition: Let A be a subset of a space X and let $x \in X$. Then $x \in \bar{A}$ iff there exists a net in A (that is a net which takes values in the set A) which converges to x in X (i.e. when regarded as a net in X).

Proof: Suppose $S : D \to A$ is a net which, when regarded as a net in X (or more precisely the net $i \circ S$ where i is the inclusion map of A into X), converges to x. Let U be any neighbourhood of x. Then there exists $n \in D$ such that for all $m \in D$, $m \geq n$ implies $S_m \in U$. But $S_m \in A$ for all $m \in D$. So $A \cap U \neq \Phi$. Thus every neighbourhood of x meets A and so $x \in \bar{A}$ by Theorem (5.2.10). Conversely suppose $x \in \bar{A}$. Then every nbd of x meets A. Let $\mathcal{N}_x$ be the neighbourhood system of x in X, directed as usual. Define $S : \mathcal{N}_x \to A$ by $S(U) =$ any point in $U \cap A$. Then S is a net in A. We claim S converges to x in X. Let G be any open set in X containing x. Then for any $U \in \mathcal{N}_x$, $U \geq G$ implies $U \subset G$ and hence that $S(U) \in U \subset G$. So S converges to x in X. ∎

We already saw (see Exercise (6.1.14)) that in case X is first countable at x, the converse part of the proposition above can be strengthened to assert the existence of a sequence in A converging to x. This is not true in general as shown by the space in Example 10, Chapter 4, Section 2 where ∞ is in the closure of $\mathbb{N} \times \mathbb{N}$ although no sequence in $\mathbb{N} \times \mathbb{N}$ converges to ∞.

(2.2) Corollary: A subset A of a space X is closed iff limits of nets in A are in A.

Proof: This follows from the last proposition and the fact that a set A is closed iff $A = \bar{A}$. ∎

(2.3) Theorem: A subset B of a space X is open iff no net in the complement $X - B$ can converge to a point in B.

Proof: This follows by applying the last corollary to $X - B$. ∎

(2.4) Corollary: Let $\mathcal{T}_1$, $\mathcal{T}_2$ be topologies on a set X such that a net in X converges to a point w.r.t. $\mathcal{T}_1$ iff it does so w.r.t. $\mathcal{T}_2$. Then $\mathcal{T}_1 = \mathcal{T}_2$.

Proof: Let $B \in \mathcal{T}_1$. Then by the theorem above, no net in $X - B$ converges to a point in B w.r.t. $\mathcal{T}_1$. But convergence w.r.t. $\mathcal{T}_2$ is given to be identical with that with respect to $\mathcal{T}_1$. So no net in $X - B$ converges to a point in B w.r.t. $\mathcal{T}_2$. Hence again by the theorem, B is open w.r.t. $\mathcal{T}_2$, i.e. $B \in \mathcal{T}_2$. So $\mathcal{T}_1 \subset \mathcal{T}_2$. Similarly $\mathcal{T}_2 \subset \mathcal{T}_1$ and so $\mathcal{T}_1 = \mathcal{T}_2$. ∎

The theorem and the corollary above show that in a topological space $(X, \mathcal{T})$, if we are given which nets converge to which points of X, then we can uniquely recover the topology $\mathcal{T}$ from this information. It is now natural to ask whether one can obtain an axiomatic characterisation of convergence. That is, suppose we are given a set X and a collection $\mathcal{C}$ of pairs (S, x) where S is a net in X and $x \in X$. What conditions on this collection will guarantee the existence of a unique topology $\mathcal{T}$ on X such that $(S, x) \in \mathcal{C}$ iff S converges to x in X w.r.t. $\mathcal{T}$? This problem has been solved fully under what is known as 'convergence classes'. However we shall not pursue it in this book.

Example 10 of Chapter 4, Section 2, shows that all the results proved so far in this section fail if 'nets' are replaced by 'sequences'. As we have pointed earlier, however, they do hold for first countable spaces. The general rule that can be laid down is that whatever can be done with nets in an arbitrary topological space, can be done with sequences in a first countable space.

Now that open sets are characterised in terms of nets, every concept in topology can be formulated in terms of nets. We conclude this section with characterisations of continuity and compactness in terms of convergence of nets.

(2.5) Theorem: Let X, Y be topological spaces, $x_0 \in X$ and $f: X \to Y$ a function. Then f is continuous at x_0 iff whenever a net S converges to x_0 in X, the net $f \circ S$ converges to $f(x_0)$ in Y.

Proof: Suppose first that f is continuous at x_0 and $S: D \to X$ is a net in X converging to x_0. Let V be any neighbourhood of $f(x_0)$ in Y. Then $f^{-1}(V)$ is a neighbourhood of x_0 in X. Since S converges to x_0, there exists $m \in D$ that for all $n \in D$, $n \geq m$ implies $S_n \in f^{-1}(V)$. But then for all $n \geq m$, $f(S_n) \in V$. Thus $f \circ S$ converges to $f(x_0)$ in Y.

Conversely suppose the given condition holds. Suppose f is not continuous at x_0. Then there exists a neighbourhood V of $f(x_0)$ such that $f^{-1}(V)$ is not a neighbourhood of x_0 in X. This means that $f^{-1}(V)$ contains no neighbourhood of x_0, or in other words that every neighbourhood of x_0 meets $X - f^{-1}(V)$. Now let $\mathcal{N}_{x_0}$ be the neighbourhood system of x_0, directed as usual. Define $S: \mathcal{N}_{x_0} \to X$ by $S(N)$ to be any chosen point in $N \cap (X - f^{-1}(V))$. Then S converges to x_0 in X. For, given any neighbourhood U of x_0, U is a member of the domain set $\mathcal{N}_{x_0}$ as well and for any $N \in \mathcal{N}_{x_0}$, $N \geq U$ implies $N \subset U$ and hence $S(N) \in N \subset U$. On the other hand the composite net $f \circ S$ takes values only in $Y - V$. So V is a neighbourhood of $f(x_0)$ which contains no points of the net $f \circ S$. Thus $f \circ S$ does not converge to $f(x_0)$, contradicting the hypothesis. So f is continuous at x_0. ∎

In Theorem (6.1.16) we had proved that the converse implication holds with 'nets' replaced by 'sequences' provided X is first countable at x_0. This is yet another instance of the general remark we made preceding the theorem above.

Finally we derive a succinct characterisation of compactness in terms of convergence of nets. Before doing it let us obtain a characterisation of compactness which, although not at all profound, is quite handy to use. First we need a definition.

(2.6) Definition: A family $\mathcal{F}$ of subsets of a set X is said to have the **finite intersection property** (abbreviated **f.i.p.**) if for any $n \in \mathbb{N}$ and $F_1, F_2, \ldots, F_n \in \mathcal{F}$, the intersection $\bigcap_{i=1}^{n} F_i$ is non-empty.

In particular, every member of a family having f.i.p. is non-empty. It does not follow however that a family having f.i.p. is always closed under finite intersections. For example the family of all closed discs in the plane which contain the origin has f.i.p. by its very definition, but it is not closed under finite intersections. Also just because a family has f.i.p. does not mean that the intersection of all its members is non-empty. As an example, consider the family of all cofinite subsets of any infinite set.

The proof of the following simple proposition (which will not be needed until the next section) is left to the reader.

(2.7) Proposition: If a family $\mathcal{F}$ of subsets of a set X is closed under finite intersections then $\mathcal{F}$ has f.i.p. iff $\Phi \notin \mathcal{F}$. ∎

We next prove a characterisation of compactness in terms of families having finite intersection property.

(2.8) Proposition: A topological space is compact iff every family of closed subsets of it, which has the finite intersection property, has a non-empty intersection.

Proof: Let X be a space. Suppose first that X is compact. Let $\mathcal{C}$ be a family of closed sets of X and assume $\mathcal{C}$ has f.i.p. We have to show that $\bigcap_{C \in \mathcal{C}} C$ is non-empty. If this intersection were empty then by De Morgan's laws, the family $\mathcal{U}$ consisting of complements of members of $\mathcal{C}$ would be a cover of X and this cover would be open since members of $\mathcal{C}$ are closed. By compactness of X, there exists a finite subcover consisting of, say, $X - C_1, X - C_2, \ldots, X - C_n$ with $C_i \in \mathcal{C}$ for $i = 1, 2, \ldots, n$. But then again by De Morgan's laws, this means that $\bigcap_{i=1}^{n} C_i = \Phi$, contradicting that $\mathcal{C}$ has finite intersection property. So $\bigcap_{C \in \mathcal{C}} C$ is non-empty. This proves the direct implication. The proof of the converse implication is similar, involving nothing more than the definition, De Morgan's laws and the fact that a set is closed iff its complement is open. ∎

Now we have all the machinery we need to prove the

(2.9) Theorem: For a topological space X, the following statements are equivalent:

(1) X is compact.

(2) Every net in X has a cluster point in X.

(3) Every net in X has a convergent subnet in X (i.e. a subnet which converges to at least one point in X).

Proof: We prove only the equivalence of (1) with (2). That of (2) with (3) follows from Theorem (1.12). So assume (1) and let $S : D \to X$ be a net in X. Suppose X has no cluster point in X. Then for each $x \in X$ there exists a neighbourhood N_x of x and an element $m_x \in D$ such that for all $n \in D$, $n \geq m_x$ implies $S(n) \in X - N_x$. Cover X by such neighbourhoods (or more precisely, by their interiors). By compactness of X, there exist $x_1, \ldots, x_k \in X$ such that $X = \bigcup_{i=1}^{k} N_{x_i}$. Let the corresponding elements in D be $m_1, m_2, \ldots, m_k$. Because D is a directed set, there exists $n \in D$ such that $n \geq m_i$ for $i = 1, 2, \ldots, k$. But then $S(n) \in \bigcap_{i=1}^{k} (X - N_{x_i}) = X - \bigcup_{i=1}^{k} N_{x_i} = \Phi$, a contradiction. So S has at least one cluster point in X. Thus (2) holds.

Conversely assume (2) holds. Let $\mathcal{C}$ be a family of closed sets of X having the finite intersection property. Let $\mathcal{D}$ be the family of all finite intersections of members of $\mathcal{C}$. Note that $\mathcal{D}$ itself is closed under finite intersections and that $\mathcal{C} \subset \mathcal{D}$. For $D, E \in \mathcal{D}$ we define $D \geq E$ to mean $D \subset E$. This makes $\mathcal{D}$ a directed set because whenever $D, E \in \mathcal{D}$, $D \cap E \in \mathcal{D}$ and $D \cap E \geq D$, $D \cap E \geq E$. Note that each member of $\mathcal{D}$ is non-empty because $\mathcal{C}$ is given to have the finite intersection property. So we can define a net $S : \mathcal{D} \to X$ by $S(D) =$ any point in D. By (2) this net has a cluster point say x in X. We claim $x \in \bigcap_{C \in \mathcal{C}} C$. For, if not there exists $C \in \mathcal{C}$ such that $x \notin C$. Then $X - C$ is a neighbourhood of x (since members of $\mathcal{C}$ are closed). Also $C \in \mathcal{D}$. So by the definition of a cluster point, there exists $D \in \mathcal{D}$ such that $D \geq C$ and $S(D) \in X - C$. But then $D \subset C$ and so $X - C \subset X - D$ contradicting that $S(D) \in D$. So $x \in \bigcap_{C \in \mathcal{C}} C$. We have thus shown that every family of closed subsets of X having f.i.p. has nonempty intersection. But by the proposition above this means that X is compact. Thus (1) holds. ∎

Exercises

2.1 Prove Proposition (2.7).

2.2 Let $S : D \to X$ be a net in a space X. For each $m \in D$ let $B_m = \bar{A}_m$ where $A_m = \{S(n) : n \in D, n \geq m\}$. Prove that the family $\{B_m : m \in D\}$ has the finite intersection property.

2.3 Give an alternate proof of the implication (1) $\Rightarrow$ (2) in Theorem (2.9) using the exercise above and Exercise (1.10).

2.4 Prove the converse implication in Proposition (2.8).

2.5 Let X be a set and $\mathcal{T}_1$, $\mathcal{T}_2$ topologies on X. Prove that $\mathcal{T}_1$ is stronger than $\mathcal{T}_2$ iff whenever a net converges to a point w.r.t. $\mathcal{T}_1$, it does so w.r.t. $\mathcal{T}_2$ also.

2.6 Prove that in a compact, first countable space, every sequence has a convergent subsequence.

2.7 Prove that a subnet of a convergent subnet of a net is again a convergent subnet of the original net.

2.8 Using the last exercise and Theorem (2.9) show that the product of two (and hence any finite number of) compact spaces is compact. (Hint: Let S be a net in $X_1 \times X_2$. Let T be a subnet of S such that $\pi_1 \circ T$ is a convergent net in X_1. Let U be a subnet of T such that $\pi_2 \circ U$ is convergent in X_2. Now show that U is a convergent subnet of S.)

2.9 What difficulty would arise in attempting to prove that compactness is a productive property using Theorem (2.9)?

Notes and Guide to Literature

Axiomatic characterisation of convergence of nets can be found in Kelley [1]. Exercise (2.8) provides an interesting application of the theory of nets in that it proves easily that compactness is finitely productive. Unfortunately, the difficulties involved in passing to arbitrary products cannot be tackled in a natural way using nets. As we shall see later in this chapter, filters can do the job instead of nets.

3. Filters and Their Convergence

The concept of a filter is essentially lattice-theoretic. It is the dual of what is known as an ideal in a lattice. However we shall have no need for such general filters. In our case, the lattice concerned will always be the power set of some set, partially ordered by the usual inclusion relation $\subset$. So we define filters only in this context.

(3.1) Definition: A **filter** on (or in) a set X is a nonempty family $\mathcal{F}$ of subsets of X such that (i) $\Phi \notin \mathcal{F}$, (ii) $\mathcal{F}$ is closed under finite intersections, and (iii) if $B \in \mathcal{F}$ and $B \subset A$ then $A \in \mathcal{F}$ for all $A, B \subset X$.

In view of Proposition (2.7) conditions (i) and (ii) imply that a filter has the finite intersection property. Condition (iii) says that a filter is closed under the operation of taking supersets of its members. It implies in particular that the set X always belongs to every filter on it. Before proving theorems about filters, it is instructive to have a few examples of them. Throughout X denotes an arbitrary nonempty set.

(1) The singleton family $\{X\}$ is a filter on X.

(2) Fix some non-empty subset A of X. Then the collection of all supersets of A (in X) is a filter on X. Such a filter is known as an **atomic filter**, the set A being called the **atom** of the filter. Note in

this case that A is the intersection of all members of the filter.

(3) In case X is infinite, the family $\mathcal{F}$ of all cofinite subsets of X is a filter on X. Note that this filter is not atomic. Such a filter is called a **cofinite filter.**

(4) Suppose $\mathcal{T}$ is a topology on X. Then for any $x \in X$, the neighbourhood system $\mathcal{N}_x$ at x is a filter. This filter is quite important for our purpose. It is called the **$\mathcal{T}$-neighbourhood filter** at x. It depends both on x and $\mathcal{T}$.

(5) Let $S : D \to X$ be a net. For each $m \in D$, let $B_m = \{S(n) : n \in D, n \geq m\}$. Let $\mathcal{F} = \{A \subset X : A \supset B_m$ for some $m \in D\}$. In other words, $\mathcal{F}$ is the collection of all supersets of sets of the form B_m for $m \in D$. Using the fact that D is a directed set, it is not hard to show that $\mathcal{F}$ is filter on X. Obviously it depends on the net S and is called the **filter associated with the net S.**

We noted that every filter has the finite intersection property. The converse is not true in general, because a family having f.i.p. need not even be closed under finite intersections. However, there is a canonical way of obtaining a filter from a family having f.i.p. We proceed to describe it.

(3.2) Definition: Let $\mathcal{F}$ be a filter on a set X. Then a sub-family $\mathcal{B}$ of $\mathcal{F}$ is said to be a **base** for $\mathcal{F}$ (or a filter base) if for any $A \in \mathcal{F}$ there exists $B \in \mathcal{B}$ such that $B \subset A$.

If $\mathcal{B}$ is a base for a filter $\mathcal{F}$ then every member of $\mathcal{F}$ is a superset of some member of $\mathcal{B}$. On the other hand if $B \in \mathcal{B}$ then $B \in \mathcal{F}$ and so any superset of B is in $\mathcal{F}$ by condition (iii) in the definition of a filter. Thus if $\mathcal{B}$ is a base for a filter $\mathcal{F}$ then $\mathcal{F}$ consists precisely of all supersets of members of $\mathcal{B}$. It follows that a filter is uniquely determined by any base for it and that no family of sets can be a base for more than one filter. As examples of bases, note that a base for the neighbourhood filter at a point is nothing but what we have been calling a local base at that point, while, by its very definition, the family $\{B_m : m \in D\}$ is a base for the filter in example (5) above.

We now characterise those families of sets which can be bases for filters.

(3.3) Proposition: Let $\mathcal{B}$ be a family of non-empty subsets of a set X. Then there exists a filter on X having $\mathcal{B}$ as a base iff $\mathcal{B}$ has the property that for any $B_1, B_2 \in \mathcal{B}$, there exists $B_3 \in \mathcal{B}$ such that $B_1 \cap B_2 \supset B_3$.

Proof: Suppose there exists a filter $\mathcal{F}$ on X having $\mathcal{B}$ as a base. Then $\mathcal{B} \subset \mathcal{F}$ and $\Phi \notin \mathcal{F}$; hence $\Phi \notin \mathcal{B}$. Also let $B_1, B_2 \in \mathcal{B}$. Then $\mathcal{B}_1, \mathcal{B}_2 \in \mathcal{F}$ (since $\mathcal{B} \subset \mathcal{F}$) and so $B_1 \cap B_2 \in \mathcal{F}$ as $\mathcal{F}$ is closed under finite intersections. So by the definition of a base, there exists $B_3 \in \mathcal{B}$ such that $B_1 \cap B_2 \supset B_3$. This proves the necessity of the condition. Conversely suppose $\mathcal{B}$ satisfies the given condition. We then construct a filter from $\mathcal{B}$ as follows. Let $\mathcal{F}$ be the family of all supersets of members of $\mathcal{B}$. Then condition (iii) in the definition of a filter automatically holds for $\mathcal{F}$. The empty set cannot be a superset of any other set. Hence it follows that $\Phi \notin \mathcal{F}$ as $\Phi \notin \mathcal{B}$. It only remains to show that $\mathcal{F}$ is closed under finite intersections. For this it suffices to show that the intersection of any two members of $\mathcal{F}$ is again in $\mathcal{F}$ for then one can apply induction. So suppose $A_1, A_2 \in \mathcal{F}$. Then there exist $B_1, B_2 \in \mathcal{B}$ such that $B_1 \subset A_1$ and $B_2 \subset A_2$. We are given that there exists $B_3 \in \mathcal{B}$ such that $B_3 \subset B_1 \cap B_2$. But then $A_1 \cap A_2$ is a superset of $B_3 \in \mathcal{B}$ and so $A_1 \cap A_2 \in \mathcal{F}$. Thus $\mathcal{F}$ is a filter on X and $\mathcal{B}$ is a base for it by its very construction. ∎

(3.4) Corollary: Any family which does not contain the empty set and which is closed under finite intersections is a base for a unique filter.

Proof: The condition in the last proposition is trivially satisfied for such a family. ∎

The reader must have been reminded of bases for topologies. It is natural to inquire if there is a corresponding notion of sub-bases for filters. The answer is in the affirmative.

(3.5) Definition: Let $\mathcal{F}$ be a filter on a set X. Then a subfamily $\mathcal{S}$ of $\mathcal{F}$ is said to be a **sub-base** for $\mathcal{F}$ if the family of all finite intersections of members of $\mathcal{S}$ is a base for $\mathcal{F}$. We also say $\mathcal{S}$ **generates** $\mathcal{F}$.

Obviously every base is a sub-base. It is easy to characterise those families that can generate filters.

(3.6) Proposition: Let $\mathcal{S}$ be a family of subsets of a set X. Then there exists a filter on X having $\mathcal{S}$ as a sub-base if and only if $\mathcal{S}$ has the finite intersection property.

Proof: If there exists a filter $\mathcal{F}$ on X, containing $\mathcal{S}$ then $\mathcal{F}$ has the f.i.p. and so does every subfamily of $\mathcal{F}$. Thus the condition is necessary. Conversely suppose $\mathcal{S}$ has the f.i.p. Let $\mathcal{B}$ be the family of all finite intersections of members of $\mathcal{S}$. Then $\Phi \notin \mathcal{B}$ and $\mathcal{B}$ is closed under finite intersections. So by the corollary above, $\mathcal{B}$ is a base for a filter $\mathcal{F}$ on X and thus $\mathcal{S}$ is a sub-base for $\mathcal{F}$. ∎

So far the treatment was purely set-theoretic, without mention of any topology on the set X in question (except in the case of example (4) of a filter above). Suppose now a topology $\mathcal{T}$ on X is given. We then define convergence and cluster points of filters w.r.t. $\mathcal{T}$ as follows.

(3.7) Definition: Let $(X, \mathcal{T})$ be a topological space and let $\mathcal{F}$ be a filter on the set X. A point x of X is said to be a **limit** of $\mathcal{F}$ w.r.t. $\mathcal{T}$ (or $\mathcal{F}$ is is said to **converge** to x w.r.t. $\mathcal{T}$) if every neighbourhood of x belongs to $\mathcal{F}$, i.e. if $\mathcal{N}_x \subset \mathcal{F}$. Also a point $y \in X$ is said to be a **cluster point** of $\mathcal{F}$ (w.r.t. $\mathcal{T}$) if every neighbourhood of y intersects every member of $\mathcal{F}$.

Evidently it suffices if the conditions are satisfied for all open neighbourhoods of x and y respectively. Note that if $\mathcal{F}$ converges to x then x is also a cluster point of $\mathcal{F}$ because given $N \in \mathcal{N}_x$, $F \in \mathcal{F}$, both N and F are in $\mathcal{F}$ and so $N \cap F \neq \emptyset$. Trivial examples of convergent filters are the neighbourhood filters. Note that if $\mathcal{F}$, $\mathcal{G}$ are filters on X with $\mathcal{F} \subset \mathcal{G}$ then whenever $\mathcal{F}$ converges to some point in X so does $\mathcal{G}$. It is probably for this reason that in such a case $\mathcal{G}$ is said to be a **subfilter** of $\mathcal{F}$, even though as subsets of $P(X)$, $\mathcal{G}$ is a superset (and not a subset) of $\mathcal{F}$. It is not hard to show that a filter $\mathcal{F}$ has x as a cluster point iff some subfilter of $\mathcal{F}$ converges to x, because if x is a cluster point of $\mathcal{F}$ then the family $\mathcal{F} \cup \mathcal{N}_x$ has the f.i.p. and so generates a filter by Proposition (3.6).

There is a canonical way of converting nets to filters and vice versa. In example (5) above we saw that any net $S : D \to X$ determines a filter having the family of sets of the form $\{S(n) : n \in D, n \geq m\}$ for $m \in D$ as a base. It may of course happen that two distinct nets determine the same filter. Conversely given a filter $\mathcal{F}$ on X, there is a net associated with it as follows. Let $D = \{(x, F) \in X \times \mathcal{F} : x \in F\}$. For $(x, F), (y, G) \in D$ define $(x, F) \geqslant (y, G)$ if $F \subset G$. It is easily seen that $\geqslant$ directs D because $\mathcal{F}$ is closed under finite intersections. Now define $S : D \to X$ by $S(x, F) = x$. Then S is a net in X. It is called the **net associated with** $\mathcal{F}$. Limits and cluster points are preserved in switching over from nets to filters and vice versa as we now prove.

(3.8) Proposition: Let $S : D \to X$ be a net and $\mathcal{F}$ the filter associated with it. Let $x \in X$. Then S converges to x as a net iff $\mathcal{F}$ converges to x as a filter. Also x is a cluster point of the net S iff x is a cluster point of the filter $\mathcal{F}$.

Proof: Assume S converges to x. Let U be any neighbourhood of x in X. Then there exists $m \in D$ such that $B_m \subset U$ where $B_m = \{S(n) : n \in D, n \geqslant m\}$. But this means $U \in \mathcal{F}$ by the definition of $\mathcal{F}$. So $\mathcal{N}_x \subset \mathcal{F}$, i.e. $\mathcal{F}$ converges to x. Conversely suppose $\mathcal{F}$ converges to x. Let U be an open nbd of X. Then $U \in \mathcal{F}$. Recalling how $\mathcal{F}$ was generated, there exists $m \in D$ such that $B_m \subset U$ where B_m is defined as above. This means that $S(n) \in U$ for all $n \in D$, $n \geqslant m$. Thus S converges to x in X. The proof of the assertion regarding cluster points is similar and left to the reader. ∎

(3.9) Proposition: Let $\mathcal{F}$ be a filter in a space X and S be the associated net in X. Let $x \in X$. Then $\mathcal{F}$ converges to x as a filter iff S converges to x as a net. Moreover, x is a cluster point of the filter $\mathcal{F}$ iff it is a cluster point of the net S.

Proof: By way of variation, this time we leave the part regarding convergence to the reader and prove the one about cluster points. First suppose that $\mathcal{F}$ has x as a cluster point. Recall that the associated net $S : D \to X$ is defined by taking $D = \{(y, G) : G \in \mathcal{F}, y \in G\}$ and putting $S(y, G) = y$. Let an open neighbourhood U of x and an element (y, G) of D be given. Then $G \cap U \neq \Phi$ by definition of a cluster point of a filter. Let $z \in G \cap U$. Then $(z, G) \in D$, $(z, G) \geqslant (y, G)$ and $S(z, G) = z \in U$. So x is a cluster point of S. Conversely suppose x is a cluster point of S. Let U be any open neighbourhood of x and let $F \in \mathcal{F}$. We have to show $F \cap U \neq \emptyset$. Let z be any point of F. Then $(z, F) \in D$. Since x is a cluster point of S, there exists $(y, G) \in D$ such that $(y, G) \geq (z, F)$ and $S(y, G) \in U$. But then,

$y \in G$, $G \subset F$ and $y \in U$ showing that $F \cap U \neq \emptyset$ as desired. Thus x is a cluster point of $\mathcal{F}$. ∎

The last two propositions allow us to translate a topological problem about filters into a corresponding problem about nets and vice versa. A solution to either one of them can also be translated into a solution of the other. However, it is more instructive to give direct, independent solutions whenever possible. As an illustration, we prove the following theorem by both the approaches.

(3.10) Theorem: A topological space is Hausdorff iff no filter can converge to more than one point in it.

Proof: In Theorem (1.4) we proved the assertion for nets. Now suppose X is a T_2 space and $\mathcal{F}$ is a filter converging to say, x and y in X. Then so does the net associated with $\mathcal{F}$ by Proposition (3.9). So $x = y$ by Theorem (1.4). Conversely suppose no filter in X converges to more than one point. Then no net in X can converge to more than one point, for if a net were to converge to more than one point, so would its associated filter by Proposition (3.8). Hence X is T_2 again by Theorem (1.4).

We also give a direct proof. Suppose X is a Hausdorff space and a filter $\mathcal{F}$ converges to x as well as y. This means $\mathcal{N}_x \subset \mathcal{F}$ and $\mathcal{N}_y \subset \mathcal{F}$. Now if $x \neq y$, then there exist $U \in \mathcal{N}_x$, $V \in \mathcal{N}_y$ such that $U \cap V = \emptyset$, which will contradict that $\mathcal{F}$ has the finite intersection property. So $x = y$. Thus limits of convergent filters in X are unique. Conversely assume that no filter in X has more than one limit in X. If X is not Hausdorff, there exist $x, y \in X$, $x \neq y$ such that every neighbourhood of x intersects every neighbourhood of y. From this it follows that the family $\mathcal{N}_x \cup \mathcal{N}_y$ has the f.i.p. (prove!) So by Proposition (3.6) there exists a filter $\mathcal{F}$ on X containing $\mathcal{N}_x \cup \mathcal{N}_y$. Evidently $\mathcal{F}$ converges both to x and y, contradicting the hypothesis. So X is a Hausdorff space. ∎

We invite the reader to prove the following theorem using both approaches.

(3.11) Theorem: For a topological space X, the following statements are equivalent:

(1) X is compact.

(2) Every filter on X has a cluster point in X.

(3) Every filter on X has a convergent subfilter. ∎

We now consider the problem of describing continuity in terms of filters. Suppose X, Y are topological spaces, $x \in X$ and $f : X \to Y$ is a function. We expect that f would be continuous at x iff whenever a filter $\mathcal{F}$ converges to x in X, then its image filter under f converges to $f(x)$ in Y. The only question is how to define the image filter. It is natural to consider the family of images of members of $\mathcal{F}$. Denote this family by $f(\mathcal{F})$, that is $f(\mathcal{F}) = \{f(A) \subset Y : A \in \mathcal{F}\}$. Unfortunately $f(\mathcal{F})$ may not be a filter by itself. In fact unless f is onto, $Y \notin f(\mathcal{F})$. Nevertheless, we can generate a filter out of $f(\mathcal{F})$.

(3.12) Proposition: Let X, Y be sets, $f : X \to Y$ a function and $\mathcal{F}$ a filter on X. Then the family $f(\mathcal{F}) = \{f(A) : A \in \mathcal{F}\}$ is a base for a filter on Y.

Proof: We show that $f(\mathcal{F})$ satisfies the condition of Proposition (3.3). Evidently $\emptyset \notin f(\mathcal{F})$. Also let $B_1, B_2 \in f(\mathcal{F}]$. Then there exist $A_1, A_2 \in \mathcal{F}$ such that $f(A_1) = B_1$ and $f(A_2) = B_2$. Then $A_1 \cap A_2 \in \mathcal{F}$ and $B_1 \cap B_2$ contains $f(A_1 \cap A_2)$ which is a member of $f(\mathcal{F})$. So by Proposition (3.3) $f(\mathcal{F})$ is a base for a unique filter on Y. ∎

(3.13) Definition: With the notation of the last proposition, the filter on Y having $f(\mathcal{F})$ as a base is called the **image filter of $\mathcal{F}$ under** f and is denoted by $f_{**}(\mathcal{F})$.

We are now in a position to characterise continuity in terms of convergence of filters.

(3.14) Proposition: Let X, Y be a topological space, $x \in X$, and $f : X \to Y$ a function. Then f is continuous at x iff whenever a filter $\mathcal{F}$ converges to x in X, the image filter $f_{**}(\mathcal{F})$ converges to $f(x)$ in Y.

Proof: Assume first that f is continuous at x and $\mathcal{F}$ is a filter which converges to x in X. We have to show that $f_{**}(\mathcal{F})$ converges to $f(x)$ in Y. Let N be a given neighbourhood of $f(x)$ in Y. Then, by continuity of f at x, $f^{-1}(N)$ is a neighbourhood of x in X. By convergence of $\mathcal{F}$ to x in X, $f^{-1}(N) \in \mathcal{F}$. So $f(f^{-1}(N)) \in f(\mathcal{F})$. But $N \supset f(f^{-1}(N))$ and so $N \in f_{**}(\mathcal{F})$. Since N was an arbitrary neighbourhood of $f(x)$, it follows that $f_{**}(\mathcal{F})$ converges to $f(x)$. Conversely suppose the condition about filters is satisfied. We have to show that f is continuous at x. If this is not so then there exists a neighbourhood N of $f(x)$ in Y such that $f^{-1}(N)$ is not a nbd of x in X. This means that every neighbourhood of x in X intersects the complement $X - f^{-1}(N)$ and hence the family $\mathcal{S} = \mathcal{N}_x \cup \{X - f^{-1}(N)\}$ has the finite intersection property. So by Proposition (3.6) $\mathcal{S}$ generates a filter $\mathcal{F}$ on X. Obviously $\mathcal{F}$ converges to x in X. However we contend that $f_{**}(\mathcal{F})$ does not converge to $f(x)$ in Y. Indeed since $X - f^{-1}(N) \in \mathcal{F}, f(X - f^{-1}(N)) \in f(\mathcal{F})$. But $Y - N$ contains $f(X - f^{-1}(N))$. So $Y - N \in f_{**}(\mathcal{F})$. Therefore $N \notin f_{**}(\mathcal{F})$, since no filter can contain both a set and its complement. But N is a neighbourhood of $f(x)$ in Y. Thus $f_{**}(\mathcal{F})$ does not converge to $f(x)$ in Y. This contradiction proves that f is a continuous of x. ∎

We conclude this section with a characterisation of convergence of filters in topological products, which will be used later.

(3.15) Theorem: Let X be the topological product of an indexed family of spaces $\{X_i : i \in I\}$. Let $\mathcal{F}$ be a filter on X and let $x \in X$. Then $\mathcal{F}$ converges to x in X iff for each $i \in I$, the filter $\pi_{i_{**}}(\mathcal{F})$ converges to $\pi_i(x)$ in X_i.

Proof: The necessity of the condition follows from the last proposition in view of continuity of π_i. For sufficiency, suppose for each $i \in I$, $\pi_{i_{**}}(\mathcal{F})$ converges to $\pi_i(x)$ in X_i. We have to show that $\mathcal{F}$ converges to x in X. Let N be a neighbourhood of x in X. Then N contains a basic open set V containing x. Let $V = \prod_{i \in I} V_i$ where each V_i is an open set in X_i and $V_i = X_i$

for all $i \in I$ except for $i = i_1, i_2, \ldots, i_n$ (say). Now $\pi_{i_{k**}}(\mathcal{F})$ converges to $\pi_{i_k}(x)$ for all $k = 1, \ldots, n$. So $V_{i_k} \in \pi_{i_{k**}}(\mathcal{F})$ and hence there exists $F_k \in \mathcal{F}$ such that $V_{i_k} \supset \pi_{i_k}(F_k)$ for $k = 1, 2, \ldots, n$. Note that $\pi_{i_k}^{-1}(V_{i_k}) \supset \pi_{i_k}^{-1}(\pi_{i_k}(F_k)) \supset F_k$ for $k = 1, 2, \ldots, n$. So $N \supset V = \bigcap_{k=1}^{n} \pi_{i_k}^{-1}(V_{i_k}) \supset \bigcap_{k=1}^{n} F_k$. But $\bigcap_{k=1}^{n} F_k$ is in $\mathcal{F}$ since $\mathcal{F}$ is closed under finite intersections, and therefore N being a superset of $\bigcap_{k=1}^{n} F_k$ is also in $\mathcal{F}$. Thus $\mathcal{N}_x \subset \mathcal{F}$, showing that $\mathcal{F}$ converges to x. ∎

Exercises

3.1 Prove that the intersection of any family of filters on a set is again a filter on that set.

3.2 If X is a set and $\mathcal{S}$ is a family of subsets of X having the finite intersection property, prove that the filter generated by $\mathcal{S}$ is the intersection of all filters on X containing $\mathcal{S}$, and hence is the smallest such filter.

3.3 Let X be a topological space and $x \in X$. Prove that the intersection of all filters on X converging to x is precisely the neighbourhood system of x in X.

3.4 Let X be a space and suppose $A \subset X$, $x \in X$. Prove that $x \in \bar{A}$ iff there exists a filter $\mathcal{F}$ on X such that $A \in \mathcal{F}$ and $\mathcal{F}$ converges to x.

3.5 Using either of the last two exercises, prove that a topology on a set is uniquely determined if we know which filters on that set converge to which points.

*3.6 Obtain an axiomatic characterisation of convergence of filters in topological spaces. In other words, given a set X and a collection $\mathcal{C}$ whose members are pairs of the form $(\mathcal{F}, x)$ where $\mathcal{F}$ is a filter on X and $x \in X$, find necessary and sufficient conditions on $\mathcal{C}$ in order that there exists a topology $\mathcal{T}$ on X such that $(\mathcal{F}, x) \in \mathcal{C}$ iff $\mathcal{F}$ converges to x w.r.t. $\mathcal{T}$.

3.7 Complete the proofs of Propositions (3.8) and (3.9).

3.8 Prove Theorem (3.11) both by passing to nets and by a direct argument.

3.9 Prove that the family $\mathcal{N}_x \cup \mathcal{N}_y$ constructed in the proof of the converse implication in Theorem (3.10) has the finite intersection property.

3.10 Generalise Theorem (3.15) to the case of convergence of filters w.r.t. the weak topology generated by a family of functions from a common domain into various topological spaces.

3.11 Suppose X, Y are sets, $f: X \to Y$ is a function and $S: D \to X$ is a net in X. Let $\mathcal{F}$ be the filter on X associated with S. Prove that the filter $f_{**}(\mathcal{F})$ coincides with the filter associated with the net $f \circ S$ in Y. Use this fact and Theorem (2.5) to give an alternate proof of Proposition (3.14).

3.12 Let X, Y be sets and suppose $\theta: P(X) \to P(Y)$ is a function which preserves finite intersections and complements (i.e. for any $A, B \in P(X)$, $\theta(A \cap B) = \theta(A) \cap \theta(B)$ and $\theta(X - A) = Y - \theta(A)$).

(a) Prove that θ also preserves finite unions. (Hint: Use De Morgan's laws.)

(b) Prove that θ is monotonic (that is, for $A, B \in X$, $A \subset B$ implies $\theta(A) \subset \theta(B)$).

(c) For any filter $\mathcal{F}$ on Y prove that $\theta^{-1}(\mathcal{F})$ is a filter on X. Note that in particular $\theta^{-1}(\{Y\})$ is a filter on X.

3.13 How are subnets related to subfilters?

Notes and Guide to Literature

The definition of a filter is due to H. Cartan. Bourbaki [1] defines topological spaces in terms of convergence of filters. The power set of a set is an example of what is known as a *Boolean algebra* and a function such as θ in Exercise (3.12) is called a *Boolean algebra homomorphism*. Theoretical aspects of Boolean algebras can be found in Bell and Slomson [1]; however it is an area which has immense practical applications as well in electrical networks; see for example, Whitestitt [1].

4. Ultrafilters and Compactness

In the last section we saw that there is a canonical way of passing from nets to filters and vice versa. As a result, any concept for nets will have an analogue for filters and vice versa. However, certain concepts appear more natural in one context than their counterparts in the other. For example the concepts of a limit and of a cluster point come more naturally for nets than for filters. On the other hand, the notion of a subfilter is defined easily by set inclusion while the corresponding notion of a subnet had a rather clumsy definition.

Wherever the filters have a leverage over the nets, it is generally because a filter on a set is described more intrinsically in terms of that set than is the case with a net in that set. The very definition of a net involves something extraneous, namely its domain, which can be any directed set whatsoever. As we have seen before, sometimes this is an advantage. But it puts us in difficulty when we want to consider a collection of nets in the same set inasmuch as the domains of these nets may be quite unrelated to one another.

The situation is better with filters. A filter on a set X is a subfamily of $P(X)$ or equivalently a member of $P(P(X))$. A collection of filters on X is thus a well-defined subset of $P(P(X))$ and is amenable to whatever additional structure we can impose on $P(P(X))$. In the present section we are concerned with the partial ordering in $P(P(X))$ (which is by inclusion). We begin with a definition.

(4.1) Definition: A filter $\mathcal{F}$ on a set X is said to be an **ultrafilter** if it is a maximal element in the collection of all filters on X, partially ordered by inclusion, that is, $\mathscr{F}$ is an ultrafilter if it is not properly contained in any filter on X.

For example, all atomic filters whose atoms are singleton sets are maximal. These are not the only examples as the following theorem shows.

(4.2) Theorem: Every filter is contained in an ultrafilter.

Proof: The proof consists of a standard application of the Zorn's lemma. Let $\mathscr{F}$ be a filter on a set X. Let G be the collection of all filters on X containing $\mathscr{F}$. Then $\mathscr{F} \in G$ and so G is non-empty. Partially order G by inclusion. We apply Zorn's lemma to G. Let $\{\mathcal{G}_i : i \in I\}$ be a non-empty chain in G. Let $\mathcal{G} = \bigcup_{i \in I} \mathcal{G}_i$. We claim $\mathcal{G}$ is a filter on X. Clearly $\Phi \notin \mathcal{G}$ because $\Phi \notin \mathcal{G}_i$ for all $i \in I$. To show that $\mathcal{G}$ is closed under finite intersections, it suffices to show that the intersection of two members of $\mathcal{G}$ is again in $\mathcal{G}$. So let $A, B \in \mathcal{G}$. Then there exists $i, j \in I$ such that $A \in \mathcal{G}_i$ and $B \in \mathcal{G}_j$. Since the collection $\{\mathcal{G}_i : i \in I\}$ is a chain (i.e. linearly ordered) under $\subset$, it follows that either $\mathcal{G}_i \subset \mathcal{G}_j$ or $\mathcal{G}_j \subset \mathcal{G}_i$. In the first case $A, B \in \mathcal{G}_j$ and so $A \cap B \in \mathcal{G}_j$ since $\mathcal{G}_j$ is a filter. Similarly in the second case $A \cap B \in \mathcal{G}_i$. In either case, $A \cap B \in \mathcal{G}$. Finally suppose $C \in \mathcal{G}$ and D is a superset of C in X. We have to show $D \in \mathcal{G}$. Now $C \in \mathcal{G}_i$ for some $i \in I$. So $D \in \mathcal{G}_i$ as $\mathcal{G}_i$ is a filter. But then $D \in \mathcal{G}$. Thus we have shown that $\mathcal{G}$ is a filter on X. Obviously $\mathcal{G}$ contains $\mathscr{F}$ as each $\mathcal{G}_i$ does. So $\mathcal{G} \in G$ and by its very construction, it is an upper bound for the chain $\{\mathcal{G}_i : i \in I\}$. We have shown that every chain in G has an upper bound in G. So by Zorn's lemma, G contains a maximal element, $\mathscr{H}$. We claim $\mathscr{H}$ is an ultrafilter, that is, $\mathscr{H}$ is also maximal in the set of all filters on X. For suppose $\mathscr{K}$ is a filter on X such that $\mathscr{H} \subset \mathscr{K}$. Then $\mathscr{F} \subset \mathscr{K}$ (since $\mathscr{F} \subset \mathscr{H}$) and so $\mathscr{K} \in G$. But $\mathscr{H}$ is maximal in G. So $\mathscr{H} = \mathscr{K}$. Thus $\mathscr{H}$ is an ultrafilter containing $\mathscr{F}$. ∎

Like other arguments based on Zorn's lemma (which is a version of the axiom of choice), the proof above is nonconstructive in that it merely asserts the existence of an ultrafilter without giving an explicit construction for it. For example, no one has yet concretely exhibited any ultrafilter containing the cofinite filter on an infinite set. Nevertheless the existence of ultrafilters is important for theoretical purposes as we shall see later in this section. First it is convenient to have a succinct characterisation of ultrafilters.

(4.3) Proposition: For a filter $\mathscr{F}$ on a set X the following statements are equivalent:

(1) $\mathscr{F}$ is an ultrafilter.

(2) For any $A \subset X$ either $A \in \mathscr{F}$ or $X - A \in \mathscr{F}$.

(3) For any $A, B \subset X$, $A \cup B \in \mathscr{F}$ iff either $A \in \mathscr{F}$ or $B \in \mathscr{F}$.

Proof: First we show (1) is equivalent to (2). Assume $\mathscr{F}$ is an ultrafilter on X and A is a subset of X. If $A \notin \mathscr{F}$ then A contains no member of $\mathscr{F}$, or equivalently every member of $\mathscr{F}$ intersects $X - A$. Thus the family $\mathscr{F} \cup \{X - A\}$ has the finite intersection property and so generates a filter

$\mathcal{G}$ by Proposition (3.6). Since $\mathcal{F}$ is maximal, no filter on X can properly contain $\mathcal{F}$. In particular $\mathcal{G} = \mathcal{F}$ which¡ forces that $X - A \in \mathcal{F}$ and so (2) holds. Conversely assume (2) holds. If $\mathcal{F}$ is not an ultrafilter then there exists a filter $\mathcal{G}$ which property contains $\mathcal{F}$. Then there exists $A \in \mathcal{G} - \mathcal{F}$. Since $A \notin \mathcal{F}$, $X - A \in \mathcal{F}$ by (2). Hence $X - A \in \mathcal{G}$. So $\mathcal{G}$ contains A as well as $X - A$ which contradicts the finite intersection property of a filter. Thus $\mathcal{F}$ is an ultrafilter.

In view of the fact that every filter contains the set X, (2) follows from (3) by taking $B = X - A$. Conversely assume (2) holds. Let $A, B \subset X$. Since $A \cup B$ is a superset of A as well as B, one way implication in (3) is immediate from the very definition of a filter. For the other way, suppose $A \cup B \in \mathcal{F}$ but neither $A \in \mathcal{F}$ nor $B \in \mathcal{F}$. Then by (2), $X - A \in \mathcal{F}$ and $X - B \in \mathcal{F}$ and so $(X - A) \cap (X - B) \in \mathcal{F}$. But $(X - A) \cap (X - B) = X - (A \cup B)$. So $\mathcal{F}$ contains $A \cup B$ as well as its complement, which is a contradiction. So (3) holds. Thus we have shown that (2) is equivalent to (3). This completes the proof. ∎

So far we dealt with ultrafilters on a set X. Suppose now there is a topology $\mathcal{T}$ on X. Convergence of ultrafilters has the following important property.

(4.4) Proposition: An ultrafilter converges to a point iff that point is a cluster point of it.

Proof: The direct implication is true for any filter, not just for an ultrafilter, as we observed in the last section after defining cluster points. For the converse, suppose X is a space and $x \in X$ is a cluster point of an ultrafilter $\mathcal{F}$ on X. If $\mathcal{F}$ does not converge to x, then there exists a neighbourhood N of x such that $N \notin \mathcal{F}$. By the characterisation of an ultrafilter just obtained, we then have that $X - N \in \mathcal{F}$. But since x is a cluster point of $\mathcal{F}$, every nbd of x intersects every member of $\mathcal{F}$, whereas $N \cap (X - N) = \Phi$. This contradiction proves that $\mathcal{F}$ converges to x in X. ∎

An important consequence of this proposition is the

(4.5) Theorem: A topological space is compact iff every ultrafilter in it is convergent.

Proof: If a space is compact then every filter in it has a cluster point by Theorem (3.11). In particular every ultrafilter has a cluster point and hence is convergent by the last proposition. Conversely suppose X is a space with the property that every ultrafilter on it is convergent. To show X is compact we again use Theorem (3.11) after showing that every filter on X has a cluster point. Suppose $\mathcal{F}$ is a filter on X. By Theorem (4.2) there exists an ultrafilter $\mathcal{G}$ containing $\mathcal{F}$. By hypothesis $\mathcal{G}$ converges to a point say x on X. Then x is also a cluster point of $\mathcal{G}$. So every nbd of x meets every member of $\mathcal{G}$ and in particular every member of $\mathcal{F}$ since $\mathcal{F} \subset \mathcal{G}$. So x is also a cluster point of $\mathcal{F}$. Thus every filter on X has a cluster point in X. As noted before this proves that X is compact. ∎

With this characterisation of compactness we are now ready to prove that compactness is a productive property. This is a celebrated theorem due to Tychonoff.

(4.6) Theorem: Let $\{X_i : i \in I\}$ be a collection of nonempty spaces and let X be its topological product. Then X is compact iff each X_i is so for $i \in I$.

Proof: The direction implication is immediate from Proposition (6.1.8) since each X_i is the image of X under the continuous function π_i, for $i \in I$. It is the converse that is really non-trivial. Suppose each X_i is compact for $i \in I$. To show that X is compact, we show that every ultrafilter on it is convergent. So let $\mathcal{F}$ be an ultrafilter on X. For $i \in I$, let $\mathcal{F}_i = \pi_{i_{**}}(\mathcal{F})$. Then $\mathcal{F}_i$ is a filter on X_i. We claim it is in fact an ultrafilter. For this we use the characterisation in Proposition (4.3). Let $A \subset X_i$. Put $B = \pi_i^{-1}(A)$. Note that $X - B = \pi_i^{-1}(X_i - A)$. Now since $\mathcal{F}$ is an ultrafilter on X, either $B \in \mathcal{F}$ or $X - B \in \mathcal{F}$. In the first case, $\pi_i(B) = A \in \pi_i(\mathcal{F}) \subset \pi_{i_{**}}(\mathcal{F}) = \mathcal{F}_i$ while in the other case we get similarly that $X - A \in \mathcal{F}_i$. So $\mathcal{F}_i$ is an ultrafilter on X_i. By compactness of X_i, $\mathcal{F}_i$ converges to x_i (say) in X_i. By Theorem (3.15) this implies that $\mathcal{F}$ converges to x where $x \in X$ is defined by $x(i) = x_i$ for $i \in I$. Thus every ultrafilter on X is convergent and so X is compact by the last theorem. ∎

No elementary proof of Tychonoff's theorem is known. We shall give another proof of it in the next chapter. But even that proof will involve very similar ideas. The advantage of the proof above is that it is broken into a series of propositions about filters and ultrafilters, each of which is fairly simple and of independent interest. Perhaps the only place a beginner is likely to feel unfamiliar is the use of Zorn's lemma in the proof of Theorem (4.2). This is the typical way in which Zorn's lemma is applied and we trust that the reader has already seen one or two applications of it in algebra or in analysis. For the benefit of the reader who has not, we give a few simple exercises as a drill in the use of Zorn's lemma.

The concept of an ultrafilter gives the theory of filters an advantage over the theory of nets. There is a parallel concept for nets and we shall define it here. But even the definition will make it clear how artificial it is.

(4.7) Definition: A net S in a set X is said to be **universal** if for each subset A of X, S is either eventually in A or eventually in $X - A$.

Properties of universal nets are parallel to those of ultrafilters and will be developed through exercises. For example it is easy to show that a universal net converges to a point iff that point is a cluster point of it. However, a little difficulty arises when one tries to prove the analogue of Theorem (4.2), namely that every net has a subnet which is universal. Since we can no longer consider the set of all nets in a set, a direct application of Zorn's lemma becomes impossible. In this case, the problem is translated into filters, solved there and the solution is translated back. It is for this reason that the notion of a universal net has no independent appeal.

Exercises

4.1 Prove that a space is Hausdorff iff every ultrafilter converges to at most one point in it.

4.2 Let G be a group and let x be an element of G other than the identity element. Prove that there exists a subgroup H of G which is maximal w.r.t. the property of not containing x (i.e. whenever K is a subgroup of G properly containing H then $x \in K$). (This and the next exercise are of no use here, they are given as a drill in the use of Zorn's lemma.)

4.3 Let (X, d) be a metric space and r a positive real number. Prove that there exists a subset of X which is maximal w.r.t. the property that every two distinct points of it are at least a distance r apart.

4.4 Prove that a non-empty family $\mathcal{F}$ of subsets of a set X is an ultrafilter on X iff it has the following properties

(i) For any $A, B \subset X$, $A \cup B \in \mathcal{F}$ iff $A \in \mathcal{F}$ or $B \in \mathcal{F}$:

(ii) For any $A \subset X$, $A \in \mathcal{F}$ iff A intersects every member of $\mathcal{F}$.

(Note: In the converse implication, it is not given beforehand that $\mathcal{F}$ is a filter on X. This needs to be established first.)

4.5 Give a direct proof of Proposition (4.4) without using the characterisation in Proposition (4.3). (Hint: Note that if x is a cluster point of a filter $\mathcal{F}$ then the family $\mathcal{F} \cup \mathcal{N}_x$ has the finite intersection property.)

4.6 Prove that the filter associated with a universal net is an ultrafilter and that the net associated with an ultrafilter is a universal net. Prove that both the converses also hold.

4.7 Prove that if a universal net has a cluster point then it converges to that point (Hint: Either convert to filters using the last exercises and apply Proposition (4.4) or give a simple, direct argument.)

4.8 Prove that every net has a subnet which is universal. (Hint: Let $S : D \to X$ be a net. Let $\mathcal{F}$ be an ultrafilter on X containing the filter associated with S. Now let $E = \{(n, A) \in D \times \mathcal{F} : S(n) \in A\}$. For $(n, A), (m, B) \in E$ define $(n, A) \geqslant (m, B)$ to mean $n \geqslant m$ in D and $A \subset B$. Show that E is a directed set and define $T : E \to X$ by $T(n, A) = S(n)$. Prove that T is universal and that it is a subnet of S.)

4.9 Prove that a space is compact iff every universal net in it is convergent (Hint: Apply the last exercises alongwith Theorem (2.9) or convert to filters again and use Theorem (4.5).)

4.10 Using the last exercise give an alternative proof of Tychonoff's theorem.

4.11 Let X be a set and Y be any singleton set. Prove that $\mathcal{F}$ is an ultrafilter on X iff there exists a function $\theta : P(X) \to P(Y)$ preserving intersections and complements such that $\mathcal{F} = \theta^{-1}(\{Y\})$. ($P(Y)$ is often denoted by $\mathbb{Z}_2 = \{0, 1\}$ with 0 corresponding to Φ and 1 corresponding to Y. $\mathbb{Z}_2$ is thus a Boolean algebra.)

4.12 Let X be a set and A, B be its subsets. Prove that $A \subset B$ iff every ultrafilter containing A also contains B. Deduce that a subset of X is uniquely determined by the collection of all ultrafilters on X to which it belongs.

Notes and Guide to Literature

The importance of filters and ultrafilters in general topology, especially where compactness is involved, is now well-established. We shall return to this topic again in the next chapter. The existence of ultrafilters in Theorem (1.2) relies crucially on the use of Zorn's lemma which is equivalent to the axiom of choice. It is apparently not clear whether Theorem (4.2) implies the axiom of choice. It is well known, however, that Tychonoff's theorem implies the axiom of choice. This was proved by Kelley [2]. Unfortunately there is a minor error in that argument, but it can easily be corrected.

Exercise (4.11) shows that ultrafilters on a set X are in one-to-one correspondence with Boolean algebra homomorphisms from $P(X)$ onto $\mathbb{Z}_2$. This also holds when $P(X)$ is replaced by an abstract Boolean algebra and thus provides a key to the representation of abstract Boolean algebras due to M. H. Stone [1].

Chapter Eleven

Compactness

We defined compact spaces as early as in Chapter 6 and proved a number of their properties in subsequent chapters. There are other topological conditions of varying strengths which resemble compactness. We shall study a few of them in this chapter. We shall also prove the well known Alexander sub-base theorem for compact spaces and apply it to give an alternate proof of Tychonoff's theorem which was proved in the last chapter using the theory of filters. In the last section we discuss compactifications.

1. Variations of Compactness

A space is compact, as we know, if every open cover of it has a finite subcover. Historically this condition was found to be a little too restrictive and it turns out that some of the pleasant consequences of compactness (especially where real-valued continuous functions are involved) continue to hold even under a weaker condition which we now define.

(1.1) Definition: A topological space is said to be **countably compact** if every countable open cover of it has a finite sub-cover.

Historically this was taken as the definition of compactness and this practice still persists to some extent in Europe. What we call as compactness, is a much stronger condition and used to be called bicompactness. It is immediate from definition that a space is compact (in our sense) iff it is countably compact and Lindelöff. The standard example of a space which is countably compact but not compact is not hard but involves the use of ordinals and is therefore deferred. However, another example will be given in the exercises.

Many results about compactness continue to hold for countable compactness. For example, the following propositions can be proved by arguments essentially similar to those for Propositions (6.1.8) and (6.1.10).

(1.2) Proposition: A continuous image of a countably compact space is countably compact. ∎

(1.3) Proposition: Countable compactness is a weakly hereditary property. ∎

We now obtain a few characterisations of countable compactness in T_1-

spaces.

(1.4) Theorem: For a T_1 topological space X the following statements are equivalent:

(1) X is countably compact.

(2) every countable family of closed subsets of X which has the finite intersection property has non-empty intersection.

(3) Every infinite subset of X has an accumulation point.

(4) Every sequence in X has a cluster point.

(5) Every infinite open cover of X has a proper subcover (i.e. a subcover which fails to contain at least one member of the original cover).

Proof: The equivalence of (1) with (2) follows by a simple application of De Morgan's law (cf. the proof of Proposition (10.2.8)). We shall prove that (1), (3) and (4) imply each other in a cyclic order. Finally we shall establish the equivalence of (3) with (5).

Assume (1) and suppose A is an infinite subset of X. If A has no accumulation point in X then its derived set is empty and so A (and in fact every subset of A) is closed in X (see Theorem (5.2.9)). Now A contains a countably infinite subset, say, B. For each $b \in B$, there exists an open subset V_b containing b such that V_b contains no point of A other than b. Now the family $\{V_b : b \in B\}$ is a countable open cover of B. By Proposition (1.3) B itself is countably compact, since it is a closed subset of a countably compact space. But any finite subfamily of $\{V_b : b \in B\}$ will cover only finitely many points of the infinite set B. Thus the countable open cover $\{V_b \cap B : b \in B\}$ of B (in its relative topology) has no finite subcover. This contradiction establishes (3).

Next assume (3) and suppose $\{x_n\}$ is a sequence in X. If the sequence assumes some value, say x, infinitely often, then clearly this value is a cluster point of the sequence. Otherwise the set A of distinct values assumed by the sequence $\{x_n\}$ (or in other words, the range of the sequence) is an infinite subset of X. By (3), A has an accumulation point, say x, in X. We contend that x is a cluster point of the sequence $\{x_n\}$. For if this were not so then there would exist a neighbourhood U of x and an integer m such that $x_n \in X - U$ for all $n \geq m$. Then U is a neighbourhood of x which contains at most m elements of the set A. In a T_1-space this contradicts that x is an accumulation point of A (see Proposition (7.1.4)). Thus x is a cluster point of $\{x_n\}$ in either case and so (4) holds.

Now assume (4). Let $\mathcal{U} = \{U_1, U_2, \ldots, U_n, \ldots\}$ be a countable open cover of X. For each $n \in \mathbb{N}$, let $V_n = \bigcup_{k=1}^{n} U_k$. Then $\mathcal{U}$ has a finite subcover iff for some $n \in \mathbb{N}$, $X = V_n$. Suppose $\mathcal{U}$ has no finite subcover. Then we shall derive a contradiction as follows. We may evidently assume $U_1 \neq \emptyset$. Pick any $x_1 \in U_1$. Now $V_1 = U_1 \neq X$. Set $n_1 = 1$. Let n_2 be the smallest integer greater than n_1 such that U_{n_2} contains an element, say x_2, not in V_1 (such an integer exists because $V_1 \neq X$ and $\mathcal{U}$ is a cover of X). Now $V_{n_2} \subsetneq X$. So again, let n_3 be the smallest integer greater than n_2 such that U_{n_3} contains an element, say x_3 in $X - V_{n_2}$. Continuing in this manner inductively, we get integers $n_1 < n_2 < n_3 < \ldots$ and a sequence $\{x_k\}$ in X such that for each $k \in \mathbb{N}$, $x_k \in U_{n_k}$ and $x_k \notin V_n$ for $n < n_k$. Now by (4), $\{x_k\}$ has a cluster point, say, x in X. Then there exists $n \in \mathbb{N}$ such that $x \in V_n$. Given such n, there exists $r \in \mathbb{N}$ such that $n_r > n$. But then V_n is an open neighbourhood of x which contains no terms of the sequence $\{x_k\}$ for $k \geq r$, contradicting that x is a cluster point of $\{x_k\}$. Thus $\mathcal{U}$ has a finite subcover and so X is countably compact.

Finally we prove the equivalence of (3) with (5). Note first that a cover $\mathcal{U}$ of a set X fails to have a proper subcover iff every member of $\mathcal{U}$ contains a point which does not belong to any other member of $\mathcal{U}$. Now assume (3) holds and let $\mathcal{U}$ be an infinite open cover of X. If $\mathcal{U}$ has no proper subcover then by what we just observed, there exists an infinite subset A of X such that each member of $\mathcal{U}$ intersects A in only one point. By (3), A has an accumulation point, say, x in X. Then $x \in U$ for some $U \in \mathcal{U}$. But U contains only one element of A, which in a T_1-space contradicts that x is an accumulation point of A (once again see Proposition (7.1.4)). Thus (5) holds. Conversely suppose (5) holds. Suppose A is an infinite subset of X. If A has no accumulation point in X then A is closed in X. Moreover, for each $x \in A$, there exists, an open set U_x which contains no point of A other than x. The family $\{U_x : x \in A\}$ may cover X. In that case we call it $\mathcal{U}$. If not, let $\mathcal{U} = \{U_x : x \in A\} \cup \{X - A\}$. In either case, $\mathcal{U}$ is an open cover of X which fails to have a proper subcover by its very construction, contradicting (5). So (5) implies (3). This completes the proof that all the five statements are equivalent. ∎

We remark that the T_1-axiom was used only in the implications (3) ⇒ (4) and (3) ⇒ (5). Even without the T_1-condition, (1), (2) and (4) remain equivalent, for it is not hard to prove (4) directly from (2), bypassing (3). Nevertheless, property (3) is important because it was one of the most classically observed facts about the unit interval [0, 1]. Using it we get the following important result.

(1.5) Theorem: Every countably compact metric space is second countable.

Proof: We know that in a metric space, separability implies second countability (see Exercise (5.1.11)). So it suffices to prove that every countably compact, metric space is separable. Let (X, d) be such a space. For each positive real number r, let A_r be a subset of X which is maximal w.r.t. the property that every two distinct points of it are at least a distance r apart. Such a set exists by the result of Exercise (10.4.3). We claim that for each $r > 0$, the set A_r is finite. For if not then it would have an accumulation point, say, x in X, by property (3) of the last theorem. Now X being a T_1-space, evey neighbourhood of x and in particular the open ball $B\left(x, \frac{r}{2}\right)$ contains infinitely many points of A_r. However, any two distinct points of $B\left(x, \frac{r}{2}\right)$ are less than r apart, a contradiction. Thus each A_r for $r > 0$ is a finite set. Let $D = \bigcup_{n=1}^{\infty} A_{1/n}$. Then D is a countable set. We contend that D is dense in X. For this it suffices to show that D meets every open ball in X. Let $B(x, r)$ be such a ball. Choose $n \in \mathbb{N}$ such that $1/n < r$. Now if $B(x, r) \cap A_{1/n} = \emptyset$ then the distance between any two distinct points of the set $A_{1/n} \cup \{x\}$ is at least $1/n$, contradicting that $A_{1/n}$ is maximal w.r.t. this property. Thus $B(x, r) \cap A_{1/n} \neq \emptyset$ and hence $B(x, r) \cap D \neq \emptyset$. So D is a countable dense subset of X. As noted before, this proves that X is second countable. ∎

(1.6) Corollary: A metric space is compact iff it is countably compact.

Proof: Compactness always implies countable compactness whether the space is metrisable or not. For the converse, we just saw that every countably compact metric space is second countable and hence Lindelöff by Theorem (6.1.4). But in presence of the Lindelöff property, countable compactness implies compactness. ∎

As an application of the last corollary, we can prove the following result which is a generalisation of Theorem (6.1.6).

(1.7) Proposition: Every continuous, real-valued function on a countably compact space is bounded and attains its extrema.

Proof: Let $f : X \to \mathbb{R}$ be continuous and suppose X is a countably compact space. Then the range $f(X)$ is countably compact by Proposition (1.2). But $\mathbb{R}$ is a metric space and so is each of its subspaces. So by the corollary above, $f(X)$ is in fact compact. Now every compact subset of $\mathbb{R}$ is bounded and contains both its supremum and infimum as it is also closed. This fact, applied to the set $f(X)$, implies the desired conclusion about f. ∎

One of the characterisations of countably compact spaces obtained above was that in such a space, every sequence has a cluster point. By Theorem (10.1.12), there exists a subnet converging to this cluster point. But a subnet of a sequence need not always be a subsequence. Thus it does not follow that in a countably compact space, every sequence has a convergent subsequence. In fact this is not true even under the stronger hypothesis of compactness. We give a counter-example. Let $\mathbb{Z}_2$ as usual be the space $\{0, 1\}$ with the discrete topology. Let $P(\mathbb{N})$ be the power set of the set of positive integers. We let $P(\mathbb{N})$ be the indexset, and define X to be the power space $\mathbb{Z}_2^{P(\mathbb{N})}$. X is then compact by Theorem (10.4.6). Elements of X are functions from $P(\mathbb{N})$ into $\mathbb{Z}_2$. We define a sequence $\{f_n\}$ in X as follows. For $n \in \mathbb{N}$ and $A \in P(\mathbb{N})$ we let $f_n(A) = 1$ or 0 according as $n \in A$ or $n \notin A$ (note that A is a subset of $\mathbb{N}$). We claim that the sequence $\{f_n\}$ has no convergent subsequence in X. For suppose $\left\{f_{nk}\right\}_{k=1}^{\infty}$ is subsequence. Now let A be the set $\{n_1, n_3, n_5, n_7, \ldots, n_{2k+1}, \ldots\}$ (we are assuming that for $p < q$, $n_p < n_q$). Then $f_{n_k}(A) = 1$ if k is odd and 0 if k is even. So the sequence $f_{n_k}(A)$ does not converge in the coordinate space $\mathbb{Z}_2$. By Proposition (8.2.9) it follows that the sequence $\{f_{n_k}\}$ cannot converge in X. So the sequence $\{f_n\}$ has no convergent subsequence in X.

The existence of convergent subsequences is especially desirable in certain aspects of analysis, such as those where a solution to a certain problem is found by finding better and better approximations. It is therefore important to isolate this property.

(1.8) Definition: A space is said to be **sequentially compact** if every sequence in it has a convergent subsequence.

The example given above shows that in general compactness does not imply sequential compactness. Nevertheless we have,

(1.9) Proposition: A first countable, countably compact space is sequentially compact.

Proof: Suppose X is a first countable and countably compact space. Let $\{x_n\}$ be a sequence in X. By statement (4) of Theorem (1.4), $\{x_n\}$ has a cluster point, say x, in X. Now X is first countable at x and so by the result of Exercise (10.1.11), there exists a subsequence of $\{x_n\}$ which converges to x. Thus X is a sequentially compact. ∎

(1.10) Theorem: In a second countable space, compactness, countable compactness and sequential compactness are all equivalent to each other.

Proof: We already know the equivalence of compactness with countable compactness in the presence of second countability. Also sequential compactness always implies countable compactness, with the converse holding for first countable spaces as we just saw. Since every second countable space is first countable, the assertion of the theorem is clear. ∎

(1.11) Corollary: In a metric space all the three forms of compactness are equivalent.

Proof: In a metric space any one of the three forms of compactness implies second countability. The result now follows from the last theorem. ∎

Easy properties of sequentially compact spaces will be developed through the exercises. There is one property which is of particular interest. It deals with the products of sequentially compact spaces. In the example given above, the space $\mathbb{Z}_2^{P(\mathbb{N})}$ was not sequentially compact although each coordinate space $\mathbb{Z}_2$ is obviously sequentially compact. So sequential compactness is not a productive property. However, it is easy to show that it is finitely productive. Suppose X, Y are sequentially compact spaces and $\{(x_n, y_n)\}$ is a sequence in the product space $X \times Y$. Then $\{x_n\}$ and $\{y_n\}$ are sequences in X and Y respectively. However we cannot hastily take their convergent subsequence and get a convergent subsequence of $\{(x_n, y_n)\}$. The trouble is that the indices of the subsequences may not match. For example, the convergent subsequence of $\{x_n\}$ might be $\{x_1, x_3, x_5, x_7, \ldots\}$ while that of $\{y_n\}$ might be $\{y_2, y_4, y_6, y_8, \ldots\}$. In this case we cannot put them together.

However, we can overcome the difficulty if we are just a little careful. First, since X is sequentially compact, $\{x_n\}$ has a convergent subsequence say $\left\{x_{n_k}\right\}_{k=1}^{\infty}$. Now consider the sequence $\left\{y_{n_k}\right\}_{k=1}^{\infty}$ in Y. Since Y is sequentially compact, this sequence has a convergent subsequence, say $\left\{y_{n_{k_r}}\right\}_{r=1}^{r=\infty}$. Now the sequence $\left\{x_{n_{k_r}}\right\}_{r=1}^{r=\infty}$, being a subsequence of the convergent sequence $\left\{x_{n_k}\right\}_{k=1}^{\infty}$ is convergent. Since the sequence $\{x_{n_{k_r}}\}$ converges in X and $\{y_{n_{k_r}}\}$ converges in Y, we see that the sequence $\left\{(x_{n_{k_r}}, y_{n_{k_r}})\right\}_{r=1}^{r=\infty}$ converges in $X \times Y$ in the product topology because the convergence in the product topology is coordinatewise. So $X \times Y$ is sequentially compact. One can now apply induction and show that the product of any finite number of sequentially compact spaces is sequentially compact. A direct argument is also possible but the notations for the subsequences become monstrous.

A question that arises naturally is whether sequential compactness is countably productive. The answer is in the affirmative and we shall prove it not so much for the sake of the result as for the sake of an important technique in the proof. For notational simplifications we write sequences as functions f etc. from $\mathbb{N}$ instead of the usual form $\{x_n\}$ etc. (actually, by very definition, a sequence is just such a function. So we are going back to the definition by adopting the new notation). If f is a sequence in a set X then a subsequence of f will be a sequence in X of the form $f \circ g$ where $g : \mathbb{N} \to \mathbb{N}$ is a strictly monotonically increasing function. Note that $g(n) \geq n$ for all $n \in \mathbb{N}$.

(1.12) Theorem: Sequential compactness is a countably productive property.

Proof: We already dealt with the case of the product of finitely many spaces. Suppose now that $\{x_k : k \in \mathbb{N}\}$ is an indexed collection of sequentially compact spaces and let X be their topological product. Let $f : \mathbb{N} \to X$ be a sequence in X. Consider the sequence $\pi_1 \circ f$ in X_1. Since X_1 is sequentially compact, there exists a subsequence say $\pi_1 \circ f \circ g_1$ of this which converges in X_1. Let f_1 be the sequence $f \circ g_1$. Then f_1 is a subsequence of f and $\pi_1 \circ f_1$ is convergent. Now consider $\pi_2 \circ f_1$. This is a sequence in X_2. So by the sequential compactness of X_2, it has a convergent subsequence say $\pi_2 \circ f_1 \circ g_2$. Let $f_2 = f_1 \circ g_2$. Then f_2 is a subsequence of f_1 and $\pi_2 \circ f_2$ is convergent in X_2. Note that $\pi_1 \circ f_2$ is also convergent in X_1 because it is a subsequence of the convergent sequence $\pi_1 \circ f_1$. Similarly, considering the sequence $\pi_3 \circ f_2$ in X_3, we get a subsequence f_3 of f_2 such that $\pi_3 \circ f_3$ converges in X_3. Continuing in this manner inductively, we get a sequence of sequences $\{f_1, f_2, f_3, \ldots, f_n, \ldots\}$ such that (i) f_1 is a subsequence of f, (ii) for each $n \in \mathbb{N}$, f_{n+1} is subsequence of f_n and (iii) for each $n \in \mathbb{N}$, $\pi_n \circ f_n$ is convergent in X_n. Now define $g : \mathbb{N} \to X$ by $g(n) = f_n(n)$ for $n \in \mathbb{N}$.

We contend that g is a subsequence of f. Let $g_n : \mathbb{N} \to \mathbb{N}$ be the function such that $f_n = f_{n-1} \circ g_n$ for $n \geq 2$. g_1 was already defined above. For $n \in \mathbb{N}$, let $h(n) = g_1 \circ g_2 \circ \ldots \circ g_n(n)$. Then the function $h : \mathbb{N} \to \mathbb{N}$ is strictly monotonically increasing. For, $h(n+1) = g_1 \circ g_2, \ldots \circ g_n \circ g_{n+1}(n+1) = g_1 \circ g_2 \circ \ldots \circ g_n(g_{n+1}(n+1))$. Now $g_{n+1}(n+1) > n$ and so $h(n+1) > h(n)$ because each function $g_1, g_2, \ldots, g_n$ is strictly monotonically increasing. Evidently, by very construction, $g = f \circ h$. Thus g is a subsequence of f. We claim g is convergent in X. By Proposition (8.2.9) it suffices to show that for each $n \in \mathbb{N}$, $\pi_n \circ g$ is convergent in X_n. Now for a given $n \in \mathbb{N}$, all except the first $n - 1$ terms of the sequence $\pi_n \circ g$ are from the sequence $\pi_n \circ f_n$ (why?). Thus $\pi_n \circ g$ can be regarded as a subsequence of $\pi_n \circ f_n$ except for the first $n - 1$ terms. (We leave it to the reader to state this fact more precisely). But by construction, $\pi_n \circ f_n$ is convergent. Therefore $\pi_1 \circ g$ is also convergent in X_n. Since this holds for each $n \in \mathbb{N}$, g is a convergent subsequence of f in X. So X is sequentially compact. ∎

The sequence g in the proof above was defined by taking the nth term of the nth sequence for $n \in \mathbb{N}$. This technique is known as **diagonalisation,** a name that fits naturally if we write down each sequence as a row of its terms. A reader familiar with the well known Arzela-Ascoli theorem about families of equicontinuous real-valued functions on a separable metric space will recall that diagonalisation is used there too. A couple of simple illustrations of this technique will be given through the exercises.

Exercises

1.1 Prove Propositions (1.2) and (1.3).

1.2 In Theorem (1.4) prove that the statement (2) implies (4) directly, even without the T_1 axiom. (Hint: Given a sequence $\{x_n\}$, for $m \in \mathbb{N}$, let $B_m = \bar{A}_m$ where $A_m = \{x_n : n \geq m\}$. Prove that the family $\{B_m : m \in \mathbb{N}\}$ has the finite intersection property and apply the result of Exercise (10.1.10)).

1.3 Prove that sequential compactness is a weakly hereditary property.

1.4 Prove that sequential compactness is preserved under continuous functions.

1.5 Prove the Lebesgue covering lemma (Theorem (6.1.7)) using the sequential compactness of a compact metric space.

1.6 Prove that a continuous function from a compact metric space into another metric space is uniformly continuous, using the fact that the domain space is sequentially compact.

1.7 Prove that if X is a first countable Hausdorff space and A is a countably compact subset of X then A is closed in X.

1.8 Let I be an uncountable set and let X be the power space $\mathbb{Z}_2^I$. Let $Y = \{x \in X : x(i) = 0$, for all except countably many $i \in I\}$.

(a) Prove that Y is dense in X and hence that Y is not compact.

*(b) Prove that Y is sequentially compact and hence also countably compact. (Hint: Given a sequence $\{x_n\}$ in Y, for each $n \in \mathbb{N}$, let

$J_n = \{i \in I : x_n(i) \neq 0\}$ and let $J = \bigcup_{n=1}^{\infty} J_n$. Then J is a countable set and so $\mathbb{Z}_2^J$ is sequentially compact. The sequence $\{x_n\}$ can now be regarded as a sequence in $\mathbb{Z}_2^J$ a singelton space.)

1.9 In the proof of Theorem (1.2), show how $\pi_n \circ g$ can be regarded as a subsequence of $\pi_n \circ f_n$ 'except for the first $n - 1$ terms'.

1.10 Assuming that every positive real number has a unique, non-terminating decimal expansion, prove that the set of real numbers in the interval (0, 1) is uncountable. (Hint: If not, arrange it as a sequence $\{a_1, a_2, \ldots, a_n, \ldots\}$. Now construct a real number b in (0, 1) which differs from a_n in the nth decimal place).

1.11 Prove that there exists a function $F: \mathbb{R} \times \mathbb{R} \to \mathbb{R}$ with the property that for any continuous function $f: \mathbb{R} \to \mathbb{R}$ there exists $y \in \mathbb{R}$ such that $F(x, y) = f(x)$ for all $x \in \mathbb{R}$. In other words, the family of restrictions of F to horizontal lines exhausts the set of all continuous functions from $\mathbb{R}$ to $\mathbb{R}$. (Hint: Show that the set of all continuous functions from $\mathbb{R}$ to $\mathbb{R}$ has the same cardinality as $\mathbb{R}$.)

1.12 Prove, however, that a function F with the property in the last exercise can never be continuous. (Hint: Consider the function $g: \mathbb{R} \to \mathbb{R}$ defined by $g(x) = F(x, x) + 1$.)

Notes and Guide to Literature

Compactness of a closed and bounded subset of $\mathbb{R}$ was discovered by Heine and Borel whereas its sequential compactness is due to Bolzano and Weierstrass. These two are among the most classic results about the real numbers. Each is equivalent to the completeness of the real line. Yet another equivalent statement is that every infinite bounded subset of $\mathbb{R}$ has at least one limit point. Each of these three properties led to the definition of a form of compactness.

The proper subcover characterisation of countable compactness for T_1 spaces is due to Arens and Dugundji [1]. A proof of the Arzela-Ascoli theorem can be found, for example, in Goffmann [1]. Exercises (1.10) and (1.12) illustrate the use of diagonalisation arguments.

2. The Alexander Sub-base Theorem

In discussing continuity of functions we saw that although it requires that the inverse image of every open set be open, it actually suffices if the inverse image of every member of some sub-base for the codomain is open. Similarly for convergence of nets, it suffices to check the condition in the definition only for all sub-basic neighbourhoods of the point (see Exercise (10.1.3)). It is natural to inquire if something similar holds for compactness too. That is, in order to check that a space is compact, instead of showing that every open cover of it has a finite subcover, will it suffice to show that this is the case for certain special covers, say for example, those covers whose members come from a given base or a sub-base for the space? The first result in this direction is quite simple.

(2.1) Proposition: A topological space X is compact if and only if there exists a base $\mathcal{B}$ for it such that every cover of X by members of $\mathcal{B}$ has a finite subcover.

Proof: The necessity of the condition is obvious. For sufficiency assume that $\mathcal{B}$ is a base for X with the property that every cover of X by members of $\mathcal{B}$ has a finite subcover. Now let $\mathcal{U}$ be any open cover of X, not necessarily by members of $\mathcal{B}$. For each $U \in \mathcal{U}$ there exists a subfamily $\mathcal{B}_U$ of $\mathcal{B}$ such that $U = \bigcup_{B \in \mathcal{B}_U} B$. Let $\mathcal{V} = \bigcup_{U \in \mathcal{U}} \mathcal{B}_U$. Then $\mathcal{V}$ is a cover of X (since $\mathcal{U}$ is a cover of X) and moreover $\mathcal{V} \subset \mathcal{B}$. So $\mathcal{V}$ has a finite subcover say $\{V_1, V_2, \ldots, V_n\}$. For each $i = 1, 2, \ldots, n$ there exists $U_i \in \mathcal{U}$ such that $V_i \in \mathcal{B}_{U_i}$. But then clearly $V_i \subset U_i$ and so $\{U_1, U_2, \ldots, U_n\}$ is a subcover of $\mathcal{U}$. Thus X is compact. ∎

This result can be paraphrased by saying that a space is compact iff every basic open cover of it (w.r.t. some base) has a finite subcover. It is now natural to inquire if a similar result holds if we replace basic open covers by sub-basic open covers. In other words, suppose $\mathcal{S}$ is a sub-base for a space X and every cover of X by members of $\mathcal{S}$ has a finite subcover. Does it follow that X is compact? The affirmative answer is due to Alexander. The proof, however, is surprisingly more involved than that of the last proposition. The argument will resemble those used in Section 4 of the last chapter. We shall need Zorn's lemma once again. In order to apply it in a natural manner, it is convenient to translate the whole problem from open covers to families of closed sets. Of course, this translation, as we have seen before, involves nothing more than the use of De Morgan's laws.

(2.2) Definition: Let X be a topological space and $\mathcal{C}$ a family of closed subsets of X. Then $\mathcal{C}$ is said to be a **closed base** (or **a closed sub-base**) for X if the family of complements of members of $\mathcal{C}$ is a base (respectively a sub-base) for X.

The following proposition is an immediate consequence of the definitions and De Morgan's laws.

(2.3) Proposition: A family $\mathcal{C}$ of closed subsets of a space X is a closed sub-base for X if and only if the family consisting of all finite unions of members of $\mathcal{C}$ is a closed base for X. ∎

Proposition (2.1) can now be translated as follows:

(2.4) Proposition: A topological space X is compact iff it has a closed base $\mathcal{C}$ such that every subfamily of $\mathcal{C}$ having the finite intersection property has a non-empty intersection. ∎

The theorem we are after can now be stated.

(2.5) Theorem: A topological space X is compact if and only if there exists a closed sub-base $\mathcal{C}$ for X such that every subfamily of $\mathcal{C}$ having the finite intersection property has a non-empty intersection.

Proof: Once again the necessity of the condition is trivial. For sufficiency, suppose $\mathcal{C}$ is a closed sub-base for a space X with the given property. Let $\mathcal{D}$ consist of all finite unions of members of $\mathcal{C}$. Then by Proposition (2.3) above, $\mathcal{D}$ is a closed base for X and in view of the last proposition, if we could prove that every subfamily of $\mathcal{D}$ having the f.i.p. has a non-empty intersection, then it would follow that X is compact. So suppose $\mathcal{F}$ is a subfamily of $\mathcal{D}$ having the f.i.p. We have to show that $\bigcap_{F \in \mathcal{F}} F$ is nonempty.

First using Zorn's lemma, there exists a subfamily $\mathcal{G}$ of $\mathcal{D}$ such that $\mathcal{F} \subset \mathcal{G}$ and $\mathcal{G}$ is maximal w.r.t. the finite intersection property. This is a fairly standard application of the Zorn's lemma and we leave it to the reader as an exercise. Since $\mathcal{F} \subset \mathcal{G}$, if we could show that $\bigcap_{G \in \mathcal{G}} G \neq \emptyset$, then it would follow that $\bigcap_{F \in \mathcal{F}} F \neq \emptyset$, as desired. Now let $\mathcal{H} = \mathcal{G} \cap \mathcal{C}$. Then $\mathcal{H}$ is a sub-family of $\mathcal{C}$ and has f.i.p. because it is also a sub-family of $\mathcal{G}$ which has the finite intersection property. So by the hypothesis, there exists $x \in X$ such that $x \in H$ for all $H \in \mathcal{H}$. We assert that $x \in \bigcap_{G \in \mathcal{G}} G$.

Let $G \in \mathcal{G}$. Since $\mathcal{G} \subset \mathcal{D}$, we have $G \in \mathcal{D}$. By the construction of $\mathcal{D}$, there exist $C_1, C_2, \ldots, C_n \in \mathcal{C}$ such that $G = \bigcup_{i=1}^{n} C_i$. Note that each C_i is also in $\mathcal{D}$ since $\mathcal{C} \subset \mathcal{D}$. We claim that for some $i = 1, 2, \ldots, n$, $C_i \in \mathcal{G}$. For otherwise, by maximality of $\mathcal{G}$ w.r.t. the finite intersection property, for each $i \in \{1, 2, \ldots, n\}$, the family $\mathcal{G} \cup \{C_i\}$ would fail to have the f.i.p. Then for each i, there would exist $G_{i,1}, G_{i,2}, \ldots, G_{i,k_i} \in \mathcal{G}$ such that $\left(\bigcap_{r=1}^{k_i} G_{i,r}\right) \cap C_i = \emptyset$. But then the finite sub-family $\{G, G_{1,1}, G_{1,2}, \ldots, G_{1,k_1}, G_{2,1}, \ldots, G_{2,k_2}, \ldots, G_{n,1}, \ldots, G_{n,k_n}\}$ of $\mathcal{G}$ has empty intersection since $G = \bigcup_{i=1}^{n} C_i$. This contradicts the finite intersection property of $\mathcal{G}$. Thus $C_i \in \mathcal{G}$ for some $i = 1, 2, \ldots, n$. Since $\mathcal{H} = \mathcal{G} \cap \mathcal{C}$ we have $C_i \in \mathcal{H}$ and hence $x \in C_i$. So $x \in G$ since $C_i \subset G$.

Thus we have shown that $x \in G$ for all $G \in \mathcal{G}$ or in other words that $\bigcap_{G \in \mathcal{G}} G \neq \emptyset$ and as noted before this implies that X is compact. ∎

The following corollary is known as the **Alexander Sub-base Theorem.**

(2.6) Corollary: A topological space X is compact iff it has a sub-base $\mathcal{S}$ with the property that every cover of X by members of $\mathcal{S}$ has a finite subcover.

Proof: This is merely a De Morgan translation of the theorem above the way proposition (2.1) is a translation of Proposition (2.4). ∎

The Alexander sub-base theorem can be used to give an alternate and more direct proof of the Tychonoff's theorem that products of compact spaces are compact.

(2.7) Theorem: Let $\{X_i : i \in I\}$ be an indexed family of nonempty compact spaces and let X be their topological product. Then X is compact.

Proof: Let $\mathcal{S}$ be the standard sub-base for the product topology on X, that is, $\mathcal{S}$ consists of all sets of the form $\pi_i^{-1}(V_i)$ where $i \in I$ and V_i is open in X_i. To show that X is compact, it suffices to prove that every cover of X by members of $\mathcal{S}$ has a finite subcover. Let $\mathcal{U}$ be such a cover. For each $i \in I$ let $\mathcal{U}_i = \{V \subset X_i : \pi_i^{-1}(V) \in \mathcal{U}\}$. Note that if $V \in \mathcal{U}_i$ then V is open in X_i because $V = \pi_i(\pi_i^{-1}(V))$ and the projection functions are onto. If for some $i \in I$, $\mathcal{U}_i$ covers X_i then by compactness of X_i we get a finite subcover of $\mathcal{U}_i$ and the inverse images of its members under π_i will yield a finite subcover of $\mathcal{U}$. Otherwise, we have that for each $i \in I$ there exists $x_i \in X_i$ such that $x_i \notin \bigcup_{V \in \mathcal{U}_i} V$. Let $x \in X$ be the point defined by $x(i) = x_i$ for $i \in I$. Then $x \in U$ for some $U \in \mathcal{U}$ because $\mathcal{U}$ is a cover of X. U is of the form $\pi^{i^{-1}}(V_i)$ for some $i \in I$ and some open set V_i in X_i. But then $x_i = \pi_i(x) \in V_i$ and $V_i \in \mathcal{U}_i$, a contradiction. So only the first possibility holds, namely, that for some $i \in I$, $\mathcal{U}_i$ is a cover of X_i and we have already shown how it implies that $\mathcal{U}$ has a finite subcover. ∎

Another interesting application of the sub-base theorem involves the order topology on a linearly ordered set X. Recall that, by definition, this is the topology which has a sub-base consisting of sets of the form L_a and R_a for $a \in X$ where $L_a = \{x \in X : x < a\}$ and $R_a = \{x \in X : a < x\}$.

(2.8) Theorem: A linearly ordered set is compact (w.r.t. the order topology) iff the order is complete and bounded.

Proof: Let $<$ be a linear order on the set X. Assume X is compact w.r.t. the order topology induced by $<$. First we show that X is bounded. If X is not bounded above then the family $\{L_b : b \in X\}$ would be an open cover of X but would have no finite subcover, because if, $\{L_{b_1}, L_{b_2}, \ldots, L_{b_n}\}$ were such a finite subcover then we may assume without loss of generality that $b_1 < b_2 < \ldots < b_n$; but then $b_n \notin L_{b_i}$ for $i = 1, 2, \ldots, n$. Thus X is bounded above, i.e. has a greatest element. Similarly it has a smallest element. To show the order is complete, let S be a non-empty subset of X. We make S into a directed set by letting $a \geq b$ in S mean by $b = a$ or $b < a$ in X. Let $D : S \to X$ be the inclusion function. Then the net D has a cluster point say x in X by compactness of X. We claim x is the least upper bound of S. First $s \leq x$ for all $s \in S$, for otherwise there would exist $s \in S$ such that $x < s$, but then the open set L_s contains no term of the net D_t for $t \geq s$, contradicting the definition of a cluster point. So x is an upper bound of S. Now if x is not the least upper bound of S then there exists $y \in X$ such that y is an upper bound of S and $y < x$. The set R_y contains x but no term of the net D, a contradiction again. So x is the least upper bound of S. Thus every nonempty set has a least upper bound in X, whence X is order complete.

For the converse, assume X is bounded and order complete. Let α be the smallest element of X. Now let $\mathcal{S}$ be the defining sub-base for the order topology on X. Elements of $\mathcal{S}$ are of the form L_a for $a \in X$ or R_b for some $b \in X$. Now suppose $\mathcal{U}$ is a sub-family of $\mathcal{S}$ covering X. Write $\mathcal{U} = \mathcal{U}_1 \cup \mathcal{U}_2$ where elements of $\mathcal{U}_1$ are of the form L_a while those of $\mathcal{U}_2$ are of the form R_b. Note that α cannot be in any R_b. Hence $\alpha \in L_a$ for some $L_a \in \mathcal{U}_1$. So the set $A = \{a \in X : L_a \in \mathcal{U}_1\}$ is nonempty. Let β be its least upper bound, which exists because X (and hence A) is bounded above and X is order complete. Then $\beta \notin L_a$ for any $a \in A$. Thus no member of $\mathcal{U}_1$ contains β and so there exists $b \in X$ such that $R_b \in \mathcal{U}_2$ and $\beta \in R_b$. This means $b < \beta$ and so, β being the supremum of A, there exists $a \in A$ such that $b < a \leq \beta$. Clearly then $X = L_a \cup R_b$ with $L_a \in \mathcal{U}_1$, $R_b \in \mathcal{U}_2$. Thus $\mathcal{U}$ has a finite subcover (in fact a subcover consisting of only two members). So every sub-basic open cover of X has a finite subcover whence X is compact by the Alexander sub-base theorem. ∎

In particular we see that the unit interval is compact in the usual topology because the usual topology on it coincides with the order topology. In Chapter 6 we established the compactness of $[0, 1]$ through its connectedness.

Exercises

2.1 Prove Propositions (2.3) and (2.4).

2.2 Prove the part involving Zorn's lemma in the proof of Theorem (2.5).

2.3 Prove that a space is Tychonoff iff it can be embedded into a compact, Hausdorff space.

2.4 Prove that a subset of a Euclidean space is compact iff it is closed and bounded. (Hint: Every bounded subset is contained in a box whose sides are closed and bounded intervals.)

2.5 Let X be a compact space and let $\mathcal{F}$ be a family of closed subsets of X. Suppose $\bigcap_{F \in \mathcal{F}} F$ is contained in some open subset U of X. Prove that there exist $F_1, F_2, \ldots, F_n \in \mathcal{F}$ such that $\bigcap_{i=1}^{n} F_i \subset U$.

2.6 Let X be a compact, Hausdorff space and $\mathcal{F}$ be a family of closed subsets of X. Prove that if the intersection of every finite subfamily of $\mathcal{F}$ is connected then so is the intersection of all members of $\mathcal{F}$. (Hint: Use the last exercise.)

2.7 Let X be the cartesian product of the unit interval $[0, 1]$ with itself. Put the lexicographic ordering on X (see Example 8, Chapter 4, Section 2). Prove that in the order topology, X is compact, connected and Hausdorff. Also show that X is first countable but not separable. Is X metrisable?

2.8 Prove that a topological space is Lindelöff iff it has a base with the property that every basic open cover has a countable subcover.

*2.9 Let $\mathcal{S}$ be the set of all intervals of the form $[a, b]$ for $a < b \in \mathbb{R}$. Prove that $\mathcal{S}$ generates the discrete topology on $\mathbb{R}$. Show that every cover of $\mathbb{R}$ by members of $\mathcal{S}$ has a countable subcover although the space $\mathbb{R}$ with the discrete topology is not Lindelöff.

Notes and Guide to Literature

Exercises (2.5) to (2.7) are taken directly from Kelley [1], Chapter 5, where the reader will also find many other interesting and important facts about compactness. Exercise (2.9) shows that the analogue of the Alexander sub-base theorem fails if compactness is replaced by the Lindelöff property. It will be instructive for the reader to find out precisely where the proof breaks down.

3. Local Compactness

In this section we localise the concept of compactness.

(3.1) Definition: A topological space X is said to be **locally compact** at a point $x \in X$ if x has a compact neighbourhood in X. X is called **locally compact** if it is locally compact at every point.

An alert reader will notice that this definition is in contrast with the comments we made about localisation of a topological property when we defined local connectedness at a point (Chapter 6, Section 3). There it was required that the point in question possess not just one connected neighbourhood but a whole bunch of arbitrarily small connected neighbourhoods. It would therefore be more logical to say that X is locally compact at a point x, if given any neighbourhood N of x there exists a compact neighbourhood M of x such that $M \subset N$. This would be much stronger than our definition. According to our definition any compact space is locally compact. We shall see in the next section that there exists a compact space $\mathbb{Q}^*$ which contains $\mathbb{Q}$ (with the usual topology) as an open subspace. Then at any point of $\mathbb{Q}$, $\mathbb{Q}^*$ would be locally compact according to our definition although there exists no compact neighbourhood of a point of $\mathbb{Q}$ which is contained in $\mathbb{Q}$. This example shows that in general, our requirement is strictly weaker than requiring that the point in question possess arbitrary small compact neighbourhoods. However, it turns out that for a fairly large class of topological spaces, the weaker version implies the stronger one.

(3.2) Proposition: If a space X is regular and locally compact at a point $x \in X$ then x has a local base consisting of compact neighbourhoods.

Proof: Let X be regular and locally compact at x. Then x has a compact neighbourhood C in X. Now let U be any neighbourhood of x. By regularity, there exists a closed neighbourhood M of x such that $M \subset U$ (see Proposition (7.1.15)). Now $M \cap C$ is a neighbourhood of x contained in U. Also $M \cap C$ is a closed subset of M (since C is a closed subset of X). Since M is compact, $M \cap C$ is compact by Proposition (6.1.10). So x has arbitrarily small compact neighbourhoods in X, or in other words such neighbourhoods form a local bases at x. ∎

A slight modification of the argument above is applicable if we replace regularity by the Hausdorff property.

(3.3) Proposition: Assume X is Hausdorff and locally compact at a point x in X. Then the family of compact neighbourhoods of x is a local base at x.

Proof: Once again let C be a compact neighbourhood of x and let U be a given neighbourhood of x in X. Now C, in the relative topology, is a compact, Hausdorff space and is therefore regular by Theorem (7.2.6). $C \cap U$ is a neighbourhood of x in X and hence also in C. So by regularity of C, there exists a closed neighbourhood D of x in C such that $D \subset C \cap U$. D being a closed subset of the compact set C, is itself compact. Moreover it is a neighbourhood of x relative to X as well because C itself is a neighbourhood of x relative to X (Prove!). Clearly $D \subset U$ and this completes the proof. ∎

The following corollary is often useful.

(3.4) Corollary: Every locally compact, Hausdorff space is regular.

Proof: Suppose X is locally compact and Hausdorff. Let $x \in X$. By the proposition above, the family of all compact neighbourhoods of x is local base at x. But in a Hausdorff space, every compact set is closed by Corollary (7.2.2). So x has a local base consisting of closed neighbourhoods. By Proposition (7.1.15) this means that X is regular. ∎

Useful as the corollary is, it is not the best possible result. It can be proved that a locally compact, Hausdorff space is in fact completely regular. We could do so right away, but this would follow much more naturally from a result of the next section. It is however not true that a locally compact, Hausdorff space is necessarily normal.

As examples of locally compact spaces, we already noted that every compact space is locally compact. As other examples all discrete spaces are locally compact (from which it follows that local compactness is not always preserved under continuous functions, for every space can be expressed as the continuous image of a discrete space). The real line and in fact any euclidean space is locally compact. However, the infinite dimensional product space $\mathbb{R}^{\mathbb{N}}$, is not locally compact. The space of rationals with the usual topology is not locally compact at any point. Thus local compactness is not a hereditary property. It is easy to check, however, that it is weakly hereditary. The following proposition gives another source of locally compact spaces.

(3.5) Proposition: An open subspace of a locally compact, regular space is locally compact.

Proof: Let X be locally compact and regular and suppose Y is open in X. Let $y \in Y$. Then Y is a neighbourhood of y in X and so by Proposition

(3.2) there exists a compact neighbourhood C of y such that $C \subset Y$. But then, Y is locally compact at y. Since y was an arbitrary point of Y, it follows that Y is locally compact in the relative topology on it. ∎

There is a sort of converse to the last proposition in case the ambient space is Hausdorff and the subspace is dense.

(3.6) Proposition: Let X be a Hausdorff space and Y be a dense subset of X. If Y is locally compact in the relative topology on it, then Y is open in X.

Proof: Let $y \in Y$ and let C be a compact neighbourhood of y in Y. Then C is of the form $D \cap Y$ where D is a neighbourhood of y in X. Let G be the interior of D in X. Clearly $y \in G$. We claim $G \subset Y$. First note that $G \subset \overline{G \cap Y}$ (the closure being in X), for if $z \in G$ and H is any open neighbourhood of z in X then $G \cap H$ is a non-empty open subset of X and so $G \cap H \cap Y$ is non-empty as Y is dense in X. But this means that every open nbd of z meets $G \cap Y$ and so $z \in \overline{G \cap Y}$. Now, $G \cap Y \subset D \cap Y = C$. But C, being a compact subset of the Hausdorff space X is closed in X and so $\overline{G \cap Y} \subset C$. So $G \subset C \subset Y$. Thus we have shown that Y contains an open neighbourhood of each of its points. Therefore Y is open in X. ∎

Putting the last two propositions together, we get the

(3.7) Theorem: A subspace of a locally compact, Hausdorff space is locally compact iff it is open in its closure.

Proof: Let X be a locally compact, Hausdorff space and Y be a subspace of X. Put $Z = \overline{Y}$. Then Z itself is locally compact for it is easy to show that local compactness is weakly hereditary. Also Z is Hausdorff and therefore regular by Corollary (3.4). Now if Y is open in Z then by Proposition (3.5) Y is locally compact. Conversely if Y is locally compact, then by the last proposition applied to Z, we see that Y is open in Z. ∎

This theorem is useful in telling whether a given subspace of a euclidean space is locally compact or not. For example, we see immediately that the space of rationals with the usual topology is not locally compact.

Exercises

3.1 Prove that local compactness is preserved under continuous, open functions and also that it is a weakly hereditary property.

3.2 Prove that a non-empty product of topological spaces is locally compact iff each coordinate space is locally compact and all except finitely many of them are compact.

3.3 Prove that in a locally compact, regular space, every point has an open neighbourhood whose closure is compact.

3.4 Let X be an infinite set with a distinguished point x_0. Let $\mathcal{T}$ consist of the empty set and all subsets of X containing x_0. Prove that $(X; \mathcal{T})$ is a locally compact space although no point has a neighbourhood whose closure is compact.

3.5 Let A be a compact subset of a locally compact, regular space X and let U be an open set containing A. Prove that there exists a closed, compact subset V of X such that $A \subset \text{int}(V)$ and $V \subset U$.

*3.6 Prove that every locally compact, regular space is completely regular.

3.7 Give an example of a locally compact, Hausdorff space which is not normal. (Hint: Consider the space Z in Exercise (8.2.11).)

3.8 Let $X = ((0, 1] \times [0, 1]) \cup \{(0, 0)\}$ with the relativised topology of the plane. Find at what points X is locally compact.

3.9 Prove that a countable, connected, T_2-space with at least two points cannot be locally compact. (Hint: see Exercise (7.3.8).)

Notes and Guide to Literature

Some authors, for example Hocking and Young [1], require local compactness at a point to mean the existence of a neighbourhood whose closure is compact. As exercise (3.4) shows, this condition is strictly stronger than our definition. However, for regular spaces, the two notions coincide.

4. Compactifications

Given a topological space X, it is often convenient to embed X into a compact space Y. Properties of the larger space sometimes give us a new insight into those of X itself. Although theoretically anything that can be done with the help of Y can also be done without it, the use of the larger space often simplifies the argument. A reader familiar with complex analysis knows that in order to study the behaviour of a complex analytic function $f(z)$ on the complex plane $\mathbb{C}$, as $z \to \infty$, it is convenient to think of $\mathbb{C}$ as a subspace of the Riemann sphere S^2, with the 'North pole' corresponding to the 'point at infinity'. S^2 is obtained by adjoining ∞ to $\mathbb{R}^2$, or by augmenting $\mathbb{R}^2$. More generally, we have seen that S^n can be obtained by adjoining one point to $\mathbb{R}^n$. This is effectively the content of Exercise (6.2.4).

We now generalise this construction to any arbitrary topological space $(X, \mathcal{T})$ instead of $\mathbb{R}^n$. Let ∞ be any symbol not in X and let $X^+ = X \cup \{\infty\}$. We define a topology $\mathcal{T}^+$ on X^+ as follows. Let $\mathcal{T}^+$ consist of all members of $\mathcal{T}$ and those subsets G of X^+ such that $\infty \in G$ and $X^+ - G$ is a compact and closed subset of X (w.r.t. $\mathcal{T}$). We verify that $\mathcal{T}^+$ is in fact a topology on X^+ and study some of the properties of the space $(X^+, \mathcal{T}^+)$.

(4.1) Proposition: With the notation above, $\mathcal{T}^+$ is a topology on X^+. $\mathcal{T}^+/X$ is the topology $\mathcal{T}$ on X. The space $(X^+, \mathcal{T}^+)$ is compact. X is dense in X^+ iff X is not compact.

Proof: Let $\mathcal{T}_1 = \{G \subset X^+ : \infty \in G \text{ and } X^+ - G \text{ is closed and compact in } X\}$. Then $\mathcal{T}^+ = \mathcal{T} \cup \mathcal{T}_1$. Clearly the empty set $\emptyset \in \mathcal{T}^+$ since $\emptyset \in \mathcal{T}$. Also $X^+ \in \mathcal{T}_1$ because $X^+ - X^+$ is the empty set which is compact and closed in X. So $X^+ \in \mathcal{T}^+$. To show that $\mathcal{T}^+$ is closed under finite intersections, it suffices to show that whenever $G, H \in \mathcal{T}^+$, $G \cap H \in \mathcal{T}^+$. There are several possibilities. If both $G, H \in \mathcal{T}$ then clearly $G \cap H \in \mathcal{T}$ and so

$G \cap H \in \mathcal{T}^+$. If one of them, say, $G \in \mathcal{T}$ and $H \in \mathcal{T}_1$ then $G \cap H = G \cap (H - \{\infty\})$ because $\infty \notin G$. Now $H - \{\infty\}$ is an open subset of X since its complement in X equals $X^+ - H$ which is given to be closed in X. So $H - \{\infty\} \in \mathcal{T}$ and thus $G \cap H \in \mathcal{T} \subset \mathcal{T}^+$. Finally suppose both $G, H \in \mathcal{T}_1$. Then $\infty \in G \cap H$. Also $X^+ - (G \cap H) = (X^+ - G) \cup (X^+ - H)$ which is closed and compact in X, as $X^+ - G$, $X^+ - H$ are so. Thus in this case, $G \cap H \in \mathcal{T}_1 \subset \mathcal{T}^+$. In any case $G \cap H \in \mathcal{T}^+$. Lastly, suppose $\{V_i : i \in I\}$ is an indexed family of members of $\mathcal{T}^+$ and $V = \bigcup_{i \in I} V_i$. If $V_i \in \mathcal{T}$ for all $i \in I$ then $V \in \mathcal{T}$, because $\mathcal{T}$ is closed under arbitrary unions. Otherwise there exists $j \in I$ such that $V_j \in \mathcal{T}_1$. Now $V_j \subset V$ and so $\infty \in V$. Also $X^+ - V = \bigcap_{i \in I} (X^+ - V_i) = \bigcap_{i \in I} (X - V_i)$ since $\infty \notin X - V$. Each $X - V_i$ is closed in X whether $\infty \in V_i$ or not. So $X^+ - V$ is a closed subset of X. But $X^+ - V \subset X^+ - V_j$ which is given to be compact. So $X^+ - V$ is itself compact, showing $V \in \mathcal{T}_1 \subset \mathcal{T}^+$. Thus we have shown that $\mathcal{T}^+$ is closed under arbitrary unions. This completes the verification that $\mathcal{T}^+$ is a topology on X^+. If $H \in \mathcal{T}$ then clearly $H = H \cap X$ while if $H \in \mathcal{T}_1$ then $\infty \in H$ and so $X \cap H = H - \{\infty\}$ which is open in X as we had seen above. Thus $H \cap X \in \mathcal{T}$ for any $H \in \mathcal{T}^+$. So $\mathcal{T}^+/X$ is precisely $\mathcal{T}$.!

Now we prove the compactness of X^+. Let $\mathcal{U}$ be an open cover of X. Then $\infty \in U$ for some $U \in \mathcal{U}$. $X^+ - U$ is a compact subset of X and hence of X^+. So $X^+ - U$ can be covered by finitely many numbers of $\mathcal{U}$. These, along with U give a finite subcover of $\mathcal{U}$. Thus X^+ is compact. Note that the set $\{\infty\}$ is open in X^+ iff X is compact. Since $\{\infty\}$ is the only non-empty set not meeting X, it follows that X is dense in X iff the set $\{\infty\}$ is not open, i.e. iff X is not compact. ∎

We remarked after the proof of Theorem (5.2.6) that sometimes the topology can be described more easily by specifying the neighbourhoods of various points. The present construction provides a good example of this. Here, the $\mathcal{T}^+$-neighbourhoods of points of X are the same as their $\mathcal{T}$-neighbourhoods (along with their supersets obtained by adding the point ∞) while the neighbourhoods of ∞ are those subsets whose complements in X^+ are closed and compact in X.

(4.2) Definition: The space X^+ is called the **one-point** or **Alexandroff compactification** of X.

There is a minor logical difficulty with this definition. In defining X^+ we let ∞ be any symbol not in X. Thus X^+ depends on the choice of this symbol and unless a particular choice is specified the definition is ambiguous. There are two ways of handling this. It is one of the axioms of set theory that no set can belong to itself. So we can always let ∞ be X itself. This canonical choice makes the definition free of ambiguity. Alternatively, it is easy to show that if α, β are two symbols not in X then the spaces $X \cup \{\alpha\}$ and $X \cup \{\beta\}$ with the topologies defined in the manner above are homeomorphic to each other in such a way that points of X correspond to themselves while α corresponds to β. In other words, the one-point compactification is well-defined (i.e. independent of the choice of the added symbol) upto a homeomorphism.

It is easy to show that the one-point compactification of the euclidean space $\mathbb{R}^n$ is homeomorphic to S^n. More generally we have

(4.3) Proposition: If Y is a compact, Hausdorff space and y_0 is any point of Y then Y is homeomorphic to the Alexandroff compactification of the space $Y - \{y_0\}$.

Proof: Let X be $Y - \{y_0\}$ with the subspace topology and let X^+ be the Alexandroff compactification of X. Define $f: Y \to X^+$ by $f(x) = x$ for $x \in X$ and $f(y_0) = \infty$. Evidently f is a bijection. We contend it is a homeomorphism. Now f/X is obviously continuous and since X is an open subset of Y, we see that f is continuous at each point of Y other than y_0. To establish continuity at y_0, we take an open set, say, G containing $\infty = f(y_0)$ in X^+. Then $Y - f^{-1}(G) = f^{-1}(X^+ - G) = X - G$ is a compact subset of X and hence closed in Y as Y is a T_2 space. So $f^{-1}(G)$ is open in Y. This completes the proof of continuity of f. To show f is open, let G be an open subset of Y. If $y_0 \notin G$ then $G \subset X$ and $f(G) = G$ is open in X and hence also in X^+ by its construction. If $y_0 \in G$ then $\infty \in f(G)$ and $X^+ - f(G) = Y - G$ which is closed and so compact since Y is compact. So $f(G)$ is open in X^+. In any case $f(G)$ is open. Hence f is a homeomorphism. ∎

An important case where this proposition is applicable is when Y is a compact metric space. Removal of any one point from Y gives a space whose Alexandroff compactification is homeomorphic to Y. As a concrete example, if we remove the 'junction' point from the figure eight curve we get a space homeomorphic to the coproduct of two copies of the real line, or the space $\mathbb{R} \times \mathbb{Z}_2$. Thus we see that the Alexandroff compactification of $\mathbb{R} \times \mathbb{Z}_2$ is the figure eight curve.

It is evident that the topological properties of a space X and its one-point compactification X^+ are intimately related to each other. For example, it is clear that X is T_1 iff X^+ is T_1. The following theorem tells precisely when X^+ is T_2.

(4.4) Theorem: The one-point compactification of a space is Hausdorff iff the space is locally compact and Hausdorff.

Proof: Let X^+ be the one-point compactification of a space X. Then X is a subspace of X^+ and so whenever X^+ is T_2 so is X. Further, X is an open subspace of X^+. In case X^+ is T_2, it is a locally compact, Hausdorff space and so X is locally compact by Proposition (3.5). This establishes the necessity of the condition. Conversely suppose that X is locally compact and Hausdorff. Let x, y be distinct points of X. If neither x nor y equals ∞, then $x, y \in X$ and so there exist open sets U, V in X such that $x \in U$, $y \in V$ and $U \cap V = \Phi$. But U, V are also open sets in X^+. So x, y can be separated in X^+. If one of them, say $x = \infty$, then $y \in X$. By local compactness of X, there exists a compact neighbourhood C of y in X. Since X is Hausdorff, C is closed in X. So the set $X^+ - C$ is an open neighbourhood of x. We let $U = X^+ - C$, $V =$ interior of C in X. Then V is open in X and also in X^+. Finally $U \cap V = \emptyset$. Thus in any case we have shown that x and y can be separated in X^+ by mutually disjoint open sets. Hence X^+ is a Hausdorff space. ∎

As a corollary we get

(4.5) Corollary: Every locally compact, Hausdorff space is completely regular (and hence a Tychonoff space).

Proof: Let X be a locally compact, T_2-space and let X^+ be its Alexandroff compactification. By the theorem above, X^+ is a Hausdorff space. Since X^+ is also compact, it is a T_4 space by Corollary (7.2.10) and hence also completely regular by Corollary (7.3.6). Since complete regularity is a hereditary property, it follows that X is completely regular. ∎

It is interesting to note that although the Alexandroff compactification X^+ appeared in the proof, the statement of the corollary does not involve it at all. As we remarked in the last section, it is possible to prove this corollary directly. But the proof above illustrates the convenience gained by embedding the given space into a compact space. There are many other ways of embedding a space into a compact space, and we give them a general name.

(4.6) Definition: Let X be a topological space. Then a **compactification** of X is a pair (e, Y) where Y is a compact space and $e: X \to Y$ is an embedding such that $e(X)$ is dense in Y (in other words, e is an embedding of X onto a dense subspace of Y). If $Y - e(X)$ consists of n points, (e, Y) is called an ***n*-point compactification.**

The requirement that $e(X)$ be dense in Y is not a stringent one, because when this is not the case, one can replace Y by $\overline{e(X)}$ which is also a compact space. What happens in $Y - \overline{e(X)}$ is, in certain respects, irrelevant as far as X is concerned and the condition that $e(X)$ be dense in Y is there merely to weed out these irrelevant happenings.

When X is a non-compact space, we know that it is a dense subspace of X^+ and so the pair (i, X^+) where $i: X \to X^+$ is the inclusion map is a compactification of X. Thus the Alexandroff compactification is a foremost example of a compactification. Strictly speaking, the term 'compactification' should not be used for X^+ when X is compact, because in such a case X is not dense in X^+. Nevertheless the term stays as a matter of convention. Anyway, the Alexandroff compactification of a space rarely comes into the picture when the space itself is compact.

The main advantage of the Alexandroff compactification is that it is defined for all spaces and is easy to describe in terms of the original space. There is another compactification called the Stone-Čech compactification which, although defined only for a rather restricted class of topological spaces, plays a crucial role in the theory of real-valued functions. We proceed to define it.

Let X be a Tychonoff space and let $\mathscr{F}$ be the family of all continuous functions from X into $[0, 1]$. In Chapter 9, we saw that $\mathscr{F}$ distinguishes points and points from closed sets and consequently that the evaluation map $e : X \to [0, 1]^{\mathscr{F}}$ defined by $e(x)(f) = f(x)$ for $f \in \mathscr{F}$, $x \in X$ is an embedding. The cube $[0, 1]^{\mathscr{F}}$ is compact by Theorem (2.7). So $\overline{e(X)}$ is a compact space. We denote it by $\beta(X)$ and we regard e as a map from X into $\beta(X)$ (although strictly speaking, when the codomain is changed, the function changes even though its graph remains the same.)

(4.7) Definition: The pair $(e, \beta(X))$ where X is a Tychonoff space and $\beta(X)$ is as above is called the **Stone-Čech compactification** of X.

The Stone-Čech compactification is defined for all Tychonoff spaces. In case X is itself compact, $e(X)$ is compact and hence a closed subset of $[0, 1]^{\mathscr{F}}$ which is a Hausdorff space. Thus in such a case $\beta(X) = e(X)$ which is homeomorphic to X itself. However, when X is not compact, $\beta(X)$ is in general much larger than $e(X)$. The remainder, that is the space $\beta(X) - e(X)$, has been an area of much mathematical activity.

Our main interest in the Stone-Čech compactification stems from an important extension property it possesses. Let X be a Tychonoff space, Y some other space and suppose $f : X \to Y$ is continuous. Let $(e, \beta(X))$ be the Stone-Čech compactification of X. By means of the embedding e, we identify X with $e(X)$. f can then be regarded as a map from $e(X)$ into Y and the problem is to extend it continuously to a map from $\beta(X)$ to Y. If we do not identify X with $e(X)$ then the problem amounts to asking whether there exists a map $g : \beta(X) \to Y$ such that $g \circ e = f$.

The general remarks we made about extension problems in Chapter 5 Section 3 and in Chapter 7, Section 4 apply here too. In general the extension problem has no solution. However, in the special case where Y is a compact, Hausdorff space, a solution exists and is unique. We shall prove this first for the case where Y is the unit interval $[0, 1]$.

(4.8) Lemma: Let X be a Tychonoff space, $(e, (\beta(X))$ its Stone-Čech compactification and suppose $f : X \to [0, 1]$ is continuous. Then there exists a map $g : \beta(X) \to [0, 1]$ such that $g \circ e = f$, i.e. g is an extension of f to $\beta(X)$, if we identify X with $e(X)$.

Proof: Let $\mathscr{F}$ be the family of all continuous functions from X into $[0, 1]$. Then $\beta(X) \subset [0, 1]^{\mathscr{F}}$. We define g on the entire cube $[0, 1]^{\mathscr{F}}$ by $g(\lambda) = \lambda(f)$ for $\lambda \in [0, 1]^{\mathscr{F}}$. This is well-defined because an element of $[0, 1]^{\mathscr{F}}$ is a function from $\mathscr{F}$ into $[0, 1]$ and can be evaluated at f since $f \in \mathscr{F}$. Equivalently g is nothing but the projection π_f from $[0, 1]^{\mathscr{F}}$ onto $[0, 1]$, and hence is continuous. Now if $x \in X$ then, by the definition of the evaluation map, $e(x) \in [0, 1]^{\mathscr{F}}$ is the function $e(x) : \mathscr{F} \to [0, 1]$ such that $e(x)(h) = h(x)$ for $h \in \mathscr{F}$. Now $g \circ e(x) = g(e(x)) = e(x)(f) = f(x)$ for all $x \in X$. So $g \circ e = f$. Thus we extended f not only to $\beta(X)$ but to the entire cube $[0, 1]^{\mathscr{F}}$. Its restriction to $\beta(X)$ proves the lemma. ∎

We are now ready for the main theorem due to Stone and Čech.

(4.9) Theorem: Any continuous function from a Tychonoff space into a compact, Hausdorff space can be extended continuously over the Stone-Čech compactification of the domain. Moreover such an extension is unique.

Proof: Let X be a Tychonoff space, $\beta(X)$ its Stone-Čech compactification and $f : X \to Y$ a map where Y is a compact, Hausdorff space. Let $\mathcal{F}$, $\mathcal{G}$ be respectively the families of all continuous functions from X, Y respectively to the unit interval $[0, 1]$ and let e, e' be the embeddings of X, Y into $[0, 1]^{\mathcal{F}}$ and $[0, 1]^{\mathcal{G}}$ respectively. For any $g \in \mathcal{G}$, let $\pi_g : [0, 1]^{\mathcal{G}} \to [0, 1]$ be the corresponding projection. Then $\pi_g \circ e' \circ f$ is a map from X into $[0, 1]$ and so by the lemma above, it has an extension say θ_g to $\beta(X)$. Then $\theta_g \circ e = \pi_g \circ e' \circ f$. Now consider the family $\{\theta_g : g \in \mathcal{G}\}$ of maps from $\beta(X)$ into $[0, 1]$. Let $\theta : \beta(X) \to [0, 1]^{\mathcal{G}}$ be the evaluation map determined by this family. We claim that $\theta \circ e = e' \circ f$. For, let $x \in X$. Then $\theta(e(x))$ is an element of $[0, 1]^{\mathcal{G}}$ given by $\theta(e(x))(g) = \theta_g(e(x))$ (by the definition of the evaluation function). But $\theta_g(e(x)) = \pi_g(e'(f(x)) = e'(f(x))(g)$. Thus for all $g \in \mathcal{G}$, $[\theta \circ e(x)](g) = [e' \circ f(x)](g)$ and so $\theta \circ e = e' \circ f$ as claimed. Now $\theta(e(X)) = e'(f(X)) \subset e'(Y)$. Since Y is compact, $e'(Y)$ is compact and hence a closed subset of $[0, 1]^{\mathcal{G}}$. So $\overline{\theta(e(X))} \subset e'(Y)$. But since θ is continuous, $\theta(\beta(X)) = \theta(\overline{e(X)}) \subset \overline{\theta(e(X))}$. Thus we see that θ maps $\beta(X)$ into $e'(Y)$. Since e' is an embedding there exists a map $e_1 : e'(Y) \to Y$ which is an inverse to e' regarded as a map from Y onto $e'(Y)$. Then $e_1 \circ \theta$ is a desired extension of f. For $e_1 \circ \theta \circ e = e_1 \circ e' \circ f = f$. Uniqueness of the extension is immediate in view of the fact that Y is a Hausdorff space and $e(X)$ is dense in $\beta(X)$ (see Exercise (7.1.9)). ∎

The argument above may appear a little intricate at first sight. However the essential idea is simple. We embed Y into a certain cube. The composition of the given function with each projection is first extended to $\beta(X)$ by the lemma. This way we get a function from $\beta(X)$ into the cube and using compactness of Y, we show that the range of this function stays within $\beta(Y)$, which is just a copy of Y.

It is instructive to reformulate the theorem above. For this we introduce an order and an equivalence relation on the class of all possible compactifications of a space.

(4.10) Definition: Let (e, Y) and (f, Z) be two compactifications of a topological space X. Then (e, Y) is said to be **greater than** (f, Z) (written $(e, Y) \geq (f, Z)$) if there exists a map $g : Y \to Z$ such that $g \circ e = f$.

(4.11) Definition: Let (e, Y) and (f, Z) be two compactifications of a space X. Then they are said to be **topologically equivalent** if there exists a homeomorphism $h : Y \to Z$ such that $h \circ e = f$.

The following proposition, whose simple proof is left as an exercise summarises a few properties of these relations.

(4.12) Proposition: The relation $\geqq$ is reflexive and transitive while the relation of topological equivalence of compactifications is indeed an equivalence relation. Moreover it is compatible with $\geqq$ in the sense that if (e, Y) is topologically equivalent to (e', Y') and (f, Z) to (f', Z') then $(e, Y) \geqq (f, Z)$ iff $(e', Y') \geqq (f, Z')$. ∎

Thus it follows that we can define $\geqq$ on the set of all equivalence classes of compactifications of a space. Even then $\geqq$ need not be an anti-symmetric relation. However, as far as Hausdorff compactifications are concerned we can prove the following result.

(4.13) Proposition: If (e, Y) and (f, Z) are Hausdorff compactifications (that is, compactifications with Y, Z Hausdorff) of a space X and $(e, Y) \geqq (f, Z) \geqq (e, Y)$ then (e, Y) is topologically equivalent to (f, Z).

Proof: By hypothesis, there exist continuous functions $h : Y \to Z$ and $k : Z \to Y$ such that $h \circ e = f$ and $k \circ f = e$. Then, $k \circ h \circ e = e$ which means that the two continuous functions $k \circ h$ and id_Y from Y into itself agree on the subset $e(X)$ of Y. But $e(X)$ is dense in the domain Y, while the range space (which is also Y) is Hausdorff. Hence $k \circ h = id_Y$ (use the result of Exercise (6.1.9)). Similarly $h \circ k = id_Z$. Thus h and k are continuous inverses of each other and hence are homeomorphisms. As we already have that $h \circ e = f$, it follows that (e, Y) is topologically equivalent to (f, Z). ∎

Now we are in a position to reformulate Theorem (4.9).

(4.14) Theorem: Among all Hausdorff compactifications of a Tychonoff space, the Stone-Čech compactification is the largest compactification upto a topological equivalence.

Proof: Let X be a Tychonoff space and $(e, \beta(X))$ its Stone-Čech compactification. Since $\beta(X)$ is a Hausdorff space, $(e, \beta(X))$ is a Hausdorff compactification. Now let (f, Y) be any Hausdorff compactification of X. Then Y is a compact, T_2 space and so by Theorem (4.9) the map $f : X \to Y$ extends to $\beta(X)$, that is, there exists $g : \beta(X) \to Y$ such that $g \circ e = f$. By definition this means that $(e, \beta(X)) \geqq (f, Y)$. Thus the Stone-Čech compactification is larger (or greater) than every Hausdorff compactification. On the other hand, the last proposition says that as far as Hausdorff compactifications are concerned, the relation $\geqq$ is anti-symmetric upto topological equivalence. Therefore upto topological equivalence, the largest compactification is unique. This is the desired conclusion. ∎

We conclude this section with a discussion of yet another compactification due to Wallman. It is defined for all T_1 spaces and throughout the remainder of this section we assume that all spaces are T_1 spaces. Given a space X and a point $x \in X$, let $\mathcal{F}_x$ be the collection of all subsets of X which contain x. Then evidently $\mathcal{F}_x$ is a filter, and in fact an ultrafilter on X. Moreover this filter converges to x w.r.t. any topology on X. It is obvious that if $x \neq y$ in X then $\mathcal{F}_x$ cannot converge to y (here we are using that X is a T_1 space). Let $F(X)$ be the collection of all such filters $\mathcal{F}_x$ for $x \in X$.

Then elements of $F(X)$ correspond uniquely to points of X and we may therefore identify X with $F(X)$. The Wallman compactification is obtained by adjoining to $F(X)$ certain other filters on X and then by defining a suitable topology on this larger collection of filters. Recall that a space is compact if and only if every ultrafilter on it is convergent. A non-convergent ultrafilter in a space may intuitively be thought of as a hole in that space and the extra filters added to $F(X)$ serve to fill up these holes.

As an analogy, the reader may recall here the construction of real numbers as Dedekind cuts of rational numbers. Every rational number determines a Dedekind cut of $\mathbb{Q}$, but because $\mathbb{Q}$ is not order-complete, there are many 'holes' in it, that is to say, there are Dedekind cuts of $\mathbb{Q}$ which do not correspond to any rational number. Any such cut, when regarded as a real number patches the hole corresponding to it. It is a commonly adopted device in mathematics. Whenever we want to complete something, we fill up the holes by adjoining suitable extra elements. We shall study another instance of this device in the next chapter where we shall discuss complete metric spaces.

With this introduction, we now construct the Wallman compactification of a space. Throughout, X denotes a T_1 space.

(4.15) Definition: A filter on X is said to be **closed** if it has a base consisting of closed subsets of X. A closed filter is said to be **ultraclosed** if it is a maximal element in the set of all closed filters on X, i.e. if it is not properly contained in any closed filter on X.

Open filters can similarly be defined although we shall have no occasion to use them. For $x \in X$, it is clear that the filter $\mathcal{F}_x$ defined above is closed because it has a base consisting of the singleton set $\{x\}$ (which is closed since X is T_1). It is also an ultraclosed filter because it is maximal not only in the collection of all closed filters but in fact in the larger collection of all possible filters on X. It is obvious that a space X is regular iff for each $x \in X$, the neighbourhood filter $\mathcal{N}_x$ is a closed filter. The existence of ultraclosed filters is guaranteed by the following proposition.

(4.16) Proposition: Every closed filter is contained in an ultra-closed filter.

Proof: The proof resembles that of Theorem (10.4.2). Let $\mathcal{F}$ be a closed filter on a space X and G the collection of all closed filters on X containing it. If $\{\mathcal{G}_i : i \in I\}$ is a non-empty chain in G, with $\mathcal{B}_i$ a base for $\mathcal{G}_i$, then $\bigcup_{i \in I} \mathcal{B}_i$ is a base for the filter $\bigcup_{i \in I} \mathcal{G}_i$ and Zorn's lemma applies. ∎

Now let $W(X)$ be the collection of all ultra-closed filters on X. As observed above the function $\Phi : X \to W(X)$ defined by $\Phi(x) = \mathcal{F}_x$ is one-to-one. We shall define a topology on $W(X)$ such that $(\Phi, W(X))$ will become a compactification of X. This topology will obviously be related to the topology on X. For a subset A of X, let A^+ be the set $\{\mathcal{F} \in W(X) : A \in \mathcal{F}\}$. Then A^+ is a subset of $W(X)$ and we have a dual relationship that for $A \subset X$ and $\mathcal{F} \in W(X)$, $A \in \mathcal{F}$ iff $\mathcal{F} \in A^+$. Clearly $\Phi^+ = \Phi$ and $X^+ = W(X)$. The transition from A to A^+ preserves finite intersections as we now show.

(4.17) Proposition: For any $A, B \subset X$, $(A \cap B)^+ = A^+ \cap B^+$.

Proof: Let $\mathcal{F} \in W(X)$. If $\mathcal{F} \in (A \cap B)^+$ then $A \cap B \in \mathcal{F}$ whence $A \in \mathcal{F}$ since A is a superset of $A \cap B$. So $\mathcal{F} \in A^+$. Similarly $\mathcal{F} \in B^+$, and so $\mathcal{F} \in A^+ \cap B^+$. This shows that $(A \cap B)^+ \subset A^+ \cap B^+$. Conversely if $\mathcal{F} \in A^+ \cap B^+$ then $\mathcal{F} \in A^+$ and $\mathcal{F} \in B^+$. This means both $A, B \in \mathcal{F}$ and hence $A \cap B \in \mathcal{F}$, showing $\mathcal{F} \in (A \cap B)^+$. ∎

Note that the proof used merely that members of $W(X)$ are filters on X and not that they are ultraclosed filters, or even that they are closed filters. However, when it comes to unions, the topology of X comes into the picture.

(4.18) Proposition: For any open sets A, B in X, $(A \cup B)^+ = A^+ \cup B^+$.

Proof: The inclusion $A^+ \cup B^+ \subset (A \cup B)^+$ is simple and hinges only on the fact that $A \cup B$ contains A as well as B. For the other way inclusion, assume $\mathcal{F} \in (A \cup B)^+$. Then $A \cup B \in \mathcal{F}$. We have to show that either $A \in \mathcal{F}$ or $B \in \mathcal{F}$. Now if $A \notin \mathcal{F}$ then no subset of A is in $\mathcal{F}$ and hence $X - A$ intersects every member of $\mathcal{F}$. But then the family $\mathcal{F} \cup \{X - A\}$ has the finite intersection property and hence generates a filter on X, which is easily seen to be a closed filter since $\mathcal{F}$ is a closed filter and $X - A$ is a closed set. By maximality of $\mathcal{F}$ among all closed filters on X, it follows that $X - A \in \mathcal{F}$. Similarly if $B \notin \mathcal{F}$ then $X - B \in \mathcal{F}$. So if neither A nor B is in $\mathcal{F}$ then $(X - A) \cap (A - B) \in \mathcal{F}$. But its complement, $A \cup B$ is also in $\mathcal{F}$, a contradiction. Thus either $A \in \mathcal{F}$ or $B \in \mathcal{F}$. In other words either $\mathcal{F} \in A^+$ or $\mathcal{F} \in B^+$. ∎

We are now in a position to define the desired topology on $W(X)$. Let $\mathcal{B} = \{U^+ \subset W(X) : U \text{ open in } X\}$. By Proposition (4.17), $\mathcal{B}$ is closed under finite intersections and so by corollary (4.3.7), is a base for a unique topology on $W(X)$. The properties of this topology are proved below.

(4.19) Theorem: The set $W(X)$ along with the topology having $\mathcal{B}$ as a base is a compact space. The function $\Phi : X \to W(X)$ defined by $\Phi(x) = \mathcal{F}_x$ is an embedding onto a dense subset of $W(X)$.

Proof: In view of Proposition (2.1), compactness of $W(X)$ will follow if it can be shown that every cover of $W(X)$ by elements of the base $\mathcal{B}$ has a finite subcover. So let $\{U_i : i \in I\}$ be an indexed family of open sets of X such that $\{U_i^+ : i \in I\}$ is a cover of $W(X)$. Now either X is covered by finitely many of U_i's or else the family $C = \{X - U_i : i \in I\}$ has the finite intersection property (by De Morgan's laws). In the first case the corresponding U_i^+'s will cover $X^+ = W(X)$ by Proposition (4.18) and we would get a desired finite subcover. In the second case we derive a contradiction as follows. The family C generates a filter $\mathcal{F}$ on X which is obviously closed since members of C are closed in X. By Proposition (4.16) $\mathcal{F}$ is contained in an ultraclosed filter $\mathcal{G}$ on X. Then $\mathcal{G} \in W(X)$. Since $\{U_i^+ : i \in I\}$ is a cover of $W(X)$, there exists $j \in I$ such that $\mathcal{G} \in U_j^+$ that is, $U_j \in \mathcal{G}$. But by the very construction of $\mathcal{G}$, $X - U_j \in \mathcal{G}$ a contradiction. Thus only the first possibility holds and $\{U_i^+ : i \in I\}$ has a finite subcover.

Continuity of $\Phi : X \to W(X)$ will follow if it is shown that the inverse image of every basic open set in $W(X)$ is open in X. So let $U^+ \in \mathcal{B}$ where U is an open subset of X. Now $\Phi(x) = \mathcal{F}_x \in U^+$ iff $U \in \mathcal{F}_x$ which is the case iff $x \in U$. Thus $\Phi^{-1}(U^+)$ is precisely U which is open in X. So Φ is continuous. We already noted before that Φ is one-to-one. Moreover Φ is open when regarded as a function from X onto $\Phi(X)$ because it is easy to show that if V is open in X then $\Phi(V) = V^+ \cap \Phi(X)$. Hence Φ is an embedding of X into $W(X)$. It only remains to prove that $\Phi(X)$ is dense in $W(X)$. For this, it again suffices to show that $\Phi(X)$ meets every non-empty member of the base $\mathcal{B}$. Let U be any non-empty open set in X. Then for any $x \in U$, $\mathcal{F}_x \in U^+$ i.e. $\Phi(x) \in U^+$ and so $\Phi(X) \cap U^+ \neq \Phi$, completing the proof. ∎

We are now justified in making the following definition.

(4.20) Definition: For a T_1-space X, the pair $(\Phi, W(X))$ as defined above is called the **Wallman compactification of** X.

The main advantage of the Wallman compactification is that it is more general than the Stone-Čech compactification which is defined only for Tychonoff spaces. Moreover, the standard base for $W(X)$ is described in a manner which is both simple and directly related to the topology on X. Properties of the Wallman compactification will be developed through exercises.

Exercises

4.1 Prove that the Alexandroff compactification of a non-empty, connected space is connected iff that space is not compact.

4.2 Prove that the one-point compactification is a topological construction in the sense that if two spaces are homeomorphic to each other then so are their one-point compactifications. Prove similar results for the other two compactifications discussed in this section.

4.3 If the Alexandroff compactifications of two spaces X and Y are homeomorphic to each other, does it follow that X is homeomorphic to Y?

4.4 Prove Proposition (4.12).

*4.5 Prove that if X is a locally compact and T_2 space then its Alexandroff compactification is the smallest among all its Hausdorff compactifications.

4.6 Prove that if (f, Z) is a Hausdorff compactification of a space X with the property that every bounded continuous real valued function on X can be extended continuously over Z then (f, Z) is equivalent to $(e, \beta(X))$. (Hint: Note first that the existence of a Hausdorff compactification of X implies that X is a Tychonoff space. By an argument similar to that of Theorem (4.9), get a map $h : Z \to \beta(X)$ such that $h \circ f = e$. Finally use Proposition (4.13).)

4.7 Let $\mathbb{N}$ be the set of positive integers with the discrete topology. Prove that $\beta(\mathbb{N})$ is not sequentially compact. In fact show that the sequence $\left\{e(n)\right\}_{n=1}^{\infty}$ has no convergent subsequence in $\beta(\mathbb{N})$.

4.8 Prove that the one-point compactification of an uncountable discrete space is not first countable and does not satisfy the countable chain condition (see Definition (6.1.17)).

4.9 If X is embeddable in Y, does it follow that X^+ is embeddable in Y^+? What about the converse?

4.10 Prove that Proposition (4.3) is no longer true if the space Y is compact and T_1. (Hint: Consider an infinite set with the cofinite topology. Also see the next exercise.)

4.11 Let $(X, \mathcal{T})$ be a T_1 space and y_0 a point not in X. Let $Y = X \cup \{y_0\}$. Let $\mathcal{T}_1$ be the collection of all cofinite subsets of Y containing y_0. Prove that $\mathcal{T} \cup \mathcal{T}_1$ is a topology on Y making it a compact T_1 space. If X is infinite, this gives a one-point compactification of X, which is in general not equivalent to the Alexandroff compactification. [Thus a T_1 space may have two one-point compactifications which are not mutually equivalent. It is therefore erroneous to refer to the Alexandroff compactification as *the* one-point compactification. By the proof of Proposition (4.3), however, a locally compact, T_2 space has only one one-point, T_2 compactification upto equivalence).

4.12 Obtain a 2-point, Hausdorff compactification of the real line with the usual topology and prove that it is unique upto equivalence among all such compactifications.

*4.13 Prove that the real line with the usual topology has no 3-point, Hausdorff compactification.

4.14 Let $\mathcal{F}$ be an ultraclosed filter on a space X. Prove that if A is a closed subset of X which meets every member of $\mathcal{F}$ then $A \in \mathcal{F}$. Prove also that this property characterises ultraclosed filters among closed filters.

4.15 Prove that a space is compact iff every ultraclosed filter in it is convergent. [Thus a non-convergent ultraclosed filter is like a 'hole' in a space.]

4.16 If X is a T_4 space, prove that its Wallman compactification is Hausdorff. (Hint: Given two distinct ultraclosed filters $\mathcal{F}$, $\mathcal{G}$, show that there exist closed sets, A, B such that $A \in \mathcal{F}$, $B \in \mathcal{G}$ and $A \cap B = \emptyset$.)

4.17 Prove that every continuous function from a T_1 space into a compact, Hausdorff space can be continuously extended to the Wallman compactification of the domain space. (Hint: Let $f: X \to Y$ be such a function. If $\mathcal{F}$ is an ultraclosed filter on X, prove by a standard compactness argument for Y that the image filter $f_{\#}(\mathcal{F})$ is convergent. Define $g(\mathcal{F})$ to be the unique limit of this filter. Prove that g extends f. For continuity of g observe that for an open subset V of Y, $g^{-1}(V) = (f^{-1}(V))^+$.)

4.18 If X is a space for which $W(X)$ is Hausdorff then prove that X is a

Tychonoff space and, that its Stone-Čech compactification is equivalent to its Wallman compactification. (Hint: Use the last exercise with Exercise (4.6).)

4.19 Let $f: X \to Y$ be a closed map where X, Y are T_1 spaces. Prove that f induces a map $f_*: W(X) \to W(Y)$ which extends f regarded as a map from X to $W(Y)$.

Notes and Guide to Literature

The three compactifications considered here are all very classic. There are many other ways of compactifying a topological space, the motivation in each case being to fill up the lacunas in the original space in some sense. For a deeper study of the Stone-Čech compactification, see Nagata [2], Smirnov [2] and Wallace [1].

The construction in the Wallman compactification can be generalised by taking, instead of all ultra-closed filters on the space, certain subcollections of them. A recent reference in this direction is Reed [1].

The standard notation for the one-point compactification of a space X is X^*. We have used X^+ for typographical convenience.

n-point compactifications are discussed in Magill [1].

Chapter Twelve

Complete Metric Spaces

Completeness of a metric is not, strictly speaking, a topological invariant, for it is sometimes possible to induce the same topology by two metrics one of which is complete while the other is not. Nevertheless some topological conditions such as compactness ensure completeness of a metric and in turn the existence of a complete metric inducing the topology of a space has interesting topological consequences. Moreover, complete metrics arise naturally and frequently in applications to analysis. In view of this, a study of topological aspects of complete metrics has a place in general topology.

1. Complete Metrics

A reader familiar with analysis has undoubtedly come across the definition of a complete metric. Nevertheless we state it here formally for the sake of completeness.

(1.1) Definition: A sequence $\{x_n\}$ in a metric space $(X; d)$ is said to be a **Cauchy sequence** if for every $\epsilon > 0$, there exists $n_0 \in \mathbb{N}$ such that for all $m, n \in \mathbb{N}$, $m \geq n_0$ and $n \geq n_0$ implies that $d(x_m, x_n) < \epsilon$.

Whether a sequence is a Cauchy sequence or not depends as much on the sequence as on the metric. For example if X is the open interval $(0, 1)$ the sequence $\left\{\frac{1}{n}\right\}_{n=1}^{\infty}$ is a Cauchy sequence w.r.t. the usual metric on X. But if we define a metric d on X by $d(x, y) = \left|\frac{1}{x} - \frac{1}{y}\right|$ then the same sequence is not a Cauchy sequence. Note that both these metrics induce the same topology on $(0, 1)$. Thus the concept of a Cauchy sequence, unlike that of a convergent sequence, is not topological in character. In other words, it is impossible to define Cauchy sequences solely in terms of open sets, without involving the metric explicitly. Nevertheless, a simple application of the triangle inequality shows that if a sequence converges w.r.t. a metric then it is also a Cauchy sequence w.r.t. that metric. That the converse is false is shown by the sequence $\{1/n\}$ in the usual metric on $(0, 1)$. Although this sequence converges to 0 in the space $[0, 1]$ it has no limit in the space $(0, 1)$. More generally, we can construct a non-convergent Cauchy sequence by taking a sequence $\{x_n\}$ in any metric space $(X; d)$, which converges to a point x in X such that no term of the sequence equals x. Then when regarded as a sequence in $X - \{x\}$, the sequence $\{x_n\}$ is a Cauchy sequence which fails to converge. Intuitively a non-convergent Cauchy sequence corresponds to a 'hole' in a metric space. Completeness of a metric space means the absence of such holes. More formally,

(1.2) Definition: A metric space $(X; d)$ is said to be **complete** if every Cauchy sequence (w.r.t. d) in X is convergent. We also say sometimes that the metric d is complete.

Inasmuch as Cauchy sequence is not a topological concept it is to be expected that completeness is not a topological invariant either. For example the open interval (0, 1) with the usual metric is not complete. However, it is not hard to show that the other metric d defined above on (0, 1) is complete. The proof depends upon the completeness of the usual metric on the real line. This is a standard fact from analysis and it is easily seen to be equivalent to many other properties of the real line, for example its order completeness.

Although completeness of a metric is not a topological property, there are some topological conditions which ensure that metrics on certain spaces are complete. As an example we have

(1.3) Proposition: Every compact metric space is complete.

Proof: Before actually starting the proof, let us see carefully what the statement says. Suppose d is a metric on a set X. Then the proposition says that if the topology induced by d renders X a compact space, then d is complete. In other words, the hypothesis is topological while the conclusion is not.

So, assume $(X; d)$ is a metric space which is compact. Then by Corollary (11.1.11), X is also sequentially compact. Now let $\{x_n\}$ be a Cauchy sequence in $(X; d)$. By sequential compactness, there exists a subsequence $\left\{x_{n_k}\right\}_{k=1}^{\infty}$ converging to some point, say y, in X, in the topology induced by d. We assert that the original sequence $\{x_n\}$ converges to y, w.r.t. d. Let $\epsilon > 0$ be given. Then since $\{x_n\}$ is a Cauchy sequence, there exists $p \in \mathbb{N}$ such that for all $m, n \geq p$, $d(x_m, x_n) < \epsilon/2$. Moreover since $x_{n_k} \to y$ as $k \to \infty$, there exists $r \in \mathbb{N}$ such that $d(x_{n_k}, y) < \epsilon/2$ for all $k \geq r$. We may assume r is large enough so that $n_r \geq p$. Then for any $n \geq n_r$, $d(x_n, y) \leq d(x_n, x_{n_r}) + d(x_{n_r}, y) < \epsilon/2 + \epsilon/2 = \epsilon$. Thus $\{x_n\}$ converges to y. Since this holds for every Cauchy sequence in $(X; d)$, $(X; d)$ is a complete metric space. ∎

The converse of the result above is false. For example, the usual metric on the real line is complete but the topology induced by it does not make $\mathbb{R}$ a compact space. With an additional hypothesis, completeness implies that the space is compact. We introduce it now.

(1.4) Definition: Let $(X; d)$ be a metric space. A subset A of X is said to be **totally bounded** (w.r.t. d) if for every $\epsilon > 0$, A can be covered by finitely many open balls of radii less than ϵ each. The metric d is said to be **totally bounded** if the entire set X is totally bounded w.r.t. it.

It is obvious that a subset of a totally bounded set as well as a finite union of totally bounded sets is totally bounded. Note that the definition does not require that the centres of the balls lie in the set itself. But this can always be assumed to be the case as we now show.

(1.5) Proposition: Let A be a subset of a metric space $(X; d)$. Then A is totally bounded w.r.t. d iff for every $\epsilon > 0$, A can be covered by finitely many open balls with centres in A and of radii less than ϵ each. Consequently A is totally bounded w.r.t. d iff it is so w.r.t. the restriction of the metric d to A.

Proof: The sufficiency part requires no proof. For the necessity part, suppose A is totally bounded w.r.t. d according to our definition. Let $\epsilon > 0$ be given. First cover A by finitely many balls, say, $\{B(x_i, r_i) : x_i \in X, i = 1, 2, \ldots, n\}$ of radius less than $\epsilon/2$ each. The centres x_i may lie possibly outside A. Evidently we may assume each ball $B(x_i, r_i)$ meets A, for otherwise we merely throw it out. So for each i, pick $a_i \in A \cap B(x_i, r_i)$. Now $B(a_i, 2r_i)$ has radius less than ϵ and contains the ball $B(x_i, r_i)$ by the triangle inequality. So the family $\{B(a_i, 2r_i) : i = 1, 2, \ldots, n\}$ is a desired family of open balls covering A. The last statement of the proposition is easily seen to be an immediate consequence of this. ∎

It is easy to show that every totally bounded metric is bounded. However, total boundedness is a much stronger concept than boundedness. If we define a metric d on $\mathbb{R}$ by $d(x, y) = \min \{1, |x - y|\}$ we get an example of a metric which is bounded but not totally bounded. Note that this metric is complete because the usual metric on $\mathbb{R}$ is complete. More generally, if d_1, d_2 are metrics on a set X and if λ is a positive constant with the property that for any $x, y \in X$, $d_1(x, y) \leqslant \lambda$ iff $d_2(x, y) \leqslant \lambda$, then $(X; d_1)$ is complete iff $(x; d_2)$ is so. The proof of this simple fact is left as an exercise. In loose but suggestive terms it says that completeness of a metric depends only on small distances.

Compactness of a metric space can now be characterised as follows:

(1.6) Theorem: A metric space is compact iff it is complete and totally bounded.

Proof: Once again, we are dealing with two sets of properties one of which is topological while the other is not. Let $(X; d)$ be a metric space. Then the theorem says that X with the induced metric topology is compact iff d is complete and totally bounded.

Assume first that X is compact. By Proposition (1.3) above, d is complete. Total boundedness of d also follows from compactness of X. Given $\epsilon > 0$, we simply cover X by all open balls of radii less than ϵ and extract a finite subcover.

For the converse, suppose d is complete and totally bounded. Let $\mathcal{U}$ be an open cover of X. If $\mathcal{U}$ has no finite subcover we shall get a contradiction as follows. Set $\epsilon = 1$ and cover X by finitely many open balls of radii less than 1, using total boundedness of d. If each of these balls can be covered by finitely many members of $\mathcal{U}$ then $\mathcal{U}$ would have a finite subcover, contradicting the hypothesis. So at least one of these balls say $B(x_1, r_1)$ cannot be covered by finitely many members of $\mathcal{U}$. Now $B(x_1, r_1)$ is also totally bounded. Setting $\epsilon = 1/2$, we cover it by finitely many open balls of radii less than 1/2 each and having centres in $B(x_1, r_1)$. Then, by an argument similar to the above, at least one of these balls say $B(x_2, r_2)$ cannot be covered by finitely many members of $\mathcal{U}$. We then repeat the argument using total boundedness of $B(x_2, r_2)$ setting $\epsilon = 1/4$ this time. Continuing in this manner, we get a sequence $\{x_n\}$ in X_n and a sequence $\{r_n\}$ of positive real numbers such that for all n, $x_{n+1} \in B(x_n, r_n)$, $r_n < 2^{1-n}$ and $B(x_n, r_n)$ cannot be covered by finitely many members of $\mathcal{U}$. The first two conditions imply that $\{x_n\}$ is a Cauchy sequence, because for $m, n \in \mathbb{N}$ with $m < n$ (say) we can write

$$\begin{aligned} d(x_m, x_n) &\leq d(x_m, x_{m+1}) + d(x_{m+1}, x_{m+2}) + \ldots + d(x_{n-1}, x_n) \\ &< \sum_{k=m}^{n-1} r_k \\ &< \sum_{k=m}^{n-1} 2^{1-k} \end{aligned}$$

and use the fact that the geometric series $\sum_{1}^{\infty} 2^{-k}$ is convergent. So by completeness of d, $\{x_n\}$ converges, say, to y in X. Then there exists $U \in \mathcal{U}$ and $r > 0$ such that $B(y, r) \subset U$. Now for all sufficiently large n we have $d(x_n, y) < r/2$. If n is further chosen so large that $r_n < r/2$ then we see that $B(x_n, r_n)$ is contained in $B(y, r)$ which is contained in a single member, U, of $\mathcal{U}$ contradicting that $B(x_n, r_n)$ cannot be covered by finitely many members of $\mathcal{U}$. This contradiction shows that $\mathcal{U}$ must have a finite subcover. Hence X is compact. ∎

We conclude this section with a discussion of some well-known complete metric spaces. We noted earlier that the usual metric on the real line is complete. From this it is easy to show that the Pythagorean metric on the euclidean space $\mathbb{R}^n$ is complete for all n. There are other ways of putting a complete metric on $\mathbb{R}^n$. Indeed any of the L^p metrics for $p \geq 1$, defined by $d((x_1, \ldots, x_n), (y_1, y_2, \ldots, y_n)) = \left\{ \sum_{k=1}^{n} |x_k - y_k|^p \right\}^{1/p}$ for $(x_1, \ldots, x_n), (y_1, \ldots, y_n) \in \mathbb{R}^n$ is complete. The proof requires, of course, that the usual metric on $\mathbb{R}$ is complete. Note that elements of $\mathbb{R}^n$ are functions from the index set $\{1, 2, \ldots, n\}$ into $\mathbb{R}$ and if we view them as such then the formula for the metric reduces to $d(f, g) = \left\{ \sum_{k=1}^{n} |f(k) - g(k)|^p \right\}^{1/p}$. It is natural to inquire if we could replace the index set $= \{1, 2, \ldots, n\}$ by some other index set, say S. Evidently there is little difficulty when S is finite. When S is infinite and f, g are functions from S into $\mathbb{R}$ then we may try to define

$d(f, g)$ as $\{\sum_{x \in S} |f(x) - g(x)|^p\}^{1/p}$. The trouble is how to define the sum. When S is countable, we may look at it as an infinite series. In other cases, a more general limiting process, namely, integration must be used. For integration to make sense the set S must be equipped with some measure μ and instead of taking all functions from S to $\mathbb{R}$ we must take only those that are integrable, or more precisely those functions $f : S \to \mathbb{R}$ for which $\int_S |f|^p \, d\mu$ is defined and finite. Restricted to such functions, the formula for d becomes $d(f, g) = \left\{\int_S |f - g|^p \, d\mu\right\}^{1/p}$. The resulting metric space is called the **L^p-space** associated with the measure space (S, μ) and is denoted by $L^p(S, \mu)$. It is a nontrivial and classic theorem of functional analysis that the L^p-spaces are complete. This theorem is called the **Riesz-Fisher theorem.**

There is yet another way of putting a metric on the set of functions. Here it is unnecessary that the range space be the real line. Instead, any metric space (X, d) would do. Let S be any set and let Y be the set of all bounded functions from S into X, that is, functions whose ranges are bounded subsets of X. For $f, g \in Y$ we define $e(f, g) = \sup \{d(f(s), g(s)) : s \in S\}$. This is a well-defined real number because f, g are bounded. It is easy to show that e defines a metric on Y, cf. Exercise (3.3.3). Note that in case d itself is a bounded metric then Y is merely the power set X^S.

Evidently the properties of the metric e will depend on those of d. As an example, we have

(1.7) Proposition: With the notations above, the metric e is complete iff the metric d is so.

Proof: Assume first that e is complete. Let $\{x_n\}$ be a Cauchy sequence in (X, d). For $n \in \mathbb{N}$, let $f_n : S \to X$ be the constant function, $f_n(s) = x_n$ for all $s \in S$. Then clearly $f_n \in Y$ and $e(f_m, f_n) = d(x_m, x_n)$ for all $m, n \in \mathbb{N}$. Thus $\{f_n\}$ is a Cauchy sequence in (Y, e). By completeness of e, $\{f_n\}$ converges to a bounded function $f : S \to X$, w.r.t. e. We assert that f is a constant function. For, otherwise, there would exist in S, points, s, t such that $f(s) \neq f(t)$. Let $r = \frac{1}{2}d(f(s), f(t))$. Then $r > 0$. Since $f_n \to f$ w.r.t. e, there exists $n \in \mathbb{N}$ such that for all $m \geq n$, $e(f_m, f) < r$. In particular, $e(f_n, f) < r$. Since $e(f_n, f)$ is the supremum of the set $\{d(f_n(u), f(u)) : u \in S\}$, in particular we have $d(f_n(s), f(s) < r$ and $d(f_n(t), f(t)) < r$. So $d(f(s), f(t)) \leq d(f_n(s), f(s)) + d(f_n(s), f_n(t)) + d(f_n(t), f(t)) < r + 0 + r = d(f(s), f(t))$ a contradiction. So f is a constant function, say $f(s) = x$, for all $s \in S$. Clearly for all $n \in \mathbb{N}$, $d(x_n, x) = e(f_n, f)$ showing that $x_n \to x$ in X since $f_n \to f$ in Y.

Conversely suppose that (X, d) is complete. Let $\{f_n\}$ be a Cauchy sequence in $(Y; e)$. For each $s \in S$ and $m, n \in \mathbb{N}$ we have that $d(f_m(s), f_n(s)) \leq e(f_m, f_n)$ showing that for each $s \in S$ the sequence $\left\{f_n(s)\right\}_{n=1}^{\infty}$ is a Cauchy sequence in (X, d). So by completeness of d, $\{f_n(s)\}$ converges to a point in X. This point is unique, because limits of sequences are unique in a metric space (and in fact in all Hausdorff spaces). We denote the limit of $\{f_n(s)\}$ in X by

$f(s)$. This defines a function f from S into X. We first show that $f \in Y$, i.e. that f is a bounded function. First, choose $p \in \mathbb{N}$ such that for all $m, n \geq p$, $e(f_m, f_n) < 1$. Such p exists because $\{f_n\}$ is given to be a Cauchy sequence in $(Y; e)$. For each $s \in S$, we have $d(f_p(s), f_n(s) < 1$ and so letting $n \to \infty$ we have $d(f_p(s), f(s)) \leq 1$ by continuity of the metric function d (see Exercise (3.3.10)). For all $s, t \in S$ we can now write

$$\begin{aligned} d(f(s), f(t)) &\leq d(f(s), f_p(s)) + d(f_p(s), f_p(t)) + d(f_p(t), f(t)) \\ &\leq 2 + d(f_p(s), f_p(t)) \end{aligned}$$

showing that f is bounded as f_p is bounded. Thus $f \in Y$.

It remains to prove that $\{f_n\}$ converges to f in $(Y; e)$. Let $\epsilon > 0$ be given. First find $p \in \mathbb{N}$ such that for all $m, n \geq p$, $e(f_m, f_n) < \epsilon/2$. Then for any $m \geq p$ and $s \in S$, we have once again, $d(f_m(s), f(s)) = \lim_{n\to\infty} d(f_m(s), f_n(s)) \leq \epsilon/2$. So $e(f_m, f) \leq \epsilon/2$ since $e(f_m, f) = \sup \{d(f_m(s), f(s)) : s \in S\}$. Thus $\{f_n\}$ converges to f in $(Y; e)$. So every Cauchy sequence in $(Y; e)$ is convergent, that is, $(Y; e)$ is complete. ∎

The topology induced by e on Y is called the **topology of uniform convergence** for obvious reasons. In general this topology depends on the metric d and not merely on the topology induced by d. Throughout our discussion so far, S was any arbitrary set. In applications, S often has an additional structure such as a topology on it and instead of taking the entire metric space $(Y; e)$ one considers certain subspaces of it.

Exercises

1.1 Let A be a subset of a metric space $(X; d)$ such that A is complete w.r.t. the metric induced on it. Prove that A is closed in X.

1.2 Prove that a closed subset of a complete metric space is complete w.r.t. the induced metric.

1.3 Prove that every totally bounded metric space is bounded.

1.4 Show by an example that total boundedness is not topologically invariant.

1.5 Let d_1, d_2 be metrics on a set X, λ a positive real number with the property that for all $x, y \in X$, $d_1(x, y) \leq \lambda$ iff $d_2(x, y) \leq \lambda$. Prove that a sequence $\{x_n\}$ is a Cauchy sequence w.r.t. d_1 iff it is so w.r.t. d_2. Deduce that $(X; d_1)$ is complete iff $(X; d_2)$ is complete.

1.6 Let $(X_1; d_1)$, (X_2, d_2) be complete metric spaces. Prove that $X_1 \times X_2$ is complete w.r.t. any of the metrics put on it as given in Exercise (3.3.4).

*1.7 Let X be the set of all infinite sequences $x = \left\{x_n\right\}_{n=1}^{\infty}$ of real numbers for which $\sum_{n=1}^{\infty} x_n^2$ is convergent. For $x, y \in X$ define $d(x, y) = \left\{\sum_{i=1}^{\infty} (x_i - y_i)^2\right\}^{1/2}$. Prove that $(X; d)$ is a complete metric space.

1.8 Obtain the space (X, d) in the exercise above as the L^2-space of a suitable measure space.

1.9 Let S be an infinite set and $(X; d)$ a bounded metric space. Then the set Y defined just before Proposition (1.7) is the entire power set X^S. Show by an example that the topology on Y induced by the metric e need not coincide with the product topology on X^S. Which topology is always stronger?

1.10 Let S be the unit interval $[0, 1]$ with the usual topology and let $(X; d)$ be $\mathbb{R}$ with the usual metric. Let Z be the set of all continuous functions from S into X. Prove that $Z \subset Y$ and that Z with the restriction of the metric e on Y is complete. (Hint: Show that Z is a closed subset of Y). Here Y is the set of all bounded functions from $[0, 1]$ to $\mathbb{R}$.

1.11 Let $(X; d)$ be a metric space and $f: X \to \mathbb{R}$ a continuous function. Define $e: X \times X \to \mathbb{R}$ by $e(x, y) = d(x, y) + |f(x) - f(y)|$. Prove that e is a metric on X and that it induces the same topology on X as d does.

1.12 Prove that on every non-compact metric space, there exists an unbounded continuous real-valued function. (Hint: Such a space must contain a countably infinite subset A having no points of accumulation. Enumerate A as $\{a_1, a_2, a_3, \ldots, a_n, \ldots\}$ and define $f: A \to \mathbb{R}$ by $f(a_n) = n$. Then f is continuous. Also A is closed in X. Apply Tietze extension theorem.)

1.13 Let X be a metrisable space. Prove that X is compact iff every metric which induces the given topology on X is bounded. (Hint: One way implication is easy. For the other way, if X is not compact then use the last two exercises to get an unbounded metric on X.)

Notes and Guide to Literature

Completeness of the real line is due to Cauchy. Although it is equivalent to many other well-known properties of real numbers, this particular formulation has the advantage that it generalises readily to abstract metric spaces. Such a generalisation is definitely worthwhile as we shall see in the next two sections.

Although the converse of Proposition (1.3) is false, the converse of its strengthened version is true. Namely, a metrisable space is compact iff every metric on it is complete. See Hausdorff [2]. The analogous result for boundedness, given in Exercise (1.13) is due to Levine [1].

A proof of the Riesz-Fisher theorem can be found in almost any treatise on functional analysis or in Royden [1]. The L^p-spaces have an additional structure of a vector space over the field of real numbers. Moreover, the metric on them is induced by a norm compatible with this algebraic structure. With this data, the L^p spaces are the foremost examples of what are called Banach spaces, a most fertile area of mathematical research.

2. Consequences of Completeness

As remarked in the last section, the existence of a complete metric has important topological consequences. In this section we shall prove the most celebrated result of this type due to Baire. First we need a few definitions.

(2.1) Definition: A subset A of a topological space is said to be **nowhere dense** in X if the interior of its closure is empty that is if int $(\bar{A}) = \emptyset$.

As examples, in the real line with the usual topology, the empty set, any finite set, the set of all integers, the set $\{1/n : n \in \mathbb{N}\}$ are all nowhere dense sets. The Cantor set is an example of an uncountable set which is nowhere dense in $\mathbb{R}$. On the other hand the set of rationals is not nowhere dense in $\mathbb{R}$, nor is its complement. In the plane $\mathbb{R}^2$, any line, circle or any simple closed curve is a nowhere dense set. More generally in $\mathbb{R}^n$, any hyperplane and the sphere S^{n-1} are nowhere dense. Note that these are closed sets with no interior points. In other words these sets do not contain any solid open ball. Intuitively speaking, they are 'thin' sets as compared to the ambient euclidean space $\mathbb{R}^n$. Note the role of the ambient space in the definition of a nowhere dense subset. Unlike compactness, it is not an intrinsic property of the subset but depends as much on what the whole space is. For example, although a line is a nowhere dense subset of $\mathbb{R}^2$, it is no longer nowhere dense in itself. In the terminology introduced in Chapter 6, this means that nowhere denseness is a relative and not an absolute property of a subset.

A subset of a nowhere dense subset as well as a finite union of nowhere dense subsets is clearly nowhere dense again. However a countable union of nowhere dense subsets need not be so as shown by $\mathbb{Q}$ in $\mathbb{R}$, for we can write $\mathbb{Q}$ as a countable union of singleton sets each of which is nowhere dense in $\mathbb{R}$. There is a name for those sets which can be written as countable unions of nowhere dense sets.

(2.2) Definition: A subset A of a space is said to be of **first category** (in X) if there exist nowhere dense subsets $F_1, F_2, \ldots, F_n, \ldots$ of X such that $A = \bigcup_{n=1}^{\infty} F_n$. Otherwise A is said to be of **second category** in X.

Once again the role of the ambient space X is vital although where it is understood we may suppress its explicit mention. As we just saw $\mathbb{Q}$ is a subset of first category in $\mathbb{R}$. More generally if X is a T_1 space with no isolated points (that is, points x for which $\{x\}$ is open) then every countable subset of X is of first category. The set of irrationals is of second category. This is not obvious but will follow from the theorem we are about to prove. A set of first category is sometimes called a **meagre set**; the term suggests that the set in question is topologically small. It should be noted that such a set need not be small with respect to other attributes such as cardinality or measure. For example the Cantor set has cardinality c although it is of first category and a modification of it yields a subset of $[0, 1]$ of Lebesgue measure 1 which is of first category.

We are now ready to prove the famous theorem of Baire.

(2.3) Theorem: Let $(X; d)$ be a complete metric space. Then a subset of first category (in X) cannot have any interior points.

Proof: Let A be a subset of first category and write A as $\bigcup_{i=1}^{\infty} F_i$ where F_i's are nowhere dense subsets of X. Note that the assertion of the theorem is not trivial, because although each F_i (in fact $\overline{F_i}$) has empty interior, the interior operator does not commute with unions and so we cannot immediately conclude that int $(A) = \emptyset$. To show that A has no interior points we proceed by a contradiction. Suppose x_0 is an interior point of A. Then there exists $r_0 > 0$ such that $B(x_0; r_0) \subset A$. By taking r_0 sufficiently small we may even assume that $C(x_0, r_0) \subset A$ where $C(x_0, r_0)$ denotes the closed ball with centre x_0 and radius r_0.

Now since $\overline{F_1}$ has no interior points, $X - \overline{F_1}$ is an open dense subset and so $B(x_0, r_0) \cap (X - \overline{F_i})$ is a non-empty open set. Hence there exists $x_1 \in X$ and $r_1 > 0$ such that $C(x_1, r_1) \subset B(x_0, r_0) \cap (X - \overline{F_1})$. We also choose r_1 smaller than $r_0/2$. Next consider $X - \overline{F_2}$ which is also an open dense set. So $B(x_1, r_1) \cap (X - \overline{F_2})$ is a non-empty open set and hence contains $C(x_2, r_2)$ for some $x_2 \in X$ and for some $0 < r_2 < r_1/2$. Continuing in this manner we get a sequence $\{x_n\}$ in X and a sequence $\{r_n\}$ of positive real numbers such that for each $n \geq 1$, $C(x_n, r_n) \subset B(x_{n-1}, r_{n-1}) \cap (X - \overline{F_n})$ and $r_n < r_0 2^{-n}$. By an argument similar to that in the proof of Theorem (1.6), it follows that $\{x_n\}$ is a Cauchy sequence. By completeness of d, $\{x_n\}$ converges to a point, say y, of X. Each term of the sequence $\{x_n\}$ is in $C(x_0, r_0)$ which is a closed set. So $y \in C(x_0, r_0)$ and hence $y \in A$. On the other hand for each $n \in \mathbb{N}$ all except the first $n - 1$ terms of the sequence are in $C(x_n, r_n)$ which is also a closed set. So $y \in C(x_n, r_n) \subset X - \overline{F_n}$. Thus $y \in \bigcap_{n=1}^{\infty} (X - \overline{F_n}) = X - \bigcup_{n=1}^{\infty} \overline{F_n} \subset X - \bigcup_{n=1}^{\infty} F_n = X - A$, a contradiction. This contradiction establishes that A has no interior points in X. ∎

(2.4) Corollary: In a complete metric space the intersection of countably many open dense sets is dense.

Proof: A set is dense iff its complement has no interior points. The complement of an open dense set is nowhere dense. With these observations, the result follows from the last theorem in view of DeMorgan's laws. ∎

(2.5) Corollary: A non-empty complete metric space is of second category (in itself).

Proof: This follows directly from the theorem, for the entire space certainly has a non-empty interior and therefore cannot be of the first category. ∎

In particular, the real line with the usual topology on it is of the second category, because this topology is induced by the usual metric which is complete. It now follows that the set of irrationals is of the second category for otherwise its union with $\mathbb{Q}$ (which is already known to be of the first category) would be of the first category.

Note that although the hypothesis of Theorem (2.3) is non-topological, its conclusion is topological and remains unchanged under homeomorphisms. This fact enlarges the applicability of the theorem. For example, the usual metric on the open interval (0, 1) is not complete. Still the space is homeomorphic to $\mathbb{R}$ whose topology is induced by a complete metric. So the conclusion of Theorem (2.3) (and its corollaries) remains valid for the space (0, 1). Alternatively we can translate the metric on $\mathbb{R}$ to a complete metric on (0, 1) and then apply Theorem (2.3) to it. Thus the existence of at least one complete metric inducing the topology makes the theorem applicable. There is a name for such spaces.

(2.6) Definition: A topological space is said to be **metrically topologically complete** if there exists a complete metric inducing the given topology on it.

Obviously this is a topological property. The real line as well as all euclidean spaces, all compact metric spaces are metrically topologically complete. The space of rationals (with usual topology) is not metrically topologically complete. For $\mathbb{Q}$ is a countable set and since there are no isolated points, $\mathbb{Q}$ is of first category in itself. So by Corollary (2.5), there cannot exist any complete metric inducing the topology on $\mathbb{Q}$, although it is a metrisable space. Note that all the results above are applicable for metrically topologically complete spaces.

We now show that an important class of spaces is metrically topologically complete. First we prove

(2.7) Proposition: An open subspace of a metrically topologically complete space is metrically topologically complete.

Proof: Let $(X; d)$ be a complete metric space and G an open subspace of X. In general the restriction of d to G will not be complete. However we can replace d on G by another metric as follows. Define $f : G \to \mathbb{R}$ by $f(x) = 1/d(x, X - G)$. Then f is continuous on G, (f is not defined if $X - G$ is empty but then there is nothing to prove). For $x, y \in G$ define $e(x, y) = d(x, y) + |f(x) - f(y)|$. Then e is a metric on G inducing the same topology on G as d (or rather its restriction) does (see Exercise (1.11)). We contend that e is a complete metric. For suppose $\{x_n\}$ is a Cauchy sequence in G w.r.t. e. Then $\{x_n\}$ is also a Cauchy sequence w.r.t. d because for any $m, n \in \mathbb{N}$, $d(x_m, x_n) \leq e(x_m, x_n)$. By completeness of d, $\{x_n\}$ converges to a point say y, in X. We claim that $y \in G$. Otherwise, for each $n \in \mathbb{N}$ we have that $d(x_n, y) \geq d(x_n, X - G)$ and so $f(x_n) \geq \dfrac{1}{d(x_n, y)}$ showing that $f(x_n) \to \infty$ as $n \to \infty$. But on the other hand $|f(x_m) - f(x_n)| \leq e(x_m, x_n)$ and $e(x_m, x_n)$ can be made sufficiently small for large m and n. It follows that $\{f(x_n)\}$ is a bounded sequence. This contradiction shows that $y \in G$. Since we already know that $\{x_n\}$ converges to y w.r.t. d, it does so w.r.t. e. Thus e is a complete metric for the relative topology on G and hence G is metrically topologically complete. ∎

Using the embedding lemma, this result can be extended as follows.

Recall from Exercise (7.4.5) that a G_δ-set in a space means a set which can be expressed as the intersection of a countable family of open sets.

(2.8) Theorem: A G_δ-set in a complete metric space is metrically topologically complete.

Proof: Let $(X; d)$ be a complete metric space and G be a G_δ set in X, that is, we can write G as $\bigcap_{n=1}^{\infty} G_n$ where each G_n is open in X. By the last proposition there exists a complete metric d_n on G_n and we may assume that d_n is bounded by 2^{-n} (see Proposition (8.4.1) with Exercise (1.5)). Let H be the topological product $\prod_{n=1}^{\infty} G_n$. Then by the construction in the proof of Theorem (8.4.2) we can put a metric on H and we leave it to the reader to verify that the metric so defined is complete. Thus H is a metrically topologically complete space.

Now, for each $n \in \mathbb{N}$ let $f_n : G \to G_n$ be the inclusion map. The family $\{f_n : n \in \mathbb{N}\}$ evidently distinguishes points and points from closed sets. So by the embedding lemma (Theorem 9.2.3)) the evaluation map $e : G \to H$ is an embedding. Obviously the image of e is the diagonal ΔG in H, that is the set of those elements of $\prod_{n=1}^{\infty} G_n$ all of whose entries are equal and belong to G. It is easy to show that ΔG is a closed subset of H (cf. Exercise (7.1.8)). So the restriction of a complete metric on H is a complete metric on ΔG by Exercise (1.2). Thus ΔG is metrically topologically complete and so is G which is homeomorphic to it. ∎

A space which is of second category when regarded as a subset of itself is called a **Baire** space. The Baire category theorem asserts that every complete metric space is such a space. There is yet another source of Baire spaces as we now prove.

(2.9) Theorem: In a locally compact, Hausdorff space a subset of a first category can have no interior points. Consequently, every such space is a Baire space, except when it is the empty space.

Proof: Let X be a locally compact, Hausdorff space and let $A = \bigcup_{n=1}^{\infty} F_n$ where each F_n is nowhere dense in X. To show that A has no interior points in X, we proceed as in the proof of Theorem (2.3) except that instead of closed balls we use compact neighbourhoods (which exist in abundance in view of Proposition (11.3.3)). We thus get a sequence $\{x_n\}$ in X and a sequence $\{C_n\}$ of compact subsets of X such that for each $n \in \mathbb{N}$, C_n is a neighbourhood of x_n and $C_n \subset C_{n-1} \cap (X - \bar{F}_n)$, C_0 is a compact neighbourhood of x_0 contained in A. Now the family $\{C_n : n \in \mathbb{N}\}$ of closed subsets of the compact set C_0 has the finite intersection property and so $\bigcap_{n=0}^{\infty} C_n$ is non-empty. Let $y \in \bigcap_{n=0}^{\infty} C_n$. Then $y \in C_0 \subset A$. But on the other hand $y \in$

$X - \bar{F}_n \subset X - F_n$ for all $n \in \mathbb{N}$ and so $y \in \bigcap_{n=1}^{\infty} (X - F_n) = X - A$, a contradiction. So A has no interior points in X. If now X is nonempty, then it certainly has a non-void interior and so must be of second category, i.e. must be a Baire space. ∎

We conclude this section with another consequence of completeness which is of a different spirit than the Baire category theorem. It asserts that certain mappings from a complete metric space into itself have fixed points. This theorem, popularly known as the **contraction fixed point theorem,** has profound applications in analysis and one such result will be given in the next section. Apart from its importance in applications, we prove the contraction fixed point theorem because it is the only non-trivial fixed point theorem to be proved in this book. There are well-known fixed point theorems due to Brouwer and Lefschetz but the methods they require are far beyond our scope.

We begin by defining what are contractions. Our intuition demands that a contraction should bring the points of the space closer and closer. A precise formulation is:

(2.10) Definition: A function T of a metric space $(X; d)$ into itself is said to be a **contraction** if there exists a number k, $0 < k < 1$ such that for all $x, y \in X$, $d(T(x), T(y)) \leq k\, d(x, y)$. T is said to be a **weak contraction** if for all $x, y \in X$, $d(T(x), T(y)) \leq d(x, y)$.

Obviously every contraction is a weak contraction. The converse is false. For example, let X be the set $[1, \infty)$ and d the usual metric on it. Define $T : X \to X$ by $T(x) = x + 1/x$. Then for all $x, y \in X$ we have, $d(T(x), T(y)) = \dfrac{xy - 1}{xy} |x - y|$ from which it follows that T is a weak contraction. However T cannot be a contraction because although the ratio $\dfrac{xy - 1}{xy}$ is less than 1 for all $x, y \in X$, it approaches 1 as x, y become large. Thus given any k, $0 < k < 1$, we can always find some $x, y \in X$ for which the inequality $d(T(x), T(y)) \leq k\, d(x, y)$ fails.

Simple examples of contractions are mappings $T : \mathbb{R} \to \mathbb{R}$ of the form $T(x) = kx$ where k is a fixed constant of absolute value less than 1. More generally this is true of certain linear transformations $T : \mathbb{R}^n \to \mathbb{R}^n$ all whose eigenvalues are less than 1 in absolute value. As a consequence of the well-known Lagrange mean value theorem in calculus, it follows that if $f : \mathbb{R} \to \mathbb{R}$ is continuously differentiable and there exists $k > 0$ such that $|f'(x)| \leq k$ for all $x \in \mathbb{R}$, then f is a weak contraction if $k \leq 1$ and is actually a contraction if $k < 1$. Note incidentally that whether a function is a contraction depends as much on the metric as on the function itself. For example, let $X = (0, \infty)$ and let $T : X \to X$ be $T(x) = x/2$ for all $x \in X$. Then T is a contraction w.r.t. the usual metric on X. However, if we put the metric e on X defined by $e(x, y) = \left|\dfrac{1}{x} - \dfrac{1}{y}\right|$ for $x, y \in X$ then T is no longer a contraction, even though e induces the same topology on X as the usual metric.

Thus contraction is a non-topological concept. Nevertheless, it has topological consequences. For example, it is trivial to show that a contraction (or even weak contraction) is continuous.

We now prove the contraction fixed point theorem due to Banach.

(2.11) Theorem: Every contraction of a complete metric space into itself has a unique fixed point.

Proof: Let $(X; d)$ be a metric space and $T : X \to X$ a contraction. There is then a number k, $0 < k < 1$ such that for all $x \in X$, $d(T(x), T(y)) \leq kd(x, y)$. Now let x_1 be any point of X and for $n > 1$, recursively define $x_n = T(x_{n-1})$. In other words $x_n = T^{n-1}(x_1)$ where T^{n-1} is the $(n-1)$-th power of T, i.e. the composite of T with itself taken $(n-1)$ times. Consider the sequence $\{x_n\}$ in X. We claim that it is a Cauchy sequence (w.r.t. d). First for any $r \in \mathbb{N}$, note that $d(x_r, x_{r+1}) = d(T^{r-1}(x_1), T^{r-1}(T(x_1))) \leq k^{r-1}\, d(x_1, x_2)$. Now for $m < n$ we have

$$\begin{aligned} d(x_m, x_n) &\leq \sum_{r=m}^{n-1} d(x_r, x_{r+1}) \\ &\leq \sum_{r=m}^{n-1} k^{r-1}\, d(x_1, x_2) = d(x_1, x_2) \sum_{r=m}^{n-1} k^{r-1} \end{aligned}$$

which can be made as small as we like for sufficiently large m, n in view of the fact the geometric series $\sum_{r=1}^{\infty} k^r$ is convergent (as $|k| < 1$). This shows that the sequence $\{x_n\}$ is a Cauchy sequence. By completeness of d, this sequence converges to, say, $x_0 \in X$. We assert that x_0 is a fixed point of T; i.e. that $T(x_0) = x_0$. This is easy, because by continuity of T we have

$$T(x_0) = \lim_{n \to \infty} T(x_n) = \lim_{n \to \infty} x_{n+1} = x_0$$

It remains to show that x_0 is the only fixed point of T. This is also easy, because if y_0 is any fixed point of T. Then $d(x_0, y_0) = d(T(x_0), T(y_0)) \leq kd(x_0, y_0)$ which can hold only when $d(x_0, y_0) = 0$ (since $k < 1$). This forces y_0 to equal x_0, thus establishing the uniqueness of the fixed point. ∎

This proof does a little more than proving what the statement of the theorem claims. For it not only shows the existence of a fixed point but also tells how to trace it. Note that the point x_1 in the proof was arbitrary. Thus we could start with any point in X, apply the contraction T to it repeatedly and the resulting sequence will converge to the fixed point of T, regardless of the point we started with. This procedure is useful when we have to content ourselves with approximations to the fixed point. It sometimes so happens that we know that a sequence is convergent and we can compute its n-th term by an algorithmic process and yet we have no way to find its limit. In such a case we approximate the limit by the n-th term of the sequence for large n.

Exercises

2.1 Let μ denote the usual Lebesgue measure on the unit interval $[0, 1]$.

Prove that for every $\epsilon > 0$, there exists a nowhere dense subset A of $[0, 1]$ such that $\mu(A) > 1 - \epsilon$. (Hint: Start as in the construction of the Cantor set given in Exercise (8.2.9), except that the sum of the lengths of the removed intervals should not exceed ϵ.)

2.2 With μ as above prove that there exists a subset of first category in $[0, 1]$ whose measure is 1. Prove however that a subset of measure 1 cannot be nowhere dense.

2.3 Let $\{X_i : i \in I\}$ be a family of non-empty, Hausdorff spaces. Prove that if infinitely many of them are non-compact then every compact subset of the topological product ΠX_i is nowhere dense.

2.4 Show by an example that the result above fails if the spaces are not assumed to be Hausdorff. (Hint: Consider an infinite power of the space X in Exercise (11.3.4).

2.5 Prove that every complete metric space with no isolated points is uncountable.

2.6 Prove that metrical topological completeness is a countably productive property. (This fact was essentially used in the proof of Theorem (2.8).).

2.7 In the proof of Theorem (2.8) verify ΔG is indeed a closed subset of the product $\prod_{n=1}^{\infty} G_n$.

2.8 (a) Prove that there exists a complete metric which induces the usual topology on the set of irrational numbers.
*(b) Give an explicit formula for such a metric.

*2.9 Let Y be a Hausdorff space and suppose X is a dense subspace of Y having a complete metric d on it (Y itself need not be metrisable). Prove that X is a G_δ set in Y. (Hint: For each $n \in \mathbb{N}$, let $G_n = \{y \in Y$: there exists a nbd M of y in Y such that $M \cap X$ has diameter less than 2^{-n} w.r.t. $d\}$. Prove that $X = \bigcap_{n=1}^{\infty} G_n$.)

2.10 Prove that a non-empty space cannot be expressed as the union of a finite number of nowhere dense subsets.

2.11 Show by examples that the contraction fixed point theorem does not hold for weak contractions; nor for contractions of an incomplete metric space.

2.12 Prove that if $T: (X; d) \to (X; d)$ is such that some power of T, say T^m $(m > 1)$, is a contraction, then T has a unique fixed point (Caution : A fixed point of T^m need not necessarily be a fixed point of T.)

Notes and Guide to Literature

The Baire category theorem is one of the most classic results. Its power in applications will be apparent in the next section. Theorem (2.7) is due to Alexandroff [1] and Hausdorff [1], while the result of Exercise (2.9) was discovered by Sierpinski [2]. They are taken here from Kelley [1]. Exercise (2.3) is also claimed as a theorem in Kelley [1] (p. 145) but the proof given is erroneous as the spaces are not assumed to be Hausdorff.

For a detailed study of the correlationship between category and measure see the book by Oxtoby [1]. For variations and many applications of the contraction fixed point theorem, see Hille [1].

3. Some Applications

In this section we give a few applications of the theorems proved in the last section. The more important applications of the Baire category theorem such as the open mapping theorem and the closed graph theorem come from functional analysis and are beyond the scope of this book. In topology too there are important applications, for example in proving certain embedding theorems. But to discuss these will take us too far afield. We shall, therefore, content ourselves in giving a few applications which will illustrate the underlying techniques even though the results obtained are of little use in the sequel. We also discuss a classic application of the contraction fixed point theorem to solving a differential equation.

We begin with an old problem of showing that there exists a continuous real-valued function on $[0, 1]$ which is not differentiable at any point of $[0, 1]$. Here differentiability at an end point means the appropriate right or left handed differentiability. For a function to be differentiable at an interior point, both the right and left handed derivatives must exist (as finite real numbers) and be equal. We shall show that there exists a continuous function on $[0, 1]$ which has no finite right-handed derivative at any point of $[0, 1]$ (right derivative at the end point 1 is not defined). In fact we shall show that the set of such functions is quite big.

First we set up some notation. Let C denote the set of all continuous real-valued functions on the unit interval $[0, 1]$. Since each such function is bounded, we can introduce a metric d on C defined by $d(f, g) = \sup\{|f(x) - g(x)| : x \in [0, 1]\}$. In Exercise (1.10) we asked the reader to show that (C, d) is a complete metric space and this fact will be needed crucially. Let R be the set of those functions in C which have a finite right-handed derivative at at least one point in $[0, 1)$. We show that R is a set of first category in C. For this we need an elementary fact from calculus, whose proof is left to the reader.

(3.1) Lemma: Let $f: [0, 1] \to \mathbb{R}$ be continuous and suppose f is right differentiable at some point $x_0 \in [0, 1)$. Then there exists a positive integer k such that for all $x \in [x_0, 1]$, $|f(x) - f(x_0)| \leq k(x - x_0)$. ∎

Now for each $k \in \mathbb{N}$, let $R_k = \{f \in C :$ there exists some $x_0 \in [0, 1)$ such that for all $x \in [x_0, 1]$, $|f(x) - f(x_0)| \leq k(x - x_0)\}$. Then the lemma shows that $R \subset \bigcup_{k=1}^{\infty} R_k$. So in order to show that R is of first category it would suffice to show that each R_k is of first category in C. (Here we are using that a countable union of sets of first category as well as any of its subsets is of first category. We leave the proof of this easy fact to the reader.)

So fix $k \in \mathbb{N}$ and consider R_k. For each $m \in \mathbb{N}$, let $R_{km} = \{f \in C$. there exists $x_0 \in [0, 1 - 1/m]$ such that for all $x \in [x_0, 1]$, $|f(x) - f(x_0)| \leq k(x - x_0)\}$. Clearly $R_k = \bigcup_{m=1}^{\infty} R_{km}$. We claim that each R_{km} is nowhere dense in C. First we find its closure.

(3.2) Lemma: The set R_{km} is closed in C and so $\bar{R}_{km} = R_{km}$.

Proof: Since C is a metric space, it suffices to show that the limit, if any, of a sequence in R_{km} is in R_{km}. Let $\{f_n\}$ be a sequence in R_{km} and suppose $\{f_n\}$ converges to a function f in C. As we noted earlier this means that $\{f_n\}$ converges to f uniformly on $[0, 1]$. Now for each f_n there exists $x_n \in [0, 1 - 1/m]$ such that for all $x \in [x_n, 1]$, $|f(x) - f(x_n)| \leq k(x - x_n)$. The sequence $\{x_n\}$ has a convergent subsequence, say $\{x_{n_r}\}$, converging to $y \in [0, 1 - 1/m]$. Let $y < x \leq 1$. Then $x > x_{n_r}$ for all sufficiently large r and so $|f_{n_r}(x) - f_{n_r}(x_{n_r})| \leq k(x - x_{n_r})$. It is easy to show that $f_{n_r}(x_{n_r}) \to f(y)$ as $r \to \infty$. So taking limits as $r \to \infty$ we have $|f(x) - f(y)| \leq k(x - y)$. For $x = y$ equality holds. So for all $x \in [y, 1]$ we have shown that $|f(x) - f(y)| \leq k(x - y)$ and since $y \in [0, 1 - 1/m]$ it follows that $f \in R_{km}$. Thus R_{km} is closed or in other words $\bar{R}_{km} = R_{km}$. ∎

In order to show that R_{km} is nowhere dense we now have to prove that its interior is empty or equivalently that its complement is dense in C. Since $C - R_k \subset C - R_{km}$ for all m, it will suffice to show that for each $k \in \mathbb{N}$, $C - R_k$ is dense in C. Note that a function whose right-handed derivative at every point is greater than k in absolute value is in $C - R_k$. A very simple class of such functions consists of those whose graphs consist of chains of line segments with slopes greater than k in absolute value. A typical such graph is pictured in Fig. 1. Let S_k be the set of such functions. We shall show that every function in C can be approximated by a function in S_k to any desired degree of approximation. In other words given $f \in C$

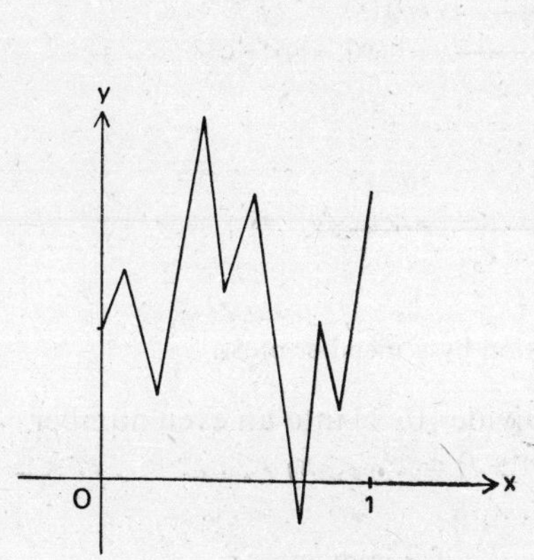

Fig. 1. A function in S_k.

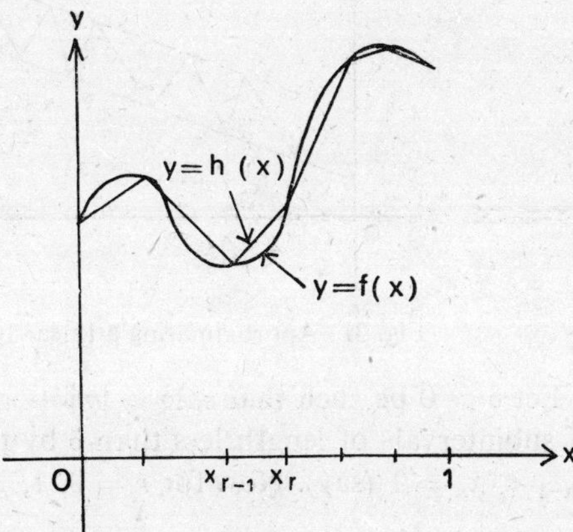

Fig. 2. Approximation by a piecewise linear function.

and $\epsilon > 0$, we shall find $g \in S_k$ such that $d(f, g) \leq \epsilon$. This approximation will be done in two stages, first we shall find a continuous, piecewise linear approximation h i.e. a function whose graph consists of straight line segments and then we shall approximate h by a function in S_k.

(3.3) Lemma: Given $f \in C$ and $\epsilon > 0$, there exists a continuous piecewise linear function h on $[0, 1]$ such that $d(f, h) \leq \epsilon/2$.

Proof: By uniform continuity of f, there exists $\delta > 0$ such that for all $x, y[0, 1]$, $|x - y| < \delta$ implies $|f(x) - f(y)| < \epsilon/2$. Divide $[0, 1]$ into subintervals of length less than δ by points $0 = x_0 < x_1 < x_2 <, \ldots < x_n = 1$. For $r = 1, 2, \ldots, n$, define h on $[x_{r-1}, x_r]$ as a linear function whose graph is the line segment joining the points $(x_{r-1}, f(x_{r-1}))$ and $(x_r, f(x_r))$, that is for $x \in [x_{r-1}, x_r]$ we define $h(x) = (1 - m)f(x_{r-1}) + mf(x_r)$ where $m = \dfrac{x - x_{r-1}}{x_r - x_{r-1}}$. This gives a well-defined continuous function on $[0, 1]$. Also for $x \in [x_{r-1}, x_r]$, $|f(x) - h(x)| = |(1 - m)f(x) + mf(x) - h(x)| \leq (1 - m)|f(x) - f(x_{r-1})| + m|f(x) - f(x_r)| \leq (1 - m)\epsilon/2 + m\epsilon/2 = \epsilon/2$. As this holds for all $r = 1, 2, \ldots, n$ we have that $\sup \{|f(x) - h(x)| : x \in [0, 1]\} \leq \epsilon/2$ or that $d(f, h) \leq \epsilon/2$ as desired. ∎

Next, we show that h can be approximated by a member of S_k. For simplicity we give the proof for the case where h is linear on the entire interval $[0, 1]$. For the general case where h is only piecewise linear one simply applies the argument on each subinterval over which h is linear.

(3.4) Lemma: Given a linear function h on $[0, 1]$, $\epsilon > 0$ and $k \in \mathbb{N}$ there exists a piecewise linear function g on $[0, 1]$ such that $d(h, g) \leq \epsilon/2$, the right hand derivative of g at each point exceeds k numerically, and $g(0) = h(0)$, $h(1) = g(1)$.

Proof: Suppose $h(x) = mx + c$, where m, c are constants.

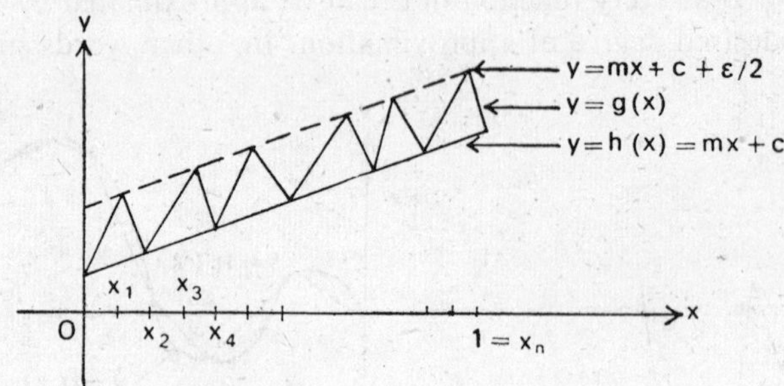

Fig. 3. Approximating a linear function by a member of S_k.

Let $\delta > 0$ be such that $\epsilon/2\delta > |m| + k$. Divide $[0, 1]$ into an even number of subintervals of length less than δ by points $0 = x_0 < x_1 < \ldots < \ldots < x_{n-1} < x_n = 1$ (say). Now for $r = 0, 1, 2, \ldots, n$

$$\text{define } g(x_r) = \begin{cases} mx_r + c & \text{if } r \text{ is even} \\ mx_r + c + \epsilon/2 & \text{if } r \text{ is odd.} \end{cases}$$

Then extend g linearly to each subinterval. We thus get a continuous, piecewise linear function g with $g(0) = h(0)$ and $g(1) = h(1)$ (since $1 = x_n$ and n is even). The slope of the graph of g over $[x_{r-1}, x_r]$ is $m + \dfrac{\epsilon}{2(x_r - x_{r-1})}$ if r is odd and $m - \dfrac{\epsilon}{2(x_r - x_{r-1})}$ is r is even. In either case it is numerically larger than k since $\dfrac{\epsilon}{2(x_r - x_{r-1})} > \dfrac{\epsilon}{2\delta} > |m| + k$. ∎

We have now completed the sticky part of the proof. Putting together what we have done so far, we see that the set R of all those functions in C which have a finite right hand derivative at at least one point in $[0, 1)$ is of first category in C. By similar arguments, it follows that the set L of those functions in C which are left differentiable at at least one point in $(0, 1]$ is of first category in C. So $R \cup L$ is a set of first category in C. But C is a complete metric space. So by Theorem (2.3), $R \cup L$ has no interior points or equivalently its complement is dense in C. But an element of $C - R \cup L$ is a function which is neither left nor right differentiable at any point of $[0, 1]$. Thus we have completed the proof of the

(3.5) Theorem: There exists a continuous real-valued function on $[0, 1]$ which is neither left nor right differentiable at any point of $[0, 1]$. In fact the set of such functions is dense in the space of all continuous real-valued functions on $[0, 1]$. ∎

Although the second statement of the theorem is stronger than the first, it is the first statement that is more celebrated. This is a peculiarity shared by almost all other instances of the application of the Baire category theorem. A typical application merely asserts the existence of something although in fact what we are looking for not only exists but exists in abundance.

As our second application of the Baire category theorem, we make good a promise given in Chapter 7 about Exercise (7.2.7). There we considered $\mathbb{R}$ with the semi-open interval topology and the problem was to show that the space $\mathbb{R} \times \mathbb{R}$ is not normal. For this we consider the sets $A = \{(x, -x) : x \in \mathbb{Q}\}$ and $B = \{(x, -x) : x \in \mathbb{R} - \mathbb{Q}\}$. These are closed in $\mathbb{R} \times \mathbb{R}$ and we claim that if U is any open set in $\mathbb{R} \times \mathbb{R}$ containing A then $\overline{U} \cap B \neq \emptyset$. This would imply that A and B have no mutually disjoint neighbourhoods and hence that $\mathbb{R} \times \mathbb{R}$ is not normal.

So suppose U is an open set in $\mathbb{R} \times \mathbb{R}$ containing A. Let $V = \mathbb{R} \times \mathbb{R} - \overline{U}$. Then V is an open set and we have to show that B is not contained in V. We assume $B \subset V$ and derive a contradiction as follows. For $x \in \mathbb{R}$ and $\epsilon > 0$, let $R(x; \epsilon)$ denote the rectangle $[x, x + \epsilon) \times [-x, -x + \epsilon)$. Note that this rectangle is a basic open set in the product topology on $\mathbb{R} \times \mathbb{R}$ (although it is not open in the usual topology on $\mathbb{R} \times \mathbb{R}$). Let I be the set of all irrational numbers. Then for each $x \in I$, $(x, -x) \in B \subset V$ and since V is open, there exists $\epsilon > 0$ (depending upon x) such that $R(x; \epsilon) \subset V$. Now for each $k \in \mathbb{N}$ let $I_k = \left\{x \in I : R\left(x; \frac{1}{k}\right) \subset V\right\}$. Then $I = \bigcup_{k=1}^{\infty} I_k$. The set of irrationals is of second category in the usual topology on $\mathbb{R}$. So there exists some $k \in \mathbb{N}$ such that I_k is not nowhere dense in $\mathbb{R}$ w.r.t. the usual topology on $\mathbb{R}$. So $\bar{I}_k$ contains an open interval (in the usual topology) and hence a rational number, say, x. Thus for some $k \in \mathbb{N}$, there exists a sequence $\{x_n\}$ in I_k converging to a rational number x. It is now easy to see what goes wrong. Since $(x, -x) \in A$ and A is contained in the open set U, we have that $R(x; \epsilon) \subset U$ for all sufficiently small $\epsilon > 0$. Fix one such ϵ and choose n large enough so that $|x_n - x| < \min\{\epsilon, 1/k\}$. Then the rectangles $R(x; \epsilon)$ and $R(x_n; 1/k)$ are not mutually disjoint. This is a contradiction since $R(x; \epsilon) \subset U$, $R(x_n, 1/k) \subset V$ and $U \cap V = \emptyset$. We have thus proved

(3.6) Theorem: The space $\mathbb{R} \times \mathbb{R}$ where each coordinate space has the semi-open interval topology is not normal. ∎

A few more applications of the Baire category theorem will be indicated in the exercises. We conclude the present section with an application of the contraction fixed point theorem. The problem here is to prove the existence and uniqueness of a solution of a certain first order differential equation. We solve it by showing that such a solution must be a fixed point of a certain contraction. In the familiar form such a differential equation looks like $\frac{dy}{dx} = F(x, y)$. F is a continuous real-valued function of two variables. A solution of this differential equation over an interval means a function $y = f(x)$ for which $f'(x) = F(x, f(x))$ for all x in that interval. A reader familiar with the elementary theory of differential equations would know that in general the solutions constitute what is known as a one-parameter family of curves. A particular solution is obtained by requiring that the function f assume a specified value for a specified value of x. Such a requirement is called an **initial condition.**

Our concern here is not with the methods of solving such a differential equation but rather with proving the existence of a solution. For simplicity we assume that the initial condition is $f(0) = 0$ and that the domain on which F is defined is a closed rectangle $D = \{(x, y) \in \mathbb{R} \times \mathbb{R}: |x| \leq a, |y| \leq b\}$ where a, b are some positive real numbers. Besides continuity, we need one more condition on the function F, the so-called **Lipschitz condition.** This means that there exists a positive constant K (not necessarily less than 1) such that for all $(x, y_1), (x, y_2) \in D$, the inequality

$$|F(x, y_1) - F(x, y_2)| \leq K\,|y_1 - y_2|$$

holds true. We then say that F **satisfies the Lipschitz condition in y with constant K.** This is not a very stringent requirement. Indeed, if the partial derivative $\frac{\partial F}{\partial y}$ exists and is bounded on D, then a simple application of the Lagrange mean value theorem from calculus shows that F satisfies the Lipschitz condition with a suitable constant.

With this set-up, we can now prove

(3.7) Theorem: Let D be the closed rectangle $\{(x, y) \in \mathbb{R}^2 : |x| \leq a, |y| \leq b\}$. Assume F is a continuous real-valued function on D satisfying the Lipschitz condition in the variable y with constant K. Then there exists $r > 0$ such that the differential equation $\frac{dy}{dx} = F(x, y)$ with the initial condition $y(0) = 0$ has a unique solution in the interval $[-r, r]$.

Proof: As remarked before we convert the problem to that of finding a fixed point of a mapping. Let M be any positive constant such that $|F(x, y)| \leq M$ for all $(x, y) \in D$. The existence of such M is guaranteed by continuity of F and by compactness of D. Let r be any positive real number less than $\min\left\{a, \frac{b}{M}, \frac{1}{K}\right\}$.

Now consider the space X of all continuous functions from $[-r, r]$ to $[-b, b]$ which vanish at 0, i.e. $X = \{g : [-r, r] \to [-b, b] \mid g \text{ is continuous and } g(0) = 0\}$. We put the usual metric $d(h, g) = \sup\{|h(x) - g(x)| : x \in [-r, r]\}$ on X. It is easily seen that d is a complete metric on X. Now for $g \in X$, let $T(g) : [-r, r] \to \mathbb{R}$ be the function defined by $T(g)(x) = \int_0^x F(s, g(s))\, ds$ for $-r \leq x \leq r$. Because of continuity of g and F the integrand is a continuous function of s and so $T(g)(x)$ is well defined. Further, we have $|T(g)(x)| \leq Mr < b$ for all $x \in [-r, r]$ by choice of M and r. Thus $T(g)$ takes values in $[-b, b]$. Properties of functions defined by definite integrals show that $T(g)$ is continuous (in fact differentiable) in x. Finally, $T(g)(0) = 0$. Putting it all together, $T(g)$ is a member of X. Thus we get a well-defined function $T : X \to X$. We assert that T is a contraction w.r.t. the metric d on X.

Let $g_1, g_2 \in X$. Then for each $x \in [-r, r]$ we have

$$\begin{aligned}
|T(g_1(x) - T(g_2)(x)| &= \left|\int_0^x F(s, g_1(s)) - F(s, g_2(s))\, ds\right| \\
&\leq \int_0^x |F(s, g_1(s)) - F(s, g_2(s))|\, ds \\
&\leq \int_0^x K|g_1(s) - g_2(s)|\, ds \\
&\qquad\qquad \text{(by Lipschitz condition)} \\
&\leq K\, d(g_1, g_2)\, |x| \\
&\leq Kr\, d(g_1, g_2).
\end{aligned}$$

Taking supremum of the left hand side as x varies over $[-r, r]$ we see that $d(T(g_1), T(g_2)) \leq K \cdot r \cdot d(g_1, g_2)$ for all $g_1, g_2 \in X$. By choice of r, Kr is a fixed positive real number less than 1. So T is a contraction.

We now apply the contraction fixed point theorem. This gives us a unique function $f \in X$ satisfying $T(f) = f$ i.e. $f(x) = \int_0^x F(s, f(s))\, ds$ for all $x \in [-r, r]$. Applying the fundamental theorem of calculus, we get that f is differentiable (by continuity of the integrand) and that $f'(x) = F(x, f(x))$ for all $x \in [-r, r]$. Moreover $f(0) = 0$. This is precisely what is meant by saying that f is a solution of the differential equation $\frac{dy}{dx} = F(x, y)$ with initial condition $y(0) = 0$. This proves the existence and uniqueness of the solution. ∎

Exercises

3.1 Prove Lemma (3.1).

3.2 Suppose X is a metric space and $\{f_n\}$ is a sequence of functions from X into $\mathbb{R}$ (or into another metric space). If $\{f_n\}$ converges to a function f uniformly and a sequence $\{x_n\}$ converges to x in X then prove that $\{f_n(x_n)\}$ converges to $f(x)$ in $\mathbb{R}$. (This type of convergence is called the **diagonal convergence**.)

3.3 A subset of $\mathbb{R}^2$ is said to be **in general position** if not three of its points are collinear. Prove that $\mathbb{R}^2$ contains a dense subset in general position. Extend the definition and the result to higher dimensional euclidean spaces. (Hint: Let D be any countable dense subset of $\mathbb{R}^2$. Enumerate D as $\{a_1, a_2, \ldots, a_n, \ldots\}$. Choose $b_n \in \mathbb{R}^2$ inductively so that $d(a_n, b_n) < 1/n$ and b_n is not collinear with any pair of distinct points in $\{b_1, \ldots, b_{n-1}\}$. Consider the set $\{b_n : n \in \mathbb{N}\}$.)

3.4 Prove that a countable union of sets of first category as well as a subset of a set of first category are of first category.

3.5 Let $\mathcal{F}$ be a family of real-valued continuous functions on a complete metric space X and suppose for each $x \in X$ the set $\{f(x) : f \in \mathcal{F}\}$ is bounded (the bound could conceivably depend on x). Prove that there is a non-empty open set G in X and a constant K such that $|f(x)| \leq K$ for all $x \in G$ and $f \in \mathcal{F}$. $\Big($Hint: For $m \in \mathbb{N}$ let $E_m = \{x \in X : |f(x)| \leq m$, for all $f \in \mathcal{F}\}$. Prove that E_m is closed and note that $X = \bigcup_{m=1}^{\infty} E_m$.$\Big)$

3.6 A countable union of closed sets is said to be **a set of type F_σ or an F_σ-set**. Obviously a set is of type F_σ iff its complement is a G_δ.

Let $f : X \to \mathbb{R}$ be a function, where X is a metric space. Prove that the set of points of X where f is not continuous is an F_σ set. $\Big($Hint: Let D be the set in question. For $n \in \mathbb{N}$ let $D_n = \{x \in X :$ every nbd of x contains points y, z such that $|f(y) - f(z)| \geq 1/n\}$. Prove that each D_n is closed in X and $D = \bigcup_{n=1}^{\infty} D_n$.$\Big)$

3.7 (a) Prove that there is no function $f: \mathbb{R} \to \mathbb{R}$ which is continuous precisely at every rational.

(b) Construct a function $f: \mathbb{R} \to \mathbb{R}$ which is continuous precisely at every irrational.

3.8 Let F be any countable subset of $\mathbb{R}^2$. Given $p \in \mathbb{R}^2$ and $\epsilon > 0$ prove that there exists $q \in \mathbb{R}^2$ such that $d(p, q) < \epsilon$ and no circle through q cuts F in more than two points.

3.9 Prove that the space X in the proof of Theorem (3.7) is in fact a complete metric space w.r.t. d.

3.10 Generalise Theorem (3.7) to vector-valued functions of a real variable x.

3.11 Let X be the unit square $[0, 1]\times[0, 1]$. Define $f : X \to \mathbb{R}$ by $f(x, y) = \dfrac{xy}{x^2 + y^2}$ if $(x, y) \neq (0, 0)$ and $f(0, 0) = 0$. Prove that f is not continuous at $(0, 0)$. Prove, however, that for each $x \in [0, 1]$, the function $f_x : [0, 1] \to \mathbb{R}$ defined by $f_x(y) = f(x, y)$ for $y \in [0, 1]$ is continuous and similarly for each $y \in [0, 1]$ the function $f_y : [0, 1] \to \mathbb{R}$ defined by $f_y(x) = f(x, y)$ for $x \in [0, 1]$ is continuous. (In other words, continuity of a function in each variable does not imply joint continuity.)

*3.12 Let X be as above and suppose $f : X \to \mathbb{R}$ is a function which is separately continuous in each argument. Prove that if $f^{-1}(\{0\})$ is dense in X, then f vanishes identically on X. (In other words, although f need not be jointly continuous, it shares some properties of continuous functions.)

Notes and Guide to Literature

Weierstrass gave the first known example of a continuous nowhere differentiable function. A simple example can be found in Sift [1]. Proofs of the open mapping theorem and the closed graph theorem can be found in Royden [1]. The result of Exercise (3.5) is also important in analysis. It is called the **uniform boundedness principle**.

An application of the Baire category theorem to show that under certain conditions there exists an embedding of a space in a euclidean space (and, as always, a great many of such embeddings) may be found in Hurewicz Wallman [1].

Theorem (3.6) is due to Sorgenfrey [1]. Another counter-example where the fact that the set of irrationals is of the second category is used, is the classic example due to Knaster and Kuratowski [1] of a connected subset of the plane having a dispersion point. Exercise (3.12) is a problem posed by Montgomery [1].

4. Completions of a Metric

In the last chapter we studied ways of compactifying a topological space. There is analogously a way to complete a given metric space. The idea is to fill up certain holes in the given space where a 'hole', as we noted in Section 1, corresponds to a non-convergent Cauchy sequence. The construction resembles closely Cantor's construction of real numbers from the rationals, with which the reader is probably already familiar. For this reason, we shall be fairly sketchy in the presentation, leaving a good many technical details to the reader. Throughout, $(X; d)$ will denote a metric space.

As we noted, a non-convergent Cauchy sequence corresponds to a hole. It may of course happen that two Cauchy sequences determine the same hole. They should therefore be regarded as equivalent. This leads to the following definition.

(4.1) Definition: Two Cauchy sequences $\{x_n\}$ and $\{y_n\}$ in $(X; d)$ are said to be **equivalent** if $d(x_n, y_n) \to 0$ as $n \to \infty$.

Let S denote the set of all Cauchy sequences in $(X; d)$. We then leave it to the reader that the definition above indeed defines an equivalence relation on S. We denote elements of S by circumflexed lower case letters. Thus the Cauchy sequences $\{x_n\}$ and $\{y_n\}$ will be denoted by $\hat{x}$ and $\hat{y}$ respectively and we write $\hat{x} \sim \hat{y}$ to mean that they are equivalent in the sense defined above. Let $\hat{X}$ denote the set of all equivalence classes of S under $\sim$, that is $\hat{X} = S/\sim$. For $\hat{x} \in S$, we denote its equivalence class in $\hat{X}$ by $[\hat{x}]$.

We now make $\hat{X}$ into a metric space.

(4.2) Definition: For $[\hat{x}], [\hat{y}] \in \hat{X}$, we define $e([\hat{x}], [\hat{y}]) = \lim_{n\to\infty} d(x_n, y_n)$ where $\hat{x} = \{x_n\}$ and $\hat{y} = \{y_n\}$.

It must of course be verified that this gives a well-defined metric on X. It must first be shown that if $\{x_n\}$ and $\{y_n\}$ are Cauchy sequences in the metric space $(X; d)$ then $d(x_n, y_n)$ is a convergent sequence of real numbers. Secondly it must be verified that if $\hat{x} \sim \hat{z}$ and $\hat{y} \sim \hat{w}$ then $\lim_{n\to\infty} d(x_n, y_n) = \lim_{n\to\infty} d(z_n, w_n)$. And finally, the axioms of a metric must be established for e. We leave all these verifications to the reader and work with the metric space $(\hat{X}; e)$.

We now show that the metric space $(\hat{X}, e)$ is an extension of the original metric space $(X; d)$ in the sense that the latter can be isometrically (that is, with preservation of the metric) embedded into the former. Note that for any $x \in X$, the constant sequence $\{x, x, x, x, \dots\}$ is evidently a Cauchy sequence. We denote it by $\hat{x}$ and define $h : X \to \hat{X}$ by $h(x) = [\hat{x}]$. It is then trivial to check that for any $x, y \in X$, $d(x, y) = e(h(x), h(y))$ and thus h is an isometrical embedding of X into $\hat{X}$. The real crux of the matter is that $h(X)$ is dense in $\hat{X}$ as we now show.

(4.3) Proposition: The range of the embedding $h : (X; d) \to (\hat{X}, e)$ is a dense subset of $\hat{X}$ in the metric topology induced by e.

Proof: It suffices to show that every element of $\hat{X}$ can be expressed as the limit of a sequence in $h(X)$. Let $[\hat{x}]$ be any such element where $\hat{x} = \{x_n\}$ is a Cauchy sequence in $(X; d)$. We claim that the sequence $h(x_n)$ converges to $[\hat{x}]$ in $(\hat{X}; e)]$ Let $\epsilon > 0$ be given. First, find $p \in \mathbb{N}$ such that for all m,

$n \geq p$, $d(x_m, x_n) < \epsilon/2$. This implies that for all $m \geq p$, $\lim\limits_{n\to\infty} d(x_m, x_n) \leq \epsilon/2$ or that $e([\hat{x}_m], [\hat{x}]) \leq \epsilon/2 < \epsilon$, by the definition of e. Hence $e(h(x_m), [\hat{x}]) < \epsilon$ for all $m \geq p$, showing that $h(x_m)$ converges to $[\hat{x}]$ in $(\hat{X}; e)$. ∎

Finally we come to the most important part of our construction, namely to show that $(\hat{X}, e)$ is complete. For this, let $\{[\hat{x}_n]\}$ be a Cauchy sequence in (X, e) where for each n, $\hat{x}_n$ is itself a Cauchy sequence in $(X; d)$ whose entires will be denoted by means of superscripts, thus $\hat{x}_n = \{x_n^m\}_{m=1}$. We want a Cauchy sequence $\hat{y} = \{y_n\}$ in $(X; d)$ such that $\{[\hat{x}_n]\}$ will converge to $[\hat{y}]$ in $(\hat{X}; e)$. We shall define $\hat{y}$ by picking its nth term from the nth sequence $\left\{x_n^m\right\}_{m=1}^{\infty}$ and to this extent the argument will resemble a diagonalisation argument. However, a strict diagonalisation would require us to let $y_n = x_n^n$ and then $\{y_n\}$ need not be a Cauchy sequence even though each x_n and $\{[\hat{x}_n]\}$ are Cauchy sequences. The difficulty can be overcome by choosing for each n, a sufficiently large $p = p(n)$ and setting $y_n = x_n^{p(n)}$. Just how large this p should be we now specify.

Fix $n \in \mathbb{N}$. Since $\left\{x_n^m\right\}_{m=1}^{\infty}$ is a Cauchy sequence, there exists $p = p(n)$ such that for all $k, m \geq p$, $d(x_n^m, x_n^k) < 1/n$. Fix one such $p(n)$ for each $n \in \mathbb{N}$ and set $y_n = x_n^{p(n)}$. We claim that $\hat{y} = \{y_n\}$ is a Cauchy sequence in $(X; d)$. For, let $\epsilon > 0$ be given. We have to find q so that for all $r, s \geq q$, $d(y_r, y_s) < \epsilon$.

First choose $q_1 \in \mathbb{N}$ so that $1/q_1 < \epsilon/3$. Also since $\{[\hat{x}_n]\}$ is a Cauchy sequence, there exists q_2 such that for all $r, s \geq q_2$, $e([\hat{x}_r], [\hat{x}_s]) < \epsilon/3$. Set $q = \max\{q_1, q_2\}$. Now for any $r, s \geq q$ we have

$$\begin{aligned} d(y_r, y_s) &= d(x_r^{p(r)}, x_s^{p(s)}) \\ &\leq d(x_r^{p(r)}, x_r^m) + d(x_r^m, x_s^m) + d(x_s^m, x_s^{p(s)}) \end{aligned}$$

for all $m \in \mathbb{N}$. In particular for $m \geq \max\{p(r), p(s)\}$ the first and the last term on the right hand side are each less than $\epsilon/3$. As $m \to \infty$ the second term tends to $e([\hat{x}_r], [\hat{x}_s])$ which is less than $\epsilon/3$. So for all sufficiently large m, each of the three terms is smaller than $\epsilon/3$, and hence $d(y_r, y_s) < \epsilon$ for all $r, s \geq q$ as desired. Thus $\{y_n\}$ is a Cauchy sequence in $(X; d)$.

It remains to prove that $\{[\hat{x}_n]\}$ converges to $[\hat{y}]$ in $(\hat{X}; e)$. Note that $e([\hat{x}_n], [\hat{y}]) = \lim\limits_{m\to\infty} d(x_n^m, y_m)$. Now

$$\begin{aligned} d(x_n^m, y_m) &\leq d(x_n^m, x_n^{p(n)}) + d(x_n^{p(n)}, y_m) \\ &= d(x_n^m, x_n^{p(n)}) + d(y_n, y_m). \end{aligned}$$

The second term can be made arbitrarily small for all sufficiently large m, n because $\{y_n\}$ is a Cauchy sequence as we have just shown. The first term is smaller than $1/n$ for all $m \geq p(n)$. Hence it follows that $e([\hat{x}_n], [\hat{y}])$ can be made arbitrarily small for all sufficiently large n. We leave it to the reader to make this argument precise, for doing so would help him master the construction. ∎

The argument given for completeness of e is admittedly a little sticky. It is possible to do this part much more elegantly as will be pointed out in the exercises. However, we have preferred the present argument because it stays closer to the construction of $\hat{X}$.

Putting together what we have proved so far, we get

(4.4) Theorem: Every metric space can be isometrically embedded as a dense subspace of a complete metric space. ∎

(4.5) Definition: The space $(\hat{X}, e)$ is often called the **completion** of (X, d).

Obviously if $(X; d)$ is complete to start with then $\hat{X}$ will coincide with $h(X)$ and conversely. In general, the complement of $h(X)$ in $\hat{X}$ will be a measure of lack of completeness of the metric d. If we identify X with $h(X)$ then X becomes a subspace of $\hat{X}$ and we can inquire about the extendability of a function from X over $\hat{X}$. One such extension result will be given in the exercises.

Now that we know that every metric space has a completion we are in a position to give a characterisation of metrically topologically complete spaces defined in Section 2. First we need a definition.

(4.6) Definition: A topological space is said to be an **absolute G_δ** if it is metrisable and is a G_δ in every metric space in which it is topologically embedded.

(4.7) Theorem: A topological space is metrically topologically complete if and only if it is an absolute G_δ.

Proof: Let X be a topological space. Assume first that X is an absolute G_δ. Then X is metrisable and moreover X is a G_δ set in its completion $\hat{X}$. So by Theorem (2.8) X itself can be topologised by a complete metric. In other words X is metrically topologically complete.

Conversely assume X is metrically topologically complete. Let Y be any metric space into which X is topologically embedded. We regard X as a subspace of Y. We have to show that X is a G_δ in Y. Let Z be the closure of X in Y. Then X is dense in Z and we first show that X is a G_δ in the relative topology on Z. We are given that there exists a complete metric say d on X. (Caution: Y is also given to be metrisable, but the restriction of a metric on Y to X may not coincide with d.) For a subset A of X, by $\delta(A)$ we shall mean its diameter w.r.t. d. Now for each $n \in \mathbb{N}$ let $G_n = \{z \in Z :$ there exists some open nbd V of z in Z such that $\delta(X \cap V) < 2^{-n}\}$. Clearly each G_n is an open set in Z. We assert that $X = \bigcap_{n=1}^{\infty} G_n$. Clearly $X \subset G_n$ for all $n \in N$ and so $X \subset \bigcap_{n=1}^{\infty} G_n$. For the other way inclusion, suppose $z \in G_n$ for all $n \in \mathbb{N}$. For each $n \in \mathbb{N}$ fix an open nbd V_n of z in Z such that $\delta(X \cap V_n) < 2^{-n}$. We may suppose $V_n \supset V_{n+1}$ for all n, for otherwise we can replace V_n by $V_1 \cap V_2 \cap \ldots \cap V_n$. Since X is dense in Z, for each n we can find $x_n \in X \cap V_n$. Note that $d(x_n, x_{n+1}) \leq 2^{-n}$ for all $n \in \mathbb{N}$ since $\delta(X \cap V_n) < 2^{-n}$. From this it follows that $\{x_n\}$ is a Cauchy sequence in $(X; d)$ (see the argument given in the proof of Theorem (1.6)). By completeness of d, $\{x_n\}$ converges to some point say x in X. We claim $x = z$. If this were not so, then since Z is a Hausdorff space, x and z would have disjoint neighbourhoods say U and V in Z. Then for each n we could pick $y_n \in X \cap V \cap V_n$ and the sequence $\{y_n\}$ would also converge to x since $d(x_n, y_n) < 2^{-n}$ for all $n \in \mathbb{N}$. This would be a contradiction because the nbd U of x contains no terms of the sequence $\{y_n\}$. Thus $x = z$ and so $z \in X$. This completes the proof that $X = \bigcap_{n=1}^{\infty} G_n$ and so X is a G_δ set in Z. Note that the only topological property of Z used in this argument was that it is a Hausdorff space. (We have thus given the solution of Exercise (2.9).)

We still have to show that X is a G_δ in the larger space Y. We first write each G_n above as $H_n \cap Z$ where H_n is open in Y. Now Z is a closed subset of Y which is a metric space. Since all metric spaces are perfectly normal (see Exercise (7.4.6)), Z is a G_δ set in Y, say $Z = \bigcap_{n=1}^{\infty} K_n$ where each K_n is open in Y. Now for $n \in \mathbb{N}$, let $L_n = H_n \cap K_n$. Then L_n is open in Y and $X = \bigcap_{n=1}^{\infty} L_n$ showing that X is a G_δ in Y. Thus X is a G_δ in every metric space into which it is topologically embedded. So X is an absolute G_δ. ∎

Exercises

4.1 Verify that $\sim$ is an equivalence relation on the set of all Cauchy sequences in $(X; d)$.

4.2 Prove that the function $e : \hat{X} \times \hat{X} \to \mathbb{R}$, defined in Definition (4.2) is well-defined.

4.3 Prove that the function e is a metric on $\hat{X}$.

4.4 Write in a precise form the argument given at the end of the proof of Theorem (4.4) to show that $\{[\hat{x}_n]\}$ converges to $[\hat{y}]$ in $(\hat{X}; e)$.

4.5 Prove that a metric space X is complete if and only if it contains a dense subset D such that every Cauchy sequence in D has a limit in X. Use this result to give an alternate proof of completeness of the metric in the construction in Theorem (4.4).

4.6 Let $(X; d)$ be a metric space and $(\hat{X}, e)$ its completion. Prove that for any complete metric space $(Y; f)$, any uniformly continuous function $g : (X; d) \to (Y; f)$ can be extended uniquely to a uniformly continuous function from $(\hat{X}; e)$ to $(Y; f)$.

4.7 Prove that the extension property of $(\hat{X}; e)$ in the last exercise characterises the completion upto an isometry in the following sense. Suppose $(Z; k)$ is a complete metric space containing $(X; d)$ as a metric subspace and having the property that any uniformly continuous function from $(X; d)$ into a complete metric space can be uniquely extended to a uniformly continuous function from $(Z; k)$ into that space. Then there exists a unique isometry $\theta : (\hat{X}, e) \to (Z, k)$ such that $\theta(x) = x$ for all $x \in X$. (Hint: The result is really simple. The spirit of the proof is the same as that of Theorem (9.1.3).)

4.8 Prove that every complete metric space is the completion of any of its dense metric subspaces.

4.9 Prove or disprove that completion is a topological construction. In other words, if two metric spaces are homeomorphic, must their completions be also so?

4.10 Prove that the completion of a totally bounded metric space is compact and hence second countable.

4.11 A space is said to be **topologically totally bounded** if its topology can be induced by a totally bounded metric. Prove that a metrisable space is topologically totally bounded if and only if it is second countable.

4.12 Let $(X; d)$ be a metric space and $C(X)$ the space of all bounded real-valued functions on X with the metric e defined by $e(f, g) = \sup. \{|f(x) - g(x)| : x \in X\}$ for $f, g \in C(X)$. Fix $a \in X$. For $x \in X$ define $h_x : X \to \mathbb{R}$ by $h_x(u) = d(x, u) - d(a, u)$. Prove that $h_x \in C(X)$. Define $h : X \to C(X)$ by $h(x) = h_x$. Prove that h is an isometric embedding of X into $C(X)$.

Notes and Guide to Literature

The construction given here for completion is a standard one. With obvious modifications it applies to the construction of real numbers from rational numbers. (One obvious modification would be that to start with, in the definition of a Cauchy sequence of rational numbers, we must take only rational ϵ's.) The details (along with a proof of the equivalence of the real numbers so obtained with those obtained by the Dedekind construction) may be found, for example, in Goffman [1].

However, if we know the completeness of the usual metric on real numbers beforehand (as we indeed do) then Exercise (4.12) gives a very easy construction for completion, for the space $\overline{h(X)}$ in that exercise is a desired completion. This construction is due to Hajek [1].

See Sierpinski [1] for further results about complete metric spaces as well as other non-topological attributes of metrics.

Chapter Thirteen

Category Theory

In Chapter 3, we remarked that certain concepts in topology (such as that of a homeomorphism) have remarkable resemblance to those from other branches in mathematics (e.g., a congruence in geometry, a bijection in set theory and an isomorphism in group theory). The theory of categories seeks to isolate what is common to these various branches of mathematics. Its compass is virtually all of mathematics. The price one pays for its generality is the lack of depth in its most widely applicable results. Nevertheless, the theory is useful because it puts constructions in one branch into broader perspective and thereby often inspires similar constructions in other branches of mathematics.

In this Chapter, we shall study the fundamentals of category theory, illustrating them with topological spaces. Incidentally, 'category' used here has nothing to do with the Baire category in Chapter 12.

1. Basic Definitions and Examples

In set theory we deal with sets and functions that 'go' from one such set to another. In topology we deal with topological spaces and continuous functions that go from one such space to another. In group theory we deal with groups and homomorphisms that go from one such group to another. It follows that if category is to encompass all three, it should have some 'objects' and some way of going from one object to another. There should, in addition, be a law of composition. These are the basic ingredients of a category. More precisely,

(1.1) Definition: A **category** $\mathcal{C}$ consists of the following data:

(i) a class, denoted by $\mathcal{O}b(\mathcal{C})$, whose members are called **objects** of $\mathcal{C}$

(ii) a set $M(X, Y)$ for every pair of objects X, Y of $\mathcal{C}$ whose members are called **morphisms** or **$\mathcal{C}$-morphisms** from X to Y, ($M(X, Y)$ is also denoted by Mor (X, Y) or by $\text{Mor}_{\mathcal{C}}(X, Y)$)

(iii) a function called **composition,** $o_{X,Y,Z} : M(X, Y) \times M(Y, Z) \rightarrow M(X, Z)$ for every triple of objects X, Y, Z in $\mathcal{C}$ (if $f \in M(X, Y)$ and $g \in M(Y, Z)$, then $o_{X,Y,Z}(f, g)$ will be denoted by $g \circ f$)

(iv) an element, $1_X \in M(X, X)$ for each object X of $\mathcal{C}$, called the **identity morphism** on X,

such that the following conditions are satisfied:

(a) the morphism sets $M(X, Y)$ and $M(Z, W)$ are mutually disjoint **unless** $X = Z$ and $Y = W$ for all objects X, Y, Z, W

(b) the composition is associative, that is, for any objects X, Y, Z, W and for any $f \in \text{Mor}(X, Y)$, $g \in \text{Mor}(Y, Z)$ and $h \in \text{Mor}(Z, W)$, $o_{X,Z,W}(o_{X,Y,Z}(f, g), h) = o_{X,Y,W}(f, o_{Y,Z,W}(g, h))$, or in simpler notation, $h \circ (g \circ f) = (h \circ g) \circ f$

(c) the identity morphism acts as a two-sided identity, that is, for any objects, X, Y, Z of $\mathcal{C}$ and any $f \in \text{Mor}(X, Y)$ and any $g \in \text{Mor}(Y, Z)$, $1_Y \circ f = f$ and $g \circ 1_Y = g$.

Despite the length of the definition, it should be obvious that what it demands is the minimum possible. Some points require a little comment before we give examples. The totality of all objects in a category need not form a set, it is only a class; however for any two objects, the morphisms from one to the other must form a set. This distinction is important in order to avoid certain logical difficulties that would result in its absence. However, we shall not dwell on this point. If two objects X, Y belong to more than one category at a time, say $\mathcal{C}$ and $\mathcal{D}$, then the respective morphism sets from X to Y are denoted by $M_{\mathcal{C}}(X, Y)$ and $M_{\mathcal{D}}(X, Y)$. However, where the category in question is understood, it may be suppressed from the notation. If $f \in \text{Mor}(X, Y)$ then X is called the **domain** and Y the **codomain** of f. In view of condition (a) these are uniquely defined. Instead of writing $f \in \text{Mor}(X, Y)$ we also write $f : X \to Y$ or $X \xrightarrow{f} Y$. Note that X and Y need not be sets and even when they are so, f need not necessarily be a function from X to Y. Although this will be the case in many categories, we caution the reader against thinking that 'morphism' is just another name for a function. The best intuition for a morphism is that of an arrow; the domain and the codomain then correspond naturally to the tail and the head of the arrow. In fact many authors do call morphisms as arrows.

Let us now give some examples of categories. The foremost is the category of sets and functions. It is commonly denoted by $\mathbb{E}$ns (from 'ensemble' which in French means a collection or a set) or by $\mathcal{S}$. Its objects are all sets. For any two sets X, Y, a morphism from X to Y is simply a function from X to Y. In order to satisfy condition (a), it is necessary here that a function from X to Y be defined as a triple (X, Y, f) where f is a subset of $X \times Y$ with the property that for each $x \in X$, there is a unique $y \in Y$ such that $(x, y) \in f$. The composition function is the usual composition and for each set X, l_X is the identity function from X to itself. Conditions (a) to (c) are trivially verified and thus $\mathbb{E}$ns is a category. It is not necessary to take all sets, or to take all functions from one set to another. As long as conditions (b) and (c) are obeyed, we can consider some selected functions and still get a category. The category so obtained will be a subcategory of $\mathbb{E}$ns, a term we now define.

(1.2) Definition: A category $\mathcal{D}$ is said to be **subcategory** of a category $\mathcal{C}$ if

(i) $\mathcal{Ob}(\mathcal{D}) \subset \mathcal{Ob}(\mathcal{C})$, that is, objects of $\mathcal{D}$ are also objects of $\mathcal{C}$.

(ii) for any objects X, Y of $\mathcal{D}$, $\mathrm{Mor}_{\mathcal{D}}(X, Y) \subset \mathrm{Mor}_{\mathcal{C}}(X, Y)$

(iii) for any objects X, Y, Z of $\mathcal{D}$, and $f \in \mathrm{Mor}_{\mathcal{D}}(X, Y)$, $g \in \mathrm{Mor}_{\mathcal{D}}(Y, Z)$, $g \circ f$ is the same morphism from X to Z whether the composition is carried out in $\mathcal{D}$ or $\mathcal{C}$.

Note that we are not requiring separately that for any object X of $\mathcal{D}$, the identity morphism on X in $\mathcal{D}$ coincides with that in $\mathcal{C}$; this will follow as a result of the uniqueness of the identity morphism, which is easy to establish and left to the reader. We see from the definition that the category of sets and bijections is a subcategory of $\mathbb{E}$ns. If in condition (ii) above, we have equality instead of inclusion, then $\mathcal{D}$ is said to be a **full subcategory** of $\mathcal{C}$. As an example, the category of all finite sets and functions is a full subcategory of $\mathbb{E}$ns.

Another important category is the category of topological spaces and continuous functions. It is commonly denoted by $\mathbb{T}$op or by $\mathcal{T}$. Its objects are topological spaces and morphisms from one space to another are continuous functions. Composition is the usual one. Since compositions of continuous functions are continuous and the identity function from each space to itself is continuous, it follows that $\mathbb{T}$op is indeed a category. Again, if we take only homeomorphisms instead of all continuous functions, or restrict the spaces to be, say, compact, connected or Hausdorff etc. we get subcategories of $\mathbb{T}$op.

There is then $\mathcal{G}$, the category of groups and homomorphisms, with $\mathcal{Ab}$, the category of abelian groups as a full subcategory. Similarly we can have categories of rings, of fields or of some other algebraic structures and corresponding homomorphisms. In such cases one can of course consider all set-theoretic functions from one object to another and still get a category. Thus, for example, we may consider a category whose objects are groups and whose morphisms are any functions (not necessarily group homomorphisms) from one group to another. However such a category is of little interest. So, when we say the category of groups, it is understood that the morphisms are group homomorphisms. Often categories are specified merely by their objects, the morphisms being understood to be the ones that are most obvious and interesting in that context. Thus 'category of topological spaces' means the one whose objects are topological spaces and whose morphisms are continuous functions. Where the morphisms are functions (with or without some additional properties) the rule of composition is almost exclusively the usual one and is therefore not explicitly stated.

In addition to the examples given above, many structures familiar in mathematics can be viewed as categories. We present two examples. Recall that a monoid is a set with an associative binary operation having an identity element. Let $(M, *)$ be such a monoid with $e \in M$ as the identity element. Let X be any symbol. Consider a category $\mathcal{M}$ whose sole object is X and $\mathrm{Mor}(X, X)$ is the set M with composition being defined by $*$. The defining axioms of a monoid then make $\mathcal{M}$ into a category. Conversely given any category with just one object, the set of morphisms from that object to itself forms a monoid, the binary operation being defined by the composition. For this reason, monoids are often defined as categories with only one object.

As our second example, let X be a set and suppose $\leq$ is a transitive and reflexive binary relation on X. Consider a category $\mathcal{X}$ whose objects are elements of X. For $x, y \in X$ we let $\text{Mor}_{\mathcal{X}}(x, y)$ be the set $\{(x, y)\}$ if $x \leq y$ and the empty set otherwise. Transitivity of $\leq$ gives a unique composition function for every triple of objects of $\mathcal{X}$ while its reflexivity ensures the existence of identity morphisms. Conversely any category in which there is at most one morphism between any two objects gives rise to a reflexive and transitive relation on the class of its objects.

For theoretical purposes it is convenient to construct certain new categories from the old ones. We mention a few such constructions.

Let $\mathcal{C}$ be any category. We define its **dual** or **opposite** category $\mathcal{C}^{\text{op}}$ as follows. The objects of $\mathcal{C}^{\text{op}}$ are the same as those of $\mathcal{C}$. However, for two objects X and Y we define $\text{Mor}_{\mathcal{C}^{\text{op}}}(X, Y)$ to be $\text{Mor}_{\mathcal{C}}(Y, X)$. In other words, we interchange the domain and the codomain of every morphism of $\mathcal{C}$ and get a morphism in $\mathcal{C}^{\text{op}}$. The law of composition is also changed accordingly. Thus if $f \in \text{Mor}(X, Y)$ and $g \in \text{Mor}(Y, Z)$ in $\mathcal{C}^{\text{op}}$, then $g \in \text{Mor}_{\mathcal{C}}(Z, Y)$ and $f \in \text{Mor}_{\mathcal{C}}(Y, X)$ and we define $g \circ f$ in $\mathcal{C}^{\text{op}}$ to be $f \circ g$ in $\mathcal{C}$. The conditions in the definition are trivially verified. Intuitively the arrows in $\mathcal{C}^{\text{op}}$ go in the opposite directions of those in $\mathcal{C}$ and hence the name. It is clear that the dual of $\mathcal{C}^{\text{op}}$ is the original category, $\mathcal{C}$. As an example of dual categories, we see that if $(X, \leq)$ is a partially ordered set regarded as a category then its dual is the partially ordered set $(X, \geq)$.

Given two categories $\mathcal{C}$, $\mathcal{D}$ we can define their product $\mathcal{C} \times \mathcal{D}$ in the obvious way. Objects of $\mathcal{C} \times \mathcal{D}$ are all ordered pairs (X, Y) with $X \in \mathcal{O}b(\mathcal{C})$ and $Y \in \mathcal{O}b(\mathcal{D})$. If (X_1, Y_1), (X_2, Y_2) are two objects of $\mathcal{C} \times \mathcal{D}$ we define $\text{Mor}((X_1, Y_1), (X_2, Y_2))$ as $\text{Mor}_{\mathcal{C}}(X_1, X_2) \times \text{Mor}_{\mathcal{D}}(Y_1, Y_2)$. The composition is defined 'co-ordinatewise'. The resulting category $\mathcal{C} \times \mathcal{D}$ is called the **product** of the categories $\mathcal{C}$ and $\mathcal{D}$. More generally one can apply the method to form products of arbitrary families of categories.

We mention yet another construction which may appear rather clumsy at this stage. However, its importance will be borne out when we shall discuss products and coproducts. Let $\mathcal{C}$ be any category and $\{X_i : i \in I\}$ some fixed indexed family of its objects. By a **cone** over this family we mean a pair $(X, \{\lambda_i\}_{i \in I})$ where $X \in \mathcal{O}b(\mathcal{C})$ and $\{\lambda_i\}_{i \in I}$ is a collection of morphisms in $\mathcal{C}$ such that for each $i \in I$, $\lambda_i \in \text{Mor}_{\mathcal{C}}(X, X_i)$. The term cone is used here simply because, it can be graphically represented by a family of arrows all

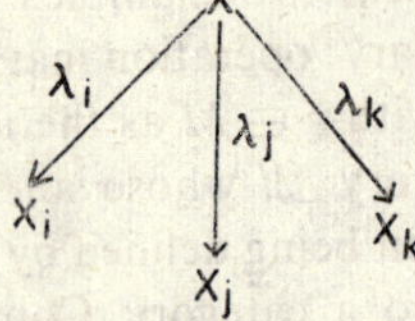

Fig. 1. Cone over. $\{X_i\}_{i \in I}$

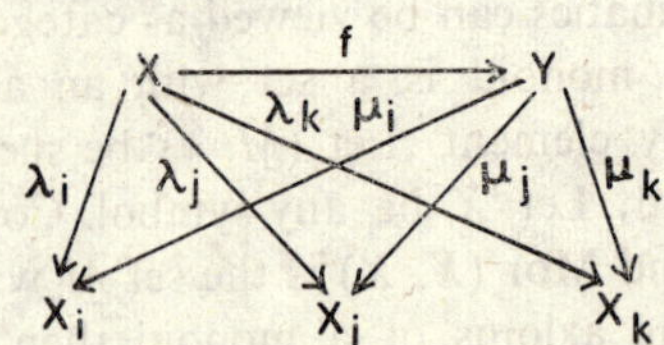

Fig. 2. Morphisms of cones.

originating at the 'vertex', which is their common domain.

We now define a category $\mathscr{D}$ whose objects are such cones. Given two such cones $(X, \{\lambda_i\})$ and $(Y, \{\mu_i\})$, a morphism from $(X, \{\lambda_i\})$ to $(Y, \{\mu_i\})$ is a morphism f in $\mathcal{C}$ from X to Y such that for each $i \in I$, $\mu_i \circ f = \lambda_i$, or equivalently the triangle

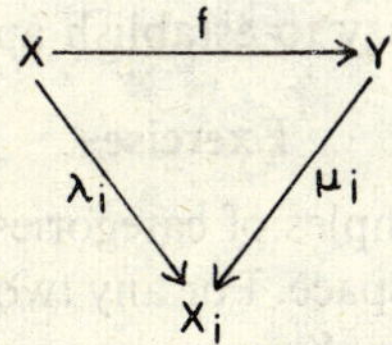

commutes for each $i \in I$. It is easy to check that $\mathscr{D}$ is a category. We can similarly define a **co-cone** over $\{X_i\}_{i \in I}$ to be a pair $(X, \{\lambda_i\})$ where $\lambda_i \in$ Mor (X_i, X) for each $i \in I$ and form a category of such co-cones. It is easy to see that a co-cone in $\mathcal{C}$ is precisely a cone in the dual category $\mathcal{C}^{\text{op}}$ and vice versa. The notions of a cone and a co-cone are therefore said to be **dual** to each other. It will be seen that every notion in category theory has a dual notion and along with every theorem there is a dual theorem. A statement holds in a category if and only if its dual holds in the dual category.

We conclude this section with a few definitions. Throughout, $\mathcal{C}$ will denote an arbitrary, but fixed category, and $f : X \to Y$ a morphism in $\mathcal{C}$.

(1.3) Definition: f is said to be an **epimorphism** if for any object Z of $\mathcal{C}$ and any morphisms $g_1, g_2 : Y \to Z$ in $\mathcal{C}$ $g_1 \circ f = g_2 \circ f$ implies that $g_1 = g_2$. Dually f is said to be a **monomorphism** if for any object Z of $\mathcal{C}$ and any two morphisms $h_1, h_2 : Z \to X$ in $\mathcal{C}, f \circ h_1 = f \circ h_2$ implies that $h_1 = h_2$.

In $\mathbb{E}$ns, the category of sets, an epimorphism is merely a surjective function and a monomorphism is an injective function. This is not the case in general. In fact, in an arbitrary category, morphisms need not be functions and so terms such as injective or surjective functions need not even make sense. If morphisms are functions (with or without additional properties) in a category $\mathcal{C}$, then it is easy to see that every morphism which is surjective as a function is an epimorphism. The converse does not hold. For example, in the category of Hausdorff topological spaces, if X is a dense subspace of a Hausdorff space Y then the inclusion map $i : X \to Y$ is an epimorphism although it need not be a set-theoretic surjection.

(1.4) Definition: f is said to a **retraction** if it is right invertible, that is, if there exists a morphism $g : Y \to X$ in $\mathcal{C}$ such that $f \circ g = 1_Y$. A **coretraction** is defined dually.

It is easy to be check that every retraction is an epimorphism. The converse holds in the category of sets but not in general. Note that a retraction in the topological sense (see Exercise (7.4.1)) is also a retraction in the categorical sense in $\mathbb{T}$op.

Finally we define an equivalence in a category.

(1.5) Definition: f is said to be an **equivalence** (or sometimes an **isomorphism**) in $\mathcal{C}$, if it is invertible, that is, if there exists a morphism $g : Y \to X$ such that $g \circ f = 1_X$ and $f \circ g = 1_Y$. Two objects are said to be **equivalent** in $\mathcal{C}$ if there exists an equivalence between them.

We now see that an equivalence in $\mathbb{E}$ns is a bijection, one in $\mathbb{T}$op is a homeomorphism while one in $\mathcal{G}$ is an isomorphism. Results about equivalences, retractions etc. are easy to establish and will be given as exercises.

Exercises

1.1 Give at least three examples of categories other than those given here.

1.2 Let X be a topological space. For any two points x, y of X let Mor (x, y) be the set of all paths in X from x to y. If we define composition of paths by concatenation (see Chapter 6, Section 3), why can we not get a category whose objects are points of X and morphisms path in X?

1.3 With the notation as above, call two paths, α, β in X from x to y as **equivalent** if there exists a map $H : I \times I \to X$ (I being the unit interval) such that $H(s, o) = \alpha(s)$. $H(s, 1) = \beta(s)$, $H(o, s) = x$ and $H(1, s) = y$ for all $s \in I$. Prove that this defines an equivalence relation on the set of paths from x to y. Prove moreover that this relation is compatible with concatenation.

*1.4 Let X be a space. For $x, y \in X$, let Mor (x, y) be the set of all equivalence classes of paths in X from x to y. Prove that if composition is defined by concatenation of classes of paths then we get a category whose objects are points in X. (Hint: For $x \in X$, the equivalence class of the constant path at x is the identity morphism. The associativity requirement boils down to showing that if α, β and γ are paths in X from (say) x to y, y to z and z to w respectively then the two paths $(\alpha * \beta) * \gamma$ and $\alpha * (\beta * \gamma)$ are equivalent in the sense above. The proof is routine once you decipher Figure 3 correctly.

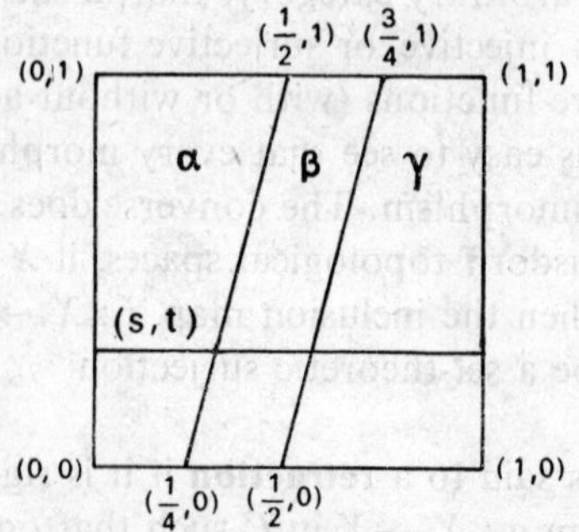

Fig. 3. Associativity of concatenation.

1.5 Prove that every retraction is an epimorphism and that every coretraction is a monomorphism. Prove that the converses hold in $\mathbb{E}$ns but not in general.

1.6 Let $\mathcal{D}$ be a subcategory of a category $\mathcal{C}$ and f a morphism in $\mathcal{D}$ (and hence in $\mathcal{C}$ as well). Prove that if f is an epimorphism in $\mathcal{C}$, so it is in $\mathcal{D}$. Show by an example that the converse is false. Do the same for monomorphisms.

1.7 With $\mathcal{C}$, $\mathcal{D}$ and f as above, prove that if f is a retraction in $\mathcal{D}$ it is so in $\mathcal{C}$. Prove that the converse need not be true.

1.8 Prove that a morphism is an equivalence if and only if it is both a retraction and a coretraction.

1.9 Prove that the composite of two monomorphisms is a monomorphism. Prove similar results for epimorphisms, retractions, coretractions and equivalences.

1.10 Let $\mathcal{C}$ be a category, X, Y, Z three of its objects and $f: X \to Z$, $g: Y \to Z$ two morphisms in $\mathcal{C}$. Consider a triple (W, h, k) where $W \in \mathcal{O}b(\mathcal{C})$ and $h: W \to X$, $k: W \to Y$ are morphisms in $\mathcal{C}$ such that $f \circ h = g \circ k$, i.e., such that the following diagram is commutative:

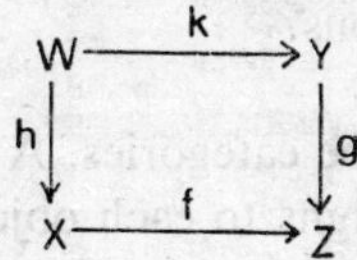

Given two such triples (W_1, h_1, k_1) and (W_2, h_2, k_2), define a morphism from (W_1, h_1, k_1) to (W_2, h_2, k_2) to be a morphism $\lambda \in \text{Mor}_{\mathcal{C}}(W_1, W_2)$ such that $h_2 \circ \lambda = h_1$ and $k_2 \circ \lambda = k_1$. Prove that this way we get a category of all such triples.

1.11 Dualise the construction in the last exercise.

Notes and Guide to Literature

Category theory is a relatively recent development in mathematics. Categories were defined by Eilenberg and MacLane around 1950. Standard references are MacLane [1], Mitchell [1] and Freud [1].

In the definition of a category, if in addition we require that each morphism set be an abelian group and that the composition functions be bilinear (i.e. linear in each argument) then we get what is known as an abelian category. The category of abelian groups is an example of an abelian category. It is a theorem of Mitchell that every abelian category is isomorphic (in a sense which we shall define in the next section) to a full subcategory of the category of abelian groups. For abelian categories, the morphism sets are often denoted by Hom instead or Mor.

2. Functors and Natural Transformations

It often happens in mathematics that to each object of a category we associate an object of another category which reflects the properties of the original object. As an example, to each topological space we associate $C(X)$, the set of all continuous real-valued functions on X. Also to each map $f : X \to Y$, we associate the function $f^{**} : C(Y) \to C(X)$ defined by $f^{**}(\lambda) = \lambda \circ f$ for $\lambda \in C(Y)$. Note here that for each space X, $C(X)$ can be given the structure of a commutative ring with identity if we define addition and multiplication of functions pointwise. In other words given $\lambda, \mu \in C(X)$ we define $\lambda + \mu$ to be the function from X to $\mathbb{R}$ defined by $(\lambda + \mu)(x) = \lambda(x) + \mu(x)$ for all $x \in X$. Similarly we define $\lambda\mu$. It is now clear that if $f : X \to Y$ is continuous then $f^{**} : C(Y) \to C(X)$ is a ring homomorphism. In this way, to each topological space we associate a commutative ring and to each map from one space to another a ring homomorphism between the corresponding rings. The advantage of such an association is that information about one category can lead to information about another category. In this section we shall study a most general and systematic way of going from one category to another. Inasmuch as the basic features of a category are compositions and identity morphisms, it is but natural to require that they be preserved under a transition from one category to another. This leads to the following basic definitions.

(2.1) Definition: Let $\mathcal{C}$, $\mathcal{D}$ be categories. A **covariant functor** F from $\mathcal{C}$ to $\mathcal{D}$ is an association which assigns to each object X of $\mathcal{C}$ an object $F(X)$ of $\mathcal{D}$ and to each morphism $f : X \to Y$ in $\mathcal{C}$ a morphism $F(f) : F(X) \to F(Y)$ in $\mathcal{D}$ in such a way that

(i) for any $f : X \to Y$ and $g : Y \to Z$ in $\mathcal{C}$,

$$F(g \circ f) = F(g) \circ F(f) \text{ in } \mathcal{D}$$

and

(ii) for any object X of $\mathcal{C}$, $F(1_X) = 1_{F(X)}$ in $\mathcal{D}$

(2.2) Definition: Let $\mathcal{C}$, $\mathcal{D}$ be categories. A **contravariant functor** F from $\mathcal{C}$ to $\mathcal{D}$ is an association which assigns to each object X of $\mathcal{C}$ an object $F(X)$ of $\mathcal{D}$ and to each morphism $f : X \to Y$ in $\mathcal{C}$ a morphism $F(f) : F(Y) \to F(X)$ in $\mathcal{D}$ in such a way that

(i) for any $f : X \to Y$ and $g : Y \to Z$ in $\mathcal{C}$,

$$F(g \circ f) = F(f) \circ F(g)$$

and

(ii) for any object X of $\mathcal{C}$, $F(1_X) = 1_{F(X)}$ in $\mathcal{D}$.

Conditions (i) and (ii) in either case are called **functorial properties**. It is immediate that the composition of two covariant or two contravariant functors is a covariant functor, while the composition of a covariant and a contravariant functor is contravariant. Note that a contravariant functor reverses the direction of the 'arrows'. Apart from this, there is no difference between covariant and contravariant functors. Indeed either can be converted to another by a simple trick. Let $\mathcal{C}^{op}$ be the opposite category of $\mathcal{C}$. Then there is a contravariant functor, $I : \mathcal{C} \to \mathcal{C}^{op}$ defined by $I(X) = X$ for all objects X of $\mathcal{C}$ and $I(f) = f$ for all morphisms f. Given any functor F from $\mathcal{C}$ to a category $\mathcal{D}$, there is a unique functor $F^{op} : \mathcal{C}^{op} \to \mathcal{D}$ such that $F^{op} \circ I = F$. Clearly F is covariant iff F^{op} is contravariant. Since the properties of $\mathcal{C}$ are dual to those of $\mathcal{C}^{op}$, it follows that properties of F and F^{op} are dual to each other as far as the role of the domain category is concerned. In practice we come across both types of functors. However, we shall generally prove theorems for covariant functors, leaving their analogues for contravariant functors to the reader.

An example of a contravariant functor was already initiated above. Let $\mathcal{C} = \mathbb{T}\text{op}$, the category of topological spaces and $\mathcal{D}$ = category of all rings and ring homomorphisms. Define $F : \mathcal{C} \to \mathcal{D}$ by $F(X) = C(X)$, the ring of continuous real-valued functions on X, and $F(f) = f^{**}$ as defined above. It is easy to check that f^{**} is a ring homomorphism (recall that the addition and multiplication of functions is pointwise). The functorial properties of F are trivial to establish. Let $f : X \to Y$ and $g : Y \to Z$ be morphisms in $\mathbb{T}\text{op}$. Then we have to show $(g \circ f)^{**} = f^{**} \circ g^{**}$ as morphisms from $C(Z)$ to $C(X)$. Now let $\lambda \in C(Z)$, i.e. λ is a continuous function from Z to $\mathbb{R}$. Then $(g \circ f)^{**}(\lambda) = \lambda \circ (g \circ f)$ which equals $(\lambda \circ g) \circ f$ by associativity of composition. But $(\lambda \circ g) \circ f$ is precisely $f^{**}(g^{**}(\lambda))$. As this holds for all $\lambda \in C(Z)$, we get the desired equality. The verification of the second functorial property is even simpler, requiring only that the composition with an identity morphism leaves a morphism unchanged.

A similar construction gives a contravariant functor F from any category $\mathcal{C}$ to $\mathbb{E}\text{ns}$, the category of sets. Let Z be a fixed object of $\mathcal{C}$. Define $F(X) = \text{Mor}_{\mathcal{C}}(X, Z)$. For $f : X \to Y$ in $\mathcal{C}$, define $F(f) : F(Y) \to F(X)$ to be f^{**}, where $f^{**}(\lambda) = \lambda \circ f$ for $\lambda \in \text{Mor}_{\mathcal{C}}(Y, Z)$. The fact that F is indeed a functor is easy to establish. We get a covariant functor $G : \mathcal{C} \to \mathbb{E}\text{ns}$ if we let $G(X) = \text{Mor}_{\mathcal{C}}(Z, X)$ and for $f : X \to Y$ in $\mathcal{C}$, $G(f)$ to be $f_{**} : G(X) \to G(Y)$ defined by $f_{**}(\lambda) = f \circ \lambda$ for $\lambda : Z \to X$. The functors F, G are often denoted respectively by $\text{Mor}_{\mathcal{C}}(-, Z)$ and $\text{Mor}_{\mathcal{C}}(Z, -)$.

We now give various examples of functors. In each case the verification of functorial properties is routine and left to the reader. Moreover, in most cases, once the action of the functor on the objects of the domain category is specified, it is generally obvious how it behaves for morphisms of the domain category. For example, let Z be a fixed set and consider the functor $F : \mathbb{E}\text{ns} \to \mathbb{E}\text{ns}$ defined by $F(X) = X \times Z$. In this case, it is obvious that if $f : X \to Y$ is a morphism in $\mathbb{E}\text{ns}$ then $F(f) : X \times Z \to Y \times Z$ should be the function $f \times 1_Z$ defined by $(f \times 1_Z)(x, z) = (f(x), z)$. In such cases it is customary to specify the action of the functor on objects only.

The simplest to remember are the so called 'forgetful functors'. It often happens that objects of one category are obtained by putting some additional structure on those of some other category. For example, a topological space is obtained by putting a topology on a set, that is an object of $\mathbb{E}\text{ns}$. A ring is obtained from an abelian group by defining an operation of multiplication, while an abelian group itself is obtained from a set. In all such cases one can define a functor from the first category to the second which simply 'forgets' the additional structure. As a concrete example define $F : \mathbb{T}\text{op} \to \mathbb{E}\text{ns}$ by $F(X; \mathcal{T}) = X$, the underlying set of the space $(X; \mathcal{T})$. Here F forgets the topology $\mathcal{T}$. If $f : (X, \mathcal{T}) \to (Y, \mathcal{U})$ is a morphism in

Top, then f is a function from X to Y which is $\mathcal{T}$-$\mathcal{U}$ continuous. We ignore its continuity and define $F(f) = f$, regarded as a set-theoretic function from X to Y. We can similarly define forgetful functors from the category $\mathcal{G}$ of groups to $\mathbb{E}$ns and from the category of rings to the category of sets. We can also define functors which are only 'partially' forgetful, in that they do not ignore the entire structure on the objects of the domain category but only a part of it. For example, there is a functor from the category of rings to the category of abelian groups which assigns to a ring $(R, +, \cdot)$ the abelian group $(R, +)$.

Sometimes it is interesting to reverse the situation, that is to say, we pass from one category to another whose objects carry more structure than those of the first category. For example, we define a functor F from $\mathbb{E}$ns to Top as follows. For each set X, let $(X, \mathcal{D}_X)$ be the discrete space with underlying set X. Note that any set-theoretic function $f: X \to Y$ is continuous if we put the discrete topologies on both X and Y. So we get a well-defined functor $F: \mathbb{E}\text{ns} \to \text{Top}$ if we let $F(X)=(X; \mathcal{D}_X)$. We could have as well chosen to put the indiscrete topology on every set and gotten a functor from $\mathbb{E}$ns to Top. Another important example of this sort of construction comes from algebra. For each set X, let $F(X)$ be the free abelian group on X. This means that $F(X)$ consists of all formal sums of the form $\sum_{i=1}^{k} n_i x_i$ with $n_i \in \mathbb{Z}$, $x_i \in X$, which are added by adding the coefficients of x_i's. If $f: X \to Y$ is any function then there is a well-defined group homomorphism $f_*: F(X) \to F(Y)$ given by $f_*(\Sigma n_i x_i) = \Sigma n_i f(x_i)$. We then define a functor $F: \mathbb{E}\text{ns} \to \mathcal{A}b$ by $F(X) =$ the free abelian group on the set X and $F(f) = f_*$ for $f: X \to Y$ in $\mathbb{E}$ns.

Many topological constructions are functorial in nature. For example, let us consider the Alexandroff compactification. If $f: X \to Y$ is a map from one space to another then we define $f^+: X^+ \to Y^+$ by $f^+(x) = f(x)$ for $x \in X$ and $f^+(\infty) = \infty$ where the same symbol ∞ is used in the construction of X^+ as well as Y^+. Continuity of f^+ at a point of X is no problem. Unfortunately f^+ need not be continuous at the point ∞ of X. A simple counterexample is obtained by taking $X = \mathbb{R}$, $Y = S^1$ and $f(x) = e^{2\pi i x}$ for $x \in \mathbb{R}$. If f satisfies a certain condition then it would follow that f^+ is continuous even at ∞.

(2.3) Definition: A map $f: X \to Y$ is said to be **proper** if the inverse image of each compact subset of Y is compact.

If X is compact and Y is Hausdoroff then every map from X to Y is proper. It is easy to see that the composite of two proper maps is again a proper map. We therefore get a category $\mathcal{P}$ Top whose objects are all topological spaces and morphisms are proper maps. This category is of course a sub-category of Top. It is easy to check that if $f: X \to Y$ is a proper map then $f^+: X^+ \to Y^+$ is continuous. Thus we get a covariant functor $F: \mathcal{P}\text{ Top} \to \text{Top}$ defined by $F(X) = X^+$ and $F(f) = f^+$. We leave it to the reader to check that the Stone-Čech compactification defines a functor from the category of Tychonoff spaces to the category of compact Hausdorff spaces.

As the final example, consider the category $\mathcal{GP}$ whose objects are pairs of groups (G, K) where G is a group and K is a normal subgroup of G. If (G, K) and (H, L) are two such pairs, by a morphism $f : (C, K) \to (H, L)$ we mean a group homomorphism $f : G \to H$ satisfying $f(K) \subset L$. It is then easy to check that $\mathcal{GP}$ is in fact a category. We define two functors from $\mathcal{GP}$ to $\mathcal{G}$. The first functor F forgets the normal subgroup. Thus $F(G, K) = G$ for a group pair (G, K). However, the second functor Q is interesting. It is called the quotient group functor. For an object (G, K), we set $Q(G, K) = G/K$, the quotient group of G by K, that is, the group of all cosets of K in G. A morphism $f : (G, K) \to (H, L)$ induces a well defined homomorphism $f_* : G/K \to H/L$ (since $f(K) \subset L$) and we define $Q(f)$ to be f_*. It is easy to check that Q is a functor. Note that for each group pair (G, K) there is a 'canonical' group homomorphism from G to G/K, i.e. from $F(G, K)$ to $Q(G, K)$. Its significance will be fully discussed later in this section.

The importance of functors in mathematics stems from the following simple result.

(2.4) Proposition: Let $\mathcal{C}$, $\mathcal{D}$ be categories, $F : \mathcal{C} \to \mathcal{D}$ a functor and $f : X \to Y$ a morphism in $\mathcal{C}$. Then,

(i) if F is covariant and f is a retraction (or a coretraction) so is $F(f)$ in $\mathcal{D}$,

(ii) if F is contravariant and f is a retraction (or a coretraction) then $F(f)$ is a coretraction (respectively a rectraction) in $\mathcal{D}$,

(iii) if f is an equivalence in $\mathcal{C}$ then $F(f)$ is an equivalence in $\mathcal{D}$, whether F is covariant or contravariant.

Proof: The arguments are very straightforward. By way of illustration, we prove (ii). So assume F is contravariant. Then $F(f) : F(Y) \to F(X)$ in $\mathcal{D}$. If f is a retraction in $\mathcal{C}$ then there exists $g : Y \to X$ such that $f \circ g = 1_Y$. Then we have $F(g) \circ F(f) = F(f \circ g) = F(1_Y) = 1_{F(Y)}$ by functorial properties of F. Thus $F(f)$ is a coretraction in $\mathcal{D}$. A similar proof shows that if f is a coretraction in $\mathcal{C}$ then $F(f)$ is a retraction in $\mathcal{D}$. ∎

Note that the converses of the statements proved above are false. For example, let $F : \mathbb{T}\mathrm{op} \to \mathbb{E}\mathrm{ns}$ be the forgetful functor. Let X be any set and $\mathcal{T}$, $\mathcal{U}$ topologies on X such that $\mathcal{T} \supset \mathcal{U}$. Let $f : (X; \mathcal{T}) \to (X; \mathcal{U})$ be the map, $f(x) = x$ for $x \in X$. (Caution: f is not the identity morphism; why?) Then if $\mathcal{T}$ is strictly stronger than $\mathcal{U}$, f is not an equivalence in $\mathbb{T}\mathrm{op}$, although $F(f)$ is the identity function on the set X, which is an equivalence in $\mathbb{E}\mathrm{ns}$. Thus the fact that $F(f)$ is an equivalence says nothing about f. However, what the proposition does say is that if $F(f)$ is not an equivalence, then f cannot be an equivalence. In particular, if X, Y are objects of $\mathcal{C}$ and $F(X)$ and $F(Y)$ are not equivalent objects in $\mathcal{D}$ then X and Y cannot be equivalent in $\mathcal{C}$. Similarly if $F : \mathcal{C} \to \mathcal{D}$ is covariant and $F(X)$ cannot be a retract of $F(Y)$ in $\mathcal{D}$ then it follows that X cannot be a retract of Y in $\mathcal{C}$.

Thus the proposition is useful in proving some non-existence theorems.

A striking application of this procedure is in proving the well-known Brouwer fixed point theorem which asserts that the closed unit ball D^n in $\mathbf{R}^n$ has the fixed point property. It is not hard to show that this assertion is equivalent to showing that the unit sphere S^{n-1} cannot be a retract of D^n. Intuitively this seems obvious enough. For if such a retraction exists then each point of D^n would have to move to a point on the boundary S^{n-1} while those points which are on S^{n-1} itself stay where they are. Because there are no 'holes' in D^n and because the retraction is to be continuous, there would have to be a point in the interior of D^n which 'would not know where to go'! However, a rigorous proof is far from trivial. Although some combinatorial proofs are known, a truly elegant proof is based on the use of functors. Suppose $r : D^n \to S^{n-1}$ were a retraction in the topological sense and hence a retraction in the categorical sense in the category Top. With a good deal of hard work it can be proved that there exists a covariant functor $H : \text{Top} \to \mathcal{A}b$ such that $H(S^{n-1})$ is isomorphic to $\mathbb{Z}$, the additive group of integers while $H(D^n)$ is isomorphic to the trivial group 0. By the proposition above, $H(r)$ would be a retraction. But it is easy to show that there can be no retraction in $\mathcal{A}b$ from 0 to $\mathbb{Z}$ and thus we get a contradiction, thereby establishing the Brouwer fixed point theorem. Of course the construction of the functor H is by no means easy and well beyond our scope.

So far we considered functors of 'one variable'. We can also consider functors of several variables. For example, if X, Y are objects of a category $\mathcal{C}$ then we can define $F(X, Y) = \text{Mor}_{\mathcal{C}}(X, Y)$ which is an object in $\mathbb{E}$ns. If $f : Z \to X$ and $g : Y \to W$ are morphisms in $\mathcal{C}$ then we get a function $F(f, g) : F(X, Y) \to F(Z, W)$ defined by $F(f, g)(\lambda) = g \circ \lambda \circ f$ for $\lambda \in \text{Mor}(X, Y)$. This way we get a functor from the product category $\mathcal{C} \times \mathcal{C}$ into $\mathbb{E}$ns. Note that this functor is contravariant in the first variable and covariant in the second. If we hold either variable as fixed, we get functors of one variable which were already considered above. As another example of a functor of two variables we have the product functor from $\text{Top} \times \text{Top}$ to Top which associates to a pair (X, Y) of spaces their topological product $X \times Y$. This functor is covariant in both the variables.

Functors are means of passing from one category to another and resemble functions in many respects. It is tempting to consider a category whose objects are categories and whose morphisms are functors from one category to another. This procedure is feasible where the functors from one category to another constitute a set and not just a class; otherwise difficulties arise similar to those in attempting to form 'a set of all sets'. Nevertheless we can define concepts such as isomorphism of categories by means of functors.

Sometimes we have two functors from the same category to the same category. For example we had two functors F and Q from the category $\mathcal{GP}$ of group pairs to $\mathcal{G}$, the category of groups. Given a group G and a normal subgroup K, in group theory there is a 'canonical' or 'natural' quotient homomorphism $\theta : G \to G/K$ defined by $\theta(g) = Kg$, the coset of K in G containing g. Here 'canonical' means simple, while 'natural' means something which depends only on G and K and not on anything else, for example on a particular set of generators for G. The idea is that the construction of the quotient homomorphism is the same for all group pairs and is not something confined to a particular group pair. If (H, L) is another such group pair and $f : (G, K) \to (H, L)$ is a morphism, then the following diagram is commutative:

$$\begin{array}{ccc} F(G, K) = G & \xrightarrow{\theta_1} & G/K = Q(G, K) \\ \downarrow f = F(f) & & \downarrow f_* = Q(f) \\ F(H, L) = H & \xrightarrow{\theta_2} & H/L = Q(H, L) \end{array}$$

where θ_1, θ_2 denote the respective quotient homomorphisms.

This is the precise meaning of naturality of the quotient homomorphism. If its definition involved something like a particular set of generators of G (or H) then it is extremely unlikely that the diagram above would be commutative, inasmuch as there is no guarantee that f would transform the particular set of generators for G to the one for H. A reader familiar with linear algebra may recall here that if V is a finite-dimensional vector space over a field F then V is isomorphic to its dual V^* (i.e. the vector space of all linear transformations from V to F). However, the construction of an isomorphism from V to V^* is based on a particular choice of a basis for V. So the isomorphism from V to V^* is not a natural one.

We now formalise the idea of passing from one functor to another in a 'natural' way. For convenience we carry out the procedure for covariant functors and leave to the reader the modifications needed for contravariant functors.

(2.5) Definition: Let $\mathcal{C}$, $\mathcal{D}$ be categories and $F, G : \mathcal{C} \to \mathcal{D}$ covariant functors. Then a **natural transformation** θ from F to G assigns to each object X of $\mathcal{C}$ a morphism θ_X in $M_{\mathcal{D}}(F(X), G(X))$ in such a way that for any morphism $f : X \to Y$, the following diagram of morphisms in $\mathcal{D}$ is commutative:

$$\begin{array}{ccc} F(X) & \xrightarrow{\theta_X} & G(X) \\ \downarrow F(f) & & \downarrow G(f) \\ F(Y) & \xrightarrow{\theta_Y} & G(Y) \end{array}$$

The fact that θ is a natural transformation from F to G is denoted by $\theta : F \rightrightarrows G$ or by $F \overset{\theta}{\rightrightarrows} G$, the double arrows stressing the fact that θ does not consist of just one arrow but a whole bunch of arrows, one for each object of $\mathcal{C}$. If each θ_X is an equivalence, then θ is called a **natural equivalence** and the functors F and G are said to be **naturally equivalent** to each other.

We already gave an example of a natural transformation above. Whenever we say that a certain construction is 'natural' it can usually be interpreted in terms of a natural transformations. For example, consider the statement that every set can be naturally regarded as a subset of its power set. Let X be a set and $P(X)$ its power set. Define $i_X : X \to P(X)$ by $i_X(x) = \{x\}$ for $x \in X$. Consider the power set functor $P : \mathbb{E}\text{ns} \to \mathbb{E}\text{ns}$, where if $f : X \to Y$ is a function, we define $P(f) : P(X) \to P(Y)$ by $P(f)(A) = f(A)$ for $A \in P(X)$. It is then easy to see that i is a natural transformation from the identity functor $I : \mathbb{E}\text{ns} \to \mathbb{E}\text{ns}$ to the power set functor P. In the exercises we ask the reader to make precise a few statements where the phrases 'in a natural way' or 'canonically' are used.

Exercises

2.1 Show by an example that the range of a functor need not be a subcategory of the codomain category.

2.2 Prove or disprove that a covariant functor takes a monomorphism to a monomorphism.

2.3 For a set X let $\mathcal{T}_X$ be the cofinite topology on X. Can we define a functor $F : \mathbb{E}\text{ns} \to \mathbb{T}\text{op}$ in such a way that $F(X) = (X; \mathcal{T}_X)$ and $F(f) = f$ for $f : X \to Y$ in $\mathbb{E}\text{ns}$?

2.4 Let $(X, \leq)$ and $(Y, \leq)$ be partially ordered sets. Regard each as a category by the method given in the last section. Interpret a functor from X to Y in terms of the order relation.

2.5 Prove that the Stone-Čech compactification gives a functor from the category of Tychonoff spaces to the category of compact Hausdorff spaces and that the Wallman compactification gives a functor from the category of T_1 spaces and closed maps to the category of compact T_1 spaces (see Exercise (11.4.19)).

2.6 Prove that the composite of two natural transformations defined in the obvious way, is a natural transformation.

2.7 Interpret the following statements in terms of natural transformations of suitable functors.

(i) For any two spaces X and Y, the spaces $X \times Y$ and $Y \times X$ are canonically homeomorphic to each other.

(ii) Every vector space over a field can be canonically embedded into its double dual (i.e. the dual of its dual).

(iii) For each set X there is a canonical bijection between its power set $P(X)$ and the set $\mathbb{Z}_2^X$ of all functions from X to $\mathbb{Z}_2 = \{0, 1\}$. (Here regard the power set functor as a contravariant functor; if $f : X \to Y$ is a function then $P(f) : P(Y) \to P(X)$ is defined by $P(f)(B) = f^{-1}(B)$ for $B \in P(Y)$.)

(iv) Complementation is a natural unary operation in the power set of any set.

(v) For any three sets X, Y, Z there is a canonical bijection between $Z^{X \times Y}$ and $(Z^X)^Y$ (The bijection was already defined in Theorem (8.1.9). The problem here is to prove its naturality in each of the three variables X, Y, Z.)

2.8 Let $\mathcal{C}$, $\mathcal{D}$, $\mathcal{E}$ be categories and F, G functors from $\mathcal{C}$ to $\mathcal{D}$ and H a functor from $\mathcal{D}$ to $\mathcal{E}$. Let θ be a natural transformation from F to G. For each $X \in \mathcal{O}b(\mathcal{C})$, define $\psi_X = H(\theta_X) : H(F(X)) \to H(G(X))$. Prove that ψ is a natural transformation from $H \circ F$ to $H \circ G$. ψ is often denoted by $H \circ \theta$.

2.9 Let $\mathcal{C}$, $\mathcal{D}$, $\mathcal{E}$ be categories; F, G functors from $\mathcal{D}$ to $\mathcal{E}$ and H a functor from $\mathcal{C}$ to $\mathcal{D}$. Let θ be natural transformation from F to G. For each $X \in \mathcal{O}b(\mathcal{C})$ define $\psi_X : F(H(X)) \to G(H(X))$ to be $\theta_{H(X)}$. Prove that ψ is a natural transformation. ψ is often denoted by $\theta \circ H$.

2.10 Let $\mathcal{C}$, $\mathcal{D}$ be categories and $F, G : \mathcal{C} \to \mathcal{D}$ be functors. Prove that a natural transformation $\theta : F \rightrightarrows G$ is a natural equivalence if and only if there exists a natural transformation $\psi : G \rightrightarrows F$ such that $\psi \circ \theta : F \rightrightarrows F$ and $\theta \circ \psi : G \rightrightarrows G$ are the respective identity transformations.

2.11 Let x_0 be any point in the interior of the closed unit ball D^n in $\mathbb{R}^n$. Prove that there exists a retraction from $D^n - \{x_0\}$ onto S^{n-1}. (Hint: For $y \neq x_0$, there is a unique half-ray from x_0 through y. Let $r(y)$ be the point where this ray cuts S^{n-1}. Then r is a desired retraction. It is called the **radial retraction** from x_0. It is impossible to extend r continuously to x_0.)

2.12 Prove that a retract of a space having the fixed point property also has the fixed point property.

*2.13 Prove that the Brouwer fixed point theorem (that is, the assertion that the closed unit ball D^n has the fixed point property) is equivalent to the assertion that S^{n-1} cannot be a retract of D^n. (Hint: Since S^{n-1} obviously fails to have the fixed point property, one-way implication follows from the last exercise. For the other way implication use the following standard argument. Let, if possible, $f : D^n \to D^n$ be a map without any fixed point. Then for each $x \in D^n$, let $r(x)$ be the point of intersection of S^{n-1} with the half ray from $f(x)$ passing through x (see Figure 4). Prove that r is a retraction of D^n onto S^{n-1}. For continuity of r, express it analytically.)

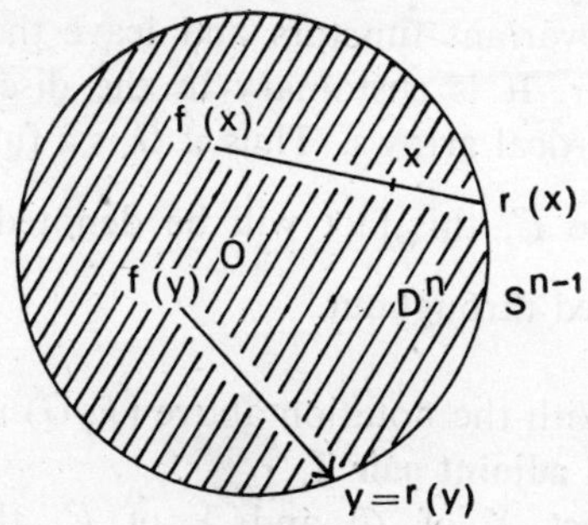

Fig. 4. Brouwer's fixed point theorem.

*2.14 Assuming Brouwer's fixed point theorem, prove the following result.

Let $f: D^n \to \mathbb{R}^n$ be a map such that $d(x, f(x)) < 1$ for all $x \in S^{n-1}$, where d is the usual metric on $\mathbb{R}^n$. Then f has at least one zero in D^n, that is, there exists $x_0 \in D^n$ such that $f(x_0) = 0$. Conversely show that this result implies that S^{n-1} is not a retract of D^n and hence the Brouwer fixed point theorem.

Notes and Guide to Literature

There are numerous examples of functors and of natural transformations in all branches of mathematics. It has been said that if there were no functors, there would be no point in studying categories and that if there were no natural transformations, there would be no point in studying functors.

Although the converses of the statements in Proposition (2.4) do not hold in general, sometimes the converse of statement (iii) does hold. For example, if C is the contravariant functor from the category of compact Hausdorff spaces to the category of rings which assigns to a compact T_2 space X, the ring $C(X)$ of continuous real-valued functions on X then it is true that $f: X \to Y$ is a homeomorphism if and only if $C(f): G(Y) \to C(X)$ is an isomorphism of rings. A proof of this fact can be found in Gillman and Jerison [1].

We remarked that a proof of the Brouwer's fixed point theorem can be given by finding a suitable functor from Top to $\mathcal{Ab}$. Algebraic topology deals with many such functors. Although many results of algebraic topology were known before category theory, a functorial formulation of the former gave it a new impetus. The first and the most influential treatment of algebraic topology in a functorial form is due to Eilenberg and Steenrod [1].

3. Adjoint Functors

We often have two categories $\mathcal{C}$ and $\mathcal{D}$ and a pair of functors one going from $\mathcal{C}$ to $\mathcal{D}$ and the other from $\mathcal{D}$ to $\mathcal{C}$. Sometimes an important relationship holds between two such functors. In this section we study such a relationship, called adjunction. Again, for the sake of simplicity, we carry out everything for covariant functors and leave the case of contravariant functors to the reader. It is customary in the discussion of adjunction to denote functors by vertical arrows. Thus if G is a functor from $\mathcal{C}$ to $\mathcal{D}$ and F a functor from $\mathcal{D}$ to $\mathcal{C}$, this fact will be denoted by $F \overset{\mathcal{C}}{\underset{\mathcal{D}}{\uparrow\!\downarrow}} G$. This notation will be maintained throughout.

(3.1) Definition: With the notation above (F, G) is said to be a pair of **adjoint functors** or an **adjoint pair** if

(i) for every object X of $\mathcal{D}$ and Y of $\mathcal{C}$, there exists a bijection $\theta_{X,Y}: \mathrm{Mor}_{\mathcal{C}}(F(X), Y) \to \mathrm{Mor}_{\mathcal{D}}(X, G(Y))$,

(ii) this bijection is natural in both the variables X and Y; in other words, if $f \in \mathrm{Mor}_{\mathcal{D}}(Z, X)$ and $g \in \mathrm{Mor}_{\mathcal{C}}(Y, W)$ then the following diagram is commutative:

$$\begin{array}{ccc} \mathrm{Mor}_{\mathcal{C}}(F(X), Y) & \xrightarrow{\theta_{X,Y}} & \mathrm{Mor}_{\mathcal{D}}(X, G(Y)) \\ \downarrow p & & \downarrow q \\ \mathrm{Mor}_{\mathcal{C}}(F(Z), W) & \xrightarrow{\theta_{Z,W}} & \mathrm{Mor}_{\mathcal{D}}(Z, G(W)) \end{array}$$

where the vertical arrows p and q are defined by means of suitable compositions in $\mathcal{C}$ and $\mathcal{D}$ respectively, that is $p(\lambda) = g \circ \lambda \circ F(f)$ for $\lambda \in \mathrm{Mor}_{\mathcal{C}}(F(X), Y)$ and $q(\mu) = G(g) \circ \mu \circ f$ for $\mu \in \mathrm{Mor}_{\mathcal{D}}(X, G(Y))$.

If (F, G) is an adjoint pair then F is called a **left adjoint** of G and G is called a **right adjoint** of F. The bijection θ (or more precisely the collection of bijections $\theta_{X,Y}$ as X ranges over $\mathcal{O}b(\mathcal{D})$ and Y over $\mathcal{O}b(\mathcal{C})$ is called an **adjunction** from F to G. (Note that if (F, G) is an adjoint pair then it does not follow that (G, F) is also one.)

The terminology probably comes from adjoint linear transformations in the theory of Hilbert spaces. If $T: V \to W$ is a linear transformation of Hilbert spaces then its adjoint T^* is a linear transformation from W to V such that for all $x \in W$ and $y \in V$, $\langle T^*x, y\rangle = \langle x, Ty\rangle$ where $\langle -, -\rangle$ denotes the inner products in V as well as in W. Of course, the analogy between adjoint functors and adjoint linear transformations is purely notational and not conceptual.

Before we give examples of adjoint functors, we warn the reader not to confuse them with functors which are inverses to each other. If F and G are indeed inverses to each other, then it is easy to check that they form an adjoint pair, for in such a case if $X \in \mathcal{O}b(\mathcal{D})$ and $Y \in \mathcal{O}b(\mathcal{C})$ then we have $\mathrm{Mor}_{\mathcal{C}}(F(X), Y) = \mathrm{Mor}_{\mathcal{C}}(F(X), F(G(X)))$ which is in bijection with $\mathrm{Mor}_{\mathcal{D}}(X, G(X))$ under the action of the functor F. However, adjunction is a much weaker concept than inversion. Indeed, in the most interesting examples of adjoint functors, the two categories are far from isomorphic.

In many examples of adjoint pairs, the right adjoint G is a forgetful functor. Let us take one such example. Let $G : \mathbb{T}\mathrm{op} \to \mathbb{E}\mathrm{ns}$ be the forgetful functor. Define $F: \mathbb{E}\mathrm{ns} \to \mathbb{T}\mathrm{op}$ by $F(X) = (X, \mathcal{T}_X)$ where $\mathcal{T}_X$ is the discrete topology on X. Now, if $(Y; \mathcal{U})$ is a topological space then $G(Y; \mathcal{U}) = Y$ and a morphism in $\mathbb{E}\mathrm{ns}$ from X to $G(Y; \mathcal{U})$ is any set-theoretic function $f: X \to Y$. If X is given the discrete topology, then f becomes continuous regardless of the topology $\mathcal{U}$ on Y. Hence f is a morphism in $\mathbb{T}\mathrm{op}$ from $(X; \mathcal{T}_X)$ to $(Y; \mathcal{U})$, i.e., from $F(X)$ to Y. This correspondence sets up a bijection $\theta_{X,(Y;\mathcal{U})}$ from $\mathrm{Mor}_{\mathbb{E}\mathrm{ns}}(X; G(Y; \mathcal{U})$ to $\mathrm{Mor}_{\mathbb{T}\mathrm{op}}((X; \mathcal{T}_X), (Y; \mathcal{U}))$. Naturality of θ is trivial to verify; because neither θ nor the functors F, G change the morphisms drastically. θ and F take a set theoretic function to the same function viewed as a map between topological spaces while G takes away the topological structure and converts maps to set-theoretic functions. The commutativity of a diagram as in (ii) requires nothing more than associativity of composition of functions. Thus we see that F is a left adjoint of G.

An analogous situation arises for the forgetful functor $G: \mathcal{A}b \to \mathbb{E}\text{ns}$. Let $F: \mathbb{E}\text{ns} \to \mathcal{A}b$ be the functor defined by $F(X) =$ free abelian group on the set X. If Y is an abelian group then every set-theoretic function f from X to Y (regarded as a set) determines a unique group homomorphism from $F(X)$ to Y and conversely. So there is a bijection between the sets $\text{Mor}_{\mathcal{A}b}(F(X), Y)$ and $\text{Mor}_{\mathbb{E}\text{ns}}(X, G(Y))$. Again the naturality of θ is routine to establish. (This will generally be the case for adjoint functors; once the adjunction is defined, the very nature of its definition would make the naturality requirement routine to verify. Henceforth we shall content ourselves with the definition of the adjunction, leaving its naturality to the reader). Thus the free abelian group functor from $\mathbb{E}\text{ns}$ to $\mathcal{A}b$ is a left adjoint to the forgetful functor from $\mathcal{A}b$ to $\mathbb{E}\text{ns}$. In this case note that $\mathcal{A}b$ is a sub-category of $\mathcal{G}$, the category of all groups. Thus we may also regard F as a functor from $\mathbb{E}\text{ns}$ to $\mathcal{G}$. However, when so regarded, it is not a left adjoint to the forgetful functor from $\mathcal{G}$ to $\mathbb{E}\text{ns}$, although the latter is merely an extension of the forgetful functor from $\mathcal{A}b$ to $\mathbb{E}\text{ns}$. It can be shown that the forgetful functor from $\mathcal{G}$ to $\mathbb{E}\text{ns}$ indeed has a left adjoint. It is obtained by associating to each set the free group on it (which is in general not the same as the free abelian group on it). However, the construction of a free group on a set is considerably more complicated.

This is perhaps a good spot to pause and comment upon the utility of category theory. Discrete spaces, free abelian groups and free groups were known long before adjoint functors were discovered and no new theorems about them have come up by realising them as left adjoints to certain forgetful functors. This is to be expected, because 'adjoint functors' is a very general concept and there are not many deep results that can be proved about them. However, realising the familiar constructions of discrete spaces and free abelian groups as left adjoint functors, makes one wonder whether every forgetful functor has a left adjoint; and if so, how to find it. Since there is an inexhaustible source of forgetful functors in mathematics, this inquiry opens a vast area of research. Thus one can consider forgetful functors from the categories of rings, fields, vector spaces, algebras, metric spaces etc., or from their various subcategories into the category of sets. One can also consider partially forgetful functors, for example, the one from the category of rings to the category of abelian groups. In each case one can look for a left adjoint and sometimes this search yields some constructions which are new and which probably would have never been thought of otherwise.

Another important class of adjoint pairs is the one where the right adjoint is the inclusion functor, from a subcategory to a category. As an example, let $\mathcal{C}$ be the category of all compact Hausdorff spaces and let $\mathcal{D}$ be the category of all Tychonoff spaces. Then $\mathcal{C}$ is a full subcategory of $\mathcal{D}$ and let $G: \mathcal{C} \to \mathcal{D}$ be the inclusion functor. Then a left adjoint for G is the familiar functor $F: \mathcal{D} \to \mathcal{C}$ given by the Stone-Čech compactification (see Exercise (2.5)). If X is a Tychonoff space, $F(X)$ is its Stone-Čech compactification and Y is any compact, Hausdorff space then we know from Theorem (11.4.9) that every map from X to Y determines a unique map from $F(X)$ to Y (namely its extension). Similarly a map from $F(X)$ to Y determines (by restriction) a map from X to Y; i.e., from X to $G(Y)$. Thus we get a bijection between $\text{Mor}_{\mathcal{C}}(F(X), Y)$ and $\text{Mor}_{\mathcal{D}}(X, G(Y))$. So F is a left adjoint to the inclusion functor G. Again, this knowledge is of little help in proving any significant results about the Stone-Čech compactification. But it provokes one to look for left adjoints to other inclusion functors.

As the last example, we consider a functor from $\mathbb{E}\text{ns}$ to $\mathbb{E}\text{ns}$. Fix a set Y, and define $G : \mathbb{E}\text{ns} \to \mathbb{E}\text{ns}$ by $G(Z) = \text{Mor}_{\mathbb{E}\text{ns}}(Y, Z) = Z^Y$. G is a covariant functor. A left adjoint F for G can be defined by $F(X) = Y \times X$ for a set X. That F is a functor is easy to check. We have already established an adjunction in Theorem (8.1.9). We recall the definition once more. For any sets X and Z we have to find a bijection from $\text{Mor}_{\mathbb{E}\text{ns}}(F(X), Z)$ to $\text{Mor}_{\mathbb{E}\text{ns}}(X, G(Z))$, i.e. from $\text{Mor}_{\mathbb{E}\text{ns}}(Y \times X, Z)$ to $\text{Mor}_{\mathbb{E}\text{ns}}(X, Z^Y)$ i.e., from $Z^{Y \times X}$ to $(Z^Y)^X$. We define $\lambda : Z^{Y \times X} \to (Z^Y)^X$ by $\lambda(f)(x)(y) = f(y, x)$ for $y \in Y$, $x \in X$ and $f : Y \times X \to Z$.

We now derive some consequences of adjuction. Suppose $F \overset{\mathcal{C}}{\underset{\mathcal{D}}{\uparrow\downarrow}} G$ is an adjoint pair with adjunction $\theta_{X,Y} : \text{Mor}_{\mathcal{C}}(F(X), Y) \to \text{Mor}_{\mathcal{D}}(X, G(Y))$ for each $X \in \mathcal{O}b(\mathcal{D})$, $Y \in \mathcal{O}b(\mathcal{C})$. In particular we put $Y = F(X)$. Then $\theta_{X, F(X)}$ is a bijection from $\text{Mor}_{\mathcal{C}}(F(X), F(X))$ to $\text{Mor}_{\mathcal{D}}(X, G \circ F(X))$. $\text{Mor}_{\mathcal{C}}(F(X), F(X))$ contains $1_{F(X)}$, the identity morphism on $F(X)$. Let $\epsilon_X \in \text{Mor}(X, G \circ F(X))$ be the image of $1_{F(X)}$ under $\theta_{X, F(X)}$. Then ϵ_X is a morphism in $\mathcal{D}$ from $I_{\mathcal{D}}(X)$ to $G \circ F(X)$ where $I_{\mathcal{D}}$ is the identity functor from $\mathcal{D}$ to itself. We assert that it is natural in X.

(3.2) Proposition: With the notation above, ϵ is a natural transformation from $I_{\mathcal{D}}$ to $G \circ F$.

Proof: Let $f : X \to Z$ be a morphism in $\mathcal{D}$. We have to show that the following diagram is commutative:

$$\begin{array}{ccc} X & \xrightarrow{\epsilon_X} & G \circ F(X) \\ {\scriptstyle f}\downarrow & & \downarrow{\scriptstyle G \circ F(f)} \\ Z & \xrightarrow{\epsilon_Z} & G \circ F(Z) \end{array}$$

i.e. to show that $\epsilon_Z \circ f = G(F(f)) \circ \epsilon_X$.

Now, by naturality of θ, we certainly have that the following diagram is commutative:

$$\begin{array}{ccc} \text{Mor}_{\mathcal{C}}(F(X), F(X)) & \xrightarrow{\theta_{X, F(X)}} & \text{Mor}_{\mathcal{D}}(X, G \circ F(X)) \\ {\scriptstyle p}\downarrow & & \downarrow{\scriptstyle q} \\ \text{Mor}_{\mathcal{C}}(F(X), F(Z)) & \xrightarrow{\theta_{X, F(Z)}} & \text{Mor}_{\mathcal{D}}(X, G \circ F(Z)) \end{array}$$

where the vertical arrows p and q are defined by $p(\lambda) = F(f) \circ \lambda$ for $\lambda \in$ Mor $(F(X), F(X))$ and $q(\mu) = G(F(f)) \circ \mu$ for $\mu \in$ Mor $(X, G \circ F(X))$. In particular, applying the commutativity to $1_{F(X)}$ in $\text{Mor}_{\mathcal{C}}(F(X), F(X))$ we get that $q(\theta_{X, F(X)}(1_{F(X)})) = \theta_{X, F(Z)}(p(1_{F(X)}))$. But by definition, $\epsilon_X = \theta_{X, F(X)}(1_{F(X)})$. So we get

$$G(F(f)) \circ \epsilon_X = \theta_{X, F(Z)}(F(f)) \tag{I}$$

We apply naturality of θ once again to get another commutative diagram:

$$\begin{array}{ccc} \text{Mor}_{\mathcal{C}}(F(Z), F(Z)) & \xrightarrow{\theta_{Z, F(Z)}} & \text{Mor}_{\mathcal{D}}(Z, G \circ F(Z)) \\ \downarrow r & & \downarrow s \\ \text{Mor}_{\mathcal{C}}(F(X), F(Z)) & \xrightarrow{\theta_{X, F(Z)}} & \text{Mor}_{\mathcal{D}}(X, G \circ F(Z)) \end{array}$$

where the vertical arrows are given by $r(\lambda) = \lambda \circ F(f)$ and $s(\mu) = \mu \circ f$. In particular starting with $1_{F(Z)}$ in $\text{Mor}_{\mathcal{C}}(F(Z), F(Z))$ we get, $s(\theta_{Z, F(Z)}(1_{F(Z)})) = \theta_{X, F(Z)}(r(1_{F(Z)}))$. Since $\theta_{Z, F(Z)}(1_{F(Z)}) = \epsilon_Z$ by definition, we get

$$\epsilon_Z \circ f = \theta_{X, F(Z)}(F(f)) \tag{II}$$

From (I) and (II) the result follows. ∎

The proof above was merely a play with commutativity of diagrams. This is generally true of all proofs where the naturality of something is to be established. The reader is well advised to master the handling of commutative diagrams until it becomes a routine matter.

(3.3) Definition: The natural transformation ϵ is called the **unit** of the adjoint pair (F, G).

The unit ϵ often has a concrete interpretation in particular examples of adjunctions. For example, in the case of the Stone-Čech compactification, it is the canonical embedding of a Tychonoff space into its compactification. Also in the example in which the discrete space functor was a left adjoint to the forgetful functor from Top to $\mathbb{E}$ns, ϵ_X is simply the identity morphism 1_X for each set X.

Dually, there is the **co-unit** η of an adjoint pair (F, G). It is obtained by putting $X = G(Y)$ for an object Y of $\mathcal{C}$. Thus $\eta_Y \in \text{Mor}_{\mathcal{C}}(F \circ G(Y), Y)$ is the morphism $\theta^{-1}_{G(Y), Y}(1_{G(Y)})$. We leave it to the reader to check that η is a natural transformation from $F \circ G$ to $I_{\mathcal{C}}$, the identity functor on $\mathcal{C}$.

The unit and the co-unit are related to each other as follows:

(3.4) Proposition: For any object X of $\mathcal{D}$, $\eta_{F(X)} \circ F(\epsilon_X) = 1_{F(X)}$ and for any object Y of $\mathcal{C}$, $G(\eta_Y) \circ \epsilon_{G(Y)} = 1_{G(Y)}$.

Proof: Once again, the proof is an exercise in commutative diagrams. We prove only the first part, leaving the other to the reader. For an object X of $\mathcal{D}$, $\epsilon_X : X \to GF(X)$ is a morphism in D and so by the naturality of the adjunction θ, we have a commutative diagram:

$$\begin{array}{ccc} \text{Mor}_{\mathcal{C}}(FGF(X), F(X)) & \xrightarrow{\theta_{GF(X), F(X)}} & \text{Mor}_{\mathcal{D}}(GF(X), GF(X)) \\ \downarrow p & & \downarrow q \\ \text{Mor}_{\mathcal{C}}(F(X), F(X)) & \xrightarrow{\theta_{X, F(X)}} & \text{Mor}_{\mathcal{D}}(X, GF(X)) \end{array}$$

where, as usual, p and q are defined by $p(\lambda) = \lambda \circ F(\epsilon_X)$ and $q(\mu) = \mu \circ \epsilon_X$. Applying commutativity to the element $\eta_{F(X)} \in \text{Mor}\,(FGF(X), F(X))$, we get

$$\theta_{X, F(X)}(\eta_{F(X)} \circ F(\epsilon_X)) = \theta_{GF(X), F(X)}(\eta_{F(X)}) \circ \epsilon_X \qquad \text{(I)}$$

Now, by very definition of η, $\theta_{GF(X), F(X)}(\eta_{F(X)} = 1_{GF(X)})$ and so the right hand side of (I) is simply ϵ_X, which gives $\theta_{X, F(X)}(\eta_{F(X)} \circ F(\epsilon_X)) = \theta_{X, F(X)}(1_{F(X)})$, by definition of ϵ_X.

But $\theta_{X, F(X)}$ is a bijection and so in particular one-to-one. Then $\eta_{F(X)} \circ F(\epsilon_X) = 1_{F(X)}$ as was to be proved. ∎

If we use the notations of Exercises (2.8) and (2.9) the last result can be paraphrased as follows:

(3.5) Proposition: The composite natural transformation $F \overset{F \circ \epsilon}{\Longrightarrow} F \circ G \circ F \overset{\eta \circ F}{\Longrightarrow} F$ equals the identity natural transformation from F to itself. Similarly the composite $G \overset{\epsilon \circ G}{\Longrightarrow} G \circ F \circ G \overset{G \circ \eta}{\Longrightarrow} G$ is the identity natural transformation. ∎

It is interesting to note that the unit, the co-unit and the property in the last proposition determine the adjunction. More precisely, if $F \underset{\mathcal{D}}{\uparrow\!\downarrow} G$ isa pair of functors, $\epsilon : I_{\mathcal{D}} \rightrightarrows G \circ F$ and $\eta : F \circ G \rightrightarrows I_{\mathcal{C}}$ are natural transformations satisfying the property in the last proposition, then there exists an adjunction from F to G whose unit is ϵ and co-unit η. The proof will be a good exercise to the reader. The significance of this fact is that adjunction can be described in terms of natural transformations.

We conclude this section with a proof of uniqueness of adjoints.

(3.6) Theorem: Any two left adjoints of the same functor are naturally equivalent to each other. Similarly any two right adjoints of the same functor are naturally equivalent to each other.

Proof: We shall prove only the first assertion. Suppose $G : \mathcal{C} \to \mathcal{D}$ is a functor and $F_1, F_2 : \mathcal{D} \to \mathcal{C}$ are both left adjoints of G. Let θ_1, θ_2 be the respective adjunctions. For any $X \in \mathcal{O}b(\mathcal{D})$ and any $Y \in \mathcal{O}b(\mathcal{C})$, we have bijections $\text{Mor}_{\mathcal{C}}(F_1(X), Y) \xrightarrow{\theta_1} \text{Mor}_{\mathcal{D}}(X, G(Y)) \xrightarrow{\theta_2^{-1}} \text{Mor}_{\mathcal{C}}(F_2(X), Y)$. Now put $Y = F_2(X)$. Then there exists a morphism say $\psi_X \in \text{Mor}_{\mathcal{C}}(F_1(X), F_2(X))$ such that $\theta_2^{-1} \circ \theta_1(\psi_X)$ equals $1_{F_2(X)} \in \text{Mor}\,(F_2(X), F_2(X))$. Similarly putting $Y = F_1(X)$, we get a morphism ρ_X from $F_2(X)$ to $F_1(X)$. That ψ and ρ define natural transformations which are inverses to each other is straightforward and left to the reader. ∎

The adjoint functor problem seeks to find left (or right) adjoint to a given functor. The theorem above shows that in case the solution exists, it is unique upto an equivalence. However, finding necessary and sufficient conditions for a functor to have a left (or right) adjoint is not easy and beyond the scope of this book.

Exercises

3.1 In each of the following, a functor $G : \mathcal{C} \rightarrow \mathcal{D}$ is given. Find a left adjoint to it. Describe the unit of the adjunction in each case.

(i) The inclusion functor from the category of complete metric spaces to the category of all metric spaces (the morphisms in each category being uniformly continuous functions).

(ii) The inclusion functor from $\mathcal{Ab}$ into $\mathcal{G}$. (The solution requires the commutator subgroup of a group).

(iii) The functor from $\mathcal{Ab}$ to $\mathcal{Ab}$ defined by $G(Z) = \text{Hom}(Y, Z)$ where Y is a fixed abelian group. (The solution requires knowledge of the tensor product of two abelian groups.)

(iv) The inclusion functor from the category of compact spaces to the category of all spaces, the morphisms being proper maps in both the categories.

(v) The inclusion functor from $\mathcal{C}$ to $P(X)$ where X is a topological space, $\mathcal{C}$ the family of all its closed subsets and the power set $P(X)$ is regarded as a category by means of the usual partial ordering on it defined by inclusion.

(vi) The forgetful functor from the category of all vector spaces over a field (the morphisms being linear transformations) to the category of sets.

3.2 Find a right adjoint to the forgetful functor from Top to Ens.

3.3 Define the co-unit of an adjunction and prove that it is a natural transformation.

3.4 Prove the other half of Proposition (3.4).

3.5 Prove that if $F \underset{\mathcal{D}}{\overset{\mathcal{C}}{\uparrow\downarrow}} G$ is a pair of functors, $\epsilon : I_{\mathcal{D}} \rightrightarrows G \circ F$ and $\eta : F \circ G \rightrightarrows I_{\mathcal{C}}$ are natural transformations such that the composities $F \xrightarrow{F\circ\epsilon} F \circ G \circ F \xrightarrow{\eta \circ F} F$ and $G \xrightarrow{\epsilon \circ G} G \circ F \circ G \xrightarrow{G \circ \eta} G$ are the respective identity transformations then there exists an adjunction θ from F to G whose unit and co-unit coincide with ϵ and η respectively.

3.6 Complete the proof of Theorem (3.6).

3.7 Generalise the construction in the proof of Theorem (3.6) as follows. Let $\mathcal{C}$, $\mathcal{D}$ be categories, G_1, G_2 functors from $\mathcal{C}$ to $\mathcal{D}$ and F_1, F_2 from $\mathcal{D}$ to $\mathcal{C}$. Assume F_i is a left adjoint of G_i for $i = 1, 2$. Prove that any natural transformation $\beta : G_1 \rightrightarrows G_2$ determines a unique natural transformation $\alpha : F_2 \rightrightarrows F_1$ such that for any $X \in \mathcal{O}b(\mathcal{D})$ and $Y \in \mathcal{O}b(\mathcal{C})$ the following diagram is commutative, the vertical arrows being defined by means of appropriate compositions:

$$\begin{array}{ccc} \text{Mor}(F_1(X), Y) & \xrightarrow{\theta_1} & \text{Mor}(X, G_1(Y)) \\ \downarrow & & \downarrow \\ \text{Mor}(F_2(X), Y) & \xrightarrow{\theta_2} & \text{Mor}(X, G_2(Y)) \end{array}$$

3.8 If a functor $G : \mathcal{C} \to \text{Ens}$ has a left adjoint, where $\mathcal{C}$ is any category, prove that there exists an object A of $\mathcal{C}$ such that G is naturally equivalent to the functor $H : \mathcal{C} \to \text{Ens}$ defined by $H(Y) = \text{Mor}_{\mathcal{C}}(A, Y)$. (Hint: Let F be the adjoint and let A be $F(*)$ where $*$ is any singleton set.)

3.9 Define adjunction for contravariant functors and translate the results proved here to the case of contravariant functors.

Notes and Guide to Literature

A construction of free groups is given in Lang [1]. A theorem about existence of left adjoints is due to Freud [1] where the reader will also find many other interesting results.

4. Universal Objects and Categorical Notions

In this section we shall define the universal objects in a category. They are of two types, either initial or terminal. Their importance in category theory will be apparent later on.

(4.1) Definition: An object A of a category $\mathcal{C}$ is called an **initial object** for $\mathcal{C}$ if for every object B of $\mathcal{C}$ there is exactly one morphism (in $\mathcal{C}$) from A to B. A is called a **terminal object** for $\mathcal{C}$ if for every object B of $\mathcal{C}$, there is exactly one morphism from B to A.

Obviously these two are dual concepts. An initial object for $\mathcal{C}$ is precisely a terminal object of the opposite category $\mathcal{C}^{\text{op}}$ and vice versa. In the category of sets the empty set is an initial object while any singleton set is a terminal object. In $\mathcal{G}$ or in $\mathcal{A}b$ any group with only one element in it is an initial as well as a terminal object. Not every category has such objects. For example, the category of all non-empty sets has no initial objects. However, where such objects do exist, they are unique upto an equivalence as we now show.

4.2 Proposition: Any two initial objects in a category are equivalent. Similarly any two terminal objects are equivalent. Also any object equivalent to an initial (terminal) object is itself an initial (terminal) object.

Proof: Suppose A_1, A_2 are both initial objects in a category $\mathcal{C}$. Let f be the unique morphism from A_1 to A_2 and let g be the unique morphism from A_2 to A_1. Then $g \circ f \in \text{Mor}_{\mathcal{C}}(A_1, A_1)$ which is also a singleton set. But there is already the identity morphism 1_{A_1} in $\text{Mor}_{\mathcal{C}}(A_1, A_1)$. So $g \circ f = 1_{A_1}$. Similarly $f \circ g = 1_{A_2}$. So A_1 is equivalent to A_2. An identical argument applies when

A_1 and A_2 are both terminal objects. The last assertion follows from the fact that if two objects A_1, A_2 are equivalent then there is a bijection between the sets $\text{Mor}_{\mathcal{C}}(A_1, B)$ and $\text{Mor}_{\mathcal{C}}(A_2, B)$ and also between $\text{Mor}_{\mathcal{C}}(B, A_1)$ and $\text{Mor}_{\mathcal{C}}(B, A_2)$ for any object B of $\mathcal{C}$. ∎

Although there is not much that can be proved about universal objects in general, it will be seen that many constructions in category theory are so defined that they can be looked upon as universal objects in a suitable category. As a first illustration, let us define the concept of a product. Let $\mathcal{C}$ be a category, I an index set and $\{X_i : i \in I\}$ an indexed collection of objects of $\mathcal{C}$.

(4.3) Definition: With the notation above, a **product** of the family $\{X_i : i \in I\}$ is a system $(X, \{\lambda_i\}_{i \in I})$ where X is an object of $\mathcal{C}$ and for each $i \in I$, $\lambda_i : X \to X_i$ is a morphism in $\mathcal{C}$ such that for any other such system $(Y, \{\mu_i\}_{i \in I})$, there is a unique morphism $f : Y \to X$ with the property that for each $i \in I$, $\mu_i = \lambda_i \circ f$.

The definition may appear unnecessarily complicated as compared to the definition of the cartesian product of a family of sets. If $\{X_i : i \in I\}$ is an indexed family of sets then their product is the set $\prod_{i \in I} X_i$, without any reference to the projection functions $\pi_i : \prod X_i \to X_i$. The set $\prod_{i \in I} X_i$ is defined as the set of certain functions from I into $\bigcup_{i \in I} X_i$. It is obvious that this simple-minded definition cannot apply for an arbitrary category $\mathcal{C}$, because in an abstract category the concept of union of objects is not defined. We therefore try to characterise the product $\prod_{i \in I} X_i$ in such a way that the new formulation would make sense in any abstract category. This would be the case if the characterisation involves only sets and functions which are categorical concepts and not the particular elements of a set, a notion peculiar to the category $\mathbb{E}$ns. An answer is provided by the property of the system $(\prod X_i, \{\pi_i\}_{i \in I})$ as given in Proposition (9.1.2). Theorem (9.1.3) states that this property characterises the product upto a bijection, that is, upto an equivalence in $\mathbb{E}$ns. It provides a satisfactory definition of products that would be applicable in any category. Going back to Definition (4.3), if one wants to stay as close to the products of sets as possible, one may call the object X there as the product of the family $\{X_i : i \in I\}$ and call λ_i's as the projection morphisms, however, it is now customary to make the projections as a part of the product itself. It should be noted that unlike the set theoretic product, the categorical product need not always exist, and when it exists, it is unique only upto an equivalence.

Analogues of Theorem (9.1.3) hold for the product of a family of topological spaces and also for the product of a family of groups, (see Exercises (9.1.2) and (9.1.5)). We can therefore say that products exist in the categories Top and $\mathcal{G}$. In the category of all finite sets, finite products (that is, those where the index sets are finite) exist but arbitrary products do not exist. As an example, let the index set I be the set $\mathbb{N}$ and for each $n \in \mathbb{N}$. Let X_i be the set $\mathbb{Z}_2 = \{0, 1\}$. We leave it to the reader to prove that the family $\{X_n : n \in \mathbb{N}\}$, has no product in the category of finite sets, with the warning that it does not suffice merely to show that the set $\mathbb{Z}_2^{\mathbb{N}}$ is not finite (why ?).

In Section 1 we considered the category of all cones over an indexed family $\{X_i : i \in I\}$ of objects in a category $\mathcal{C}$. It is clear from the definitions that a product of the family $\{X_i : i \in I\}$ is precisely a terminal object of the corresponding category of cones.

The definition of coproducts is completely dual.

(4.4) Definition: Let $\{X_i : i \in I\}$ be an indexed family of objects in a category $\mathcal{C}$. Then its coproduct is a system $(X, \{\lambda_i\}_{i \in I})$ such that for each $i \in I$, $\lambda_i \in \text{Mor}_{\mathcal{C}}(X_i, X)$ and for any other such system $(Y, \{\mu_i\}_{i \in I})$ there exists a unique morphism $f : X \to Y$ such that for each $i \in I$, $f \circ \lambda_i = \mu_i$.

This definition is consistent with the coproducts of a family of sets or of a family of topological spaces as defined in Chapter 8. In the category of abelian groups, a coproduct of a family is what is popularly called the direct sum of abelian groups. In the category of all groups the coproducts do exist but the construction is much too complicated to be described here. In the category of finite sets coproducts do not exist in general.

It is obvious that a coproduct is precisely an initial object in the category of all co-cones on $\{X_i : i \in I\}$. So from Proposition (4.2) they are unique upto equivalences.

These examples illustrate how to translate a notion from a particular category to general categorical terms. As another example, let us see what should be the categorical analogue of the set-theoretic notion of intersection of two sets. Given two sets A and B their intersection is the set of elements common to them. This formulation does not generalise to arbitrary categories, because although sets and functions have analogues (as objects and morphisms respectively) in an arbitrary category, the concept of an element of a set is peculiar to the particular category $\mathbb{E}$ns and has no analogue in an abstract category. So we look at the intersection of sets in another way. Assume A and B are subsets of a set X (for X we may take $A \cup B$ or any larger set). Let i_1, i_2 be the inclusion functions from A, B respectively to X. Let f_1, f_2 be the inclusions of $A \cap B$ into A and B respectively. Then $i_1 \circ f_1 = i_2 \circ f_2$, in other words the following diagram is commutative:

$$\begin{array}{ccc} A \cap B & \xrightarrow{f_1} & A \\ \Big\downarrow{\scriptstyle f_2} & & \Big\downarrow{\scriptstyle i_1} \\ B & \xrightarrow{i_2} & X \end{array}$$

Moreover for any set C and any functions $g_1 : C \to A$ and $g_2 : C \to B$ such that $i_1 \circ g_1 = i_2 \circ g_2$, there exists a unique function $h : C \to A \cap B$ such that $g_1 = f_1 \circ h$ and $g_2 = f_2 \circ h$; we simply define $h(x) = g_1(x)$ or, which is the same as, $g_2(x)$. It is clear that this property characterises the system

$(A \cap B, f_1, f_2)$ upto a bijection. Thus we have succeeded in describing the intersection of two sets, upto a bijection, in terms involving only the objects and morphisms in $\mathbb{E}$ns, and without actually touching the particular elements of any set. This leads to the

(4.5) Definition: Let $\mathcal{C}$ be a category and $k_1 : A_1 \to X$, $k_2 : A_2 \to X$ be morphisms. Then their **pull-back** is a triple (P, f_1, f_2) such that $f_i : P \to A_i$ $(i = 1, 2)$, are morphisms such that

(i) $k_1 \circ f_1 = k_2 \circ f_2$ and

(ii) for any other triple (Q, g_1, g_2) for which $k_1 \circ g_1 = k_2 \circ g_2$ there exists a unique morphism h (in $\mathcal{C}$) from Q to P such that $f_i \circ h = g_i$ for $i = 1, 2$.

A pull-back is said to be an **intersection** when all the morphisms k_1, k_2, f_1, f_2 are monomorphisms.

If we consider the category $\mathcal{D}$ defined by the method in Exercise (1.10), then a pull-back of the pair (k_1, k_2) is precisely a terminal object in $\mathcal{D}$. It follows that a pull-back is unique upto an equivalence in $\mathcal{D}$.

Pull-backs always exist for any pair of morphisms in $\mathbb{E}$ns, Top and in $\mathcal{G}$. However, there are categories in which they do not always exist. Notice that pull-backs (and also intersections) are constructions that depend on the morphisms k_1 and k_2 and not just on the objects A_1 and A_2. For example, in $\mathbb{E}$ns, let $A_1 = A_2 = \{1\}$, $X = \{0, 1\}$, $k_1(1) = 0$, $k_2(1) = 1$. Then the intersection of (k_1, k_2) is the triple (Φ, f_1, f_2) where f_1, f_2 are the inclusions of the empty set Φ into A_1 and A_2. However, the set-theoretic intersection of A_1 and A_2 is $\{1\}$ which is not the empty set. The anomaly can be explained as follows. If k_1 and k_2 were inclusion functions then their categorical intersection would indeed coincide with their set-theoretic intersection. In the present case, k_1 is not an inclusion function, although certainly it is a monomorphism. If k_1 were to be an inclusion function, then A_1 would have to be taken as $\{0\}$, in which case $A_1 \cap A_2$ is indeed the empty set.

The dual concept of a pull-back is push-out. We leave the definition to the reader. When applied to $\mathbb{E}$ns, a push-out is a generalisation of the set-theoretic union of two sets.

Most of the constructions in category theory are special cases of what are known as right and left roots. We proceed to define these concepts.

(4.6) Definition: A category is said to be **small**, if the class of its objects is a set.

This definition is of interest only if we are using the formal set theory in which a set is scrupulously distinguished from a class. As remarked after the definition of a category, we shall not dwell on this point. Its importance is that if $\mathcal{C}$ is any category, $\mathcal{D}$ is any small category and F, G are functors from $\mathcal{D}$ to $\mathcal{C}$, then the natural transformations from F to G form a set and so we can consider a category of all functors from $\mathcal{D}$ to $\mathcal{C}$, the morphisms being the natural transformations from one functor to another. Examples of small categories include any category with only finitely many objects, any category obtained from a partially ordered set and the category given in Exercise (1.4).

(4.7) Definition: Let $\mathcal{D}$ be a small category and F a functor from $\mathcal{D}$ to a category $\mathcal{C}$. Then a **right root** of F is a pair (R, θ) where R is a constant functor from $\mathcal{D}$ to $\mathcal{C}$ (i.e. R assigns the same object to every object of $\mathcal{D}$ and the identity morphism on this object to every morphism in $\mathcal{D}$) and θ is a natural transformation from F to R such that for any other such pair (S, ψ) there is a unique natural transformation $\alpha : R \rightrightarrows S$ with the property that $\alpha \circ \theta = \psi$.

Let us analyse the definition. A constant functor into $\mathcal{C}$ can of course be identified with an object of $\mathcal{C}$. Thus we look at R simultaneously as a constant functor and as an object of $\mathcal{C}$. A natural transformation θ from F to R is then merely a set $\{\theta_A : A \in \mathcal{O}b(\mathcal{D})\}$ such that $\theta_A : F(A) \to R$ and if $f : A \to B$ is a morphism in $\mathcal{D}$ then $\theta_B \circ F(f) = \theta_A$. Moreover, a natural transformation between two constant functors is merely a morphism between the objects they represent.

With these remarks it is easy to see how a coproduct in a category $\mathcal{C}$ can be looked upon as a right root of a suitable functor. Let I be an index set and $\{X_i : i \in I\}$ an indexed family of objects in $\mathcal{C}$. We put the trivial partial order $\leq I$, defined by $i \leq j$ iff $i = j$ and then regard I as a category by the method described in Section 1. We continue to denote this category by I. If we define $F(i) = X_i$ for $i \in I$ and $F(1_i) = 1_{X_i}$, we get a functor F from I into $\mathcal{C}$. I is also a small category. Clearly a cocone over $\{X_i : i \in I\}$ corresponds to a pair (R, θ) where $R : I \to \mathcal{C}$ is a constant functor and θ is a natural transformation from F to R. It is also clear that a co-product of the family $\{X_i : i \in I\}$ determines a right root of F and vice versa.

We leave it to the reader to define the dual notion of a **left root** of a functor from a small category and to interpret the two notions as universal objects in appropriate categories. Note that a pull-back of a pair of morphisms in a category $\mathcal{C}$ can be interpreted as a left root of a certain functor $F : I \to \mathcal{C}$, where I is a set $\{a_1, a_2, a_3\}$ on which the partial ordering is defined by $a_i \leq a_j$ iff $i = j$ or $i = 1, j = 3$ or $i = 2, j = 3$. Similarly a push-out can be interpreted as a right root of a suitable functor.

We conclude the section with a discussion of an important special case of left and right roots of a functor. It arises when the domain category $\mathcal{D}$ is obtained from a directed set $(D, \leq)$ in the manner described in Section 1. Recall from Chapter 10 that this means that $\leq$ is a partial order on the set D with the additional condition that for any $m, n \in D$ there is $p \in D$ such that $m \leq p$ and $n \leq p$. (In Chapter 10 we stated this fact in terms of $\geq$ rather than $\leq$; but this should make no difference). Note that a covariant functor $F : \mathcal{D} \to \mathcal{C}$ gives a family $\{X_n : n \in D\}$ of objects of $\mathcal{C}$ along with a family of morphisms $\{\lambda_{m,n} : m \leq n \text{ in } D\}$. The correspondence is obtained by setting $X_n = F(n)$ for $n \in D$ and $\lambda_{m,n} = F(k_{m,n})$ for $m \leq n$ where $k_{m,n}$ is the unique morphism in $\mathcal{D}$ from m to n. Note further that whenever $m \leq n \leq p$ we have $\lambda_{m,p} = \lambda_{n,p} \circ \lambda_{m,n}$. Conversely any family

$\{x_n\}_{n\in D}$ of objects and $\{\lambda_{m,n} : m \leq n \text{ in } D\}$ of morphisms satisfying the condition $\lambda_{m,p} = \lambda_{n,p} \circ \lambda_{m,n}$ for all $m \leq n \leq p$ determines a functor F from $\mathcal{D}$ into $\mathcal{C}$. We use the functor and the associated family interchangeably.

(4.8) Definition: Let $\mathcal{C}$ be a category, $(D, \leq)$ a directed set regarded as a category $\mathcal{D}$ and $F : \mathcal{D} \to \mathcal{C}$ a covariant functor. Let $(\{x_n\}, \{\lambda_{m,n}\})$ be the corresponding system of objects and morphisms in $\mathcal{C}$. Then a right root of F is called a **direct limit** of the system $(\{X_n\}, \{\lambda_{m,n}\})$.

This concept is important because it arises naturally in many important categories such as Ens, Top, and $\mathcal{G}$. We give a few examples. In all these examples the directed set D will be $\mathbb{N}$ and $\leq$ the usual ordering. In order to specify the system $(\{X_n\}_{n\in\mathbb{N}}, \{\lambda_{m,n}\}_{m\leq n})$ it suffices to specify what the morphisms $\lambda_{n,n+1}$ are for each $n \in \mathbb{N}$, because once they are given, $\lambda_{m,n}$ for $m \leq n$ is merely the composite $\lambda_{(n-1),n} \circ \lambda_{(n-2),(n-1)} \circ \ldots \circ \lambda_{m,m+1}$.

Example 1: In Ens, let $\{X_n\}_{n\in\mathbb{N}}$ be an ascending sequence of subsets of a set Y. Let $\lambda_{n,n+1} : X_n \to X_{n+1}$ be the inclusion function. Then the direct limit of this system is precisely the union $\bigcup_{n=1}^{\infty} X_n$ along with inclusion functions from X_n into this union.

Example 2: In Ens, let $X_n = \mathbb{R}^n$ (regarded only as a set) and let $\lambda_{n,n+1} : \mathbb{R}^n \to \mathbb{R}^{n+1}$ be the standard inclusion which takes a point $(x_1, \ldots, x_n)$ of $\mathbb{R}^n$ to $(x_1, x_2, \ldots, x_n, 0)$. In this case the direct limit is the system $(\mathbb{R}^\infty, \{\mu_n\}_{n\in\mathbb{N}})$ where $\mathbb{R}^\infty$ is the set of all sequences of real numbers having only finitely many non-zero entries, and $\mu_n : \mathbb{R}^n \to \mathbb{R}^\infty$ is defined by $\mu_n(x_1, x_2, \ldots, x_n) = (x_1, x_2, \ldots, x_n, 0, 0, 0, \ldots)$. Note that each $\mathbb{R}^n$ is also an abelian group and the functions $\lambda_{n,n+1}$ are group homomorphisms. Moreover $\mathbb{R}^\infty$ can be given a group structure in such a way that μ_n's are homomorphisms. Thus the present example is also an example of a direct limit in $\mathcal{G}$ or in $\mathcal{Ab}$.

The notion dual to that of a direct limit is an inverse limit. To define it, we start with the dual ordering on D, or what amounts to the same, we consider contravariant functors from $(D, \leq)$ into $\mathcal{C}$. Each such functor F determines a system $(\{x_n\}_{n\in D}, \{\lambda_{m,n}\}_{m\leq n})$ where for $m \leq n$, $\lambda_{m,n}$ is a morphism from X_n to X_m. A left root of such a functor F is called an **inverse limit** of the corresponding system. We illustrate it with examples.

Example 3: In Ens let $\{X_n : n \in \mathbb{N}\}$ be a descending sequence of subsets of a set Y and for each $n \in \mathbb{N}$ let $\lambda_{n,n+1} : X_{n+1} \to X_n$ be the inclusion function. The situation is dual to that in Example 1 and the inverse limit is precisely the intersection $\bigcap_{n=1}^{\infty} X_n$, with the appropriate inclusion functions from $\bigcap_{n=1}^{\infty} X_n$ to the X_n's.

Example 4: In Top, for each $n \in \mathbb{N}$, let X_n be the unit circle S^1 regarded as a subset of complex numbers with the usual topology and $\lambda_{n,n+1} : X_{n+1} \to X_n$ be the map $\lambda_{n,n+1}(z) = z^2$ for $z \in X_{n+1}(=S^1)$. The inverse limit in this case is popularly known as the **2-solenoid** or simply the **solenoid**. To construct it we consider the topological product $Z = \prod_{n=1}^{\infty} X_n$. Elements of Z can be written as sequences $\bar{z} = (z_1, z_2, \ldots, z_n, \ldots)$ where $z_n \in S^1$ for all n. Now let Y be the set of those $\bar{z}$ for which $z_n = (z_{n+1})^2$ for all $n \in \mathbb{N}$. For example, the sequence $(1, -1, i, e^{\pi i/4}, e^{\pi i/8}, \ldots)$ is in Y. Y is a closed subset of Z and hence is a compact metric space. For each n, let $f_n : Y \to X_n$ be the restriction of the projection map. Then for each n, $\lambda_{n,n+1} \circ f_{n+1} = f_n$. Moreover, let $(W, \{g_n\}_{n \in \mathbb{N}})$ be any other system where W is a space and $g_n : W \to S^1$ are maps satisfying $\lambda_{n,n+1} \circ g_{n+1} = g_n$ for all $n \in \mathbb{N}$. We define $h : W \to Y$ by $(h(w))_n = g_n(w)$. Then $g_n = f_n \circ h$ and h is the only map having this property. So the system $(Y, \{f_n\})$ is an inverse limit of the given sequence of circles and maps.

Exercises

4.1 Prove that in the category of finite sets, the family of infinitely many copies of $\mathbb{Z}_2$ has neither a product nor a coproduct.

4.2 Prove that any pair of morphisms with a common codomain in Ens has a pull-back. (Hint: Let $k_1 : A_1 \to X$, $k_2 : A_2 \to X$ be functions. Let $P = \{(x_1, x_2) \in A_1 \times A_2 : k_1(x_1) = k_2(x_2)\}$ and let $f_i : P \to A_i$ be the restriction of the projection function, $i = 1, 2$.)

4.3 Prove that pull-backs also exist in Top, $\mathcal{G}$ and $\mathcal{Ab}$.

4.4 Give an example of a category in which there is a pair of morphisms with a common codomain which has no pull-back.

4.5 Define a push-out of a pair of morphisms having a common domain. Prove that push-outs exist in Ens, Top, $\mathcal{G}$ and in $\mathcal{Ab}$. (Hint: Let $k_i : X \to A_i$, $i = 1, 2$ be morphisms in Ens. Let N be the equivalence relation on the disjoint union of A_1 and A_2 generated by all pairs $(k_1(x), k_2(x))$ for $x \in X$. Let Z be the quotient set. A similar argument works for Top; however, for $\mathcal{G}$ and $\mathcal{Ab}$ modifications are needed.)

4.6 Interpret pull-backs and products as left roots and push-outs and coproducts as right roots of suitable functors.

4.7 Interpret left and right roots of a functor as universal objects in suitable categories.

4.8 Prove that inverse limits always exist in the categories Ens, Top, $\mathcal{G}$, $\mathcal{Ab}$. (Hint: The construction of the solenoid can be generalised suitably.)

4.9 Prove that direct limits always exist in the categories Ens, Top. (Hint: Take suitable quotient sets or quotient spaces respectively of the coproducts.)

4.10 In Example 2, prove that a suitable topology to $\mathbb{R}^\infty$ can be given so that it becomes the direct limit in Top of the sequence $\{\mathbb{R}^n : n \in \mathbb{N}\}$, where each $\mathbb{R}^n$ carries the usual topology.

4.11 Let $\mathcal{D}$ be a small category and $F : \mathcal{D} \to \mathcal{C}$ a functor with a right root (R, θ). Let $G : \mathcal{C} \to \mathcal{E}$ be another functor. Show by an example that $(G \circ R, G \circ \theta)$ need not be a right root of $G \circ F$ (i.e. right roots are not always preserved by functors).

*4.12 Let X_1 be a space with one point. By induction let $X_{n+1} = X_n \times \mathbb{Z}_2$ and $\lambda_{n, n+1} : X_{n+1} \to X_n$ be the projection map for $n \geq 1$, where $\mathbb{Z}_2$ is the discrete space with two points. Prove that the inverse limit of the system $(\{X_n\}_{n \in \mathbb{N}}, \{\lambda_{m, n}\})$ in Top is homeomorphic to the Cantor ternary set with the usual topology.

4.13 Prove that there is a category whose objects are all small categories and morphisms are all covariant functors between small categories.

Notes and Guide to Literature

A proof of the existence of direct limits in the category of groups may be found in Lang [1]. The solenoid is not such an abstract space as it appears; it can be realised in $\mathbb{R}^3$ as the intersection of a descending sequence of solid tori, each one going twice 'around the hole' of the preceding one; see Hocking and Young [1], p. 330. The terms 'injective' and 'projective' limits are sometimes used for direct and inverse limits respectively. The terms 'limits' and 'colimits' are also used instead of left and right roots respectively.

The terms 'forgetful functor', 'push-out', 'pull-back' illustrate the colloquial nature of the terminology used in category theory, possibly due to its American origin! Steenrod in fact described the general category theory as abstract non-sense; apparently meaning that although the statements and settings of its theorems look grand, the proofs are generally obvious almost to the point of being trivial.

Chapter Fourteen

Uniform Spaces

Uniform spaces stand somewhere in between metric spaces on one hand and general topological spaces on the other. In this chapter we present a brief treatment of uniform spaces. The presentation will be rather sketchy. We hope the reader has by now acquired sufficient maturity to supply the details.

1. Uniformities and Basic Definitions

In Chapter 4 we defined topological spaces as a generalisation of metric spaces. This generalization has well stood the test of time and its value is now universally accepted. There are however a few aspects of metric spaces that are lost in this generalisation. For example, as we remarked in Chapter 12, the concepts of completeness, uniform continuity and uniform convergence cannot be defined for arbitrary topological spaces. It is natural to ask whether a generalisation of metric spaces is possible in which such 'uniform' concepts are still salvaged. The answer is in the affirmative and leads to the theory of uniform spaces.

The key to the transition from metric spaces to topological spaces was the concept of open sets in metric spaces. Once they were defined, it turned out that the two basic notions of convergence and continuity can be expressed exclusively in terms of open sets without involving the metric directly. This fact led to the definition of a topological space.

We follow an analogous programme for uniform spaces. The key concept here is that of an entourage as defined below.

(1.1) Definition: Let $(X; d)$ be a metric space. Then a subset E of $X \times X$ is said to be an **entourage** if there exists $\epsilon > 0$ such that for all $x, y \in X$, $d(x, y) < \epsilon$ implies $(x, y) \in E$.

The definition applies equally well for pseudometric spaces as well and we shall often deal with pseudo-metrics in this chapter. It is convenient to look at an entourage in a slightly different way. Given a metric (or a pseudometric) d on X, and $\epsilon > 0$, let U_ϵ denote the set $\{(x, y) \in X \times X : d(x, y) < \epsilon\}$. If we give X the topology induced by d and $X \times X$ the product topology then it is easy to see that U_ϵ is an open neighbourhood of the diagonal ΔX in $X \times X$. Thus every entourage in $(X; d)$ is a neighbourhood of ΔX in $X \times X$. The converse is false. If d is the usual metric on $\mathbb{R}$ then the

22 (45–90/1980)

set $\left\{(x, y) \in \mathbb{R}^2 : |x-y| < \frac{1}{1+|y|}\right\}$ is a neighbourhood of $\Delta\mathbb{R}$ in $\mathbb{R}\times\mathbb{R}$ but there is no $\epsilon > 0$ such that U_ϵ is contained in this set. Incidentally, entourage is a French word meaning environment.

We now show that the basic concepts of uniform continuity, uniform convergence and Cauchy sequences (upon which depends completeness) can all be expressed in terms of entourages without involving the metrics directly. We state three simple propositions whose proofs are left to the reader.

(1.2) Proposition: Let $(X; d)$, $(Y; e)$ be metric spaces and $f : X \to Y$ a function. Then f is uniformly continuous (w.r.t. d and e) if and only if for every entourage F of $(Y; e)$, there exists an entourage E of $(X; d)$ such that for all $(x, y) \in E$, $(f(x), f(y)) \in F$. ∎

(1.3) Proposition: Let $\{x_n\}$ be a sequence in a metric space $(X; d)$. Then $\{x_n\}$ is a Cauchy sequence (w.r.t. d) if and only if for every entourage E, there exists $p \in \mathbb{N}$ such that for all m, $n \geq p$, $(x_m, x_n) \in E$. ∎

(1.4) Proposition: Let S be any set and $\{f_n : S \to X\}_{n \in N}$ be a sequence of functions into a metric space $(X; d)$. Then $\{f_n\}$ converges uniformly to a function $f : S \to X$ (w.r.t. d) if and only if for every entourage E there exists $m \in \mathbb{N}$ such that for all $n \geq m$ and $x \in S$, $(f_n(x), f(x)) \in E$. ∎

These propositions are not at all profound. But they provide us what we need. Recall that in defining topological spaces we listed down a number of properties of open sets in a metric space (see Theorems (4.1.5) and (4.1.6)) and then selected a few of them as the axioms for a topological space. We follow precisely the same procedure in defining uniform spaces. We begin by listing down a number of properties of entourages in metric spaces. Before doing so we recall a few facts about binary relations. They are important here because every entourage is a reflexive binary relation on the underlying set. Let X be a set and U a binary relation on X (that is, U is a subset of $X \times X$). Then the inverse relation U^{-1} is the set $\{(x, y) \in X \times X : (y, x) \in U\}$. For a subset A of X, $U[A]$ is defined to be set $\{y \in X :$ there exists some $x \in A$ such that $(x, y) \in U\}$. If $x \in X$, then we denote $U[\{x\}]$ by $U[x]$ or by $U(x)$ as well. Moreover, if U, V are two binary relations on X then their composite $U \circ V$ is defined to be the binary relation $\{(x, y) \in X \times X :$ there exists $z \in X$ such that $(x, z) \in V$ and $(z, y) \in U\}$. Composition is clearly associative and for two binary relations U, V we have $(U \circ V)^{-1} = V^{-1} \circ U^{-1}$ and $(U \circ V)[A] = U[V[A]]$ for any $A \subset X$. Also for any binary relation U, $(U^{-1})^{-1} = U$. Clearly U is symmetric iff $U = U^{-1}$, U is reflexive iff $\Delta X \subset U$ and U is transitive iff $U \circ U \subset U$. Note that for any U, $U \cap U^{-1}$ is always symmetric.

It may be helpful to interprete some of these concepts geometrically. Let X be the set of real numbers. A binary relation U on X is then a subset of the plane $\mathbb{R} \times \mathbb{R}$. It is easily seen that the inverse relation U^{-1} is nothing but the mirror image of U in the diagonal $\Delta\mathbb{R}$. Let $p : \mathbb{R} \times \mathbb{R} \to \mathbb{R}$ denote the projection onto the second factor. Then for a subset A of X, $U[A]$ is precisely the set $p\,(U \cap (A \times \mathbb{R}))$ as shown in Figure 1.

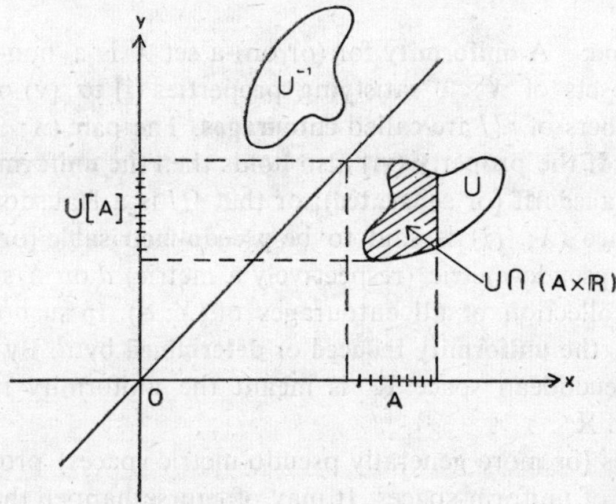

Fig. 1. **Binary Relation on ℝ.**

Unfortunately the composite of two binary relations on ℝ does not seem to admit any natural geometric interpretation.

With the terminology just introduced, we now have,

(1.5) Proposition: Let $(X; d)$ be a pseudo-metric space and $\mathcal{U}$ the family of all its entourages. Then,

(i) $\Delta X \subset U$ for each $U \in \mathcal{U}$

(ii) if $U \in \mathcal{U}$ then $U^{-1} \in \mathcal{U}$

(iii) if $U \in \mathcal{U}$ then there exists $V \in \mathcal{U}$ such that $V \circ V \subset U$.

(iv) if $U, V \in \mathcal{U}$ then $U \cap V \in \mathcal{U}$ and

(v) if $U \in \mathcal{U}$ and $U \subset V \subset X \times X$ then $V \in \mathcal{U}$.

If, moreover d is a metric then in addition to the above we also have

(vi) $\bigcap \{U : U \in \mathcal{U}\} = \Delta X$ (actually this property characterises metrics among pseudo-metrics).

Proof: (v) is immediate from the definition of an entourage. (i) follows from the fact that $d(x, x) = 0$ for all $x \in X$ while (ii) requires the symmetry of the pseudometric d. For (iii) suppose $U \in \mathcal{U}$. Then there exists $\epsilon > 0$ such that $U_\epsilon \subset U$. Put $V = U_{\epsilon/2}$. Then as a consequence of the triangle inequality it follows that $V \circ V \subset U$. For (iv), let $U, V \in \mathcal{U}$. Let ϵ, δ be such that $U_\epsilon \subset U$ and $U_\delta \subset V$. Then $U_\alpha \subset U \cap V$ where $\alpha = \min\{\epsilon, \delta\}$. So $U \cap V \in \mathcal{U}$. Finally (vi) is equivalent to saying that for $x \neq y$ in X, $d(x, y) > 0$, which is the case if and only if d is a metric. ∎

It is now obvious how to define an abstract uniform space. We start with a set X, a family $\mathcal{U}$ of binary relations on X (i.e. subsets of $X \times X$) and require that $\mathcal{U}$ satisfies some of the properties in the last proposition (or other similar properties). Once again it is a battle between generality and non-triviality. Assuming too few conditions about $\mathcal{U}$ would lead to a widely applicable but shallow theory while assuming too many would give a deep theory with a very limited ground for applications. As a compromise, the following definition is now quite standard.

(1.6) Definition: A **uniformity** for (or on) a set X is a non-empty collection $\mathcal{U}$ of subsets of $X \times X$ satisfying properties (i) to (v) of the last proposition. Members of $\mathcal{U}$ are called **entourages**. The pair $(X; \mathcal{U})$ is called a **uniform space**. If the property (vi) also holds then the uniform space $(X; \mathcal{U})$ is said to be **Hausdorff** (or **separated**); or that $\mathcal{U}$ is a Hausdorff uniformity. A uniform space $(X; \mathcal{U})$ is said to be **pseudo-metrisable** (or **metrisable**) if there exists a pseudo-metric (respectively a metric) d on X such that $\mathcal{U}$ is precisely the collection of all entourages of $(X; d)$. In such a case we also say that $\mathcal{U}$ is the **uniformity induced** or **determined** by d. By the **usual uniformity** on a euclidean space $\mathbb{R}^n$ is meant the uniformity induced by the usual metric on $\mathbb{R}^n$.

Metric spaces (or more generally pseudo-metric spaces) provide the foremost examples of uniform spaces. It may of course happen that two distinct metrics induce the same uniformity. The definition of **uniform continuity** of a function from one uniform space to another is a routine translation of Proposition (1.2) and is left to the reader. It is then easy to show that the composition of uniformly continuous functions is uniformly continuous and that the identity function is always uniformly continuous. We thus get a category whose objects are uniform spaces and morphisms are uniformly continuous functions. An equivalence in this category is called a **uniform isomorphism.** There is an evident covariant functor from the category of metric spaces (and uniformly continuous maps) to the category of uniform spaces. This functor associates to a metric space $(X; d)$, the uniform space $(X; \mathcal{U})$ where $\mathcal{U}$ is the family of all entourages in $(X; d)$.

Uniform convergence can be defined for sequences (or nets) of functions from a set into a uniform space. The concept of a Cauchy sequence (or a Cauchy net) is also the obvious one. These definitions are the 'only if' parts in the statements of Propositions (1.4) and (1.3) respectively and will not be restated.

Analogous to topologies, we have bases and sub-bases for uniformities as well.

(1.7) Definition: Let $(X; \mathcal{U})$ be a uniform space. Then a subfamily $\mathcal{B}$ of $\mathcal{U}$ is said to be a **base** for $\mathcal{U}$ if every member of $\mathcal{U}$ contains some member of $\mathcal{B}$; while a subfamily $\mathcal{S}$ of $\mathcal{U}$ is said to be a **sub-base** for $\mathcal{U}$ if the family of all finite intersections of members of $\mathcal{S}$ is a base for $\mathcal{U}$.

It is obvious that a uniformity is completely determined by any base or sub-base for it. This provides a general method for constructing uniformities. The following proposition gives a sufficient condition for a family of binary relations to generate a uniformity.

(1.8) Proposition: Let X be a set and $\mathcal{S} \subset P(X \times X)$ be a family such that

for every $U \in \mathcal{S}$ the following conditions hold:
(a) $\Delta X \subset U$,
(b) U^{-1} contains a member of $\mathcal{S}$, and
(c) there exists $V \in \mathcal{S}$ such that $V \circ V \subset U$. Then there exists a unique uniformity $\mathcal{U}$ for which $\mathcal{S}$ is a sub-base.

Proof: Let $\mathcal{B}$ be the family of all finite intersections of members of $\mathcal{S}$ and let $\mathcal{U}$ be the family of all supersets of members of $\mathcal{B}$. We assert that $\mathcal{U}$ is a uniformity on X. Conditions (i) and (v) of the definition are immediate from the construction of $\mathcal{U}$ and (a). For (ii), suppose $U \in \mathcal{U}$. Then there exist $U_1, U_2, \ldots, U_n \in \mathcal{S}$ such that $\bigcap_{i=1}^{n} U_i \subset U$. By (b) each U_i^{-1} contains some $V_i \in \mathcal{S}$. Then $\bigcap_{i=1}^{n} V_i \subset \bigcap_{i=1}^{n} U_i^{-1} \subset U^{-1}$. So $\bigcap_{i=1}^{n} V_i \in \mathcal{B}$ and $U^{-1} \in \mathcal{U}$ as desired. For condition (iii) in the definition suppose $U \in \mathcal{U}$ and suppose $U_1, \ldots, U_n \in \mathcal{S}$ are such that $\bigcap_{i=1}^{n} U_i \subset U$. For each $i = 1, 2, \ldots, n$, find $V_i \in \mathcal{S}$ such that $V_i \circ V_i \subset U_i$ by (c). Let $V = \bigcap_{i=1}^{n} V_i$. It is easily seen that $V \circ V \subset \bigcap_{i=1}^{n} (V_i \circ V_i)$. So $V \circ V \subset U$. Further $V \in \mathcal{B}$ and hence $V \in \mathcal{U}$. Thus (iii) is satisfied. It remains to verify (iv) in the definition. Let $U, V \in \mathcal{U}$. Find $U_1, \ldots, U_n$ and $V_1, \ldots, V_m \in \mathcal{S}$ such that $\bigcap_{i=1}^{n} U_i \subset U$ and $\bigcap_{j=1}^{m} V_j \subset V$. Then $\left(\bigcap_{i=1}^{n} U_i\right) \cap \left(\bigcap_{j=1}^{m} V_j\right) \in \mathcal{B}$. Since $U \cap V$ is a superset of this intersection it follows that $U \cap V \in \mathcal{U}$. Thus $\mathcal{U}$ is a uniformity for X. By its very construction $\mathcal{B}$ is a base and $\mathcal{S}$ a sub-base for $\mathcal{U}$. Uniqueness of $\mathcal{U}$ is trivial since a sub-base determines a uniformity. ∎

We can compare uniformities on a set the same way as topologies are compared. On any given set X, the strongest or the finest uniformity is evidently the family of all subsets of $X \times X$ containing ΔX. It is called the **discrete** uniformity and is induced by any discrete metric on X. Its opposite extreme is the **indiscrete** uniformity consisting of $X \times X$ alone. It is the coarsest of all uniformities on X. The union of a family of uniformities on X need not be a uniformity on X by itself, however, such a union clearly satisfies the conditions of the last proposition and therefore generates a uniformity on X. However, in sharp contrast with topologies, the intersection of two uniformities on a set need not be a uniformity. This is so because if $\mathcal{U}$, $\mathcal{V}$ are uniformities on a set X then the family $\mathcal{U} \cap \mathcal{V}$ may fail to satisfy condition (iii) in the definition (even though it certainly satisfies all other conditions.) A simple counter-example can be given as follows. Note first of all that if R is an equivalence relation on a set X then the family of all supersets of R (in $X \times X$) is a uniformity on X. If R, S are two equivalence relations on X then the intersection of their corresponding uniformities is precisely the family of all supersets of $R \cup S$. Now let $X = \{1, 2, 3, 4\}$. Let R be the equivalence relation on X whose equivalence classes are the sets $\{1, 2\}$ and $\{3, 4\}$; in other words, $R = (\{1, 2\}\times\{1, 2\}) \cup (\{3, 4\}\times\{3, 4\})$. Similarly let S be the equivalence relation whose equivalence classes are $\{1, 3\}$ and $\{2, 4\}$. Let $\mathcal{U}$, $\mathcal{V}$ be the corresponding uniformities on X, generated by $\{R\}$ and $\{S\}$ respectively. Then $R \cup S \in \mathcal{U} \cap \mathcal{V}$ and there is no $W \in \mathcal{U} \cap \mathcal{V}$ such that $W \circ W \subset R \cup S$.

Given a uniform space $(X; \mathcal{U})$ and a subset Y of X, we define the **relativised uniformity** $\mathcal{U}/Y$ as the family $\{U \cap (Y\times Y) : U \in \mathcal{U}\}$. It is easy to check that $\mathcal{U}/Y$ is in fact a uniformity for Y. $(Y; \mathcal{U}/Y)$ is said to be a **uniform subspace** of $(X; \mathcal{U})$. It is easy to see that the inclusion function $i : Y \to X$ uniformly continuous w.r.t. $\mathcal{U}/Y$ and $\mathcal{U}$. If $(X; \mathcal{U})$, $(Y; \mathcal{V})$ are two uniform spaces then a function $f : X \to Y$ is said to be an **embedding** if it is one-to-one, uniformly continuous and a uniform isomorphism when regarded as a function from $(X; \mathcal{U})$ onto $(f(X))$, $\mathcal{V}/f(X)$.

Elementary properties of embeddings and subspaces are left to the reader to state and prove.

Given an indexed collection $\{(X_i, \mathcal{U}_i) : i \in I\}$ of uniform spaces we define a product uniformity on the cartesian product $X = \prod_{i\in I} X_i$ much the same way as the product topology. The idea is to make each projection $\pi_i : X \to X_i$ uniformly continuous without making the uniformity on X unnecessarily strong. Let $\mathcal{S}$ be the family of all subsets of $X\times X$ of the form $\theta_i^{-1}(U_i)$ for $U_i \in \mathcal{U}_i$, $i \in I$ where $\theta_i : X\times X \to X_i\times X_i$ is the function $\pi_i \times \pi_i$ defined by $(\pi_i \times \pi_i)(x, y) = (\pi_i(x), \pi_i(y))$. It is easy to check that the family $\mathcal{S}$ satisfies the conditions of Proposition (1.8) and consequently generates a uniformity $\mathcal{U}$ on X. The uniform space $(X; \mathcal{U})$ is called the **product** of $\{(X_i, \mathcal{U}_i) : i \in I\}$. A routine argument shows that the system $((X, \mathcal{U}), \{\pi_i\}_{i\in I})$ is a product in the category of uniform spaces considered above.

Analogues of many other constructions for topological spaces can be carried out for uniform spaces. There is little point in stating them unless they are of independent interest. We conclude the present section by defining a certain topology associated with a given uniformity. A uniformity on a set X induces a topology on X much the same way as a metric on X does. Let d be a metric on a set X. For $r > 0$, let U_r be the set $\{(x, y) \in X\times X : d(x, y) < r\}$. We already considered such sets when we defined entourages. For $x \in X$, $U_r[x]$ is precisely the open ball $B(x; r)$. Thus a subset G of X is open in the metric topology if and only if for each $x \in G$, there is an entourage U (which may well depend on x) such that $U[x] \subset G$. Note that this statement does not involve the metric d directly and therefore the construction can be generalised for any uniform space as we are about to do.

(1.9) Theorem: For a uniform space $(X; \mathcal{U})$ let $\mathcal{T}_{\mathcal{U}}$ be the family $\{G \subset X :$ for each $x \in G$, there exists $U \in \mathcal{U}$ such that $U[x] \subset G\}$. Then $\mathcal{T}_{\mathcal{U}}$ is a topology on X.

Proof: It is trivial to check that the empty set Φ and the set X belong to $\mathcal{T}_{\mathcal{U}}$. Also from the very nature of the definition, it is clear that $\mathcal{T}_{\mathcal{U}}$ is closed under arbitrary unions. Lastly to show that $\mathcal{T}_{\mathcal{U}}$ is closed under finite intersections, let $G, H \in \mathcal{T}_{\mathcal{U}}$ and suppose $x \in G \cap H$. Then there exist U,

$V \in \mathcal{U}$ such that $U[x] \subset G$ and $V[x]. \subset H$. Let $W = U \cap V$. Then $W \in \mathcal{U}$. Also $W[x] \subset U[x] \cap V[x]$ and so $W[x] \subset G \cap H$. So $G \cap H \in \mathcal{T}_\mathcal{U}$. Thus $\mathcal{T}_\mathcal{U}$ is a topology on X. ∎

In view of the remarks preceding the theorem, it follows that if d is a pseudometric on a set X, $\mathcal{U}$ the uniformity on X induced by d then $\mathcal{T}_\mathcal{U}$ as given by the theorem above coincides with the pseudometric topology on X induced by d directly. The transition from uniformities to topologies applies not only for spaces but for maps as well. Specifically,

(1.10) Proposition: Let $(X; \mathcal{U})$, $(Y, \mathcal{V})$ be uniform spaces and $f: X \to Y$ a function which is uniformly continuous w.r.t. $\mathcal{U}$ and $\mathcal{V}$. Let $\mathcal{T}_\mathcal{U}$, $\mathcal{T}_\mathcal{V}$ be the topologies on X and Y respectively as given by the theorem above. Then f is continuous w.r.t. these topologies.

Proof: Let $G \in \mathcal{T}_\mathcal{V}$. We have to show that $f^{-1}(G)$ is open in X, that is $f^{-1}(G) \in \mathcal{T}_\mathcal{U}$. Let $x \in f^{-1}(G)$. Then $f(x) \in G$ and so by definition of $\mathcal{T}_\mathcal{V}$ there exists $V \in \mathcal{V}$ such that $V[f(x)] \subset G$. Let $U = \{(z, y) \in X \times X : (f(z), f(y)) \in V\}$. Then $U \in \mathcal{U}$ since f is uniformly continuous. Moreover $U[x] \subset f^{-1}(V[f(x)]) \subset f^{-1}(G)$. So $f^{-1}(G)$ is open in X. Hence f is continuous. ∎

(1.11) Definition: Let $(X; \mathcal{U})$ be a uniform space. Then the topology $\mathcal{T}_\mathcal{U}$ is called the **uniform topology** on X induced by $\mathcal{U}$. A topological space $(X; \mathcal{T})$ is said to be **uniformisable** if there exists a uniformity $\mathcal{U}$ on X such that $\mathcal{T} = \mathcal{T}_\mathcal{U}$.

The transition from uniform spaces to topological spaces gives a covariant functor from the category of uniform spaces to the category of topological spaces. This functor has many nice properties. For example, it preserves products in the sense that if $(X; \mathcal{U})$ is the product of a collection $\{(X_i. \mathcal{U}_i) : i \in I\}$ of uniform spaces then the space $(X; \mathcal{T}_\mathcal{U})$ is precisely the topological product of the family $\{(X_i, \mathcal{T}_{\mathcal{U}_i}) : i \in I\}$ of topological spaces. There is also an obvious relationship between bases and sub-bases for a uniformity and the corresponding concepts for a topological space. However, such results are of a routine type and will be left to the reader.

It is now clear that uniform spaces stand somewhere between metric spaces and general topological spaces. Every metric space gives rise to a uniform space and in turn, every uniform space gives rise to a topological space. It is natural to ask under what conditions the reverse passage is possible. In other words, which uniform spaces are metrisable (or pseudo-metrisable) and which topological spaces are uniformisable? These are important questions and will be taken up in the next section. However, in the exercises we derive regularity as a necessary condition for uniformis-ability.

Exercises

1.1 Let $(X; \mathcal{U})$ be a uniform space and d a pseudometric on X. (Caution: $\mathcal{U}$ may not necessarily be induced by d.) Let $\{r_n\}$ be a monotonically decreasing sequence of positive real numbers converging to 0. Put the product uniformity on $X \times X$ and the usual uniformity on $\mathbb{R}$. Prove that the function $d : X \times X \to \mathbb{R}$ is uniformly continuous if and only if for each $n \in \mathbb{N}$, the set $\{(x, y) \in X \times X : d(x, y) < r_n\}$ is a member of $\mathcal{U}$. (Hint: First show that the family of all sets, in $(X \times X)^2$ of the form $\{((x, y), (u, v)) : (x, u) \in U \text{ and } (y, v) \in U\}$ for $U \in \mathcal{U}$ is a base for the product uniformity on $X \times X$.)

1.2 Let $(X; d)$ be a pseudometric space and $\mathcal{U}$ the uniformity induced by d. On $X \times X$ put the product uniformity. Prove that the function $d : X \times X \to \mathbb{R}$ is uniformly continuous w.r.t. this uniformity and the usual uniformity on $\mathbb{R}$. Prove also that $\mathcal{U}$ is the smallest uniformity on X, having this property.

1.3 Define a Cauchy net in a uniform space. Also define uniform convergence of a net of functions from a set into a uniform space. Prove that uniform convergence implies pointwise convergence.

1.4 Prove that a uniform space is Hausdorff if and only if the associated topological space is Hausdorff.

1.5 Let $(X; \mathcal{U})$ be a uniform space and $A \subset X$. Prove that the set $\{x \in A : U[x] \subset A \text{ for some } U \in \mathcal{U}\}$ is open and is in fact the interior of A w.r.t. the uniform topology $\mathcal{T}_{\mathcal{U}}$. Deduce that for $B \subset X$, $\overline{B} = \bigcap \{U[B] : U \in \mathcal{U}\}$. (Hint: Note that each $U \in \mathcal{U}$ contains a symmetric $V \in \mathcal{U}$ such that $V \circ V \subset U$.)

1.6 Let $(X; \mathcal{U})$ be a uniform space. On $X \times X$ put the product topology. Then prove that every member of $\mathcal{U}$ is a neighbourhood of ΔX and in fact contains a symmetric open member of $\mathcal{U}$. (Hint: In all such problems, it is easier to give a proof for pseudo-metric spaces first and then to translate it to uniform spaces. Given $U \in \mathcal{U}$ find a symmetric $V \in \mathcal{U}$ such that $(V \circ V) \circ (V \circ V) \subset U$ and hence that $V \circ V \circ V \subset U$ in view of reflexivity of V. Then show that for each $(x, y) \in V$, $V[x] \times V[y] \subset U$ and hence that V is contained in the interior of U in $X \times X$.)

1.7 Let $(X; \mathcal{U})$ be a uniform space. Prove that each member of $\mathcal{U}$ contains a closed symmetric member of $\mathcal{U}$. (Hint: First prove that for any subset M of $X \times X$, $\overline{M} = \bigcap \{U \circ M \circ U : U \in \mathcal{U}\}$.)

1.8 Prove that every uniformisable space is regular.

1.9 Let X be a set; $(Y; \mathcal{V})$ a uniform space and F the set of all functions from X to Y. For each $V \in \mathcal{V}$, let $\theta(V) = \{(f, g) \in F \times F : (f(x), g(x)) \in V \text{ for all } x \in X\}$. Prove that the family $\{\theta(V) : V \in \mathcal{V}\}$ is a sub-base for a uniformity $\mathcal{U}$ on F.

1.10 In the last exercise suppose X has a topology and let $M(X, Y)$ be the set of all members of F which are continuous w.r.t. this topology and the uniform topology on Y. Prove that $M(X, Y)$ is closed in F w.r.t. the uniform topology induced by $\mathcal{U}$. (Hint: In simple language, this means that the uniform limit of continuous functions is continuous.)

1.11 Prove that the functor from the category of uniform spaces to Top preserves products.

Notes and Guide to Literature

Although the concepts of uniform convergence, uniform continuity and completeness were used in analysis for a long time, the basic concept of uniformity was isolated relatively recently. The first definitive work is attributed to Weil [1]. The treatment given here follows that in Kelley [1].

The uniformity introduced in Exercise (1.9) is called the **uniformity of uniform convergence.**

See Willard [1] for an alternate approach to uniformities and for their relationship with proximities.

2. Metrisation

In this section we tackle the two basic questions raised at the end of the last section, that is to find when a uniform space is metrisable and when a topological space is uniformisable. In obtaining these answers we shall also derive as a byproduct, a description of a uniformity in terms of a certain set of pseudometrics on the underlying set.

Let $(X; \mathcal{U})$ be a uniform space. A necessary condition for pseudo-metrisability of $\mathcal{U}$ is easy to get. Let d be a pseudometric on X inducing the uniformity $\mathcal{U}$. For each $n \in \mathbb{N}$ let $B_n = \{(x, y) \in X \times X : d(x, y) < 1/n\}$. Then the family $\mathcal{B} = \{B_n : n \in \mathbb{N}\}$ is clearly a countable base for $\mathcal{U}$. Thus the existence of a countable base is a necessary condition for pseudometrisability of a uniformity. This condition also turns out to be sufficient as we shall show. But the proof is not so obvious. We begin by getting the base elements in a certain normalised form.

(2.1) Proposition: Suppose a uniformity $\mathcal{U}$ on a set X has a countable base. Then there exists a countable base $\left\{U_n\right\}_{n=1}^{\infty}$ for $\mathcal{U}$ such that each U_n is symmetric and $U_{n+1} \circ U_{n+1} \circ U_{n+1} \subset U_n$ for all $n \in \mathbb{N}$.

Proof: Let the given countable base be $\mathcal{V} = \{V_1, V_2, \ldots, V_n, \ldots\}$. We set $U_1 = V_1 \cap V_1^{-1}$. U_1 is then a symmetric member of $\mathcal{U}$. We observe that for every member U of $\mathcal{U}$ there exists a symmetric member V of $\mathcal{U}$ such that $V \circ V \circ V \subset U$ (see the hint to Exercise (1.6)). Apply this fact to the set $U_1 \cap V_2$ which is a member of $\mathcal{U}$. Then there exists a symmetric member say U_2 of $\mathcal{U}$ such that $U_2 \circ U_2 \circ U_2 \subset U_1 \cap V_2$. Next consider $U_2 \cap V_3$ and get a symmetric member U_3 of such that $U_3 \circ U_3 \circ U_3 \subset U_2 \cap V_3$. In this manner we proceed by induction and get a sequence $\{U_n : n \in \mathbb{N}\}$ of symmetric members of $\mathcal{U}$ such that for each $n \in \mathbb{N}$, $U_{n+1} \circ U_{n+1} \circ U_{n+1} \subset U_n \cap V_{n+1}$. Then $U_n \subset V_n$ for all n and so $\{U_n\}_{n \in \mathbb{N}}$ is a base for $\mathcal{U}$ since $\{V_n : n \in \mathbb{N}\}$ is given to be a base. ∎

Having obtained a normalised countable base $\{U_n : n \in \mathbb{N}\}$ for $\mathcal{U}$, we proceed with the construction of a pseudometric d on X as follows. We set $U_0 = X \times X$. Note that $U_n \subset U_{n-1}$ for all $n \in \mathbb{N}$. The idea is to define

$d: X \times X \to \mathbb{R}$ in such a way that for each $n \in \mathbb{N}$, the set $\{(x, y) \in X \times X : d(x, y) < 2^{-n}\}$ will be very close to the set U_n. To be precise, we shall construct a pseudo-metric d on X such that for each n, the set $\{(x, y) \in X \times X : d(x, y) < 2^{-n}\}$ will be between U_n and U_{n-1}.

We begin with a function $f: X \times X \to \mathbb{R}$ defined by $f(x, y) = 2^{-n}$ in case there exists $n \in \mathbb{N}$ (which must then be unique) such that $(x, y) \in U_{n-1} - U_n$. If there exists no such n, it means $(x, y) \in \bigcap_{n=1}^{\infty} U_n$ and in that case we set $f(x, y) = 0$. Note that $f(x, y) = f(y, x)$ for all $x, y \in X$ since all the sets U_n's are symmetric. Note further that for each $n \in \mathbb{N}$, $\{(x, y) \in X \times X : f(x, y) \leq 2^{-n}\}$ is precisely the set U_{n-1}.

Now comes the key step in the construction. For $x, y \in X$ we define $d(x, y) = \inf \sum_{i=1}^{n} f(x_i, x_{i+1})$ where the infimum is to be taken over all possible finite sequences $\{x_0, x_1, \ldots, x_n, x_{n+1}\}$ in X for which $x_0 = x$ and $x_{n+1} = y$. Such a sequence will be called a **chain** from x to y with n **nodes** at $x_1, \ldots, x_n$. The number $\sum_{i=1}^{n} f(x_i, x_{i+1})$ will be called the **length** of the chain. Thus $d(x, y)$ is the infimum of the lengths of all possible chains from x to y. The idea is similar to the calculus definition of the arc length of a curve except that there we take the supremum rather than the infimum of the lengths of all possible polygonal paths joining the end points and having vertices (or 'nodes') on the curve. We first prove that the function d so defined is a pseudometric on the set X.

(2.2) Lemma: The function $d : X \times X \to \mathbb{R}$ just defined is a pseudometric on the set X.

Proof: Evidently $d(x, y) \leq f(x, y)$ for all $x, y \in X$, since $\{x, y\}$ is itself a chain from x to y. So, if $x = y$ then $d(x, y) = 0$ from the definition of f $\left(\text{recall that } \Delta X \subset \bigcap_{n=1}^{\infty} U_n\right)$. Symmetry of d follows from that of f. Only the triangle inequality remains to be established. Let $x, y, z \in X$. Let S_1, S_2 and S_3 be respectively the sets of all possible chains from x to y, from y to z and from x to z. A chain $s_1 \in S_1$ and $s_2 \in S_2$ together determine an element of S_3 by juxtaposition, which we denote by $s_1 + s_2$. Note that $\lambda(s_1 + s_2) = \lambda(s_1) + \lambda(s_2)$ where λ denotes the length of the chain. We now let $J(S_3)$ be the image of $S_1 \times S_2$ in S_3 under the juxtaposition function $+ : S_1 \times S_2 \to S_3$. Then we have,

$$\begin{aligned}
d(x, y) + d(y, z) &= \inf \{\lambda(s_1) : s_1 \in S_1\} + \inf \{\lambda(s_2) : s_2 \in S_2\} \\
&= \inf \{\lambda(s_1) + \lambda(s_2) : (s_1, s_2) \in S_1 \times S_2\} \\
&= \inf \{\lambda(s_1 + s_2) : (s_1, s_2) \in S_1 \times S_2\} \\
&= \inf \{\lambda(s_3) : s_3 \in J(S_3) \subset S_3\} \\
&\geq \inf \{\lambda(s_3) : s_3 \in S_3\} \\
&= d(x, z) \text{ by definition.}
\end{aligned}$$

The triangle inequality is thereby established and so d is a pseudometric on X. ∎

We are not through yet. Although we have shown that d is a pseudometric on X we have not shown that it induces the given uniformity $\mathcal{U}$ on X. Let $\mathcal{V}$ be the uniformity on X induced by d. A countable base for $\mathcal{V}$ is given by $\{W_n : n \in \mathbb{N}\}$ where we define W_n to be the set $\{(x, y) \in X \times X : d(x, y) < 2^{-n}\}$. In order to show that $\mathcal{U}$ equals $\mathcal{V}$ it suffices to show that every member of a base for $\mathcal{U}$ contains a member of a base for $\mathcal{V}$ and vice versa. Of this one part is easy, because it is immediate from the fact that $d(x, y) \leq f(x, y)$ and the definition of f that for each $n \in \mathbb{N}$, $U_n \subset W_n$. For the other part we claim that $W_n \subset U_{n-1}$ for each $n \in \mathbb{N}$. For this we need a lemma about the function f.

(2.3) Lemma: For any integer $k \geq 0$ and $x_0, \ldots, x_k, x_{k+1} \in X$, $f(x_0, x_{k+1}) \leq 2 \sum_{i=0}^{k} f(x_i, x_{i+1})$.

Proof: We apply induction on k. More precisely, we use the second principle of induction. The case $k = 0$ is trivial. Assume $k > 0$ and that the result holds for all possible chains with less than k nodes. Let $x_0, x_1, \ldots, x_k, x_{k+1}$ be a chain with k nodes. The idea is to break this chain into smaller chains and then to apply the induction hypothesis to each of them. Let a be the length of this chain, that is, $a = \sum_{i=0}^{k} f(x_i, x_{i+1})$. Since each term in this sum is non-negative it follows that $a \geq 0$ and that $a = 0$ if and only if $f(x_i, x_{i+1}) = 0$ for all $i = 0, 1, \ldots, k$. Moreover, none of the terms $f(x_i, x_{i+1})$ can exceed a. These simple observations will be needed crucially in the argument below where our task is to show that $f(x_0, x_{k+1}) \leq 2a$.

We make three cases depending upon how big a is.

Case I: Let $a \geq \frac{1}{4}$. Then $2a \geq \frac{1}{2}$. By the definition of f, the largest value it can take is $\frac{1}{2}$. In particular, $f(x_0, x_{k+1}) \leq \frac{1}{2} \leq 2a$ as was to be proved.

Case II: Let $a = 0$. Then $f(x_i, x_{i+1}) = 0$ for every $i = 0, 1, \ldots, k$. We have to show that $f(x_0, x_{k+1}) = 0$ or equivalently that $(x_0, x_{k+1}) \in \bigcap_{n=0}^{\infty} U_n$. We decompose the chain $x_0, x_1, \ldots, x_k, x_{k+1}$ into three chains, say, $x_0, \ldots, x_r$; $x_r, x_{r+1}, \ldots, x_s$ and $x_s, x_{s+1}, \ldots, x_k, x_{k+1}$ where r and s are any integers such that $1 \leq r \leq s \leq k$. Note that each of these three chains has length 0 and less than k nodes. So by the induction hypothesis, $f(x_0, x_r)$, $f(x_r, x_s)$ and $f(x_s, x_{k+1})$ are all 0. Hence for every $n \in \mathbb{N}$, $(x_0, x_r) \in U_n$, $(x_r, x_s) \in U_n$ and also $(x_s, x_{k+1}) \in U_n$ which means $(x_0, x_{k+1}) \in U_n \circ U_n \circ U_n$. But $U_n \circ U_n \circ U_n \subset U_{n-1}$. Thus (x_0, x_{k+1}) U_{n-1} for all $n \in \mathbb{N}$ and hence $f(x_0, x_{k+1}) = 0$.

Case III: Let $0 < a < \frac{1}{4}$. Let r be the largest integer such that $\sum_{i=0}^{r-1} f(x_i, x_{i+1}) \leq a/2$. If $f(x_0, x_1) > a/2$, this definition fails and we set $r = 0$ in this case. Then $0 \leq r \leq k$ and so each of the chains $x_0, \ldots, x_r$ and $x_{r+1}, \ldots, x_{k+1}$ has less than k nodes. The first chain has length $\leq a/2$. Note that the length of the second chain is also at most $a/2$ for otherwise the chain $x_0, \ldots, x_{r+1}$ will have length $<a/2$ contradicting the definition of r. By induction hypothesis we now get, $f(x_0, x_r) \leq a$ and $f(x_{r+1}, x_{k+1}) \leq a$ while $f(x_r, x_{r+1}) \leq a$ as was observed above. Now let m be the smallest integer such that $2^{-m} \leq a$ ($m > 2$ since $a < \frac{1}{4}$). Note that the function f only assumes values of the form 2^{-n} for $n \in \mathbb{N}$. So whenever $f(x, y) \leq a$ for $x, y \in X$, it actually means $f(x, y) \leq 2^{-m}$. In particular, $f(x_0, x_{r+1}) \leq 2^{-m}$ and hence $(x_0, x_{r+1}) \in U_{m-1}$. Similarly $(x_r, x_{r+1}) \in U_{m-1}$ and $(x_{r+1}, x_{k+1}) \in U_{m-1}$. But then $(x_0, x_{k+1}) \in U_{m-1} \circ U_{m-1} \circ U_{m-1} \subset U_{m-2}$. Hence $f(x_0, x_{k+1}) \leq 2^{-(m-1)} \leq 2a$ as was to be proved. ∎

The significance of this lemma is noteworthy. As an immediate corollary of it, it follows that for any $x, y \in X$, $f(x, y) \leq 2d(x, y)$. We thus have that $\frac{1}{2}f(x, y) \leq d(x, y) \leq f(x, y)$. $f(x, y)$ may be looked upon as a crude measure of the distance between x and y. The exact distance $d(x, y)$ is obtained by taking various chains from x to y, summing up the approximate distances between every pair of successive nodes in each case and then taking the infimum over all possible chains from x to y. The lemma says that the crude measure $f(x, y)$ is not, after all, so bad an approximation for $d(x, y)$. This is in sharp contrast with geometry, where the arc length of a curve is in general much greater than the straight line distance between its end points, even though the arc length is defined by approximating the curve by straight line segments.

We now prove that for each $n \in \mathbb{N}$, $W_n \subset U_{n-1}$. Let $(x, y) \in W_n$. Then $d(x, y) < 2^{-n}$ whence $f(x, y) < 2^{-(n-1)}$ by the lemma above. But since f takes only values of the form 2^{-k} for $k \in \mathbb{N}$, it follows that $f(x, y) \leq 2^{-n}$. So $(x, y) \in U_{n-1}$.

With this we have completed the proof of the following important theorem.

(2.4) Theorem: A uniformity is pseudo-metrisable if and only if it has a countable base. ∎

The reader can hardly fail to see the beauty of the construction employed in the proof above. The theorem is of the same spirit as the Urysohn lemma (Theorem (7.3.3)). In both the results, a real-valued function is constructed from a purely set theoretic hypothesis. It is perhaps no accident that the theorem above is due to Urysohn (jointly with Alexandroff).

In the proof above if $\mathcal{U}$ were a Hausdorff uniformity then for any $x \neq y$ in X, $f(x, y) > 0$ and so $d(x, y) > 0$ since $f(x, y) < 2d(x, y)$. Thus the pseudo-metric d is actually a metric. We have therefore obtained a characterisation of metrisable uniformities.

(2.5) Theorem: A uniformity is metrisable if and only if it is Hausdorff and has a countable base. ∎

We now turn to the second basic problem, namely, that of deciding which topological spaces are uniformisable. The answer will come through an embedding theorem, which is also of interest in its own right.

Theorem (2.4) can be roughly interpreted as saying that if the uniformity $\mathcal{U}$ on X is small (in the sense of possessing a countable base) then a single pseudometric d suffices to describe it. It leads to the expectation that the smallness condition could be dropped if we allow a set D of pseudometrics on X to replace the single pseudometric d. This guess turns out to be correct. We begin with a description of the uniformity generated by a family D of pseudometrics on a set X. For each $d \in D$ and $r > 0$, let $V_{d,r} = \{(x, y) \in X \times X : d(x, y) < r\}$. Let $\mathcal{S}$ be the collection $\{V_{d,r} : r > 0, d \in D\}$. It is easy to show that $\mathcal{S}$ is a sub-base for a uniformity $\mathcal{U}$ on X (see Proposition (1.8)). This uniformity is called the **uniformity generated** by D. It has several interesting properties which we list in the following proposition.

(2.6) Proposition: Let $\mathcal{U}$ be the uniformity generated by a family D of pseudometrics on a set X. Then

(i) Each member d of D is a uniformly continuous function from $X \times X$ to $\mathbb{R}$ where $\mathbb{R}$ has the usual uniformity and $X \times X$ has the product uniformity induced by $\mathcal{U}$. Moreover, $\mathcal{U}$ is the smallest uniformity on X which makes each member of D uniformly continuous.

(ii) Let Y be the power set X^D. For each $d \in D$ let X_d be a copy of the set X and let $\mathcal{V}_d$ be the uniformity on X_d induced by the pseudometric d. Let $\mathcal{V}$ be the product uniformity on $Y = \prod_{d \in D} X_d$. Then the evaluation function $f : X \to Y$ defined by $f(x)(d) = x$ for all $d \in D$, $x \in X$ is a uniform embedding of $(X; \mathcal{U})$ into $(Y; \mathcal{V})$.

Proof: The first statement is a generalisation of Exercise (1.2) and its proof will be omitted. For (ii), we let $\pi_d : Y \to X_d$ be the projection for $d \in D$. Note that for each $d \in D$, $\pi_d \circ f : X \to X_d$ is merely the identity function and is uniformly continuous because the uniformity on X is $\mathcal{U}$ which is stronger than $\mathcal{V}_d$, the uniformity on X_d. So by the general properties of products, $f : (X; \mathcal{U}) \to (Y; \mathcal{V})$ is uniformly continuous. Clearly f is one-to-one. Let Z be the range of f. We have to show that f is a uniform isomorphism when regarded as a function from $(X; \mathcal{U})$ to $(Z; \mathcal{V}/Z)$. As uniform continuity and injectivity of f are already known, it suffices to prove that the image of every sub-basic entourage under $f \times f$ is an entourage in $\mathcal{V}/Z$. Take the sub-base $\mathcal{S}$ for $\mathcal{U}$. A member of $\mathcal{S}$ is of the form $V_{d,r}$ for some $r > 0$ and $d \in D$. It is easy to verify that $(f \times f)(V_{d,r})$ is precisely $Z \cap (\pi_d \times \pi_d)^{-1}(V_{d,r})$ which is an entourage in the relative uniformity on Z. ∎

Statement (ii) above is an embedding theorem. It says that an abstract uniform space can be 'concretely' described as a uniform subspace of the product of pseudo-metric spaces, provided that the uniformity is generated by some set D of pseudometrics on the underlying set. The smaller the family D, the sharper is the resulting embedding. To apply this proposition to a given uniform space $(X; \mathcal{U})$, we have to look for a family D of pseudometrics on the set X which generates $\mathcal{U}$. The following proposition shows that such a search is always successful.

(2.7) Proposition: Let $(X; \mathcal{U})$ be a uniform space and D the family of all pseudometrics on X which are uniformly continuous as functions from $X \times X$ to $\mathbb{R}$, the domain being given the product uniformity induced by $\mathcal{U}$ on each factor. Then D generates the uniformity $\mathcal{U}$ on X.

Proof: Let $\mathcal{V}$ be the uniformity generated by D. By statement (i) of the last proposition $\mathcal{V}$ is the smallest uniformity on X rendering each $d \in D$ uniformly continuous. So $\mathcal{V} \subset \mathcal{U}$. For the other way inclusion, suppose $U \in \mathcal{U}$. Let U_1 be a symmetric member of $\mathcal{U}$ contained in U. Set $U_0 = X \times X$. Define $U_2, U_3, \ldots, U_n, \ldots$ by induction so that each U_n is a symmetric member of $\mathcal{U}$ and $U_n \circ U_n \circ U_n \subset U_{n-1}$ for each $n \in \mathbb{N}$. Then, duplicating the construction used in the proof of Theorem (2.4), we get a pseudometric $d : X \times X \to \mathbb{R}$ such that for each $n \in \mathbb{N}$, $U_{n+1} \subset \{(x, y) \in X \times X : d(x, y) < 2^{-n}\} \subset U_{n-1}$. It follows that d is uniformly continuous w.r.t. the product topology on $X \times X$ (see Exercise (1.1))). So $d \in D$. We do not claim, nor is it necessarily true, that d generates the uniformity $\mathcal{U}$. This would have been the case if it were known that the sets U_n's formed a base for $\mathcal{U}$; but we are not assured of it. Still, putting $n = 2$ we get $\{(x, y) \in X \times X : d(x, y) < \frac{1}{4}\} \subset U_1 \subset U$. Let $\mathcal{S}$ be the defining sub-base for $\mathcal{V}$. Then $\{(x, y) : d(x, y) < \frac{1}{4}\} \in \mathcal{S} \subset \mathcal{V}$ and so $U \in \mathcal{V}$. Thus we have shown that $\mathcal{U} \subset \mathcal{V}$, and hence that $\mathcal{U} = \mathcal{V}$. Therefore D generates the uniformity $\mathcal{U}$. ∎

Note that the family of pseudometrics in the last proposition is not a very economical one. Indeed if $d \in D$ then so does λd for any $\lambda > 0$; although in terms of the uniform structure induced on X, λd tells us nothing new than d does. Still, the last proposition is of theoretical importance, because when combined with the earlier proposition, we get the following embedding theorem.

(2.8) Theorem: Every uniform spacc is uniformly isomorphic to a subspace of a product of pseudometric spaces. ∎

The embedding theorem implies a topological property of uniform spaces. A uniform embedding of one uniform space into another induces a topological embedding of the corresponding topological spaces. Moreover, the topology induced by the product uniformity is the same as the product of the corresponding uniform topologies (see Exercise (1.11)). These facts yield the following corollary.

(2.9) Corollary: If $(X; \mathcal{U})$ is a uniform space then the corresponding topological space $(X; \mathcal{T}_{\mathcal{U}})$ is completely regular.

Proof: By what we said above, it follows from the embedding theorem that $(X; \mathcal{T}_{\mathcal{U}})$ is embeddable into a product of pseudometric spaces. It then follows that $(X; \mathcal{T}_{\mathcal{U}})$ is completely regular, by the result of Exercise (9.2.5). ∎

In other words, complete regularity is a necessary condition for uniformisability. It also turns out to be a sufficient condition. Because Exercise (9.2.5) also shows that a completely regular space $(X; \mathcal{T})$ is embeddable into a product of pseudo-metric spaces. Let these pseudometric spaces be indexed as $\{(X_i, d_i) : i \in I\}$. Let $\mathcal{U}_i$ be the uniformity induced by d_i on X_i. Let $(Y; \mathcal{U})$ be the product of the family $\{(X_i, \mathcal{U}_i) : i \in I\}$ of uniform spaces. As before, $(Y; \mathcal{T}_{\mathcal{U}})$ is the topological product of the family $\{(X_i, \mathcal{T}_{\mathcal{U}_i}) : i \in I\}$ of topological spaces. We are given an embedding $f : (X; \mathcal{T}) \to (Y; \mathcal{T}_{\mathcal{U}})$. Let Z be the range of f. It is a routine matter to check that the topology on Z induced by the relativised uniformity $\mathcal{U}/Z$ is the same as the relativisation of the topology $\mathcal{T}_{\mathcal{U}}$ to Z; in symbols $\mathcal{T}_{\mathcal{U}/Z} = (\mathcal{T}_{\mathcal{U}})/Z$. $(X; \mathcal{T})$ is homeomorphic to $(Z; (\mathcal{T}_{\mathcal{U}})/Z)$ which is the same as $(Z; \mathcal{T}_{\mathcal{U}/Z})$. So $(X; \mathcal{T})$ is uniformisable. Thus we get the following theorem which completely answers our second basic question.

(2.10) Theorem: A topological space is uniformisable if and only if it is completely regular. ∎

(2.11) Definition: The **gage** of a uniform space $(X; \mathcal{U})$ is the family of all pseudometrics on the set X which are uniformly continuous as functions from $X \times X$ to $\mathbb{R}$.

In view of statement (i) in Proposition (2.6), the uniformity is completely characterised by its gage. It is often easier to work with the gage rather than with the uniformity itself. Every concept based on uniformity can be theoretically characterised in terms of its gage. A few such results will be given as exercises.

Before concluding we remark that although we have answered the two basic questions we started with, there are still many other related questions. In mathematics, as soon as a problem is shown to have a solution, the following supplementary questions arise automatically: (i) Is the solution unique? (ii) If not, which solution is the best of all in some sense? (iii) Is there a solution which can be concretely constructed from the data of the problem? Let us discuss the two problems we have studied with reference to these supplementary questions.

First consider the metrisation problem. Given a uniform space $(X; \mathcal{U})$ we know precisely when there exists a pseudometric d on X which induces the uniformity $\mathcal{U}$. There is nothing unique about such a pseudometric. Indeed if a pseudometric d works so does λd where λ is any positive real number. Also no particular pseudometric stands out among those that work. As for constructing a pseudometric concretely in terms of the given uniformity (or rather a given countable base for it), although such a construction is possible and was in fact given in the proof above, it is only of a theoretical importance. In practice, it is hardly feasible to take the infimum over all possible chains. Thus we see that for the metrisation problem none of the three questions above has a satisfactory answer.

The situation is better for the other problem, namely, the uniformisation problem. Here we are given a topological space $(X; \mathcal{T})$ and we ask when there exists a uniformity $\mathcal{U}$ on X such that $\mathcal{T}_{\mathcal{U}}$, the uniform topology induced by $\mathcal{U}$ coincides with $\mathcal{T}$. We know this to be the case iff $(X, \mathcal{T})$ is completely regular. The question as to the uniqueness of such $\mathcal{U}$ has been answered completely. Although we shall not discuss the answer, in the next section we shall show that compactness is a sufficient condition for uniqueness. In general, however, very different uniformities may induce the same topology. For example, let X be the set of positive integers and $\mathcal{T}$ the discrete topology on X. Let d_1 be the usual metric on $\mathbb{N}$ and d_2 the metric defined by $d_2(m, n) = \left|\frac{1}{m} - \frac{1}{n}\right|$ for $m, n \in \mathbb{N}$. Let $\mathcal{U}_1$, $\mathcal{U}_2$ be the uniformities induced by d_1, d_2 respectively. Then $\mathcal{U}_i$ is quite larger than $\mathcal{U}_2$ although they induce the same topology $\mathcal{T}$ on X. It is even possible to induce the same topology on a set by two uniformities out of which one is metrisable and the other is not! An example of this will be given in the next chapter. Note however, that if $(X; \mathcal{T})$ is a topological space and $\mathcal{N}$ is the family of all neighbourhoods of the diagonal ΔX in $X \times X$, then in view of Exercise (1.6), every uniformity which induces $\mathcal{T}$ is necessarily a subfamily of $\mathcal{N}$. It is natural to inquire if $\mathcal{N}$ itself is a uniformity on X. If it were so then it would be the largest uniformity on X inducting the topology $\mathcal{T}$. It turns out that $\mathcal{N}$ is not always a uniformity. All the conditions in Definition (1.6) except (iii) are satisfied by $\mathcal{N}$. If (iii) is also satisfied then the space X must be normal. For let A, B be disjoint closed subset of X. Let $U = X \times X - (A \times B)$. Then $U \in \mathcal{N}$. If (iii) is satisfied, there would exist an open neighbourhood V of ΔX such that $V \circ V \subset U$. But then the sets $V[A]$, $V[B]$ would be disjoint open neighbourhoods of A and B in X. So X is normal.

Even though $\mathcal{N}$ itself need not be a uniformity for X, it contains a largest uniformity inducing the given topology $\mathcal{T}$. It consists of all $V \in \mathcal{N}$ for which there exists a sequence $\{V_n\}$ in $\mathcal{N}$ such that $V_1 = V$ and for each n, $V_{n+1} \circ V_{n+1} \subset V_n$. Thus we see that the third question for the uniformisation problem has at least one answer. If $(X; \mathcal{T})$ is equipped with some additional structure compatible with the topology $\mathcal{T}$, then it is sometimes possible to describe a uniformity (inducing $\mathcal{T}$) in terms of this structure. A pseudometric on X (inducing $\mathcal{T}$) is one such structure. Another important example comes from topological groups, discussed in the next chapter.

Finally, now that we can tell precisely which topological spaces are uniformisable and also which uniformities are metrisable, it is tempting to think that we have an answer to the basic problem mentioned in Chapter 4, Section 1, namely of deciding which topological spaces are metrisable. But this is not so. The trouble is that although we can tell which uniformities are metrisable, we still cannot tell which uniformisable spaces are metrisable. As remarked earlier it is possible to have a metrisable topological space $(X, \mathcal{T})$ and a non-metrisable uniformity $\mathcal{U}$ on X such that $\mathcal{T}_{\mathcal{U}} = \mathcal{T}$. Thus if we find that a topology $\mathcal{T}$ on a set X is induced by a non-metrisable uniformity $\mathcal{U}$ on X, this is no ground to suppose that the space $(X, \mathcal{T})$ is not metrisable. To show non-metrisability of $(X, \mathcal{T})$ we would have to show that *every* uniformity inducing $\mathcal{T}$ is non-metrisable. To characterise this in terms of $\mathcal{T}$ is not easy. That is why, although the results of this section were known for a long time, the general metrisation theorem was proved much later.

Exercises

2.1 Let d be a pseudo-metric on a set X. Assume that $d(x, y) < 1$ for all $x, y \in X$. For each $n \in \mathbb{N}$, let $U_n = \{(x, y) \in X \times X : d(x, y) < 2^{-2n}\}$. Set $U_0 = X \times X$. Prove that each U_n is symmetric and that $U_n \circ U_n \circ U_n \subset U_{n-1}$ for all $n \in \mathbb{N}$. Can the pseudometric d be recovered from the family $\left\{U_n\right\}_{n=1}^{\infty}$?

2.2 For the case of the usual metric on the open interval (0, 1) picture the sets U_n defined above as subsets of the open unit square. Show further that in this case the metric can be recovered from the family $\left\{U_n\right\}_{n=1}^{\infty}$.

2.3 Let D be a family of pseudometrics on a set X. Let $\mathcal{U}$ be the uniformity on X generated by D. Prove that if at least one member of D is a metric then $\mathcal{U}$ is Hausdorff. Show by an example that the converse is false.

2.4 Let $(X, \mathcal{U})$ be a uniform space and D its gage. Prove that $\mathcal{U}$ is Hausdorff if and only if for each $x \neq y$ in X, there exists $d \in D$ such that $d(x, y) > 0$.

2.5 For a pseudometric d on a set X let (X_d, d^*) be the corresponding metric space obtained by the method of Exercise (3.3.9) (see also Exercise (5.4.14). Let $h_d : X \to X_d$ denote the projection function. Do this for each member d of the gage of a uniformity $(X, \mathcal{U})$ and show that the function $h : X \to \prod_{d \in D} X_d$ defined by $h(x)(d) = h_d(x)$ is uniformly continuous w.r.t. $\mathcal{U}$ and the product uniformity on ΠX_d where the factor X_d carries the uniformity induced by the metric d^*.

2.6 Prove that a Hausdorff uniform space is uniformly isomorphic to a subspace of a product of metric spaces. [Hint: Show that the function h defined in the last exercise is one-to-one.]

In the remaining exercises, $(X; \mathcal{U})$ will be a uniform space with gage D. Prove each of the following.

2.7 For any $A \subset X$ and $x \in X$, x is in the closure of A (relative to the uniform topology $\mathcal{T}_{\mathcal{U}}$ on X) if and only if for all $d \in D$, $d(x, A) = 0$.

2.8 For any $A \subset X$ and $x \in X$, x is in interior of A if and only if there exists $d \in D$ such that $B_d(x; r) \subset A$.

2.9 A net $\{S_n : n \in E\}$ converges to s in X if and only if for each $d \in D$, the net $\{d(S_n, s) : n \in E\}$ converges to 0 in $\mathbb{R}$.

2.10 If $(Y; \mathcal{V})$ is another uniform space with gage E and $f : X \to Y$ is a function, then f is uniformly continuous (w.r.t. $\mathcal{U}$ and $\mathcal{V}$) if and only if for each $e \in E$, the composite $e \circ (f \times f)$ is a member of D.

Notes and Guide to Literature

The material in this chapter is standard and taken directly from Kelley

[1] where Theorem (2.4) is attributed to Alexanderoff and Urysohn [1] and Theorem (2.8) to Weil [1].

3. Completeness and Compactness

In Chapter 12, we studied completeness for metric spaces rather thoroughly. In the introduction to uniformities, we remarked that completeness is a uniform concept. It is then natural to develop a theory of completeness for uniformities. In this section, we briefly develop such a theory. Compactness, on the other hand, is not strictly speaking a purely uniform concept. Still, just as compactness has certain interesting implications for metric spaces, so does it for uniform spaces and we discuss a few of them.

The treatment of completeness for uniform spaces is analogous to that for metric spaces with two significant changes. First of all although sequences were adequate for metric spaces (they being first countable) they are no longer so for general uniform spaces. We therefore have to replace them with generalised structures such as nets or filters. We prefer to work with nets rather than filters. However, an equivalent theory can be developed using filters instead of nets and will be indicated in the exercises. Another point to note is that unlike metric spaces, uniform spaces need not be Hausdorff and so limits of nets need not be unique thereby requiring a little caution in handling limits.

We begin with the fundamental notion of a Cauchy net.

(3.1) Definition: A net $\{S_n : n \in D\}$ in a set X is said to be a **Cauchy net** w.r.t. a uniformity $\mathcal{U}$ on X if for every $U \in \mathcal{U}$ there exists $p \in D$ such that for all $m \geq p$, $n \geq p$ in D, $(S_m, S_n) \in U$. A uniform space $(X, \mathcal{U})$ is said to be **complete** if every Cauchy net in X (w.r.t. $\mathcal{U}$) converges to at least one point in X (w.r.t. the topology $\mathcal{T}_{\mathcal{U}}$). Many results about Cauchy sequences in metric spaces continue to hold for Cauchy nets in uniform spaces. A typical result is,

(3.2) Proposition: Every convergent net is a Cauchy net. A Cauchy net is convergent if and only if it has a cluster point.

Proof: Let $\{S_n : n \in D\}$ be a net in a uniform space $(X; \mathcal{U})$. Suppose first $\{S_n : n \in D\}$ converges to say, x in X. Let $U \in \mathcal{U}$. Then there exists a symmetric $V \in \mathcal{U}$ such that $V \circ V \subset U$. Now $V[x]$ is a neighbourhood of x in the uniform topology on X. So there exists $p \in D$ such that for all $n \geq p$ in D, $S_n \in V[x]$, i.e. $(x, S_n) \in V$. Now for any $m, n \geq p$, $(S_m, x) \in V$ and $(x, S_n) \in V$ by symmetry of V. So $(S_m, S_n) \in V \circ V \subset U$. Since $U \in \mathcal{U}$ was arbitrary, it follows that $\{S_n : n \in D\}$ is a Cauchy net. For the second assertion, the 'only if' part always holds whether the net is a Cauchy net or not. For the other way implication we show that a Cauchy net converges to each of its cluster points. Let x be a cluster point of a Cauchy net $\{S_n : n \in D\}$ in a uniform space $(X; \mathcal{U})$. We have to show that S_n converges to x in the uniform topology. Let G be a neighbourhood of x. Then there exists $U \in \mathcal{U}$ such that $U[x] \subset G$. Once again, find a symmetric $V \in \mathcal{U}$ such that $V \circ V \subset U$. Then there exists $p \in D$ such that for all $m, n \geq p$ in D, $(S_m, S_n) \in V$. Since x is a cluster point of $\{S_n : n \in D\}$, there exists $q \geq p$ in D such that $S_q \in V[x]$, i.e. $(x, S_q) \in V$. Then for all $n \geq q$, we have $(S_q, S_n) \in V$ and $(x, S_q) \in V$. So $(x, S_n) \in V \circ V \subset U$. Hence $S_n \in U[x] \subset G$, showing that $\{S_n : n \in D\}$ converges to x. ∎

The argument above runs parallel to that one would give in the case of a metric space. The similarity will be clearer if the reader writes the argument in terms of a metric and then translates it in terms of entourages. Note in particular the trick of finding, for a given $U \in \mathcal{U}$, a symmetric entourage $V \in \mathcal{U}$ such that $V \circ V \subset U$. This corresponds to a most common argument for metric spaces, when we want to show $d(x, z) < \epsilon$, we find y such that $d(x, y) < \epsilon/2$, $d(y, z) < \epsilon/2$ and apply the triangle inequality. Henceforth, when a proof for uniform spaces is a straightforward translation of the corresponding proof for metric spaces, we shall usually omit it.

(3.3) Corollary: Every compact, uniform space is complete.

Proof: Here a compact uniform space means a uniform space whose associated topological space is compact. By theorem (10.2.9), every net in a compact space has a cluster point and so the result follows immediately from the last proposition. ∎

Let $(X; \mathcal{U})$, $(Y, \mathcal{V})$ be uniform spaces and $f: X \to Y$ be uniformly continuous. Then it is easy to see that for any Cauchy net $S : D \to X$, the composite net $f \circ S$ is a Cauchy net in $(Y; \mathcal{V})$. By arguments similar to those for metrics, it follows that a closed uniform subspace of a complete uniform space is complete (w.r.t. the relativised uniformity). It is not true in general that a complete uniform subspace must be closed; however, this would be the case for Hausdorff uniformities.

Before we proceed further let us verify that the definitions of a Cauchy net and of completeness are indeed valid generalisations of the corresponding concepts already defined for metric spaces. We do so in the following proposition.

(3.4) Proposition: Let $(X; d)$ be a metric space and $\mathcal{U}$ the uniformity on X induced by d. Then a net $\{S_n : n \in D\}$ in X is a Cauchy net w.r.t. the metric d if and only if it is a Cauchy net w.r.t. the uniformity $\mathcal{U}$. Moreover $(X; d)$ is a complete metric space if and only if $(X; \mathcal{U})$ is a complete uniform space.

Proof: The statement about Cauchy nets is a trivial consequence of definitions; (we never formally defined a Cauchy net in a metric space, but the definition is the obvious extension of the definition of a Cauchy sequence (12.1.1)). It is the second statement that requires a proof. Since the metric topology on X induced by d is the same as the uniform topology induced by $\mathcal{U}$, convergence of a net w.r.t. the metric d is the same as that w.r.t. the uniformity $\mathcal{U}$. The second statement then amounts to showing that every Cauchy net in $(X; d)$ is convergent if and only if every Cauchy sequence in $(X; d)$ is convergent. Since a sequence is a special case of a net, one way implication is obvious. The other way implication requires a little argument.

Assume every Cauchy sequence in $(X; d)$ is convergent. Let $\{S_n : n \in D\}$ be a Cauchy net in $(X; d)$. In view of Proposition (3.2) it suffices to show that $\{S_n : n \in D\}$ has a cluster point in X. We set $\epsilon = 2^{-k}$ for $k = 1; 2, \ldots$ and inductively obtain elements $p_1, p_2, \ldots, p_k, \ldots$ in D such that for each $k \in \mathbb{N}$, we have (i) $p_{k+1} \geq p_k$ in D and (ii) for all $m, n \geq p_k$ in D, $d(S_m, S_n) < 2^{-k}$. Now consider the sequence $\left\{S_{p_k}\right\}_{k=1}^{\infty}$. Since $d(S_{p_k}, S_{p_{k+1}}) < 2^{-k}$ for all $k \in \mathbb{N}$, and the series $\sum_{k=1}^{\infty} 2^{-k}$ is convergent, it follows that $\left\{S_{p_k}\right\}_{k=1}^{\infty}$ is a Cauchy sequence in $(X; d)$ (see the argument used in the proof of Theorem (12.1.6)). So $\{S_{p_k}\}$ converges to a point, say, x of X by hypothesis. We claim that x is a cluster point of the net $\{S_n : n \in D\}$. Let $\epsilon > 0$ and $m \in D$ be given. We have to find $n \in D$ such that $n \geq m$ and $d(S_n, x) < \epsilon$. First choose $k \in \mathbb{N}$ so that $2^{-k} < \epsilon/2$, and $d(S_{p_k}, x) < \epsilon/2$. Since $\geq$ directs the set D, there exists $n \in D$ such that $n \geq p_k$ and $n \geq m$. For any such n we have, $d(S_n, x) \leq d(S_n, S_{p_k}) + d(S_{p_k}, x) < 2^{-k} + \epsilon/2 < \epsilon$ as desired. So x is a cluster point of the net $\{S_n : n \in D\}$ and as noted before, this completes the proof. ∎

Cauchy nets in and completeness of product uniformities can be described very simply in terms of the corresponding concepts for the co-ordinate spaces.

(3.5) Proposition: Let $(X, \mathcal{U})$ be the uniform product of a family of non-empty uniform spaces $\{(X_i, \mathcal{U}_i) : i \in I\}$. Then a net $S : D \to X$ is a Cauchy net in $(X; \mathcal{U})$ iff for each $i \in I$ the net $\pi_i \circ S$ is a Cauchy net in $(X_i, \mathcal{U}_i)$. Also $(X; \mathcal{U})$ is complete iff each $(X_i, \mathcal{U}_i)$ is so.

Proof: The 'only if' part of the first assertion follows from the fact that π_i is uniformly continuous. For the converse, let $S : D \to X$ be a net for which $\pi_i \circ S$ is a Cauchy net for each $i \in I$. Let $\mathcal{S}$ be the standard sub-base for $\mathcal{U}$, consisting of all subsets of the form $\theta_i^{-1}(U_i)$ for $U_i \in \mathcal{U}_i$, $i \in I$, where $\theta_i : X \times X \to X_i \times X_i$ is the function $\pi_i \times \pi_i$. It suffices to show that the condition in the definition of a Cauchy net is satisfied for all $U \in \mathcal{S}$ (why ?). Suppose $U = \theta_i^{-1}(U_i)$ for some $U_i \in \mathcal{U}_i$ and $i \in I$. Find $p \in D$ so that for all $m, n \geq p$ in D, $(\pi_i(S_m), \pi_i(S_n)) \in U_i$. Such p exists since $\pi_i \circ S$ is given to be a Cauchy net. But then $(S_m, S_n) \in U$ for all $m, n \geq p$. So S is a Cauchy net in $(X; \mathcal{U})$.

The 'if' part of the second assertion follows from the 'only if' part of the first assertion and the fact that the convergence in the product space is coordinate-wise (see Exercise (10.4.4). The 'only if' part, is left as an exercise. ∎

In Chapter 12, we saw that every metric space can be isometrically embedded as a dense subspace of a complete metric space. Using the same construction, it can be proved that every pseudometric space is isometric to a dense sub-space of a complete pseudometric space. Combining these facts along with the last proposition and the embedding theorem for uniform spaces, we get the existence of a completion for uniform spaces as well.

(3.6) Theorem: Every uniform space is uniformly isomorphic to a dense subspace of a complete uniform space.

Proof: Let $(X; \mathcal{U})$ be a uniform space. By Theorem (2.8), there exists a family $\{(X_i, d_i) : i \in I\}$ of pseudometric spaces and a uniform embedding $f : X \to \prod_{i \in I} X_i$ where the product ΠX_i is assigned the product uniformity, each X_i being given the uniformity induced by d_i. For each $i \in I$, let (X_i^*, d_i^*) be a complete pseudometric space containing (X_i, d_i) upto an isometry. Then ΠX_i is a uniform subspace of ΠX_i^* with the product uniformities. We regard f as an embedding of $(X; \mathcal{U})$ into ΠX_i^* which is complete by the last proposition. Let Z be the closure of $f(X)$ in ΠX_i^*. Then Z is complete with the relative uniformity. Also $f(X)$ is dense in Z. Hence the result. ∎

We conclude the section with a discussion of compact uniform spaces. We already proved in Corollary (3.3) that such spaces are complete. The converse is false. However, as we saw in the case of metric spaces (see Theorem (12.1.6)), completeness along with an additional hypothesis implies compactness. This additional requirement bears the expected name.

(3.7) Definition: A uniform space $(X, \mathcal{U})$ is said to be **totally bounded** (or **pre-compact**) if for each $U \in \mathcal{U}$, there exists $x_1, \ldots, x_n \in X$ such that $X = \bigcup_{i=1}^{n} U[x_i]$. Equivalently, $(X; \mathcal{U})$ is totally bounded iff for each $U \in \mathcal{U}$ there exists a finite subset F of X such that $U[F] = X$.

We leave it to the reader to check that a metric space is totally bounded (in the sense of Definition (12.1.4)) if and only if the associated uniform space is totally bounded in the sense just defined. It is then to be expected that the analogue of Theorem (12.1.6) should hold for uniform spaces as well. This is actually true, but the proof is surprisingly more complicated and depends upon the concept of a universal net as in Definition (10.4.7).

(3.8) Theorem: A uniform space is compact if and only if it is complete and totally bounded.

Proof: We already showed that every compact uniform space is complete. Also total boundness is an easy consequence of compactness and so the necessity part of the theorem is proved. As for sufficiency, let $(X; \mathcal{U})$ be a uniform space which is complete and totally bounded. To show that X is compact we shall use the result of Exercises (10.4.7) which states that a space is compact if and only if every universal net in it is convergent. So let $\{S_n : n \in D\}$ be a universal net in X. By definition, this means that for any $A \subset X$ either $\{S_n\}$ is eventually in A or it is eventually in $X - A$. We claim that $\{S_n : n \in D\}$ is a Cauchy net in $(X; \mathcal{U})$. Let $U \in \mathcal{U}$. Find a symmetric $V \in \mathcal{U}$ such that $V \circ V \subset U$. By total boundedness of $(X; \mathcal{U})$, there exist $x_1, x_2, \ldots, x_k \in X$ such that $X = \bigcup_{i=1}^{k} V[x_i]$. Now for at least one $i = 1, 2, \ldots, k$, $\{S_n : n \in D\}$ is eventually in $V[x_i]$; for otherwise the net will be eventually $X - V_k[x_i]$ for all $i = 1, 2, \ldots, k$ and hence will be eventually in $\bigcap_{i=1}^{k} (X - V[x_i])$ which equals the empty set. Thus for some i, $\{S_n : n \in D\}$ is eventually in $V[x_i]$. This means there exists $p \in D$ such that for all $n \geq p$, $S_n \in V[x_i]$, i.e. $(x_i, S_n) \in V$. So for all $m, n \geq p$ in D, $(S_m, S_n) \in V \circ V \subset U$ as desired. Thus $\{S_n : n \in D\}$ is a Cauchy net in $(X; \mathcal{U})$. By completeness, it is convergent in X. As noted before this completes the proof. ∎

Other results about compact metric spaces have analogues for compact uniform spaces. For example it is an easy exercise to prove the following propositions.

(3.9) Proposition: Every continuous function from a compact uniform space to a uniform space is uniformly continuous. ∎

The following result is evidently the analogue of the Lebesgue covering lemma (Theorem 6.1.7)).

(3.10) Proposition: Let $(X; \mathcal{U})$ be a compact uniform space and $\mathcal{V}$ an open cover of X. Then there exists $U \in \mathcal{U}$ such that for each $x \in X$ there exists $V \in \mathcal{V}$ such that $U[x] \subset V$.

Proof: For each $x \in X$ there exists $U_x \in \mathcal{U}$ such that $U_x[x]$ is contained in some member of $\mathcal{V}$. Hence, there exists a symmetric $V_x \in \mathcal{U}$ such that $(V_x \circ V_x)[x]$ is contained in some member of $\mathcal{V}$. The interiors of the sets $V_x[x]$ for $x \in X$, cover X. So by compactness of X, there exist $x_1, \ldots, x_n \in X$ such that $X = \bigcup_{i=1}^{n} V_i[x_i]$ where V_i denotes V_{x_i} for $i = 1, 2, \ldots, n$. Now let $U = \bigcap_{i=1}^{n} V_i$. Then $U \in \mathcal{U}$. Also let $x \in X$. Then $x \in V_i[x_i]$ for some i. So

$$\begin{aligned} U[x] &\subset V_i[x] \\ &\subset V_i[V_i[x_i]] \quad (\text{since } x \in V_i[x_i]) \\ &= (V_i \circ V_i)[x_i] \\ &\subset \text{some member of } \mathcal{V}. \end{aligned}$$

Since this holds for all $x \in X$, the result follows. ∎

Finally, there is an interesting uniqueness theorem about compact uniform spaces.

(3.11) Proposition: Let $(X; \mathcal{U})$ be a compact uniform space. Then $\mathcal{U}$ is the only uniformity on X which induces the topology $\mathcal{T}_{\mathcal{U}}$ on X.

Proof: Suppose $\mathcal{V}$ is any other uniformity on X such that $\mathcal{T}_{\mathcal{U}} = \mathcal{T}_{\mathcal{V}}$. Let $f : (X; \mathcal{U}) \to (X; \mathcal{V})$ and $g : (X; \mathcal{V}) \to (X, \mathcal{U})$ be identity functions. Then each is continuous (in fact a homeomorphism) because $\mathcal{T}_{\mathcal{U}} = \mathcal{T}_{\mathcal{V}}$. Hence by Proposition (3.9) above, both f and g are uniformly continuous. This means $\mathcal{V} \subset \mathcal{U}$ and $\mathcal{U} \subset \mathcal{V}$ respectively. So $\mathcal{U} = \mathcal{V}$ as desired. ∎

Although the previous result is interesting, it does not tell us how to describe the uniformity in terms of the induced topology. This information is provided in the following theorem.

(3.12) Theorem: Let $(X; \mathcal{U})$ be a compact, uniform space. Let $\mathcal{T}_{\mathcal{U}}$ be the uniform topology on X and give $X \times X$ the product topology. Then $\mathcal{U}$ consists precisely of all the neighbourhoods of the diagonal ΔX in $X \times X$.

Proof: In Exercise (1.6) we asked the reader to prove that every member of $\mathcal{U}$ is a neighbourhood of ΔX in $X \times X$. We then only have to prove that every neighbourhood of ΔX is a member of $\mathcal{U}$. Let V be such a neighbourhood. Without loss of generality we may suppose V is open in $X \times X$. Let $\mathcal{B}$ be the family of those members of $\mathcal{U}$ which are closed in $X \times X$. By Exercise (1.7), $\mathcal{B}$ is a base for $\mathcal{U}$. Let $W = \bigcap\{U : U \in \mathcal{B}\}$. We assert that $W \subset V$. For suppose $(x, y) \in W$. Then $y \in U[x]$ for all $U \in \mathcal{B}$. Since $\mathcal{B}$ is a base for $\mathcal{U}$ it is easy to see that the family $\{U[x] : U \in \mathcal{B}\}$ is a local base at x (w.r.t. $\mathcal{T}_{\mathcal{U}}$). In particular since $V[x]$ is an open neighbourhood of x, there exists $U \in \mathcal{B}$ such that $U[x] \subset V[x]$. Hence $y \in V[x]$, i.e. $(x, y) \in V$. Thus we have shown $W \subset V$. Note that so far we have not used the compactness of X. We put it to use as follows. Since X is compact, so is $X \times X$. The family $\mathcal{B} \cup \{X \times X - V\}$ is a family of closed subsets of $X \times X$ and in view of what has just been proved, its intersection is empty. So there exist finitely many members, say, $U_1, U_2, \ldots, U_n$ of $\mathcal{B}$ such that $U_1 \cap U_2 \cap \ldots \cap U_n \cap (X \times X - V) = \emptyset$, or equivalently $U_1 \cap U_2 \ldots \cap U_n \subset V$. But this means $V \in \mathcal{U}$, and completes the proof. (The last part could also have been done by recalling the result of Exercise (11.2.5).) ∎

We have already seen (see Theorem (2.10)) that complete regularity is a necessary and sufficient condition for uniformisability of a space. In presence of compactness, complete regularity is equivalent to regularity (the proof is left as an exercise). We thus get the following corollary of the last two results.

(3.13) Corollary: For a compact, regular topological space, there exists a unique uniformity inducing its topology. This uniformity consists of all neighbourhoods of the diagonal in the product of the space with itself. ∎

It should be noted that such uniqueness results do not hold for compact metric spaces, on which there are infinitely many distinct metrics inducing the same topology. Of course. all these metrics induce the same uniformity.

In passing we remark, that throughout this section we have preferred to avoid working with the gage of a uniform space. In the spirit of Exercises (2.7) to (2.10) of the last section, the concepts of completeness and total boundedness can be characterised in terms of gages of pseudometrics and often these characterisations can be used to give alternate proofs of the results of this section. A few indicative examples will be given as exercises.

Exercises

3.1 Verify all the statements made after Corollary (3.3) and before Proposition (3.4).

3.2 Prove that if $\mathcal{S}$ is a sub-base for a uniformity $\mathcal{U}$ on a set X, then a net is a Cauchy net in $(X; \mathcal{U})$ if and only if the condition in the definition is satisfied for all members of $\mathcal{S}$.

3.3 Prove that every Hausdorff uniform space is uniformly isomorphic to a dense subspace of a complete, Hausdorff uniform space.

3.4 Let A be a dense subspace of a uniform space $(X; \mathcal{U})$. Let $(Y; \mathcal{V})$ be a complete, Hausdorff uniform space and suppose of $f : A \to Y$ is uniformly continuous (A being given the relativised uniformity). Prove that there is a unique, uniformly continuous function $g : X \to Y$ such that $g/A = f$. [Hint: Express a point x of X as a limit of a net S in A, define $g(x)$ to be the limit of the net $f \circ S$ in Y. Show that this is well-defined.]

3.5 Find a left adjoint to the inclusion functor from the category of all complete, Hausdorff uniform spaces to the category of all Hausdorff uniform spaces.

3.6 Prove Proposition (3.9).

3.7 Prove that every compact, regular space is completely regular. [Hint: Let X be such a space, $x \in X$ and C a closed subset of X not containing x. Let $F = \overline{\{x\}}$. Prove that $F \cap C = \emptyset$. Now observe that X is normal (Theorem (7.2.8)) and apply Urysohn's lemma.]

3.8 Let $(X; \mathcal{U})$ be a uniform space and G its gage. [In the last section we denoted a gage by D but now we need D for a directed set.] Prove that a net $\{S_n : n \in D\}$ is a Cauchy net in X iff for each $d \in G$, the net $\{d(S_m, S_n) : (m, n) \in D \times D\}$ converges to 0 in the space of real numbers. Here $D \times D$ is to be given the product ordering, that is $(m, n) \geq (p, q)$ iff $m \geq p$ and $n \geq q$.

3.9 Using the characterisation in the last exercise give alternate proofs of Propositions (3.2) and (3.5).

3.10 If $(X, \mathcal{U})$ is a compact, uniform space prove that every pseudometric which is continuous on $X \times X$ is in the gage of $(X; \mathcal{U})$. Show by an example that the hypothesis of compactness is essential.

3.11 Let A be a dense subset of a uniform space $(X; \mathcal{U})$. Prove that $(X; \mathcal{U})$ is totally bounded iff $(A; \mathcal{U}/A)$ is so. Deduce that a uniform space is totally bounded iff it is uniformly isomorphic to a dense subspace of a compact uniform space. [Hence the name 'precompact'.]

3.12 A filter $\mathcal{F}$ in a uniform space is said to be a **Cauchy filter** if for any $U \in \mathcal{U}$ there exists $F \in \mathcal{U}$ such that $F \times F \subset U$. In Chapter 10, Section 3 we defined the associated filter of a net and the associated net of a filter. Prove that in a uniform space,

(i) a net is a Cauchy net iff its associated filter is a Cauchy filter,

(ii) a filter is a Cauchy filter iff its associated net is a Cauchy net.

3.13 Prove that a uniform space is complete if and only if every Cauchy filter in it is convergent.

In subsequent exercises give direct arguments about filters rather than using Exercise (3.12) to translate the problem in terms of nets.

3.14 Prove that a Cauchy filter converges to each of its cluster points.

3.15 Prove that every ultrafilter in a totally bounded uniform space is a Cauchy filter.

3.16 Give an alternate proof of Theorem (3.8) using filters.

3.17 Prove the 'only if' part of the second assertion of Proposition (3.5). (Hint: Let $j \in I$ and S be a Cauchy net in X_j. Obtain a Cauchy net T in X such that $\pi_j \circ T = S$ while $\pi_i \circ T$ is a constant net for all $i \neq j$.)

3.18 Let (e, Y) be a Hausdorff compactification of a space X. Let $\mathcal{U}$ be the unique uniformity on Y inducing the topology on Y. The relativised uniformity $\mathcal{U}/X$ (regarding X as a subset of Y) is called the **uniformity induced by the compactification (e, Y)**. Prove that two Hausdorff compactifications of a Tychonoff space are equivalent (in the sense of Definition (11.4.11)) if and only if the uniformities induced by them are equal.

3.19 Prove that the uniformity induced by the Stone-Čech compactification of a Tychonoff space is the smallest uniformity such that each bounded real-valued continuous function is uniformly continuous.

Notes and Guide to Literature

Completeness of uniform spaces is a straightforward generalisation of the same notion for metric spaces. Many results about the latter have expected analogues for the former. Unfortunately, there seems to be no analogue of the Baire category theorem for complete uniform spaces.

Chapter Fifteen

Selected Topics

Through the preceding chapters we have exposed the reader to some of the most standard aspects of general topology. Considerations of space and of the level of this book have forced us to omit certain material. Nevertheless there are many topics of which a student of topology should have a general knowledge. In this chapter we have selected four such topics. Some of them are vast enough to occupy entire books. Obviously then, we have to content ourselves with bare glimpses of these topics.

1. Function Spaces

In analysis we often come across families of functions from one space to another and functions defined on such families. As remarked earlier these latter functions are generally called operators. For example, the integral operator assigns to a real valued function its integral while the differential operator assigns to it its derived function. It is natural to ask if such operators are continuous. This is a topological question and would be meaningful only if the domain, that is, the collection of functions had a topology on it. This leads to the concept of function spaces, that is, topological spaces whose elements are themselves some functions.

Suppose X, Y are topological spaces. Let $M(X, Y)$ denote the set of all continuous functions from X to Y. The problem is to put a suitable topology on $M(X, Y)$ in a way that will reflect the topologies on X and Y. One such answer is to regard $M(X, Y)$ as a subset of the power Y^X. On the latter we can put the product topology and relativise it to $M(X, Y)$. Let $\mathcal{P}$ denote the topology so obtained. Although this topology has certain nice features (for example, convergence in it is very simple, namely pointwise), it turns out not to be strong enough for certain other purposes. Consider the evaluation function $e : X \times M(X, Y) \to Y$ defined by $e(x, f) = f(x)$ for $x \in X$, $f \in M(X, Y)$. If we fix a particular $x \in X$, then this function is merely the restriction of a projection function and hence is continuous if we put the topology $\mathcal{P}$ on $M(X, Y)$. But as a function of two variables e is not continuous in general. To make it so we need a stronger topology on $M(X, Y)$ than $\mathcal{P}$. We now proceed to define it.

First let us view $\mathcal{P}$ in a particular way. For subsets A, B of X, Y respectively, let $[A, B] = \{f \in M(X, Y) : f(A) \subset B\}$. It is easy to see that the topology $\mathcal{P}$ is generated by the family $\{[F, V] : F \text{ finite}, V \text{ open}\}$. For this reason the topology $\mathcal{P}$ is also called the **finite open topology**, the name suggesting that it is generated by subsets of $M(X, Y)$ taking finite sets to open sets.

Now to get a stronger topology, we replace finite sets by the next best thing, namely compact ones. Specifically let $\mathcal{S} = \{[C, V] : C \text{ compact}, V \text{ open}\}$. Then the topology on $M(X, Y)$ generated by $\mathcal{S}$ will be called the **compact open topology**. It is stronger than the finite open topology. The latter is the relativised product topology. The set $M(X, Y)$ with the compact open topology is called the function space of maps from X to Y.

Properties of the function space naturally depend upon those of X and Y. It is easy to show, for example, that if Y is T_0, T_1, T_2 or regular so is $M(X, Y)$ respectively. Unfortunately even with the compact open topology we cannot ensure the continuity of the evaluation function $e : X \times M(X, Y) \to Y$. As an example, let $X = \mathbb{Q}$ and $Y = \mathbb{R}$ with usual topologies. Let $f : \mathbb{Q} \to \mathbb{R}$ be the identically zero function. We assert that $e^{-1}((-1, 1))$ is not a neighbourhood of the point $(0, f)$. For if it were, then by the definition of the compact open topology, there would exist compact subsets $C_1, \ldots, C_n$ of $\mathbb{Q}$, open subsets $V_1, \ldots, V_n$ of $\mathbb{R}$ and an open subset G of $\mathbb{Q}$ such that $(0, f) \in G \times (\bigcap\limits_{i=1}^{n} [C_i, V_i]) \subset e^{-1}(-1, 1)$. Since f is identically 0 it follows that each V_i contains 0. Note also that a compact subset of $\mathbb{Q}$ cannot have any interior points. So there exists $q \in G$ such that $q \notin C_1 \cup C_2 \cup \ldots \cup C_n$. Now by complete regularity of $\mathbb{Q}$ there exists $g \in M(\mathbb{Q}, \mathbb{R})$ such that $g(q) = 2$ and g is 0 on $C_1 \cup C_2 \cup \ldots \cup C_n$. But then $(q, g) \in G \times (\bigcap\limits_{i=1}^{n} [C_i, V_i])$ and still $e(q, g) = g(q) = 2 \notin (-1, 1)$ a contradiction. In this example the space $\mathbb{Q}$ is not locally compact. The compact open topology on a set $M(X, Y)$ will be strong enough only if the space X had a large number of compact subsets. Local compactness along with regularity ensures this and under this hypothesis it is not hard to show that the evaluation function is continuous. Actually a more general result holds.

(1.1) Theorem: Let X, Y, Z be spaces with X locally compact and regular. Then the composition function from $M(Z, X) \times M(X, Y) \to M(Z, Y)$ is continuous.

Proof: Let λ be the composition function defined by $\lambda(f, g) = g \circ f$ for $f \in M(Z, X)$, $g \in M(X, Y)$. To show λ is continuous, it suffices to show that the inverse image under λ of every sub-basic open set in $M(Z, Y)$ is open. Let $[C, V]$ be such an open set where C is a compact subset of Z and V is an open subset of Y. We claim $\lambda^{-1}[C, V]$ is a neighbourhood of each of its points. Suppose $(f, g) \in \lambda^{-1}[C, V]$. This means $f(C) \subset g^{-1}(V)$. Now $f(C)$ is compact and $g^{-1}(V)$ is open in X by continuity of f and g. But X is locally compact and regular. So by the result of Exercise (11.3.5) there exists a compact subset K such that $f(C) \subset G$ and $K \subset g^{-1}(V)$ where $G = \text{int}(K)$. The set $[C, G] \times [K, V]$ is open in $M(Z, X) \times M(X, Y)$ and contains (f, g). Also for any $h \in [C, G]$ and $k \in [K, V]$ we have, $k(h(C)) \subset k(G) \subset k(K) \subset V$ showing that $(h, k) \in [C, V]$. Thus $[C, G] \times [K, V] \subset \lambda^{-1}[C, V]$ and so $\lambda^{-1}[C, V]$ is a nbd of (f, g). As this holds for every point of $\lambda^{-1}[C, V]$, the set $\lambda^{-1}[C, V]$ is open. This shows that λ is continuous. ∎

Now let Z be a space consisting of just one point. Then $M(Z, X)$ and $M(Z, Y)$ are homeomorphic to X, Y respectively in such a way that the composition function $M(Z, X) \times M(X, Y) \to M(Z, Y)$ corresponds to the evaluation function. Thus we see that if X is locally compact and regular then the evaluation function $X \times M(X, Y) \to Y$ is continuous.

In Proposition (8.2.9) we saw that the convergence of sequences in the product topology is pointwise or coordinatewise. The argument there applies equally well for nets. It is natural to ask for a similar characterisation of convergence w.r.t. the compact open topology. Although we shall not answer this question in general, the answer turns out to be particularly familiar when Y is a metric space, or more generally, a uniform space.

(1.2) Theorem: Let $(Y; d)$ be a metric space and X any space. Then a net $S : D \to M(X, Y)$ converges to a function $f \in M(X, Y)$ in the compact open topology iff for every compact subset C of X, the net of functions $\{S_n/C : n \in D\}$ converges to f/C uniformly w.r.t. d.

Proof: Assume that $S : D \to M(X, Y)$ converges to f in the compact open topology. Let C be a compact subset of X. We have to show that the functions S_n converge to f uniformly on C. Let $\epsilon > 0$ be given. Since C is compact and f is continuous, $f(C)$ is compact. So there exist points $x_1, x_2, \ldots, x_n \in C$ such that the open balls (in Y) of radii $\epsilon/4$ each with centres at $f(x_1), f(x_2), \ldots, f(x_n)$ cover $f(C)$. For each $i = 1, 2, \ldots, n$ let $C_i = C \cap f^{-1}(C(f(x_i), \epsilon/4))$ where $C(f(x_i), \epsilon/4)$ is the closed ball (in Y) of radius $\epsilon/4$ and centre $f(x_i)$. Then C_i is compact. Let $U_i = B(f(x_i), \epsilon/4)$. Then U_i is open in Y and clearly $f(C_i) \subset U_i$ for all $i = 1, 2, \ldots, n$. So $f \in \bigcap_{i=1}^{n} [C_i, U_i]$ which is an open set in the compact open topology on $M(X, Y)$. Since $\{S_n : n \in D\}$ converges to f in this topology, there exists $m \in D$ so that for all $n \geq m$ in D, $S_n \in \bigcap_{i=1}^{n} [C_i, U_i]$. Now for any $x \in C$, $x \in C_i$ for some i, whence $d(f(x_i), f(x)) < \epsilon/2$. But we also have that $S_n(x) \in U_i$ for all $n \geq m$, and hence that $d(S_n(x), f(x_i)) < \epsilon/2$. Thus by triangle inequality we get $d(S_n(x), f(x)) < \epsilon$. This holds for all $n \geq m$ and for all $x \in C$. This precisely means that $\{S_n : n \in D\}$ converges to f uniformly on C.

Conversely assume that S converges uniformly to f on every compact subset of X. Here we are given **a priori** that f and S_n for each $n \in D$ are continuous. To show that S converges to f in $M(X, Y)$ it suffices to show that the condition of convergence is satisfied for all sub-basic open sets containing f. Let $[C, U]$ be such a set where C is compact subset of X and U is an open subset of Y. Then $f(C)$ is compact and disjoint from the closed set $X - U$. So $d(f(C), X - U)$ is positive. Denote this number by ϵ (ϵ is not well-defined in case $X - U$ is empty, but then there is nothing to prove).

By uniform convergence, there exists $m \in D$ such that for all $n \geq m$ and $x \in C$, $d(S_n(x), f(x)) < \epsilon$ which implies that $S_n(x) \in U$ by the definition of ϵ. So $S_n(C) \subset U$ or $S_n \in [C, U]$ for all $n \geq m$, $n \in D$. As noted earlier this implies that $\{S_n : n \in D\}$ converges to f in the compact-open topology on $M(X, Y)$. ∎

In view of the result just proved, the compact-open topology is sometimes called the **topology of uniform convergence on compacta**. 'Compacta' is the plural of 'compactum' which simply means a compact set. In case the space X itself is compact, uniform convergence on all compacta is equivalent to uniform convergence on X. So we get the following corollary of the last theorem.

(1.3) Corollary: Let X be a compact space and $(Y; d)$ a metric space. Then with the compact open topology $M(X, Y)$ is metrisable and a metric for it is given by $e(f, g) = \sup \{d(f(x), g(x)) : x \in X\}$ for $f, g \in M(X, Y)$.

Proof: Note that every continuous function from a compact space into a metric space is bounded and so e is well defined. We have already seen that e is a metric and that convergence in the topology induced by e is the uniform convergence on X (see Chapter 12, Section 1). By the theorem above, this is equivalent to convergence in the compact-open topology. But topologies are uniquely determined by convergence of nets (see Exercise (10.2.5)). So e induces the compact open topology on $M(X, Y)$. ∎

It also follows that in case X is compact, the metric e may depend upon the particular metric d on Y; but the topology induced by e depends only on the topology on Y induced by d.

The concept of function spaces provides an alternate and a fruitful description of continuity of functions of several variables. As a typical illustration, let X, Y, Z be topological spaces and suppose $f: X \times Y \to Z$ is a function. For each $y \in Y$ we get a function $f_y : X \to Z$ defined by $f_y(x) = f(x, y)$ for $x \in X$. This function f_y may be thought of as the composite of an embedding of X into $X \times Y$ (the embedding which takes a point x of X to a point (x, y) of $X \times Y$) with f. It follows that if f is continuous then f_y is continuous for every y and this gives a well-defined function $\hat{f} : Y \to M(X, Z)$ defined by $\hat{f}(y) = f_y$ for $y \in Y$. $\hat{f}$ is called the **associate** of f. It is easy to show that if we put the compact open topology on $M(X, Z)$ then $\hat{f}$ is continuous. It suffices to show that the inverse image of each sub-basic open set is open. Let $[C, V]$ be such a set where C is a compact subset of X and V is an open subset of Z. We assert that $\hat{f}^{-1}([C, V])$ is a neighbourhood of each of its points. Let $y \in \hat{f}^{-1}([C, V])$. Then $\hat{f}(y)$, that is $f_y \in [C, V]$ and hence $f(C \times \{y\}) \subset V$ or in other words, $C \times \{y\} \subset f^{-1}(V)$. The set $f^{-1}(V)$ is open in $X \times Y$ by continuity of f. Now apply the theorem of Wallace (Theorem (7.2.11)). We then get open sets G, H in X, Y respectively such that $C \subset G$, $y \in H$ and $G \times H \subset f^{-1}(V)$. Then for all $z \in H$, we have $f(C \times \{z\}) \subset f(G \times H) \subset V$ from which it follows that $H \subset \hat{f}^{-1}([C, V])$. Thus $\hat{f}^{-1}([C, V])$ is a nbd of y. This proves that $\hat{f}^{-1}([C, V])$ is open in Y and thus establishes continuity of $\hat{f}$. So we have proved that if f is continuous, its associate $\hat{f}$ is also continuous.

We now try to go the other way. Suppose we are given a map $g : Y \to M(X, Z)$. Can we find a continuous function $f : X \times Y \to Z$ such that $g = \hat{f}$. There is little difficulty in defining f. In fact we have no choice but to let $f(x, y) = g(y)(x)$ for $x \in X$, $y \in Y$ (see the proof of Theorem (8.1.9)). The trouble is that even if each $g(y)$ is continuous and g itself is continuous, it does not follow that f is continuous. The reason is that continuity of g does not mean such unless the compact open topology on $M(X, Z)$ is sufficiently strong (recall that the stronger the topology on the codomain the stronger it is to say that a function is continuous). As is to be expected, the situation can be remedied with an additional hypothesis about X.

(1.4) Theorem: Let X be locally compact and regular and let Y, Z be any spaces. Then for a function $f : X \times Y \to Z$, f is continuous iff its associate $\hat{f} : Y \to M(X, Z)$ is continuous.

Proof: One way implication was already proved and did not require the additional hypothesis about X. For the converse assume $\hat{f}$ is continuous. We could prove continuity of f directly but a much more elegant argument is as follows. Note that f is precisely the composite $X \times Y \xrightarrow{1 \times \hat{f}} X \times M(X, Z) \xrightarrow{e} Z$ where $1 \times \hat{f}$ is defined by $(1 \times \hat{f})(x, y) = (x, \hat{f}(y))$ for $(x, y) \in X \times Y$ and e is the evaluation function. Continuity of $\hat{f}$ implies that of $1 \times \hat{f}$ while e is continuous because X is locally compact and regular as seen earlier. So f is continuous. ∎

This proof is also instructive in that it shows exactly where local compactness and regularity of X were used. A counter-example would be obtained if we let $X = \mathbb{Q}$, $Z = \mathbb{R}$ and $Y = M(\mathbb{Q}, \mathbb{R})$.

The concept of function spaces allows us to give a precise meaning to the expression 'a continuously varying family of maps'. Recall that an indexed family of maps from a space X to a space Y is function $g : I \to M(X, Y)$ where I is an index set. If the index set I is equipped with some topology and g is continuous then g is called a continuously varying family of maps from X to Y. By the theorem above, if X is locally compact and regular then this is equivalent to giving a map from $X \times I$ to Y. In most applications the index set I is the unit interval $[0, 1]$ with the usual topology.

A classic example of this situation is to be found in complex analysis. There we have an analytic function $f(z)$ defined in some non-empty open connected subset D of the complex plane. Cauchy's theorem asserts that the line integral of $f(z)$ vanishes over all closed paths 'which can be continuously deformed or shrunk to a point within D'. In elementary treatises this phrase is explained by appealing to intuition. We can express it rigorously as follows. A closed path in D corresponds to a map from S^1, the unit circle, to D, or equivalently to an element of $M(S^1, D)$. A continuous shrinking of such a path to a point simply means a map $g : [0, 1] \to M(S^1, D)$ such that $g(0)$ is the given closed path and $g(1)$ is a constant path in D. Such a map g is called a deformation and $g(t)$ is called the t-th stage of deformation for $0 \leqslant t \leqslant 1$. Since S^1 is a compact metric space, the last theorem applies and we get a map say $H : S^1 \times [0, 1] \to D$ such that $H(x, t) = g(t)(x)$ for all $x \in S^1$, $t \in [0, 1]$. Such a map H is known as a homotopy, a notion we now define.

(1.5) Definition: Let f, g be maps from a space X to a space Y. Then f is said to be **homotopic** to g if there exists a map $H : X \times [0, 1] \to Y$ such that $H(x, 0) = f(x)$ and $H(x, 1) = g(x)$ for all $x \in X$. Such a map H is called a **homotopy** from f to g. For $t \in [0, 1]$, the map $H_t : X \to Y$ defined by $H_t(x) = H(x, t)$ for $x \in X$ is called the **t-th stage** of the homotopy H.

A homotopy from f to g determines a path from f to g in the function space $M(X, Y)$. The converse holds if X is locally compact and regular. Alternatively, a homotopy from f to g determines a path in Y from $f(x)$ to $g(x)$ for each $x \in X$ and these paths vary continuously with x. The converse also holds since the unit interval is compact and regular. It is easy to show that 'being homotopic to' is an equivalence relation.

To tell whether two given maps are homotopic to each other is not always easy. It is known for example that the identity map from the n-sphere S^n to itself is not homotopic to constant map. Intuitively this means that the sphere cannot be shrunk or contracted to a point. This seems obvious but is non-trivial to prove. Note, incidentally, the role of the codomain space. If we regard the identity map of S^n and the constant map as maps from S^n to $\mathbb{R}^{n+1}$ (of which S^n is a subspace) then they are homotopic. As a matter of fact any two maps from any space X whatsoever into $\mathbb{R}^{n+1}$ are homotopic. If f, g are two such maps we simply define $H : X \times [0, 1] \to \mathbb{R}^{n+1}$ by $H(x, t) = (1 - t)f(x) + tg(x)$ for $x \in X$ and $t \in [0, 1]$. Thus we must distinguish between functions that have the same domain and the same graph if their codomains are not the same. In other areas of mathematics such a distinction is often not made because there it causes no harm. In dealing with homotopies, the codomain space is very crucial.

A few results about homotopic maps will be given as exercises.

Exercises

In exercises (1.1) to (1.5) X, Y will be topological spaces and $M(X, Y)$ the function space.

1.1 Prove that if Y is T_0, T_1, T_2 or regular, $M(X, Y)$ has the corresponding separation properties.

1.2 Prove that if X is non-empty then Y can be embedded into $M(X, Y)$. (Hint: Consider constant functions.)

1.3 If $\mathcal{S}$ is a sub-base for Y prove that the family of all sets of the form $[C, V]$ with $C \subset X$ compact and $V \in \mathcal{S}$ is a sub-base for $M(X, Y)$.

1.4 If X, Y are second countable and X is locally compact and Hausdorff then prove that $M(X, Y)$ is second countable.

1.5 If, further, Y is metrisable in addition to the hypotheses of the last exercise, prove that $M(X, Y)$ is metrisable.

1.6 For any three spaces X, Y, Z prove that $M(X, Y \times Z)$ is homeomorphic to $M(X, Y) \times M(X, Z)$. Prove that the homeomorphism can be chosen to be natural in the sense of category theory. Does this result hold for maps into arbitrary products?

1.7 For any three spaces X, Y, Z prove that $M(X+Y, Z)$ is naturally homeomorphic to $M(X, Z) \times M(Y, Z)$. Does this hold for arbitrary coproducts?

1.8 For any three spaces X, Y, Z prove that the function $M(X \times Y, Z) \to M(Y, M(X, Z))$ taking f to $\hat{f}$ for $f \in M(X \times Y, Z)$ is continuous. Interpret this in terms of natural transformation of suitable functors.

1.9 Let X, Y be topological spaces. Prove that 'being homotopic to' defines an equivalence relation on $M(X, Y)$. (Hint: The proof resembles that of Proposition (6.3.9). Why is this resemblance not unexpected?) The equivalence classes under this relation are called **homotopy classes of maps** from X to Y.

1.10 Prove that the equivalence relation in the last exercise is compatible with compositions, in the sense that if f, g are homotopic maps from X to Y and h, k are homotopic maps from Y to Z then $h \circ f$ is homotopic to $k \circ g$. Hence obtain a category whose objects are topological spaces and morphisms are homotopy classes of maps. This category is called the **homotopy category.** An equivalence in this category is called a **homotopy equivalence.**

1.11 Prove that any two constant maps into a path-connected space are homotopic to each other. A map which is homotopic to a constant map is called **null-homotopic**.

1.12 A space X is called **contractible** if the identity map $1_X : X \to X$ is null-homotopic. Prove that:

(i) A contractible space is path connected.

(ii) Every map into or from a contractible space is null-homotopic

(iii) A space is contractible iff it is equivalent in the homotopy category to a singleton space.

(iv) A product of non-empty spaces is contractible iff each co-ordinate space is so.

(v) A retract of a contractible space is contractible.

(vi) All euclidean spaces, open balls, closed balls in them are contractible. A triod (that is a space hoemorphic to the figure Y) is contractible.

1.13 Let X be a space. The space $X \times [0, 1]$ is called the **cylinder** over X. The quotient space of $X \times [0, 1]$ obtained by collapsing $X \times \{1\}$ to a point is called the **cone** over X. It is denoted by $C(X)$. Prove that:

(i) X can be canonically embedded in $C(X)$.

(ii) $C(X)$ is always contractible if X is nonempty.

(iii) A map $f : X \to Y$ is null-homotopic iff there exists a map $F : C(X) \to Y$ such that $F/X = f$, regarding X as a subspace of $C(X)$.

(iv) $C(S^n)$ is homeomorphic to D^{n+1} for all n.

(v) $C(\mathbb{R})$ is not homeomorphic to the subspace $\{(x - xy, y) : x \in \mathbb{R}, 0 \leq y \leq 1\}$ of the plane.

1.14 Prove that the assertion that the sphere S^{n-1} is not contractible is equivalent to the Brouwer fixed point theorem, namely, that the closed ball D^n has the fixed point property. (Hint: See Exercise (13.2.13).)

1.15 Let X be any space and $f, g : X \to S^n$ two maps. Suppose $d(f(x), g(x)) < 2$ for all $x \in S^n$ where d denotes the usual euclidean distance in $\mathbb{R}^{n+1}$. Prove that f is homotopic to g. (Hint: for each $x \in S^n$, there is a unique great circular arc from $f(x)$ to $g(x)$.)

1.16 A space is said to be **simply connected** if every map from S^1 to it is null-homotopic, Prove that:

(i) Every contractible space is simply connected.

(ii) A retract of a simply connected space is simply connected.

*(iii) S^n is simply connected for $n > 1$. (Hint: First prove that a non-surjective map into S^n must be null-homotopic. Then, given $\alpha : S^1 \to S^n$, use uniform continuity and the last exercise to get a map $\beta : S^1 \to S^n$ which is homotopic to x and whose range consists of finitely many great circular arcs on S^n. Since $n > 1$, β cannot be surjective.)

Notes and Guide to Literature

For further information on the compact open topology see Kelley [1] or Dugundji [1]. Under certain additional hypothesis on X and Y, the correspondence given in Exercise (1.8) is a homomorphism.

Homotopy theory concerns itself with the study of the homotopy category. In algebraic topology, one studies functors from the homotopy category to the category of groups, rings etc. Non-existence theorems about the homotopy category are then proved by proving non-existence in the algebraic category. Exercise (1.14) shows how a result in homotopy theory can be applied to get a geometric result. Standard treatises on homotopy theory and algebraic topology include Hu [2], Hilton and Wylie [1], Spanier [1] and Eilenberg and Steenrod [1].

2. Paracompactness

We remarked earlier that compactness is the next best thing to finiteness. It is natural to inquire what is the next best thing to compactness. The answer is obviously some condition weaker than compactness. Just how weak should this condition be ? It should be weak enough to be satisfied by a large class of naturally occurring spaces (for example, metric spaces). On the other hand it should not be so weak that no significant results can be proved using it.

The condition to be discussed in this section meets these demands quite well. It is called paracompactness. Before defining it, let us illustrate with an example the type of problems that can be tackled with it. Let X be a space. The space $X \times [0, 1]$ is called the **cylinder** over X. The set $X \times \{0\}$ is called the **base** of the cylinder and may be identified with X in the obvious manner. Suppose now we have an open set G in $X \times [0, 1]$ containing the base. The problem is to fit within G some 'mountain' over X. More precisely we want a function $f : X \to [0, 1]$ which is positive everywhere such that the set $\{(x, y) \in X \times [0, 1] : 0 \leq y < f(x)\}$ is contained in G. This is of course always possible. Indeed, for each $x \in X$, we have, by the very definition of product topology, that there exists an open neighbourhood V_x of x in X and a positive real number r_x such that $V_x \times [0, r_x) \subset G$. We may set $f(x) = r_x$ for all $x \in X$. But if we require f to be continuous (a most natural requirement in topology) then we are in trouble and indeed it can be shown by examples that the problem may fail to have a solution. Note that since a constant function is continuous, the problem always has a local solution at each point by what we just noted. We try to patch together these local solutions and construct a global solution. It is to this end that some additional hypothesis about X is needed. For example if X were compact we could apply a standard compactness argument (or what amounts to the same, apply Wallace's theorem) and get a constant function that will work. Without compactness, even if a solution exists, in general a constant function will not work as can be seen by taking $X = (0, 1]$ and $G = \{(x, y) : 0 < x \leq 1, 0 \leq y < x\}$.

The situation illustrated by this problem is fairly common. A problem has a local solution at every point (that is, a solution defined over some neighbourhood of that point). In other words there exists an open cover $\mathcal{U}$ and for each $U \in \mathcal{U}$, a solution to the problem over U. The problem then is how to piece together these solutions to get a global solution. The thorny part is what to do if two members, say, U and V of $\mathcal{U}$ overlap ? In such a case for points of $U \cap V$ we may take some suitable linear combination of the two solutions or take the maximum or the minimum of them. More generally this can be done for all points which are common to finitely many members of $\mathcal{U}$. But what if we had a cover $\mathcal{U}$ in which infinitely members had a point in common ? This difficulty would not arise if $\mathcal{U}$ were a finite cover and this is where compactness helps. Things are not so bad if the cover $\mathcal{U}$, even though not finite, is what is known as locally finite. Recall from Definition (9.3.3) that this means that every point has a neighbourhood which meets only finitely many members of $\mathcal{U}$. This condition is stronger than saying that every point belongs to at most finitely many members of $\mathcal{U}$. If the latter holds, $\mathcal{U}$ is called **point finite**.

From this discussion it would appear that the appropriate generalisation of compactness that we are looking for would be obtained by requiring that every open cover of the space in question has a locally finite subcover. Ironically, this requirement turns out to be equivalent to compactness! For suppose X is a space with the property that every open cover of it has a locally finite subcover. Let $\mathcal{U}$ be a given open cover of X. Fix some non-empty $V \in \mathcal{U}$ and let $\mathcal{V} = \{G \cup V : G \in \mathcal{U}\}$. Then $\mathcal{V}$ is an open cover of X. Since every member of $\mathcal{V}$ contains V, no subcover of $\mathcal{V}$ can be locally finite unless it is actually finite. But then $\mathcal{U}$ would have a finite subcover. Hence X is compact.

Thus the proposed generalisation is only an illusory one. Its failure can be analysed as follows. In the argument given above the members of the cover $\mathcal{V}$ were unnecessarily large. When the members of a cover are very large it has little chance of possessing a locally finite subcover unless it possessed a finite subcover. It is clear that if we want a really meaningful generalisation of compactness in which finite covers will be replaced by locally finite covers, we must allow their members to be small enough, perhaps, smaller than those of the given open cover. We have already encountered this concept. Recall from Exercise (10.1.2) that a family $\mathcal{U}$ of subsets of a set X is said to be a refinement of another such family $\mathcal{V}$ if every member of $\mathcal{U}$ is contained in some member of $\mathcal{V}$. A subfamily is a very special case of a refinement. If we replace finiteness by local finiteness and subcovers by open refinements we indeed get a very far-reaching generalisation of compactness which we now define.

(2.1) Definition: A space is called **paracompact** if it is regular and if every open cover of it has an open, locally finite refinement which is also a cover.

The presence of the separation axiom, regularity, in the definition requires a little comment. Because of it, it is not quite correct to say that paracompactness is a generalisation of compactness. For example an infinite set with the cofinite topology or the Alexandroff compactification of the rationals are compact. But they are not paracompact because although they satisfy trivially the condition regarding open covers, they are not regular. It turns out that regularity plays a crucial role in many important properties of paracompactness spaces. If it were not made a part of the definition, then it would have to be added everytime as a part of the hypothesis. If we take the classic approach of working exclusively with Hausdorff spaces then the difficulty does not arise as shown by the following proposition.

(2.2) Proposition: Let X be a Hausdorff space. Then X is paracompact if and only if it has the property that every open cover of it has an open, locally finite refinement which is also a cover.

Proof: Clearly only the converse implication requires a proof. So suppose X has the given property. We need only show that X is regular. Suppose $x \in X$ and C is a closed subset of X not containing x. For each $y \in C$, there exists, by the Hausdorff property, an open set U_y containing y such that $x \notin \overline{U}_y$. Let $\mathcal{U} = \{U_y : y \in C\} \cup \{X - C\}$. Then $\mathcal{U}$ is an open cover of X. Let $\mathcal{V}$ be an open, locally finite refinement of $\mathcal{U}$ which is also a cover. Every member of $\mathcal{V}$ is contained either in some U_y or in $X - C$. In the latter case it cannot intersect C. Thus if we let $\mathcal{W}$ be $\{V \in \mathcal{V} : V \cap C \neq \emptyset\}$ then every member of $\mathcal{W}$ is contained in some U_y. Now set

$G = \bigcup_{W \in \mathcal{W}} W$. Then G is open and contains C. The proof would be complete if we can show that $x \notin \bar{G}$. For this we observe that since $\mathcal{V}$ is locally finite so is $\mathcal{W}$ and so by the result of Exercise (9.3.4) $\bar{G} = \bigcup_{W \in \mathcal{W}} \bar{W} \subset \bigcup_{y \in C} \bar{U}_y$. Since $x \notin \bar{U}_y$ for all $y \in C$, the result follows. ∎

It is customary to require 'Hausdorff' instead of 'regular' in the definition of paracompactness. The proposition shows that the resulting condition would be stronger than that we have assumed. Proceeding in the same vein as the last proof we get,

(2.3) Theorem: Every paracompact space is normal.

Proof: In the last proof we replace the point x by a closed set D disjoint from C and use regularity instead of the Hausdorff property. The same proof then goes through. ∎

The argument above illustrates how local finiteness can replace finiteness. Whether a point belongs to the closure of a set or not is a local property at that point and hence local finiteness is as good as finiteness. Since continuity is also a local concept, we expect that the same should apply for continuity. This is indeed the case. Suppose $\mathcal{F}$ is a family of continuous real-valued functions on a space X such that for each $x \in X$, there exists a neighbourhood N of x with the property that all except finitely many members of $\mathcal{F}$ vanish on N. Then for each $x \in X$, the sum $\sum_{f \in \mathcal{F}} f(x)$ is really a finite sum (even though $\mathcal{F}$ itself may be infinite or even uncountable). This way we get a function $f : X \to \mathbb{R}$ called the **sum function** of $\mathcal{F}$, which is continuous because X can be covered by open sets on each of which f is continuous (being a sum of finitely many continuous functions). The set of points where a real valued function takes non-zero values is often called its **support** (occasionally the closure of this set is called the support). With this terminology, the condition about $\mathcal{F}$ can be expressed by saying that the family of supports of members of $\mathcal{F}$ is locally finite. When members of $\mathcal{F}$ take only non-negative values and the sum function f is identically equal to 1, $\mathcal{F}$ is called a **partition of unity**. (Unity is the old name for 1.)

Partitions of unity are very useful in applications as we shall illustrate. Before doing so, we proceed to prove the existence of partitions of unity with certain properties. We begin with an interesting property of normal spaces. It is not hard to show by induction that if $\{U_1, \ldots, U_n\}$ is an open cover of a normal space then there exists an open cover $\{V_1, \ldots, V_n\}$ of X such that $\bar{V}_i \subset U_i$ for all $i = 1, 2, \ldots, n$. Intuitively we may think V_i as a 'shrinking' of U_i. The theorem to come shows that such a shrinking is possible even for covers that are locally finite (actually point finiteness would suffice as the proof shows.)

(2.4) Theorem: Let $\mathcal{U}$ be an open, locally finite cover of a normal space X. Then for each $U \in \mathcal{U}$ there exists an open set $G(U)$ such that $\overline{G(U)} \subset U$ and the family $\{G(U) : U \in \mathcal{U}\}$ covers X.

Proof: Let $\mathfrak{J}$ be the topology on X. We are looking for a function $G : \mathcal{U} \to \mathfrak{J}$ with the desired properties. Let us call such a function G as a total shrinking function for $\mathcal{U}$. Let us define a partial shrinking function for $\mathcal{U}$ to be a function of the form $F : \mathcal{V} \to \mathfrak{J}$ where $\mathcal{V} \subset \mathcal{U}$, $\overline{F(V)} \subset V$ for all $V \in \mathcal{V}$ and the family $(\mathcal{U} - \mathcal{V}) \cup \{F(V) : V \in \mathcal{V}\}$ is a cover of X. Let $\mathcal{F}$ be the family of all partial shrinking functions for $\mathcal{U}$. Note that $\mathcal{F}$ is non-empty because the function with the empty domain belongs to it. Given two partial shrinking functions F, H for $\mathcal{U}$, we say $F \leq H$ if, first, domain of F is contained in that of H and secondly $F(V) = H(V)$ for all $V \in$ domain of F. It is easy to show that $\leq$ is a partial order on $\mathcal{F}$. We contend that $\mathcal{F}$ has a maximal element and that such a maximal element is a total shrinking function for $\mathcal{U}$.

The existence of a maximal element in $\mathcal{F}$ is proved by applying the Zorn's lemma in the usual manner. For convenience, if $F \in \mathcal{F}$ let us write $\mathcal{D}(F)$ for the domain of F. Now suppose $\{F_i : i \in I\}$ is a chain in $\mathcal{F}$. We construct a partial shrinking function F from this chain as follows. First let $\mathcal{D}(F) = \bigcup\limits_{i \in I} \mathcal{D}(F_i)$. Now if $V \in \mathcal{D}(F)$, then there exists $i \in I$ such that $V \in \mathcal{D}(F_i)$ and we set $F(V) = F_i(V)$. This is well-defined because if $V \in \mathcal{D}(F_i) \cap \mathcal{D}(F_j)$ for $i \neq j \in I$ then $F_i(V) = F_j(V)$ by the definition of the partial ordering on $\mathcal{F}$. Clearly $\overline{F(V)} \subset V$ for all $V \in \mathcal{D}(F)$. In order to show that F is a partial shrinking function for $\mathcal{U}$ it only remains to show that the family $(\mathcal{U} - \mathcal{D}(F)) \cup \{F(V) : V \in \mathcal{D}(F)\}$ is a cover of X. Let $x \in X$. Then there exist only finitely many members of $\mathcal{U}$ say $U_1, U_2, \ldots, U_n$ which contain x. If at least one of these is in $\mathcal{U} - \mathcal{D}(F)$ we are done. Otherwise there exist $i_1, i_2, \ldots, i_n \in I$ such that $U_r \in \mathcal{D}(F_{i_r})$ for all $r = 1, 2, \ldots, n$. Since $\{F_i : i \in I\}$ is a chain we may assume without loss of generality that $i_1 \leq i_2 \leq \ldots \leq i_n$. But then $U_r \in \mathcal{D}(F_{i_n})$ for all $r = 1, 2, \ldots, n$ and since these are the only members of $\mathcal{U}$ containing x, it follows that $x \in F_{i_n}(V)$ for some $V \in \mathcal{D}(F_{i_n})$ as we are given that the family $(\mathcal{U} - \mathcal{D}(F_{i_n})) \cup \{F_{i_n}(V) : V \in \mathcal{D}(F_{i_n})\}$ is a cover of X. We also have $F(V) = F_{i_n}(V)$ and thus we have shown that $(\mathcal{U} - \mathcal{D}(F)) \cup \{F(V) : V \in \mathcal{D}(F)\}$ is a cover of X. So F is a partial shrinking function for $\mathcal{U}$ and by its very construction it is an upper bound for the chain $\{F_i : i \in I\}$. Thus we have shown that every chain in $\mathcal{F}$ has an upper bound in $\mathcal{F}$. Zorn's lemma applies and we get a partial shrinking function, say, G for $\mathcal{U}$ which is maximal w.r.t. the ordering $\leq$.

To conclude the proof, we assert that G is a total shrinking function for $\mathcal{U}$, that is, $\mathcal{D}(G) = u$. If this is not so, let $U \in \mathcal{U} - \mathcal{D}(G)$. Let W be the union of the set $\bigcup \{G(V) : V \in \mathcal{D}(G)\}$ and the set $\bigcup (\mathcal{U} - \mathcal{D}(G) - \{U\})$. Then $X - W$ is a closed set and $X - W \subset U$ since $(\mathcal{U} - \mathcal{D}(G)) \cup \{G(V) : V \in \mathcal{D}(G)\}$ is a cover of X. Now by normality there exists an open set Q such that $X - W \subset Q$ and $\bar{Q} \subset U$. We define $H : \mathcal{D}(G) \supset \{U\} \to \mathfrak{J}$ by $H(V) = G(V)$ for $V \in D(G)$ and $H(U) = Q$. Then we get a partial shrinking function which is strictly greater than G, contradicting the maximality of G. Thus $\mathcal{D}(G) = \mathcal{U}$ and so G is a desired total shrinking function for $\mathcal{U}$. ∎

The use of Zorn's lemma in the proof above may come as a surprise but it is really not so. The idea in the proof was to go on enlarging the domain of the partial shrinking function until it exhausted the entire family $\mathcal{U}$. If $\mathcal{U}$ were finite or at least countable we could do so inductively. If $\mathcal{U}$ is uncountable we need the appropriate analogue of the method of induction called the transfinite induction. This requires the use of ordinals and is therefore beyond our scope. But the underlying principle is that every set can be well ordered. It was remarked in Chapter 2 that this is equivalent to Zorn's lemma. Hence the use of the latter is not surprising.

Combining the last theorem with Urysohn's lemma we get the following important corollary:

(2.5) Corollary: Let $\mathcal{U}$ be an open, locally finite cover of a normal space X. Then there exists a partition of unity $\{f_V : V \in \mathcal{U}\}$ such that for each $V \in \mathcal{U}, f_V$ vanishes outside V.

Proof: Let $G : \mathcal{U} \to \mathfrak{J}$ be a total shrinking function as given by the last theorem. For each $V \in \mathcal{U}$, $\overline{G(V)}$ and $X - V$ are disjoint closed sets and so there exists, by Urysohn's lemma, a map $g_V : X \to [0, 1]$ such that $g_V = 1$ on $G(V)$ and g_V vanishes outside V. Since $\mathcal{U}$ is locally finite, each point, say x of X has a neighbourhood N which meets only finitely many members of $\mathcal{U}$. But this means that all except finitely many g_V's vanish on N. So we get a map $g : X \to \mathbb{R}$ defined by $g(x) = \sum_{V \in \mathcal{U}} g_V(x)$ for $x \in X$. The only trouble is that $g(x)$ may not be identically equal to 1. This can be remedied by normalization. Since the family $\{G(V) : V \in \mathcal{U}\}$ covers X, for each $x \in X$ there exists some $W \in \mathcal{U}$ such that $g_W(x) = 1$. So $g(x)$ is strictly positive. We now merely set $f_V(x) = \dfrac{g_V(x)}{g(x)}$ for $x \in X$. The family $\{f_V : V \in \mathcal{U}\}$ is a desired partition of unity. ∎

In order to apply this corollary we must have locally finite, open covers. It is precisely at this point that paracompactness comes in. As an illustration we shall show that the problem of mountain fitting discussed at the beginning of this section has a solution when the space X is paracompact.

(2.6) Theorem: Let X be a paracompact space and G be an open set in $X \times [0, 1]$ containing $X \times \{0\}$. Then there exists a positive real-valued continuous function f on X such that for each $x \in X$, $\{x\} \times [0, f(x)) \subset G$.

Proof: As noted before, there exists an open cover $\mathcal{V}$ of X with the property that for each $V \in \mathcal{V}$, there exists $r > 0$ such that $V \times [0, r) \subset G$. By paracompactness, there exists an open, locally finite refinement $\mathcal{U}$ of $\mathcal{V}$ which also covers X. For each $V \in \mathcal{U}$, fix $r_V > 0$ such that $V \times [0, r_V] \subset G$ (such r_V exists because V is contained in some member of $\mathcal{V}$). By Theorem (2.3), X is normal. Let $\{f_V : V \in \mathcal{U}\}$ be a partition of unity as given by the last corollary. For $x \in X$, set $f(x) = \sum_{V \in \mathcal{U}} f_V(x) r_V$. This is well-defined because for each x, only finitely many $f_V(x)$'s are non-zero. We assert that f is continuous. Let $x \in X$ and let N be an open neighbourhood of x which meets only finitely many members say $V_1, \ldots, V_n$

of $\mathcal{V}$. Then for each $y \in N$, $f(y) = \sum_{i=1}^{n} r_{V_i} f_{V_i}(x)$. Thus f/N is continuous, being a linear combination of continuous real-valued functions. Since N is open, f is continuous at x. But x was arbitrary. So f is continuous on X. Finally for $x \in X$ we write $f(x)$ as $\sum_{i=1}^{n} f_{V_i}(x) r_{V_i}$. Since each $f_{V_i}(x)$ is non-negative and $\sum_{i=1}^{n} f_{V_i}(x) = 1$, it follows that $\min_{1 \leqslant i \leqslant n} \{r_{V_i}\} \leqslant f(x) \leqslant \max_{1 \leqslant i \leqslant n} \{r_{V_i}\}$. Thus $f(x)$ is positive. Also since $\{x\} \times [0, r_{V_i}) \subset G$ for all $i = 1, 2, \ldots, n$ it follows that $\{x\} \times [0, f(x)) \subset G$. This completes the proof. ∎

Having illustrated the power of paracompactness let us now see which spaces are paracompact. A compact, Hausdorff space is regular and hence paracompact. In Theorem (7.2.8) it was shown that a regular, Lindelöff space is normal. With a more elaborate argument it can be shown that such a space is in fact paracompact as we now do.

(2.7) Theorem: Every regular, Lindelöff space is paracompact.

Proof: Let X be a regular, Lindelöff space. Let $\mathcal{V}$ be an open cover of X. We may assume $\mathcal{V}$ to be countable say $\mathcal{V} = \{V_1, V_2, \ldots, V_n, \ldots\}$. Now for each $n \in \mathbb{N}$, let $W_n = V_n - \bigcup_{k=1}^{n-1} V_k$ and let $\mathcal{W} = \{W_n : n \in \mathbb{N}\}$. Note that members of $\mathcal{W}$ are in general neither open nor closed. Still $\mathcal{W}$ is a cover of X. Given $x \in X$, let n be the smallest integer such that $x \in V_n$. Then $x \in W_n$. Also $V_n \cap W_m = \emptyset$ for all $m > n$. In other words, every member of $\mathcal{V}$ meets only finitely many members of $\mathcal{W}$.

Now, by regularity of X, there exists an open cover say $\mathcal{G}$ of X such that the family of closures of members of $\mathcal{G}$ is a refinement of $\mathcal{V}$. Again by the Lindelöff property, we may assume $\mathcal{G}$ to be countable, say, $\mathcal{G} = \{G_1, G_2, \ldots, G_n, \ldots\}$. Now construct $\mathcal{H}$ from $\mathcal{G}$ the same way $\mathcal{W}$ was constructed from $\mathcal{V}$. That is $\mathcal{H} = \{H_n : n \in \mathbb{N}\}$ where $H_n = G_n - \bigcup_{r=1}^{n-1} G_r$ for $n \in \mathbb{N}$. Let $\mathcal{K} = \{\bar{H}_n : n \in \mathbb{N}\}$. Then $\mathcal{H}$ and hence $\mathcal{K}$ covers X. Also $\mathcal{H}$ and hence $\mathcal{K}$ is locally finite by Exercise (9.3.4). Also note that $\mathcal{K}$ is a refinement of $\mathcal{V}$. From this it follows that every member of $\mathcal{K}$ meets only finitely many members of $\mathcal{W}$.

The next step is to expand members of $\mathcal{W}$ to open sets. For each n, let $L_n = \bigcup\{\bar{H}_i : \bar{H}_i \subset X - W_n\}$. Then L_n is contained in $X - W_n$. Also L_n is closed since $\mathcal{K}$ and hence every subfamily of $\mathcal{K}$ is locally finite (see Exercise (9.3.3)). Thus if we set $M_n = X - L_n$, then M_n is an open set containing W_n. Note that an element of $\mathcal{K}$ meets W_n iff it meets M_n. Now for each n, set $D_n = M_n \cap V_n$ and let $\mathcal{D} = \{D_n : n \in \mathbb{N}\}$. Then $\mathcal{D}$ is a refinement of the given cover $\mathcal{V}$. Also members of $\mathcal{D}$ are open. Since $W_n \subset D_n$ for all n, it follows that $\mathcal{D}$ is a cover of X. To complete the proof we need only show that $\mathcal{D}$ is locally finite. Let $x \in X$, since $\mathcal{K}$ is locally finite, there exists an open neighbourhood N of x and $n \in \mathbb{N}$ such that $N \cap \bar{H}_m = \emptyset$ for all

$m > n$. Then $N \subset \bigcup_{i=1}^{n} \bar{H}_i$. Each $\bar{H}_i$ is an element of $\mathcal{K}$ and so meets M_r iff it meets W_r which is the case only for finitely many values of r. Thus it follows that N intersects M_r only for finitely many values of r. Since $D_r \subset M_r$ for all r, it follows that N intersects only finitely many members of $\mathcal{D}$. This proves that $\mathcal{D}$ is locally finite. Thus we have shown that every open cover of X has an open, locally finite refinement which is also a cover of X. As X is also regular, it is paracompact. ∎

As a corollary it follows that every second countable, regular space is paracompact. In particular all euclidean spaces, all subspaces of the Hilbert cube are paracompact.

But the most surprising source of paracompact spaces is metric spaces (or more generally, pseudometric spaces). A. H. Stone proved that every metric space is paracompact. We shall not prove it. But it is a truly remarkable theorem because there is nothing in the definition of a metric space which has the slightest connotations of finiteness or even of local finiteness. It is also a powerful theorem because it shows that the class of paracompact spaces is quite large. It is because of this result that paracompactness is the next best thing to compactness.

Exercises

2.1 Let $X = \mathbb{R} \times \mathbb{R}$ where each $\mathbb{R}$ carries the semi-open interval topology. For $x \in \mathbb{R}$ and $r > 0$, let $B(x; r)$ be the set $[x, x + r) \times [-x, -x + r)$. Let $\{q_1, q_2, \ldots, q_n, \ldots\}$ be an enumeration of rationals. Let A be the set $\{(x, -x) : x \in \mathbb{Q}\}$. Show that the set $((X - A) \times [0, 1]) \cup \left(\bigcup_{n=1}^{\infty} (B(q_n; 1/n) \times [0, 1/n)\right)$ is an open set in $X \times [0, 1]$ for which Theorem (2.6) fails. (Hint: Let, if possible, f be a function as asserted by the Theorem. For $m, n \in \mathbb{N}$, let $R_{m,n} = \{x \in \mathbb{R} : f \geqslant 1/n \text{ on } B(x; 1/m)\}$. Then for some m, n, $R_{m,n}$ is not nowhere dense in $\mathbb{R}$ in the usual topology.)

2.2 Prove that all discrete and all indiscrete spaces are paracompact.

2.3 Prove or disprove that paracompactness is preserved under continuous functions.

2.4 Prove that the product of two paracompact spaces need not be paracompact. (Hint: Note that the real line with the semi-open interval topology is regular and Lindelöff.)

2.5 Prove that the product of a paracompact space with a compact, regular space is paracompact. (Hint: Let X, Y be such spaces respectively and $\mathcal{U}$ an open cover of $X \times Y$. First show that there exists an open cover $\mathcal{V}$ of X such that for each $V \in \mathcal{V}$, $V \times Y$ is covered by finitely many members of $\mathcal{U}$).

2.6 Prove that paracompactness is a weakly hereditary but not a hereditary property.

2.7 A space is called **σ-compact** if it is the union of a countable family of compact subsets. Prove that a regular, σ-compact space is paracompact. (Hint: First prove that every σ-compact space is Lindelöff.)

2.8 A space X is called **hemicompact** if there exists a sequence $\{K_n\}$ of compact subsets of X such that every compact subset of X is contained in some K_n. Prove that a hemicompact space is σ-compact and that the space of rationals (with the usual topology) is σ-compact but not hemicompact. (Hint: The complement of a compact subset of $\mathbb{Q}$ is necessarily dense and hence contains points arbitrarily close to 0.)

2.9 Let X be a normal space. Prove the following statement by induction on n. For every closed subset A of X and for every open family $\{U_1, U_2, \ldots, U_n\}$ covering A, there exist open sets $V_1, V_2, \ldots, V_n$ covering A such that $\bar{V}_i \subset U_i$ for all $i = 1, 2, \ldots, n$. (The case $n = 1$ is just the definition of normality.)

2.10 Give a constructive proof of Theorem (2.4) in the case where the cover is countable. (Here 'constructive' means without appealing to the axiom of choice.)

*2.11 A real-valued function f on a topological space is called **upper (lower) semi-continuous** if for each $a \in \mathbb{R}$ the set $f^{-1}((-\infty, a))$ [respectively, $f^{-1}((a, \infty))$] is open. Prove that if X is a paracompact space and g, h are two real-valued functions on X such that g is upper semi-continuous, h is lower semi-continuous and $g(x) < h(x)$ for all $x \in X$, then there exists a continuous real-valued function f on X such that $g(x) < f(x) < h(x)$. (Hint: Show that each $x \in X$ has an open neighbourhood G such that $\sup\{g(y) : y \in G\} < \inf\{h(y) : y \in G\}$. Thus a constant function will work on G. Now piece these functions together by a partition of unity.)

2.12 Deduce Theorem (2.6) from the last exercise.

Notes and Guide to Literature

Paracompact spaces were introduced by Dieudonné. There are many characterisations of the definition. They, along with a proof of paracompactness of metric spaces may be found in Kelley [1]. The converse of Theorem (2.3) is not true. A trivial counter-example is to take a normal space which is not regular. A standard non-trivial example, where the space is T_4 but not paracompact is from ordinals. It will be given in the next section.

For transfinite induction, see Halmos [1]. A few standard applications of partitions of unity arise in proving that every paracompact manifold has a Riemannian metric or in proving that a fibre bundle over a paracompact space has the homotopy lifting property; see Husemoller [1].

For further information on σ-compact and hemicompact spaces, see Willard [1].

3. Use of Ordinal Numbers

We remarked earlier that ordinal numbers provide important counter-examples in topology. However, we have deferred their discussion as it was often possible to give the necessary counter-examples without them. In this section we consolidate the strange properties possessed by certain spaces constructed from ordinal numbers. It turns out that there is one ordinal of paramount importance as far as we are concerned. We therefore dispense with the general theory of ordinal numbers and proceed to construct only this particular ordinal, or rather, a representative of this ordinal. First we review a few basic concepts.

A well-ordering on a set is a partial order with the property that every non-empty subset has a least element. Trivially every subset of a well-ordered set is well-ordered w.r.t. the induced partial order. It is also easy to see that a well-ordering is linear and complete. If $(X, <)$ is a well-ordered set and $x \in X$ then the set of all elements of X preceding x, that is the set $\{y \in X : y < x\}$ is called the **initial segment** of X determined by x. It will be denoted by W_x.

Now to construct the desired ordinal, we start with any uncountable set, for example $\mathbb{R}$. By the well ordering principle there exists a well ordering on $\mathbb{R}$ which we denote by $<$ (the same symbol is used for the usual ordering on $\mathbb{R}$, but in the present discussion the usual ordering will not come at all). Now let ∞ be any symbol not in $\mathbb{R}$ and let $X = \mathbb{R} \cup \{\infty\}$. We extend $<$ to X by letting $x < \infty$ for all $x \in \mathbb{R}$. Then $<$ is a well-ordering on X. Note that W_∞ is $\mathbb{R}$ which is uncountable. Let $S = \{x \in X : W_x$ is uncountable$\}$. Then $\infty \in S$ and so S is nonempty. (This is the only reason for introducing ∞. Without it, it may happen that each W_s is uncountable even though $\mathbb{R}$ is uncountable. Whether this can actually happen is equivalent to the continuum hypothesis.) Let z be the least element of S. We denote W_z by Ω_0, z by Ω and $W_z \cup \{z\}$ by Ω'. Note that for each $y \in \Omega_0$, W_y is countable, for otherwise $y \in S$ contradicting that z is the least element of S. Note also that W_z has no greatest element; for if y were such element then W_z would have to equal $W_y \cup \{y\}$ which is countable. Thus it follows that Ω is the supremum of Ω_0. The basic properties of Ω_0 are given below:

(3.1) Theorem: Every countable subset of Ω_0 is bounded. If $\{x_n\}$, $\{y_n\}$ are sequences in Ω_0 such that for each $n \in \mathbb{N}$, $x_n \leq y_n \leq x_{n+1}$, then the sets $\{x_n : n \in \mathbb{N}\}$ and $\{y_n : n \in \mathbb{N}\}$ have a common supremum in Ω_0.

Proof: Let A be a countable subset of Ω_0. Let $B = \bigcup \{W_x : x \in A\}$. For each $x \in A$, W_x is countable. A is also countable and so B is countable. As a subset of Ω', A is certainly bounded above, Ω (that is z) being an upper bound for it. Let y be the supremum of A in Ω' (which exists because a well ordering is complete.) Then $W_y \subset B$. For, let $x \in W_y$. Then $x < y$ and so there is some $u \in A$ such that $x < u \leq y$. But then $x \in W_u$ and $W_u \subset B$ since $u \in A$. So $x \in B$. It follows that W_y is countable, as B is so.

Hence y cannot be z for W_z (that is Ω_0) is not countable. So $y \in \Omega_0$. Thus we have shown that A is bounded above. Every subset of a well-ordered set is bounded below. So A is a bounded subset of Ω_0. Thus every countable subset of Ω_0 has a supremum in Ω_0. The second assertion is now clear. ∎

The second assertion in the theorem is called the **interlacing lemma.** The theorem will be needed frequently.

Before proceeding further with the sets Ω_0 and Ω' let us consider two objections about their construction. The first is that they are not uniquely defined as there was considerable arbitrariness in the choice of the uncountable set at the start and also the symbol ∞. Also there could be many well-orderings on the set with which we started. This is not a very serious objection. The only property of Ω_0 that we shall need is that it is an uncountable set with a well-ordering in which every element has only countably many predecessors. It can be shown that any two well-ordered sets with this property are order equivalent and hence represent the same ordinal number. Thus we may suppose that Ω_0 is defined uniquely upto an order equivalence.

The second objection is that the construction of Ω_0 is non-constructive! It relies heavily on the well-ordering principle which is a form of the axiom of choice. As remarked in Chapter 2, no explicit well-ordering of $\mathbb{R}$ is known. This is indeed a somewhat serious objection because it is actually possible to construct the ordinals without using the axiom of choice. The only answer we can give is that we have already used the axiom of choice freely (often in the form of Zorn's lemma). At any rate our interest in Ω_0 and Ω' is in the topological counter-examples they provide and not in their construction *per se*.

On both Ω_0 and Ω' we put the order topologies (induced by the respective well orderings). Note that in general the order topology on a subset is weaker than the relativisation of the order topology on the superset. But in this case the two fortunately coincide as can easily be shown. Thus from now on we shall regard Ω_0 and Ω' as topological spaces with Ω_0 as a subspace of Ω'. Note that Ω_0 is open and dense in Ω' because Ω is the supremum of Ω_0. We shall denote the first element of Ω_0 by 0. If $y \in \Omega_0$, by $y + 1$ we shall denote its *immediate successor* that is the least element of the set $\{x \in \Omega_0 : y < x\}$. We constructed Ω_0 as a certain subset of real numbers. But the 0 and $+$ have nothing to do with the usual 0 and $+$ for $\mathbb{R}$. In fact from now on we might as well forget that Ω_0 was constructed from $\mathbb{R}$. As said earlier all the results we state will follow from the fact that Ω_0 is an uncountable well-ordered set in which W_y is countable for each $y \in \Omega_0$.

Now coming to the topological properties of Ω_0 and Ω' we first study them with reference to smallness conditions and separation axioms. The results are listed below.

(3.2) Theorem: Let Ω_0, Ω' have the order topologies. Then

(i) Ω' is first countable at every point except Ω.

(ii) Neither Ω_0 nor Ω' is separable.

(iii) Both Ω' and Ω_0 are sequentially (and hence countably) compact. Ω' is compact but Ω_0 is not so.
(iv) Both Ω' and Ω_0 are T_4-spaces.
(v) The product $\Omega_0 \times \Omega'$ is not normal.
(vi) The space Ω_0 is not paracompact.
(vii) The space Ω_0 is not Lindelöff.

Proof: (i) Let $y \in \Omega_0$. Then the family of open intervals of the form $(x, y+1)$ for $x \in W_y$ is easily seen to be a local base at y. Also this base is countable. (If $y = 0$, this family is empty, but then $\{\{0\}\}$ is a local base.) Thus Ω' is first countable at each point other than Ω. This means that the space Ω_0 is first countable. However Ω' is not first countable at Ω. Let if possible $\mathcal{L}$ be a countable local base at Ω. Without loss of generality we may suppose that the members of $\mathcal{L}$ come from the defining base for the order topology. That is to say, we may suppose $\mathcal{L} = \{(a_n, \Omega] : n \in \mathbb{N}\}$, where each $a_n \in \Omega_0$. By Theorem (3.1), the supremum of the set $\{a_n : n \in \mathbb{N}\}$ (say b) is in Ω_0. But then $(b+1, \Omega)$ is a neighbourhood of Ω which contains no member of $\mathcal{L}$, a contradiction.

(ii) Given any countable subset A of Ω_0 let b be its supremum which exists by Theorem (3.1). Then $A \cap (b, \Omega) = \emptyset$ showing that A cannot be dense in Ω_0. So Ω_0 is not separable. Since Ω_0 is open in Ω', it also follows that Ω' cannot be separable. (See Exercise (5.1.8).)

(iii) Let $\{a_n\}$ be a sequence in Ω_0. If it assumes some value infinitely often, it has a constant subsequence. Otherwise let $b \in \Omega_0$ be the supremum of the set $\{a_n : n \in \mathbb{N}\}$. Then by the definition of order topology b is a limit point of the sequence $\{a_n\}$. By first countability of Ω_0, $\{a_n\}$ has a subsequence converging to b. So Ω_0 is sequentially compact. As for sequential compactness of Ω' note that a sequence in Ω' either assumes the value Ω infinitely often (in which it has a constant subsequence converging to Ω) or else it is eventually in Ω_0 (in which case it has a convergent subsequence by the sequential compactness of Ω_0). As for compactness, note that the orders on both Ω' and Ω_0 are complete (being well orderings). But Ω' is bounded while Ω_0 is not. So by Theorem (11.2.8) Ω' is compact while Ω_0 is not.

(iv) All spaces obtained from linear orders are Hausdorff as can easily be shown. Thus Ω' is a compact, T_2 space and hence a T_4 space by Corollary (7.2.10). From this we cannot immediately say that Ω_0 is also a T_4 space because normality is not a hereditary property. Nevertheless, using normality of Ω', we can prove that of Ω_0 as follows. Let A, B be closed, mutually disjoint subsets of Ω_0. Let $\bar{A}$, $\bar{B}$ be their closures in Ω'. $\bar{A}$ will consist of A and possibly the point Ω and similarly for $\bar{B}$. We assert that Ω cannot belong to both $\bar{A}$ and $\bar{B}$. For if this happened, every neighbourhood of Ω would have to meet A as well as B. Now start with any $a_1 \in A$. Then $(a_1, \Omega]$ is a nbd of Ω and so there exists $b_1 \in B$ such that $a_1 < b_1$. But $(b_1, \Omega]$ is also a nbd of Ω. So there exists $a_2 \in A$ such that $b_1 < a_2$. Continuing in this manner, we get sequences $\{a_n\}$ and $\{b_n\}$ in A, B respectively such that for each n, $a_n < b_n < a_{n+1}$. By the interlacing lemma, these sequences converge to a common limit in Ω_0 which would belong to $A \cap B$, a contradiction. Thus Ω cannot belong to both $\bar{A}$ and $\bar{B}$. Suppose $\Omega \notin \bar{A}$. Then $\bar{A} = A$ and so A is a closed subset of Ω' disjoint from $\bar{B}$. By normality of Ω', there exist open sets G, H which separate A from $\bar{B}$. $G \cap \Omega_0$ and $H \cap \Omega_0$ then separate A from B in Ω_0.

(v) Consider $\Omega_0 \times \Omega'$. This is a subspace of $\Omega' \times \Omega'$. Let $A = \{(x, x) : x \in \Omega_0\}$. Then $A = (\Omega_0 \times \Omega') \cap \Delta\Omega'$ where $\Delta\Omega'$ is the diagonal in $\Omega' \times \Omega'$. Since Ω' is a Hausdorff space, $\Delta\Omega'$ is a closed subset of $\Omega' \times \Omega'$ (see Exercise (7.1.8)). So A is a closed subset of $\Omega_0 \times \Omega'$. Let $B = \Omega_0 \times \{\Omega\}$. Then B is also a closed subset of $\Omega_0 \times \Omega'$ since $\{\Omega\}$ is closed in Ω' which is a T_1 space. Obviouly A, B are disjoint and we contend that they cannot be separated in $\Omega_0 \times \Omega'$ by disjoint open sets. Let if possible U, V be open sets containing A, B respectively such that $U \cap V = \emptyset$. Now for each $x \in \Omega_0$, the space $\{x\} \times \Omega'$ is homeomorphic to Ω' in the obvious manner and $U \cap (\{x\} \times \Omega')$ and $V \cap (\{x\} \times \Omega')$ are mutually disjoint open sets containing the points (x, x) and (x, Ω) respectively. Now in the order topology on Ω', every neighbourhood of Ω contains all except countably many points of Ω_0. Hence $V \cap (\{x\} \times \Omega')$ contains all except countably many points of $\{x\} \times \Omega_0$. Thus for all large $y \in \Omega_0$, $(x, y) \in V$ and hence $(x, y) \notin U$. Define $f : \Omega_0 \to \Omega_0$ by $f(x)$ to be the least y in Ω_0 such that $x < y$ and $(x, y) \notin U$. (We are not asserting that f is continuous.) Now start with any $x_1 \in \Omega_0$ and define inductively $x_2 = f(x_1)$, $x_3 = f(x_2), \ldots, x_{n+1} = f(x_n), \ldots$. The sequence $\{x_n\}$ is strictly monotonically increasing and by Theorem (3.1) converges to a point say y in Ω_0. So the sequence $\{(x_n, f(x_n))\}$ in $\Omega_0 \times \Omega'$ converges to (y, y). But U is a neighbourhood of (y, y) which contains no term of this sequence, a contradiction. This contradiction shows that the closed sets A, B, have no disjoint neighbourhoods in $\Omega_0 \times \Omega'$. So $\Omega_0 \times \Omega'$ is not normal.

(vi) This follows from (v) and the result of Exercise (2.5). Still, a direct argument is instructive. Consider the open cover of Ω_0 by sets of the form W_y for $y \in \Omega_0$. We assert that this cover has no open, locally finite refinement which covers Ω_0. For suppose $\mathcal{V}$ were such a refinement. Let $x \in \Omega_0$ and let N be an open neighbourhood of x in Ω_0 which meets only finitely many members say $V_1, V_2, \ldots, V_n$ of $\mathcal{V}$. For each $i = 1, 2, \ldots, n$ there is some $y_i \in \Omega_0$ such that $V_i \subset W_{y_i}$. Let $y = \max\{y_1, \ldots, y_n\}$. Then $y \notin V_i$ for all i and so it follows that no member of $\mathcal{V}$ can contain both x and y. As in (v) we construct a function $f : \Omega_0 \to \Omega_0$ such that for each $x \in \Omega_0$, $f(x) > x$ and no member of $\mathcal{V}$ can contain both x and $f(x)$. We construct a sequence exactly as in (v) and get $y \in \Omega_0$ as the limit of this sequence. Now let V be a member of Ω_0 such that $y \in V$. Then V contains x_n for all large n. In particular it contains x_n and $f(x_n)$ $(= x_{n+1})$, a contradiction. Thus Ω_0 is not paracompact.

(vii) Ω_0 is countably compact. If it were Lindelöff then it would be compact which is not the case. Alternatively, Ω_0 is regular. If it were Lindelöff, it would be paracompact by Theorem (2.7). As yet another way it is easy to see directly that the open cover $\{W_y : y \in \Omega_0\}$ has no countable subcover. So Ω_0 is not Lindelöff. ∎

This theorem amply illustrates the power of Ω_0 in providing counter-examples. In addition, Ω_0 has some curious properties. We prove one of them.

(3.3) Theorem: Every continuous real-valued function on Ω_0 is eventually constant in the sense that given such a function there exists $y \in \Omega_0$ such that the function has the same value at all $x \geq y$.

Proof: Let $f : \Omega_0 \to \mathbb{R}$ be continuous. Assume first that f is bounded, say f takes values in $[0, 1]$. Call 0 as a_0 and 1 as b_0. We assert that f is either eventually in $[0, \frac{2}{3})$ or eventually in $(\frac{1}{3}, 1]$. For let $A = f^{-1}([\frac{2}{3}, 1])$ and $B = f^{-1}([0, \frac{1}{3}])$. Then A, B are mutually disjoint, closed subsets of Ω_0. So at least one of them is countable (this was essentially proved in proving normality of Ω_0). If A is countable then f is eventually in $[0, \frac{2}{3})$ and we set $a_1 = 0$, $b_1 = \frac{2}{3}$. If A is not countable, then B is and f is eventually in $(\frac{1}{3}, 1]$ and we set $a_1 = \frac{1}{3}$, $b_1 = 1$. In either case we get an interval $[a_1, b_1]$ of length $\frac{2}{3}$ and some $y_1 \in \Omega_0$ such that $f(x) \in [a_1, b_1]$ for all $x \geq y_1$. We now repeat this argument for $[a_1, b_1]$ instead of $[a_0, b_0]$. Continuing in this manner, we get a nested sequence $\{[a_n, b_n]\}$ of intervals and a sequence $\{y_n\}$ in Ω_0 such that for each $n \in \mathbb{N}$, $b_n - a_n = (\frac{2}{3})^n$ and $f(x) \in [a_n, b_n]$ for all $x \geq y_n$. Let y be the supremum of the set $\{y_n : n \in \mathbb{N}\}$ in Ω_0 and c be the unique real number common to all the intervals $[a_n, b_n]$. Then $f(x) = c$ for all $x \geq y$ as was to be proved.

For the general case, consider $g \circ f$ where $g : \mathbb{R} \to (0, 1)$ is a homeomorphism. ∎

(3.4) Corollary: Ω' is homeomorphic to the Alexandroff as well as to the Stone-Čech compactification and the Wallman compactification of Ω_0.

Proof: The first part follows from Proposition (11.4.3). For the second part, as a trivial consequence of the last theorem it follows that every map from Ω_0 to $[0, 1]$ can be continuously extended over Ω'. This property characterises Stone-Čech compactification (see Exercise (11.4.6)). The last part follows from Exercise (11.4.18). ∎

As for connectedness, both Ω_0 and Ω' cut a sorry figure. Although the orders on them are complete, there are too many gaps in them (see Exercise (6.2.2) for the definition of a gap.) For each $y \in \Omega_0$, there is no element of Ω_0 between y and $y + 1$. As a result it is easy to show that both Ω_0 and Ω' are totally disconnected. However, if we fill up the gap between y and $y + 1$ by putting a copy of the open interval we get a connected space called the long line or transfinite line. The idea as well as the name come from the real line, or more precisely the half-line $[0, \infty)$. Let ω be the set of non-negative integers. Then the usual ordering is a well ordering on ω. The space $[0, \infty)$ is obtained by joining each $y \in \omega$ to $y + 1$. To make this construction precise we consider the set $\omega \times [0, 1)$. On ω and on the semi-open interval $[0, 1)$ we put the usual ordering and on $\omega \times [0, 1)$ we put the lexicographic (dictionary) ordering. Then the topology induced by this ordering yields a space homeomorphic to $[0, \infty)$. An explicit homeomorphism

$h : \omega\times[0, 1) \to [0, \infty)$ is given by $h(y, t) = y + t$ for $y \in \omega$, $0 \le t < 1$.

A similar procedure is followed for constructing the long line. We let $L = \Omega_0\times[0, 1)$. On Ω_0 we have the well-ordering we have been working with so far. On $[0, 1)$ we have the usual order. On L we put the lexicographic ordering. This means that for (x, t), (y, u) in L, $(x, t) \le (y, u)$ iff either $x < y$ in Ω_0 or $x = y$ and $t \le u$ in $[0, 1)$. Note that this is not a well ordering on L. Still it is a linear order. The set L with the topology induced by this ordering is called the **long** or the **transfinite line**. There is a minor notational difficulty in dealing with the order topology on L. Points of L are ordered pairs of the form (x, t) for $x \in \Omega_0$ and $t \in [0, 1)$. On the other hand a notation like this is also needed to denote open intervals which appear so frequently in any discussion of an order topology. The difficulty can be overcome either by agreeing to denote points of L by Greek letters or by using symbols like $]a, b[$ instead of (a, b) to denote open intervals.

The long line has many interesting properties, derived from the strange properties of Ω_0. A few such properties will be given as exercises.

Exercises

3.1 Prove that every well-ordering is linear and complete. (Hint: For linearity consider a set consisting of two points.)

3.2 Let $<$ be a linear order on a set X and let Y be a subset of X which is an interval (of whatever type). Prove that the order topology on Y coincides with the relativisation of the order topology on X.

3.3 Prove that Ω_0 is locally compact. Indeed for each $x \in \Omega'$ show that $W_x \cup \{x\}$ is a compact, open neighbourhood of x.

3.4 Prove that neither Ω_0 nor Ω' satisfies the countable chain condition.

3.5 Let $X = \{1/n : n \in \mathbb{N}\} \cup \{0\}$ with the usual topology. The product $\Omega' \times X$ is called the **Tychonoff plank**. Prove that the Tychonoff plank is normal but that the subspace $\Omega' \times X - \{(\Omega, 0)\}$ is not so.

3.6 Let U be a neighbourhood of the diagonal $\Delta\Omega_0$ in $\Omega_0\times\Omega_0$. Prove that there exists $x \in \Omega_0$ such that $(y, z) \in U$ for all $y \geqslant x$, $z \geqslant x$ in Ω_0. (Hint: If not, then assuming U to be symmetric, construct a strictly increasing sequence $\{x_n\}$ in Ω_0 such that $(x_n, x_{n+1}) \notin U$ for infinitely many values of n.)

3.7 Let $\mathcal{N}$ be the family of all neighbourhoods of $\Delta\Omega_0$ in $\Omega_0\times\Omega_0$. Prove that $\mathcal{N}$ is the only uniformity on Ω_0 inducing the order topology. (Hint: Let $\mathcal{U}$ be a uniformity on Ω_0 inducing the order topology. By Exercise (14.1.6), $\mathcal{U} \subset \mathcal{N}$. Let $\mathcal{V} = \{V \subset \Omega' \times \Omega' : V \cap (\Omega_0\times\Omega_0) \in \mathcal{U}$ and $(x, \Omega]\times(x, \Omega] \subset V$, for some $x \in \Omega_0\}$. Prove that $\mathcal{V}$ is a uniformity on Ω' inducing the order topology on Ω'. Then apply Corollary (14.3.13). Finally to show $\mathcal{N} \subset \mathcal{U}$, apply the last exercise).

3.8 Prove that Ω_0 with the uniformity in the last exercise is not complete even though every Cauchy sequence in it is convergent. (Hint: Take Ω_0 itself as a directed set and show that the identify function on it is a Cauchy net.)

3.9 For $x \in \Omega_0$, let $V_x = ((x, \Omega)\times(x, \Omega)) \cup \Delta\Omega_0$. Let $\mathcal{V} = \{V_x : x \in \Omega_0\}$.

Prove that $\mathcal{V}$ is a base for a uniformity on Ω_0 which induces the discrete topology on Ω_0 even though the uniformity itself is not metrisable. (This fulfills a promise given in Section 2 of the last chapter.)

3.10 Prove that the function $\omega \times [0, 1) \rightarrow [0, \infty)$ defined above is in fact a homeomorphism.

3.11 Let $L' = L \cup \{\Omega\}$. Extend the ordering on L to L' by letting $\alpha < \Omega$ for all $\alpha \in L$. Prove that L' with the order topology is compact, connected, locally connected and first countable at each point except Ω. Prove that L, L' contain Ω_0, Ω' respectively as closed subsets. (From this most pathological properties of L, L' can be deduced from those of Ω_0 and Ω'.) How are L, L' related to each other in terms of compactifications?

3.12 Prove that L is connected and locally connected. For $\alpha, \beta \in L$ with $\alpha < \beta$ prove that $[\alpha, \beta]$ is separable and is homeomorphic to the unit interval. (Hint: Having obtained a denumerable dense subset of $[\alpha, \beta]$ use Exercise (2.2.8) and extend the order equivalence to an order equivalence between $[\alpha, \beta]$ and the unit interval.) [This exercise shows that each initial segment of L is embeddable in the real line. But neither Ω_0 nor L is embeddable in $\mathbb{R}$.]

3.13 Prove that L has the fixed point property. (Hint: If $f: L \rightarrow L$ is a map without fixed points then using connectedness show that $f(\alpha) > \alpha$ for each $\alpha \in L$. Now use the analogue of Theorem (3.1).)

3.14 Prove that the long line is not contractible. (Hint: Let if possible $H: L \times [0, 1] \rightarrow L$ be a homotopy such that H_0 is the identity map and H_1 is a constant map. Each stage of the homotopy, H_t is a map from L to L. By connectedness its range is either a bounded or unbounded interval of L. Let $A = \{t \in [0, 1] :$ range of H_t is bounded$\}$. Then $0 \notin A$, $1 \in A$. Show that A is a clopen subset of the unit interval.)

Notes and Guide to Literature

For a study of ordinals as equivalence classes of well-ordered sets, see Goffman [1]. For a systematic construction of ordinals as certain sets see either Kelley [1] or Halmos [1]. When such a construction is carried out it turns out that the element Ω is the same as the set Ω_0. As a result, in the literature the reader will often find Ω_0 denoted by Ω.

We referred to the continuum hypothesis in the construction of Ω_0. This hypothesis is the statement that the only cardinal numbers less than c are countable. It is easy to show that this amounts to saying that Ω_0 and $\mathbb{R}$ have the same cardinality, which would in turn mean that it is possible to have a well-ordering on $\mathbb{R}$ in which every initial segment is countable. If one assumes the continuum hypothesis, many interesting constructions become possible, see again Goffman [1]. Quite a few conjectures in topology have been shown to depend upon the continuum hypothesis.

Although Ω_0 is extensively used in providing counter-examples, it is also important in some theoretical constructions. Its peculiar property in Theorem (3.1) is useful in constructions where countability is of essence.

Classic examples of this sort of construction are the Borel sets and the Baire functions used in analysis. See Royden [1] or Goffman [1].

4. Topological Groups

The theory of topological groups is rich both in terms of its own profound results and also in terms of its applications. In this small section it will not be possible even to hint at them. Still we study topological groups mainly to illustrate how the introduction of an algebraic structure on a space affects and enriches its topological properties.

As the name implies, a topological group is a topological space whose underlying set is also endowed with a group structure. The topological and the algebraic structures must obviously be co-related in some way, otherwise the study of topological groups would amount to separate studies of topological spaces and of groups. The most natural correlationship between the two would be to require that the two basic functions in a group namely the group multiplication and the inversion (the function taking each element to its inverse) are continuous. It turns out that the continuity of these two functions is equivalent to the continuity of a certain single function and the formal definition of a topological group is as follows.

(4.1) Definition: A topological group is a triple $(G, \cdot, \mathfrak{J})$ where $(G, \cdot)$ is a group and $\mathfrak{J}$ is a topology on G such that the function $f : G \times G \to G$ defined by $f(x, y) = x \cdot y^{-1}$ for $x, y \in G$ is continuous.

As usual, when $\cdot$ and $\mathfrak{J}$ are understood we suppress them from notation. The product $x \cdot y$ is often written as xy. If $\bullet$ is commutative (i.e. if the group is abelian) it is customary to write $x \cdot y$ as $x + y$ and powers of x additively. If $A, B \subset G$, by AB we mean the set $\{ab : a \in A, b \in B\}$. For $x \in G$, $\{x\}A$, $A\{x\}$ are abbreviated to xA and Ax respectively. For $A \subset G$, A^{-1} will be the set $\{x^{-1} : x \in A\}$.

Trivial examples of topological groups result by taking any group and putting either the discrete or the indiscrete topology on it. These examples show that every group can be converted into a topological group. It is natural to inquire if every topological space can be converted into a topological group. The answer is in the negative. The existence of a group structure compatible with the topology of a space has strong topological consequences. The simplest is the following. Let $(G, \bullet, \mathfrak{J})$ be a topological group. For $a \in G$ define $L_a : G \to G$ and $R_a : G \to G$ by $L_a(x) = ax$, $R_a(x) = xa$ for $x \in G$. These functions are called respectively the **right** and **left translations** by a and are easily seen to be homeomorphisms. Unless a is the identity element of G, L_a and R_a have no fixed points. Thus we see that the space $(G, \mathfrak{J})$ cannot have the fixed point property, unless G consists of only one point. From this it follows, for example, that the unit interval cannot be made into a topological group. Another easy consequence of translations is that the underlying topological space of a topological group is homogeneous (see Exercise (6.3.3) for the definition).

As a deeper consequence of the algebraic structure we now show that the underlying topological space of a topological group is uniformisable. Not only that, we shall construct a uniformity concretely in terms of the group structure. This will fulfill a promise made at the end of Section 2 in the last chapter. Let e be the identity element of a topological group $(G, \cdot, \mathfrak{J})$. Let $\mathcal{N}$ be the neighbourhood system of G at e. For $N \in \mathcal{N}$, let U_N be the set $\{(x, y) \in G \times G : x^{-1}y \in N\}$. Let $\mathcal{B} = \{U_N : N \in \mathcal{N}\}$.

(4.2) Theorem: The family $\mathcal{B}$ is a base for a uniformity $\mathcal{L}$ on G. Moreover $\mathcal{L}$ induces the topology $\mathfrak{J}$ on G.

Proof: We first show that $\mathcal{B}$ satisfies the conditions of (14.1.8). First for every $N \in \mathcal{N}$, $e \in N$ and so trivially $\Delta G \subset U_N$. Secondly if $N \in \mathcal{N}$ then $N^{-1} \in \mathcal{N}$ by continuity of the inversion function (note that $e^{-1} = e$). It is easy to see that $(U_N)^{-1} = U_{N^{-1}}$. For the remaining condition, suppose $U_N \in \mathcal{B}$. Now the function $g : G \times G \to G$ defined by $g(x, y) = xy$ is continuous and so $\exists$ $J, K \in \mathcal{N}$ such that $g(J \times K) \subset N$. Let $M = J \cap K$. Then $M \in \mathcal{N}$ and $U_M \in \mathcal{B}$. Let $(x, y) \in U_M \circ U_M$. Then there exists $z \in G$ such that $(x, z) \in U_M$ and $(z, y) \in U_M$. This means $x^{-1}z \in M$ and $z^{-1}y \in M$. So $x^{-1}y = g(x^{-1}z, z^{-1}y) \in g(M \times M) \subset g(J \times K) \subset N$, i.e. $(x, y) \in U_N$. Thus we have shown that $U_M \circ U_M \subset U_N$. All the conditions of Proposition (14.1.8) are satisfied by $\mathcal{B}$. So there exists a uniformity $\mathcal{L}$ on G having $\mathcal{B}$ as a sub-base. But $\mathcal{B}$ is closed under finite intersections. For if $M, N \in \mathcal{N}$ then $M \cap N \in \mathcal{N}$ and $U_M \cap U_N = U_{M \cap N}$. Hence $\mathcal{B}$ is not only a sub-base but in fact a base for $\mathcal{L}$.

It remains to show that $\mathfrak{J}_{\mathcal{L}}$, the topology induced by $\mathcal{L}$ is the same as $\mathfrak{J}$. First suppose $H \in \mathfrak{J}$ and $x \in H$. We have to find $V \in \mathcal{L}$ such that $V[x] \subset H$. Let $N = x^{-1}H$. By openness of the translation functions, N is a neighbourhood of e. Let $V = U_N$. Then $y \in V[x]$ implies that $(x, y) \in U_N$, that is $x^{-1}y \in N$ or equivalently $y \in H$ as was to be shown. So H is open in the uniform topology. Conversely suppose $H \in \mathfrak{J}_{\mathcal{L}}$. We claim that H is a $\mathfrak{J}$-neighbourhood of each of its points. Let $x \in H$. Then by the definition of $\mathcal{L}$ and the fact that $\mathcal{B}$ is a base for $\mathcal{L}$, there exists $N \in \mathcal{N}$ such that $U_N[x] \subset H$. We assert that $xN \subset H$. For suppose $y \in N$. Then $(x, xy) \in U_N$ since $x^{-1}(xy) = y \in N$. So $xy \in U_N[x]$. But $U_N[x] \subset H$. So $xy \in H$ and hence $xN \subset H$. Again by the fact that translation functions are open, xN is a $\mathfrak{J}$-neighbourhood of x since N is a $\mathfrak{J}$-neighbourhood of e. Thus we have shown that $H \in \mathfrak{J}$. So $\mathfrak{J}_{\mathcal{L}}$ coincides with $\mathfrak{J}$. ∎

The uniformity $\mathcal{L}$ constructed above is called the **left uniformity** for G. Instead of taking the sets U_N, we can consider sets of the form $\{(x, y) \in G \times G : x \cdot y^{-1} \in N\}$ for $N \in \mathcal{N}$. Then we get a uniformity $\mathcal{R}$ on G called the **right uniformity** on G which also induces the topology $\mathfrak{J}$. In general $\mathcal{L}$ and $\mathcal{R}$ distinct. Of course for an abelian group the two coincide.

There is a category $\mathcal{G}$Top whose objects are topological groups and morphisms are continuous homomorphisms, that is group homomorphisms which are continuous with respect to the given topologies. From this category one has the (partially) forgetful functors into $\mathcal{G}$ and into Top.

The ritualistic aspects of the study of topological groups are analogous to those in the study of groups. Thus one defines subgroups, normal subgroups, quotient groups and so on. If G is a topological group and H is a subgroup of the group G then it is easy to show that H with the relative topology on it is a topological group. By G/H we shall denote the set of right cosets of H in G (that is, sets of the form Hx for $x \in G$). Since distinct cosets are disjoint, we view G/H as a decomposition of G and give it the quotient topology. (Formerly, this notation was used to denote a quotient space whose only non-trivial member was H. However in connection with topological groups it will always denote a space of cosets). Let $p : G \to G/H$ be the projection map (i.e. the quotient map). It is interesting to note that this map is open. Let V be an open subset of G. Then $p^{-1}(p(V))$ is precisely the set $H \cdot V$. Writing HV as $\bigcup_{h \in H} h \cdot V$, we see that it is open since each hV, being a translate of the of the open set V is open. So $p(V)$ is open in G/H by the definition of a quotient topology.

Topological properties of G can often be deduced from those of H and G/H. As a typical result, we have

(4.3) Theorem: If H is compact and G/H is compact so is G.

Proof: Let $\mathcal{U}$ be an open cover of G. The argument will resemble the proof of Wallace's theorem. We shall use the continuity of the group multiplication. Fix $x \in G$. Then for each $h \in H$, $hx \in U$ for some $U \in \mathcal{U}$. Hence there exist open sets V_h and W_h (which could also depend upon x) containing h and x respectively such that $V_h \cdot W_h$ is contained in some member of of $\mathcal{U}$. Cover H by the open sets V_h for $h \in H$. By compactness of H obtain a finite subcover say $\{V_{h_1}, \ldots V_{h_n}\}$. Let $W_x = \bigcap_{i=1}^{n} W_{h_i}$. Then for each i, $V_{h_i} \cdot W_x$ is contained in some member of $\mathcal{U}$ and hence $H \cdot W_x$ is contained in finitely many members of $\mathcal{U}$. Thus we have shown that for every $x \in G$, there exists an open set W_x containing x such that the set $H \cdot W_x$ can be covered by finitely many members of $\mathcal{U}$.

Now $p(H \cdot W_x)$ is an open subset of G/H since p is an open map. It contains the coset Hx. It follows that the family $\{p(H \cdot W_x) : x \in G\}$ is an open cover of G/H. Since G/H is compact, some finite subfamily say $\{p(H \cdot W_{x_1}), \ldots, p(H \cdot W_{x_m})\}$ covers G/H. But this means that $G = \bigcup_{i=1}^{n} H \cdot W_{x_i}$ Since each $H \cdot W_{x_i}$ can be covered by finitely many members of $\mathcal{U}$, so can G. ∎

The reader should note the crucial use of the group structure made in this proof. The theory of topological groups could be developed further. But that is really not our intention. Instead we proceed to give examples of some important topological groups.

1. All euclidean spaces under the usual addition are topological groups. Discrete subgroups (that is subgroups having no points of accumulation) of these groups are known as lattices. (This is the other meaning of the

word 'lattice' referred to in Chapter 2, Section 2.)

2. The non-zero real numbers or the non-zero complex numbers form a topological group under multiplication. The non-zero quaternions also form a group under multiplication. These groups contain S^0, S^1 and S^3 respectively as subgroups. The last one among these is also known as the **spin group** and is important in applications to physics. It is interesting to note that these are the only values of n for which S^n (with the usual topology) can be made into a topological group. A proof of this fact is extremely non-trivial.

3. Let X be a space and $H(X)$ the group of all homeomorphisms of X. $H(X)$ is a subset of $M(X, X)$ and we put on it the relativised compact open topology. If X is locally compact and regular then $H(X)$ becomes a topological group under composition. Often the space X has some additional structure. Then instead of all homeomorphisms if we take only those which preserve this additional structure, we get subgroups of $H(X)$. As an example, let $X = \mathbb{R}^n$ or $\mathbb{C}^n$. Then X is a vector space over $\mathbb{R}$ (or $\mathbb{C}$). If we take only those homeomorphisms which are linear transformations, we get topological groups denoted by $GL(n, \mathbb{R})$ and $GL(n, \mathbb{C})$ respectively and called the **Galois linear groups.** If we take only orthogonal (or unitary) transformations we get the so-called **orthogonal** or **unitary groups**. These groups are respectively denoted by $0(n)$ and $U(n)$. They contain as subgroups the groups of orthogonal (or unitary) transformations of determinant 1. They are called the **special orthogonal** (or **special unitary**) groups and are important in applications.

4. If $\{G_i : i \in I\}$ is a collection of topological groups, the product space $\prod_{i\in I} G_i$ can be made into a topological group under co-ordinate wise multiplication. Two special cases of this construction are important. If each G_i is a copy of the group S^1 then the product group we get is called a **torus group.** It is a compact, connected abelian group. If each G_i is a copy of $\mathbb{Z}_2$, the discrete group with two elements, then the product group is a compact, totally disconnected abelian group. In particular the Cantor set is a group (see Exercise (8.2.9)).

5. We know that continuous founctions from a topological space into the unit interval [0, 1] play a crucial role in its study. For topological groups, the role of the unit interval is played by the group S^1. A **character** of a topological group G is a continuous homomorphism from G into S^1. The characters of G are easily seen to form a subgroup of the torus group $(S^1)^G$. It is called the **character group of** G. For the study of a locally compact abelian group, its character group is an invaluable tool.

We conclude the section with a discussion of a concept very intimately related to topological groups.

(4.4) Definition: Let X be a topological space and G a topological group. Then by a **left G-action** on X we mean a map $f : G \times X \to X$ such that for all $g, h \in G$ and $x \in X$ we have $f(g, f(h, x)) = f(gh, x)$ and $f(e, x) = x$ where e is the identity element of G. We also say G acts on X from the left.

We can define a right G-action similarly. If we write gx for $f(g, x)$ then the first condition takes the form $(gh)x = g(hx)$. For a fixed $g \in G$ define $\theta_g : X \to X$ by $\theta_g(x) = f(g, x)$ for $x \in X$. Then it is easy to see that $\theta_{gh} = \theta_g \circ \theta_h$ for all $g, h \in G$ while θ_e is the identity mapping 1_X. Each θ_g is a homeomorphism since $\theta_g \circ \theta_{g^{-1}} = \theta_e = 1_X$. Thus θ gives a continuous homomorphism of G into $H(X)$. Conversely if X is locally compact and regular then a continuous homomorphism from G into $H(X)$ gives rise to a left G-action on X. In particular we see that every subgroup of $H(X)$ acts on X from the left.

If G acts from the left and $x \in X$ then the sets $\{gx : g \in G\}$ and $\{g \in G : gx = x\}$ are called respectively the **orbit** of x and the **isotropy** group of x (which is easily seen to be a subgroup of G). Two distinct orbits are disjoint and thus we get a decomposition of X into various orbits. The quotient space of this decomposition is called the **projective space** of the G-action. G is said to act **transitively** if there is only one orbit and **freely** if for each x, the isotropy group of x is trivial. A few simple results on group actions will be given as exercises.

A central problem in the theory of topological groups is how to realise (upto an isomorphism) an abstract group concretely as the group of certain transformations of a suitable space X. This is called the **representation problem**. In case X is regular and locally compact this amounts to finding suitable action of that group on X. Often X has some additional structure. For example suppose X is $\mathbb{R}^n$ (or $\mathbb{C}^n$) and G acts on X from the left in such a way that for each g the corresponding transformation θ_g is orthogonal (or unitary). Then we get an orthogonal (or unitary) representation of G.

There is a huge literature on the theory of group representations. Topological groups are also important in the theory of integration. Under certain conditions, there exists a measure called the Haar measure on a topological group. However it is beyond our scope to discuss any of these things.

Exercises

4.1 Prove that the condition in Definition (4.1) is equivalent to the continuity of group multiplication and of the inversion function.

4.2 Do we get a topological group if we put the cofinite topology on a group?

4.3 Prove that the underlying space of a topological group is homogeneous.

4.4 Prove that a T_0 topological group is a Tychonoff space.

4.5 Prove that the partially forgetful functor from $\underline{G}$ Top to $\underline{G}$ has a left adjoint. Prove also that this functor factors through the category of uniform spaces. (The only thing that requires a proof is that a continuous homomorphism is uniformly continuous w.r.t. the left uniformitites.)

4.6 Let G be topological group and let $A, B \subset G$. Prove that:

(i) If at least one of A and B is open, so is $A \cdot B$. (Hint: Express $A \cdot B$ as a union of translates of open sets.)

(ii) $A \bullet B$ need not be closed even when A, B are both closed. (Hint: In $(\mathbb{R}^2, +)$ let A be the y-axis and B the set $\{(x, y) : x > 0, y > 0, xy = 1\}$.

(iii) If one of A and B is compact and the other closed when $A \bullet B$ is closed. (Hint: Use nets.)

4.7 Let H be a subgroup of a topological group G and G/H the space of right cosets of H in G. Prove that:

(i) $\bar{H}$ is a subgroup of G. (Hint: Use nets.)

(ii) If H is open in G then it is closed in G.

(iii) There is a transitive right G-action on the space G/H. The isotropy group of each element is H.

(iv) If H is a normal subgroup of G then G/H is a topological group.

(v) If H is dense in G then G/H is an indiscrete space.

(vi) If H is normal so is $\bar{H}$.

(vii) If G is connected and H is normal and totally disconnected then every element of H commutes with every element of G. (Hint: For $a \in H$, $f(x) = x^{-1}ax$ defines a map from G to H.)

(viii) If G is compact and H is closed in G then $p : G \to G/H$ is a closed map.

(ix) If H and G/H are first countable so is G.

4.8 Prove that an additive subgroup of $\mathbb{R}$ is either dense in $\mathbb{R}$ or is discrete (that is, has no accumulation points in $\mathbb{R}$). Hence show that a closed subgroup of $\mathbb{R}$ is either countable or is $\mathbb{R}$ itself.

4.9 Prove that $\mathbb{R}/\mathbb{Z}$ is isomorphic, as a topological group, to S^1. Hence or otherwise show that a finite subgroup of S^1 is cyclic.

4.10 Let G be an abelian topological group and $\hat{G}$ its character group. Prove that:

(i) If G is discrete, $\hat{G}$ is compact.

(ii) If G is compact, $\hat{G}$ is discrete.

(iii) If G is either compact or discrete, there is a natural isomorphism from G onto $\hat{\hat{G}}$ (i.e. the character group of the character group of G).

4.11 Suppose a topological group G acts on a space from the left. Prove that distinct orbits are mutually disjoint and that G acts transitively on each orbit. For $x \in X$, let H be the isotropy group of x. Prove that there is a continuous bijection from G/H, the space of left cosets of H in G, onto the orbit of x. Is this bijection necessarily a homeomorphism? Finally prove that a group G acts freely on X iff the corresponding homomorphism $\theta : G \to H(X)$ is one-to-one.

4.12 Obtain the projective plane (see Exercise (5.4.18)) as the projective space of a suitable left action of a discrete group with two elements on the space S^2. [Hence the name.]

Notes and Guide to Literature

Numerous references can be cited. Probably the most celebrated is the book of Pontrjagin [1]. See Hewitt and Ross [1] for detailed results with applications to analysis. Exercise (4.10) points out an interesting duality between a topological group and its character group. It is due to Pontrjagin. As a result, the character group is also called the **dual group**. Part (iii) is valid for all locally compact, abelian groups.

The lattices (that is, discrete subgroups of $\mathbb{R}^n$) form the starting point of what is known as geometric number theory. This branch of mathematics is comparatively new. A reference on it is Cassels [1].

To show that the sphere S^n for $n \neq 0$, 1 and 3 cannot be made into a topological group requires deep results from algebraic topology. This problem arose originally out of number theory.

The standard reference for group actions is Montgomery and Zippin [1].

Finally, a group structure is not the only algebraic structure that can be studied along with a topological structure. One can put less restrictive structures such as semi-groups or a group structure upto homotopy. On the other side, one can put stronger structures than a group. Thus one can study vector spaces over the field of real or complex numbers along with a suitable topology. They are known as topological vector spaces and studied in Kelley, Namioka *et al.* [1].

Miscellaneous Exercises

Although we have been giving a fair number of exercises at the end of each section, there are still many exercises which could not be included so far for various reasons. We give a few such exercises here. They would provide the reader with an opportunity to review most of the material. No particular order is followed and hints are given only sparingly.

M.1 Prove that a topological space is connected iff every map from it into a discrete space with two points is constant. Use this fact to give alternate proofs of Propositions (6.2.4), (6.2.11) and (6.2.13).

M.2 Prove that a compact metric space is locally connected iff for every $\epsilon > 0$, it can be covered by finitely many open, connected sets of diameter less than ϵ each.

M.3 Prove that the topologist's sine curve has the fixed point property.

M.4 Let X be the space obtained from the coproduct of n copies of the unit interval $[0, 1]$ by identifying together the points corresponding to the end point 0 in each copy. (For $n = 3$ the resulting space is called the triod). Prove that X has the fixed point property.

M.5 Prove that a space is completely normal iff every subspace of it is normal.

M.6 Prove that a space is a T_0 space iff it can be embedded into a power of the Sierpinski space.

M.7 Prove that every sequence in a totally bounded metric space has a Cauchy subsequence. Using this fact, give an alternate proof of Theorem (12.1.6).

M.8 Let C be the Cantor ternary set. For $n \in \mathbb{N}$ and for $1 \leq k \leq 2^n$ let I_k^n be the closed interval, defined inductively as the left one-third of the interval $I_{(k+1)/2}^{n-1}$ for k odd and the right one-third of $I_{k/2}^{n-1}$ for k even, I_1^0 being the unit interval. Let $C_k^n = C \cap I_k^n$. Prove that the sets C_k^n are clopen subsets of C and that each is homeomorphic to C. For each n prove that C is the disjoint union of C_k^n's ($k = 1, 2, \ldots, 2^n$).

M.9 Let A be a non-empty closed subset of the Cantor set C. Prove that there exists a retraction of C onto A. (Hint: For each n let A_n be the set $\bigcup \{C_k^n : C_k^n \cap A \neq \emptyset\}$. Obtain inductively a retraction r_n of C onto A_n, r_0 being the identity map. Now take the uniform limit of the sequence $\{r_n\}$ of maps.)

M.10 Prove that there exists a continuous function f which maps C onto the unit interval, I. (Hint: For each n define $f_n : C \to I$ by $f_n(x) = k/2^n$ for $x \in C_k^n$. Take the uniform limit of f_n's.)

M.11 Prove that every compact metric space is a continuous image of the Cantor ternary set.

M.12 Prove that there exists a map from the unit interval onto the Hilbert cube. (These are examples of the so-called space filling curves.)

M.13 Let (X, d) be a compact metric space and $f : X \to X$ an isometry (that is, for $x, y \in X$, $d(x, y) = d(f(x), f(y))$. Prove that f is a surjection. (Hint: Let $A = f(X)$. If $X - A \neq \emptyset$, let x_0 be any point of $X - A$ and define x_n inductively as $f(x_{n-1})$ for $n \in \mathbb{N}$. Consider the sequence $\{x_n\}$).

M.14 Suppose (Z, e) is a compact metric space and $g : Z \to Z$ is a function such that $e(g(z_1), g(z_2)) \geqslant e(z_1, z_2)$ for all $z_1, z_2 \in Z$. Start with any $z_0 \in Z$ and define $z_n = g(z_{n-1})$ inductively for $n \in \mathbb{N}$. Prove that the sequence $\{z_n\}$ has a subsequence converging to z_0. (Hint: For every $\epsilon > 0$, there exist infinitely many terms of the sequence contained in some ϵ-ball. Now note that for $m, n \in \mathbb{N}$ with $m < n$, $e(z_m, z_n) \geqslant e(z_0, z_{n-m})$.)

M.15 Let (X, d) be a compact metric space and $f : X \to X$ a function such that for all $x, y \in X$, $d(f(x), f(y)) \geqslant d(x, y)$. Prove that f is an isometry and hence that it is a surjection. (Hint: Apply the last exercise with $Z = X \times X$, e a suitable metric on Z. Start with a point $(x, y) \in Z$ for which $d(f(x), f(y)) > d(x, y)$ if at all.)

M.16 Prove that for the Sierpinski space, the family of all neighbourhoods of the diagonal is a uniformity but that the topology induced by this uniformity is not the topology on the Sierpinski space.

M.17 Let $\mathcal{N}$ be the family of all neighbourhoods of the diagonal in $\mathbb{R} \times \mathbb{R}$ Prove that $\mathcal{N}$ is not uniformity on $\mathbb{R}$.

M.18 Let (X, d) be a metric space. Prove that following statements are equivalent:

(1) The uniformity induced by d coincides with the family of all neighbourhoods of ΔX in $X \times X$.

(2) Every continuous real valued function on X is uniformly continuous w.r.t. d (and the usual metric on $\mathbb{R}$).

(3) For every open cover $\mathcal{U}$ of X there exists $r > 0$ such that for each $x \in X$, $B_d(x, r)$ is contained in some member of $\mathcal{U}$. (In other words, the conclusion of the Lebesgue covering lemma holds.)

M.19 For a metrisable space X prove that the following conditions are equivalent:

(1) The collection of all neighbourhoods of ΔX in $X \times X$ is a uniformity with a countable base.

(2) The set of non-isolated points of X (that is the set $\{x \in X : \{x\}$ is not open$\}$) is compact.

(3) X has a metric satisfying conditions (2) and (3) of the last exercise.

M.20 A subset S of the euclidean space $\mathbb{R}^n$ is said to be **convex** if for any two points in S the line segment joining them is contained in S, that is for all $x, y \in S$ and $t \in [0, 1]$ we have that $(1 - t)x + ty \in S$. Prove that every non-empty convex open subset of $\mathbb{R}^n$ is homeomorphic to an open ball in $\mathbb{R}^n$.

M.21 The **antipodal map** on a sphere S^n is defined as the map $\alpha : S^n \to S^n$

such that $\alpha(x) = -x$ for all $x \in S^n$. For n odd prove that α is homotopic to the identity map on S^n. (Hint: For $n = 1$, let the t-th stage of the homotopy be a rotation through an angle πt, $t \in [0, 1]$.) This result is false for even n. The proof is non-trivial.

M.22 Let I be an index set of cardinality c. Prove that the space $\mathbb{Z}_2^I$ is separable but that it contains a dense subset no countable subset of which is dense.

M.23 Prove that a surjective map $p : X \to X$ is a quotient map iff it has the property that for any space W and a function $f : Y \to W$, continuity $f \circ p$ implies that of f.

M.24 Let $p : X \to Y$ be a quotient map and Z be a locally compact, regular space. Define $q : X \times Z \to Y \times Z$ by $q(x, z) = (p(x), z)$ for $x \in X$, $x \in Z$. Prove that q is a quotient map. (Hint: Use the last exercise along with Theorem (15.1.4).)

M.25 Let $X = \mathbb{R}$ and Y be obtained from X by collapsing $\mathbb{Z}$, the set of all integers, to a point. Let $p : X \to Y$ be the quotient map. Prove that the map $q : X \times \mathbb{Q} \to Y \times \mathbb{Q}$ defined by $q(x, z) = (p(x), z)$ for $x \in X$, $z \in \mathbb{Q}$ is not a quotient map.

M.26 For $n \in \mathbb{N}$ let C_n be the circle in the plane with centre at $\left(\frac{1}{n}, 0\right)$ and radius $\frac{1}{n}$. Let $C = \bigcup_{n=1}^{\infty} C_n$. Prove that there exists a continuous bijection from the space Y in the last exercise onto C but that Y is not homeomorphic to C. (In fact Y is not first countable and hence not embeddable in the plane.)

M.27 Let $\{f_i : X_i \to Y \mid i \in I\}$ be an indexed family of functions where each X_i is a topological space. Suppose $\mathfrak{J}$ is the strong topology on Y determined by this family. Prove that $(Y, \mathfrak{J})$ is a quotient space of the coproduct $\sum_{i \in I} X_i$, provided that the family $\{f_i(X_i) : i \in I\}$ covers Y.

M.28 Let X be a set and $\{\mathfrak{J}_i : i \in I\}$ a family of topologies on X such that for every $i \in I$, $(X; \mathfrak{J}_i)$ is locally connected. Prove that $(X; \bigcap_{i \in I} \mathfrak{J}_i)$ is locally connected. Hence show that for any topology $\mathfrak{J}$ on X there exists a smallest topology $\mathfrak{J}^*$ on X containing $\mathfrak{J}$ such that $(X; \mathfrak{J}^*)$ is locally connected.

M.29 Let $(X, \mathfrak{J})$ be a topological space and let $\mathcal{B}$ be the family of all components of open subsets of X. Prove that $\mathcal{B}$ is a base for a topology $\mathcal{U}$ on X and that $\mathcal{U}$ is contained in $\mathfrak{J}^*$, the topology given in the last exercise. (Whether the two actually coincide is apparently not known.)

M.30 Let $(X, \mathfrak{J})$, $(Y, \mathcal{U})$ be spaces and let $\mathfrak{J}^*$, $\mathcal{U}^*$ be the topologies on X, Y respectively given by Exercise (M.28). Prove that if a function $f : X \to Y$ is $\mathfrak{J} - \mathcal{U}$ continuous then it is $\mathfrak{J}^* - \mathcal{U}^*$ continuous. (Hint: Let $\mathcal{V} = \{V \subset Y : f^{-1}(V) \in \mathfrak{J}^*\}$. Prove that $\mathcal{V}$ is a topology on Y and that $(Y; \mathcal{V})$ is locally connected.)

M.31 A subcategory $\mathcal{C}$ of a category $\mathcal{D}$ is said to be **reflective (coreflective)** if the inclusion functor from $\mathcal{C}$ to $\mathcal{D}$ has a left (respectively a right) adjoint. Prove that:

(i) The category of compact, Hausdorff spaces is a reflective subcategory of the category of Tychonoff spaces.

(ii) The categories of indiscrete spaces and of T_1 spaces are reflective subcategories of Top while those of discrete spaces and locally connected spaces are coreflective subcategories of Top.

M.32 A subcategory $\mathcal{C}$ of a category $\mathcal{D}$ is said to be **closed under equivalences** if whenever X, Y are equivalent objects of $\mathcal{D}$ and X is in $\mathcal{C}$, so is Y. Suppose $\mathcal{C}$ is a full, reflective subcategory of Top and that it is closed under equivalences. Prove that 'belonging to $\mathcal{C}$' is a productive topological property. State and prove the dual result for coreflective subcategories of Top.

M33 Let $(X, \mathfrak{J})$ be a topological space and $\mathcal{F}$ the family of all continuous functions from it into the Sierpinski space. Prove that the weak topology on X generated by $\mathcal{F}$ coincides with $\mathfrak{J}$.

M.34 Interpret the results of Exercise (4.2.3) in terms of the two basic problems mentioned in Chapter 5, Section 5.

M.35 Let A, B be closed subsets of a space X. Prove that if $A \cup B$ and $A \cap B$ are connected then A, B are connected.

M.36 Let $X = ([0, 1]\times[0, 1]) - (\{\frac{1}{2}\}\times(\frac{1}{2}, 1])$ and $Y = [0, 1]$. Define $p : X \to Y$ by $p((x, y)) = x$. Prove that p is continuous, open and surjective and that for any y_1, y_2 in Y, $p^{-1}(y_1)$ is homeomorphic to $p^{-1}(y_2)$. Here Y is compact and $p^{-1}(y)$ is compact for each $y \in Y$. Still X is not compact. (This example shows that Theorem (15.4.3) is not a mere topological result in the sense that it could not have been proved merely from the facts that the quotient homomorphism is continuous, open and onto and all cosets are homeomorphic to each other. The use of the group structure is vital for the proof.)

M.37 Prove that the unit square with the topology induced by the lexicographic ordering on it is compact, Hausdorff, first countable, connected, and locally connected, but that it does not satisfy the countable chain condition. Find its path components.

M.38 Let X be the subset of the plane obtained by joining the point (0, 1) to the point (0, 0) and the points $\left(\frac{1}{n}, 0\right)$ for $n \in \mathbb{N}$. (X is homeomorphic to the broom space of Exercise (6.3.5).) Let Y be the union of X and its reflection in the plane through the point (0, 0). Prove that Y is simply connected but not contractible.

M.39 Prove that the long line is simply connected. In fact show that every map from a compact space into it is null-homotopic.

M.40 Let X, Y be spaces, A a subset of X and $f : A \to Y$ a map. Take the coproduct $X + Y$ and let R be the equivalence relation on $X + Y$ generated by f (that is, R is the smallest equivalence relation on $X + Y$ containing the set $\{(x, f(x)) : x \in A\}$). The quotient space $(X + Y)/R$ is called the **attaching space** of f. It is denoted by $X \cup_f Y$.

Prove that the map $f: A \to Y$ has a continuous extension $F: X \to Y$ iff Y is a retract of $X \cup_f Y$.

M.41 Let X be the cylinder $S^1 \times [0, 1]$ and A the subset $S^1 \times \{0, 1\}$. Let Y be S^2 minus the union of two mutually disjoint open discs on S^2. For example let $Y = S^2 - (D_1 \cup D_2)$ where D_1 and D_2 are defined by $D_1 = \{(x_1, x_2, x_3) \in S^2 : x_3 > \frac{1}{2}\}$ and $D_2 = \{(x_1, x_2, x_3) \in S^2 : x_3 < -\frac{1}{2}\}$. Let f be a homeomorphism of A onto the union of the boundaries of D_1 and D_2. Prove that the attaching space of f is homeomorphic to the torus surface. (This is an example of a general construction known as attaching handles. In the present case we say that a torus is obtained by attaching a handle to the 2-sphere.)

M.42 Let $\mathbb{C}$ be the complex plane and $X = \mathbb{C} \times \mathbb{C} - \{(0, 0)\}$. Define a binary relation $\sim$ on X by $(z_1, z_2) \sim (w_1, w_2)$ iff there exists $c \in \mathbb{C}$ such that $cz_1 = w_1$ and $cz_2 = w_2$. Prove that this is an equivalence relation. The resulting quotient space of X, denoted by $\mathbb{C}P^1$, is called the **complex projective line**. Prove that it is homeomorphic to the 2-sphere S^2.

M.43 Prove that a coproduct of paracompact spaces is paracompact. Deduce that the product of a paracompact space with a discrete space is paracompact.

M.44 If G is a topological group and H is an open subgroup then prove that G/H, the space of right cosets, is discrete and that G is homeomorphic to the product of H and G/H.

M.45 Prove that a locally compact group is paracompact. (Hint: Let U be an open neighbourhood of the identity such that $\overline{U}$ is compact. Let $V = U \cup U^{-1}$. Let $H = \bigcup_{n=1}^{\infty} V^n$ where for $n \in \mathbb{N}$, V^n consists of all n-fold products of elements of V. Prove that H is σ-compact and that it is an open subgroup of G.)

M.46 Let G be a topological group and C the component of G containing the identity element. Prove that C is a normal subgroup of G and that the quotient group G/C is totally disconnected.

M.47 Prove that a continuous bijection of $\mathbb{R}$ onto itself must be a homeomorphism. (This is also true of higher dimensional euclidean spaces but the proof is highly non-trivial.)

M.48 A T_4 space X is called an **absolute retract** if whenever it is embedded as a closed subspace of a T_4 space Y, it is a retract of Y. Prove that:

(i) A T_4 space X is an absolute retract iff it has the property that whenever A is a closed subset of a normal space Z, every map from A to X can be continuously extended to a map from Z to X. (cf. Exercise (8.3.4).)

(ii) A binormal absolute retract is contractible.

(iii) Every retract of an absolute retract is an absolute retract.

M.49 A T_4 space X is called an **absolute neighbourhood retract** if whenever it is embedded as a closed subspace of a T_4 space Y it is a neighbourhood retract of Y (that is, a retract of some neighbourhood of X in Y). Prove that:

(i) A T_4 space X is an absolute neighbourhood retract iff it has the property that whenever A is a closed subset of a normal space Z, every map from A to X can be extended to a map from some neighbourhood of A (in Z) to X.

(ii) Every neighbourhood retract of an absolute neighbourhood retract is an absolute neighbourhood retract.

(iii) For every n, S^n is an absolute neighbourhood retract.

M.50 Given any cardinal number α, prove that there exists a compact, Hausdorff, abelian topological group G and a dense subgroup H of G such that the quotient group G/H has cardinality at least α. (Hint: Consider suitable powers of $\mathbb{Z}_2$, the discrete group with two elements.)

M.51 Prove that every topological space can be obtained as a quotient space of a Hausdorff space. (Hint: Let X be a space. From the last exercise obtain groups G, H, a subset S of G/H and a bijection $f: S \to X$ (f need not be continuous.) Let $Y = p^{-1}(S)$ where $p: G \to G/H$ is the quotient homomorphism. Let $Z = \{(x, y) \in X \times Y : f(p(y)) = x\}$ with the relativised product topology. The restriction of the projection from Z to X is a desired map. This result is taken from Willard [1].)

M.52 Prove that a topological space $(X; \mathfrak{J})$ has the property that every compact subset of it is closed iff it has the property that every continuous bijection from a compact space onto X is a homeomorphism. (Hint: If C is a compact subset of X prove that the topology generated by $\mathfrak{J} \cup \{X - C\}$ makes X into a compact space.)

M.53 Let X be a space in which every compact subset is closed. Prove that X is hemicompact (see Exercise (15.2.8)) iff the Alexandroff compactification of X is first countable at the point at infinity.

M.54 Prove that a hemicompact, first countable space is locally compact. Show that the space in Example 10, Chapter 4, Section 2 is hemicompact but not locally compact.

M.55 Prove that a space having a σ-locally finite base is first countable. Is the converse true?

References

P. Alexandroff, 'Sur les ensembles de la premiere classe et les ensembles abstraits', *C. R. Acad. Sci.* (Paris), **178** (1924), 185–187.

P. Alexandroff and H. Hopf, *Topologie*, Chelsea Publishing, New York, 1965.

P. Alexandroff and P. Urysohn, 'Une condition necessarie et suffisante pour qu'une class ($\mathcal{L}$) Soit une class ($\mathcal{D}$)', *C. R. Acad. Sci.* (Paris), **177** (1923), 1274–1277.

R. Arens and J. Dugundji, 'Remark on the concept of compactness', *Portugaliae Math.*, **9** (1950), 141–143.

J. L. Bell and A. B. Slomson, *Models and Ultraproducts—An Introduction*, North-Holland, London-Amsterdam, 1969.

R. H. Bing, 'The elusive fixed point property', *Amer. Math. Monthly*, **76** (1969), 119–132.

————, 'A connected, countable, Hausdorff space', Proc. Amer. Math. Soc., **4** (1953), 474.

E. Bishop, *Foundations of Constructive Analysis*, McGraw-Hill, New York.

K. Borsuk, *Theory of Retracts*, Polska Akademia Nauk, Warsaw, 1967.

N. Bourbaki, 'Topologie Générale', *Actualities Sci. Ind. Paris*, **1084** (1949).

J. W. S. Cassels, *Introduction to Geometry of Numbers*, Springer-Verlag, 1971.

J. B. Conway, 'The inadequacy of sequences', *Amer. Math. Monthly*, **76** (1969), 68–69.

R. Crowell and R. Fox, *Introduction to Knot Theory*, Blaisdell, New York, 1963.

M. Davis, *Computability and Unsolvability*, McGraw-Hill, New York, 1958.

J. Dugundji, *Topology*, Prentice-Hall, New Delhi, 1975.

N. Dunford and J. Schwartz, *Linear Operators*, Parts I, II and III, Wiley Interscience, New York, 1971.

S. Eilenberg and N. Steenrod, *Foundations of Algebraic Topology*, Princeton University Press, Princeton, 1952.

R. Engelking, *General Topology*, Polish Scientific Publishers, Warsaw, 1977.

M. K. Fort, Jr., 'Homogeneity of infinite products of manifolds with boundary', *Pacific J. Math.*, **12** (1962), 879–884.

P. Freud, *Abelian Categories*, Harper & Row, New York, 1964.

L. Gillman and M. Jerison, *Rings of Continuous Functions*, Van Nostrand, Princeton, 1960.

C. Goffman, *Real Functions*, Prindle Weber and Schmidt, Boston, 1967.

S. W. Golomb, 'A connected topology for the integers', *Amer. Math. Monthly*, **66** (1959), 663–665.

V. Guillemin and A. Pollack, *Differential Topology*, Prentice-Hall, Englewood Cliffs, N. J., 1974.

R. Gustin, 'Countable, connected spaces', *Bull. Amer. Math. Soc.*, **52** (1946), 101–106.

O. Hájek, 'Metric completion simplified', *Amer. Math. Monthly*, **75** (1968), 62–65.

P. R. Halmos, *Naïve Set Theory*, Van Nostrand, Princeton, 1960.

F. Hausdorff, 'Die Mengen G_δ in Vollstandigen Raümen', *Fund. Math.*, **6** (1924), 146–148.

————, 'Erweiterung einer Homöomorphie', *Fund. Math.*, **16** (1930), 353–360.

E. Hewitt, 'The role of compactness in analysis', *Amer. Math. Monthly*, **67** (1960), 499–516.

E. Hewitt and K. Ross, *Abstract Harmonic Analysis*, Vol. I, Springer-Verlag, Berlin, 1963.

E. Hille, *Methods in Classical and Functional Analysis*, Addison-Wesley, Reading, 1972.

P. Hilton and S. Wylie, *Homology Theory*, Cambridge University Press, 1960.

J. Hocking and G. Young, *Topology*, Addison-Wesley, Reading, 1961.

H. B. Hoyle III, 'Connectivity maps and almost continuous functions', *Duke Math. J.*, **37** (1970), 671–680.

S. T. Hu, *Theory of Retracts*, Wayne State University Press, Detroit, 1965.

———, *Homotopy Theory*, Academic Press, New York, 1959.

H. H. Hung, 'Some metrisation theorems', Proc. Amer. Math. Soc., **54** (1976), 363–367.

W. Hurewicz and H. Wallman, *Dimension Theory*, Princeton University Press, Princeton, 1941.

D. Husemoller, *Fibre Bundles*, McGraw-Hill, New York, 1966.

R. A. Johnson, 'A compact non-metrisable space such that every closed subset is a G-delta', *Amer. Math. Monthly*, **77** (1970), 172–176.

J. L. Kelley, *General Topology*, Van Nostrand, Princeton, 1955.

———, 'The Tychonoff product theorem implies the axiom of choice', *Fund. Math.*, **37** (1950), 75–76.

J. L. Kelley, I. Namioka *et al.*, *Linear Topological Spaces*, Van Nostrand, Princeton, 1963.

J. R. Kline, 'A theorem concerning connected point sets', *Fund. Math.*, **3** (1922), 238–239.

B. Knaster and C. Kuratowski, 'Sur les ensembles connexes', *Fund. Math.*, **2** (1921), 206–255.

V. Krishnamurthy, 'On the number of topologies on a finite set', *Amer. Math. Monthly*, **73** (1966), 154–157.

C. Kuratowski, *Topologie*, Vol. 1 and 2, Academic Press, New York, 1968.

S. Lang, *Algebra*, Addison-Wesley, Reading, 1967.

E. Langford, 'Characterisation of Kuratowski 14-sets', *Amer. Math. Monthly*, **78** (1971), 362–367.

S. Leader, 'On products of proximity spaces', *Math. Ann.*, **154** (1964), 185–194.

N. Levine, 'A characterisation of compact metric spaces', *Amer. Math. Monthly*, **68** (1961), 653–655.

C. L. Liu, *Introduction to Combinatorial Mathematics*, McGraw-Hill, 1968.

M. W. Lodato, 'On topologically induced generalised proximity relations', Proc. Amer. Math. Soc. **15** (1964), 417–422.

S. MacLane, *Categories for the Working Mathematician*, Springer-Verlag, New York, 1971.

K. D. Magill Jr., '*N*-point compactifications', *Amer. Math. Monthly*, **72** (1965), 1075–1081.

H. Martin, 'Remark on the Nagata-Smirnov metrisation theorem', Topology-Proceedings of the Memphis State University Conference, Marcel Dekker, New York, 1976.

E. J. McShane, 'Partial ordering and Moore-Smith limits', *Amer. Math. Monthly*, **59** (1952), 1–11.

E. Mendelson, *Introduction to Mathematical Logic*, Van Nostrand, Princeton, 1964.

B. Mitchell, *Theory of Categories*, Academic Press, New York, 1965.

D. Montgomery and Zippin, *Topological Transformation Groups*, Interscience, New York, 1955.

H. L. Montgomery, (Problem with Solution) *Amer. Math. Monthly*, **85** (1978), 198.

E. H. Moore and L. H. Smith, 'A general theory of limits', *Amer. J. Math.*, **44** (1922), 102–121.

J. R. Munkres, *Topology—A First Course*, Prentice-Hall, New Delhi, 1978.

J. Nagata, *Modern Dimension Theory*, North-Holland, Amsterdam, 1964.

———, 'On the uniform topology of bicompactifications', *J. Inst. Polytech. Osaka City Univ.*, **1** (1950), 28–39.

S. A. Naimpally, J. G. Hocking and P. Cameron, 'Nearness—a better approach to continuity and limits' (unpublished).

S. A. Naimpally and B. Warrack, *Proximity Spaces*, Cambridge University Press, London, 1970.

J. Novak, Regular space on which every continuous function is constant, *Časopis Pěst. Mat. Fys.*, **73** (1948), 58–68.

J. C. Oxtoby, *Measure and Category*, Springer-Verlag, New York, 1971.

G. Polya, *How to Solve It—a new aspect of mathematical method*, Princeton University Press, Princeton, 1973.

L. Pontrjagin, *Topological Groups*, Princeton University Press, Princeton, 1939.

A. Ramanathan, 'Maximal Hausdorff spaces', Proc. Indian Acad. Sci. Sect. A., **26** (1947), 31–42.

E. Reed, A class of T_1 compactifications', *Pacific J. Math.*, **65** (1976), 471–484.

H. Rogers Jr., *Theory of Recursive Functions and Effective Computability*, McGraw-Hill New York, 1967.

H. Royden, *Real Analysis*, The Macmillan Company, New York, 1968.

D. Scott, 'Continuous lattices' in *Toposes, Algebraic Geometry and Logic*, Springer-Verlag, Berlin-Heidelberg, 1972.

A. Seidenberg, *Lectures in Projective Geometry*, Van Nostrand Reinhold, New York, 1962.

W. Sierpinski, *General Topology*, University of Toronto Press, Toronto, 1952.

———, Sur les ensembles complets d'un espace, *Fund. Math.*, **11** (1928), 203–205.

G. Simmons, *Introduction to Topology and Modern Analysis*, McGraw-Hill Company, New York, 1963.

Yu. M. Smirnov, 'A necessary and sufficient condition for metrizability of a topological space', *Doklady Akad. Nauk S. S.S.R.N.S.*, **77** (1951), 197–200.

———, 'On normally disposed sets of normal spaces', *Mat. Sbornik*, N.S., **29** (1951), 173–176.

R. H. Sorgenfrey, 'On the topological product of paracompact spaces', *Bull. Amer. Math. Soc.*, **53** (1947), 631–632.

E. H. Spanier, *Algebraic Topology*, McGraw-Hill, New York, 1966.

L. Steen and J. Seebach, *Counterexample in Topology*, Holt, Rinehart and Winston, New York, 1970.

A. H. Stone, 'Paracompactness and product spaces', *Bull. Amer. Math. Soc.*, **54** (1948), 977–982.

M. H. Stone, 'The theory of representation for Boolean algebras', *Trans. Amer. Math. Soc.*, **40** (1936), 37–111.

W. C. Swift, 'Simple construction of non-differentiable functions and space-filling curves', *Amer. Math. Monthly*, **68** (1961), 653–655.

J. Thomas, 'A regular space, not completely regular,' *Amer. Math. Monthly*, **76** (1969), 181–182.

W. Thron, *Topological Structures*, Holt, Rinehart and Winston, New York, 1966.

J. W. Tukey, 'Convergence and uniformity in topology', *Ann. of Math. Studies*, **2** (1940).

A. Tychonoff, 'Über einen Funktionenraum', *Math. Ann.* **111** (1935), 762–766.

———, 'Über die topologische Erweiterung von Räumen', *Math. Ann.*, **102** (1929), 544–561.

A. D. Wallace, 'Extensional invariance', *Trans. Amer. Math. Soc.*, **70** (1951), 97–102.

A. Weil, 'Sur les espaces a structure uniforme et sur la topologie générale', *Actualites Sci. Ind.*, **869** (1940).

J. E. Whitesitt, *Boolean Algebra and Its Applications*, Addison-Wesley, Reading, 1961.

A. Wilansky, 'Life without T_2' *Amer. Math. Monthly*, **77** (1970), 157–161.

——— 'Correction to "Life without T_2" ', *Amer. Math. Monthly*, **77** (1970), 728.

S. Willard, *General Topology*, Addison-Wesley, Reading, 1970.

Index

ERRATA

1. The comment on p. 309, after the definition of a sub-category is not correct. Although the identity morphism on an object is unique, this does not imply that the identity morphism in a subcategory must coincide with the identity morphism in the original category. As a simple counter-example, consider a monoid whose morphisms are real numbers (with composition defined by multiplication of real numbers) and the 'subcategory' whose only morphism is 0. It is therefore necessary to add to Definition (13.1.2), one more requirement, namely,
   (iv) for any object X of $\mathscr{D}$, the identity morphism on X as an object of $\mathscr{D}$ is the same as that in $\mathcal{C}$.
2. On p. 328, part (iv) of Exercise (13.3.1) has no solution because there can be no proper map from a non-compact space to a compact space.
3. In the notation of an ordered pair, commas and semicolons (and occasionally even colons) have been used interchangeably. For example a topological space has been denoted by $(X, \mathscr{T})$, by $(X; \mathscr{T})$ or even by $(X : \mathscr{T})$.
4. A few other corrections are listed below:

Page	*Line*	*Incorrect*	*Correct*
31	−3	triples	tuples
41	11	elements	elephants
58	−11	$f(x)$	$f(t)$
65	−13	two	three
89	13	Querry	Query
114	12	write $y \bar{\delta} A$	write $y \delta A$
122	−15	X_j	X
151	Fig. 2	(0, 1) (on x-axis)	(1, 0)
166	3	$\mathbb{R}x\{0, 1\}$	$\mathbb{R} \times \{0, 1\}$
202	−5	x_i	X_i
202	−5	Y_i	Y_j
205	−14	X	G
211	9	$\leq$	$<$
232	−5	confinal	cofinal
241	13	filter	a filter
245	4	]	)
266	2	bases	base
300	11	not	no

Page	*Line*	*Incorrect*	*Correct*
320	2	transformations	transformation
327	7	$(\eta_{F(X)} = 1_{GF(X)})$	$(\eta_{F(X)}) = 1_{GF(X)}$
327	19	$F \underset{\mathcal{D}}{\uparrow\downarrow} G$	$F \overset{\mathcal{C}}{\underset{\mathcal{D}}{\uparrow\downarrow}} G$
342	13	$(f(X)), \mathcal{V}/f(X)$	$(f(X), \mathcal{V}/f(X))$
343	1	$V[x]$.	$V[x]$
350	14	topology	uniformity
373	−9	u	$\mathcal{U}$
373	−5	$\supset$	$\cup$
393	22	uniformity	a uniformity
394	8	$p : X \to X$	$p : X \to Y$
394	9	continuity	continuity of